T0297644

# Clinical Chemistry, Immunology and Laboratory Quality Control

# Clinical Chemistry, Immunology and Laboratory Quality Control

## A Comprehensive Review for Board Preparation, Certification and Clinical Practice

### Second Edition

**Amitava Dasgupta, Ph.D, DABCC**

Professor of Pathology and Laboratory Medicine
University of Texas McGovern Medical School at Houston

**Amer Wahed, M.D, FRCPath (UK)**

Professor of Pathology and Laboratory Medicine
University of Texas McGovern Medical School at Houston

ELSEVIER

Elsevier
Radarweg 29, PO Box 211, 1000 AE Amsterdam, Netherlands
The Boulevard, Langford Lane, Kidlington, Oxford OX5 1GB, United Kingdom
50 Hampshire Street, 5th Floor, Cambridge, MA 02139, United States

**Library of Congress Cataloging-in-Publication Data**
A catalog record for this book is available from the Library of Congress

**British Library Cataloguing-in-Publication Data**
A catalogue record for this book is available from the British Library

ISBN: 978-0-12-815960-6

For information on all Elsevier publications
visit our website at https://www.elsevier.com/books-and-journals

*Publisher:* Stacy Masucci
*Acquisitions Editor:* Ana Claudia A. Garcia
*Editorial Project Manager:* Megan Ashdown
*Production Project Manager:* Sreejith Viswanathan
*Cover Designer:* Christian J. Bilbow

Typeset by SPi Global, India

# Contents

# Preface

The first edition of "Clinical Chemistry, Immunology and Laboratory Quality Control: A Comprehensive Review for Board Preparation, Certification and Clinical Practice" was published by Elsevier in 2014. In 2015, we published the first edition of "Hematology and Coagulation: A concise guide for board review, board preparation and clinical practice" (Wahed A. and Dasgupta A.). Subsequently we published the third book in the series "Microbiology and Molecular Diagnosis in Pathology: A Comprehensive Review for Board Preparation, Certification and Clinical Practice" in 2017. The last book in the series, "Transfusion Medicine for Pathologists: A Comprehensive Review for Board Preparation, Certification, and Clinical Practice," was published also by Elsevier in 2018. All of these books are co-authored by faculties in the clinical pathology division of our department. We have received good feedback on our books and recently we published the second edition of Hematology and Coagulation book to incorporate the 2017 WHO guidelines as well as to update the content of the book.

Since the publication of the first edition of clinical chemistry book which was well received by readers, there are newer guidelines for drug of abuse testing, application of cardiac markers for diagnosis of myocardial infarction, etc. Moreover, point-of-care tests are gaining acceptance for rapid diagnosis. In addition, biotin interference in immunoassays using biotinylated antibodies is becoming a serious problem. Therefore, we decided to significantly revise our first edition of clinical chemistry book, adding two new chapters; one on biotin interference and the other on point-of-care testing. We have also included structures of common drugs of abuse to complete chapters on abused drugs. However, this book is a study guide, not a replacement for several excellent textbooks on clinical chemistry.

The aim of the second edition is to provide a strong foundation for students, residents, and fellows embarking on the journey of mastering clinical chemistry. It is expected that the book will also act as a valuable resource for residents preparing for the clinical pathology board exam. At the end of each chapter, we have included a section denoted as "key points" as in the first edition. We hope that this section will be a good resource for reviewing information, when time at hand is somewhat limited.

We hope that readers will find the second edition useful and, if so, our hard work will be duly rewarded.

**Amitava Dasgupta, Ph.D, DABCC**
*Professor of Pathology and Laboratory Medicine*
*University of Texas McGovern Medical School at Houston*

**Amer Wahed, M.D, FRCPath (UK)**
*Professor of Pathology and Laboratory Medicine*
*University of Texas McGovern Medical School at Houston*

# CHAPTER 1

# Instrumentation and analytical methods

## Instrumentation and analytical methods: An introduction

Various analytical methods are used in clinical laboratories (Table 1). Spectro-photometric detection is a common method of analysis where an analyte is detected and quantified using a visible (400–800 nm) or ultraviolet wavelength (below 380 nm). Atomic absorption and emission spectroscopy and fluorescence spectroscopy also fall under this broad category of spectrophotometric detection. Chemical sensors such as ion selective electrodes and pH meters are also widely used in clinical laboratories. Ions elective electrodes are the method of choice for detecting various ions such as sodium, potassium, and related electrolytes in serum or plasma. In blood gas machines, sensors capable of detecting hydrogen ion (pH meter) and partial pressure of oxygen during blood gas measurements are used. Another analytical method used in clinical laboratories is chromatographic method, but this method is utilized less frequently than other methods such as immunoassays, enzymatic assays, and colorimetric assays, which can be easily adopted on automated chemistry analyzers.

## Spectrophotometry and related techniques

Spectroscopic methods utilize measurement of a signal at a particular wavelength or a series of wavelengths. Spectrophotometric detections are used in many assays including atomic absorption, colorimetric assays, enzymatic assays, immunoassays, as well as in detecting elution of the analyte of interest from a column during high-performance liquid chromatography using ultraviolet-visible range spectrophotometric detector (UV detector).

Colorimetry is based on measuring the intensity of color after a chemical reaction so that the concentration of an analyte can be determined using the absorption by the colored compound. Use of Trinder reagent to measure salicylate level in serum is an example of a colorimetric assay. In this assay, salicylate reacts with ferric nitrate to form a purple complex which is measured in visible

1

**Table 1** Assay principles and instrumentation in clinical chemistry laboratory.

| Detection method | Various assays/analytical instrument |
| --- | --- |
| Spectrophotometric detection | Colorimetric assays<br>Atomic absorption<br>Enzymatic assays<br>Various immunoassays<br>High-performance liquid chromatography<br>with ultraviolet (HPLC-UV) or fluorescence detection |
| Chemical sensors | Various ion selective electrodes and oxygen sensors |
| Flame ionization detection | Gas chromatography |
| Mass spectrometric detection | Gas chromatography/mass spectrometry<br>(GC/MS), High-performance liquid chromatography/mass spectrometry<br>(LC/MS) or Tandem mass spectrometry<br>(LC-MS/MS)<br>Inductively coupled plasma mass spectrometry (ICP-MS) |

wavelength, but due to interferences from endogenous compounds such as bilirubin, this assay has been mostly replaced by more specific immunoassays [1].

Spectrophotometric measurements are based on Beer's law or sometimes referred as Beer-Lambert law. When a monochromatic light beam (light with a particular wavelength) is passed through a cell containing a specimen in a solution, a part of the light is absorbed and the rest is passed through the cell and reaches the detector. If "$I_o$" is the intensity of the light beam going through the cell and "$I_s$," the intensity of light beam coming out of the cell (transmitted light), then "$I_s$" should be less than "$I_o$." However, a part of light may be scattered by the cell or absorbed by the solvent in which the analyte is dissolved or even absorbed by the material of the cell. To correct this, one light beam of the same intensity and wavelength is passed through a reference cell containing solvent only and another through the cell containing analyte of interest. If "$I_r$" is intensity of the light beam coming out of the reference cell, its intensity should be close to "$I_o$." Transmittance ($T$) is defined as $I_s/I_o$. Therefore, correcting for scattered light and other nonspecific absorption, we can assume that transmittance of the analyte in solution should be $I_s/I_r$. In spectrophotometry, often, transmittance is measured as absorption "$A$" because there is a linear relation between absorbance and concentration of the analyte in the solution.

$$A = -\log T = -\log I_s/I_r = \log I_r/I_s$$

Usually transmittance is expressed as percent. For example, if 90% of the light is absorbed, then only 10% of the light is being transmitted where "$I_r$" is 100 (we are assuming that no light was absorbed when the beam passes through the reference cell, i.e., $I_o = I_r$), and $I_s$ value is 10.

Therefore, $A = \log 100/10 = \log 10 = 1$

If only 1% light is transmitted, then "$I_r$" is 100 and $I_s$ value is 1 and value of absorbance is

$A = \log 100/1 = \log 100$
$\quad = 2$. Therefore, the scale of absorbance is from 0 to 2, where zero value is for no absorbance.

Beer's law states that absorption of the light also depends on the concentration of the analyte in the solvent and the length of the cell path.

$A = \log I_r/I_s = a \cdot b \cdot c$

In this equation, "$a$" is proportionality constant termed as "absorptivity," "$b$" is the length of the cell path and "$c$" is the concentration of the analyte. Therefore, if "$b$" is 1 cm and concentration of the analyte is expressed as mol/L, then "$a$" is termed as "molar absorptivity" and often designated as "$\varepsilon$" (epsilon). The value of "$\varepsilon$" is a constant for a particular compound for a particular wavelength under prescribed condition of pH, solvent, and temperature.

Therefore $A = \varepsilon \cdot b \cdot c$ or $\varepsilon = A/b \cdot c$

For example, if "$b$" is 1 cm and concentration of the compounds is 1 mol/L, then

$A = \varepsilon$

Therefore, from measured absorbance value, concentration of the analyte can be easily calculated using know molar absorptivity and length of the cell:

$A = \varepsilon \cdot b \cdot c$ or concentration "$c$" $= A/\varepsilon \cdot b$

However, the direct proportionality between absorption and concentration must be established experimentally for a given analyte and instrument because this relation is valid up to a certain concentration of the analyte (upper end of calibration curve). Therefore, at a concentration higher than the upper limit of calibration curve, the analyte no longer obeys Beer's law and the relationship between concentration and absorption becomes nonliner. Therefore, only the concentration range where an analyte obeys Beer's law (calibration curve) can be used for measuring an analyte (analytical measurement or AMR) which is based on the calibration curve of the analyte (please see Chapter 2).

## Atomic absorption

Atomic absorption spectrophotometric techniques are widely used in clinical chemistry laboratories for analysis of various metals, although this technique is capable of analyzing many elements including trace elements which can be transformed into atomic form after vaporization. Although many elements can be measured by atomic absorption method, in clinical laboratories, lead, zinc, copper, and trace elements are commonly measured in blood or urine using atomic absorption. In atomic absorption spectrophotometry, following steps are followed:

- Sample applied (whole blood, serum, urine, etc.) to the sample cup.
- Liquid solvent is evaporated and dry sample is vaporized to a gas or producing droplets.
- Components of gaseous sample is converted into free atoms. This can be achieved by a flame or by a flameless manner using a graphite chamber that can be heated after application of the sample.
- A hollow cathode lamp containing an inert gas like argon or neon at a very low pressure is used as a light source. Inside the lamp, there is a metal cathode that contains the same metal as analyte of analysis. For example, for analysis of copper, a hollow copper cathode lamp is needed. For analysis of lead, a hollow lead cathode lamp is required.
- Atoms in the ground state then absorb a part of the light emitted by the hollow cathode lamp to be in the excited state. Therefore, a part of the light beam is absorbed, resulting in a net decrease in the intensity of the beam arriving at the detector. Applying principles of Beer's law, concentration of the analyte of interest can be measured.
- Zimmerman correction is often applied in flameless atomic absorption spectrophotometry in order to correct for background noise. This produces a more accurate result.

Since atoms for most elements are not in vapor state at room temperature, flame or heat must be applied to the sample to produce droplets or vapor and breaking the molecular bonds to produce atoms of the element for further analysis. An exception is mercury because mercury vapor can be formed at room temperature. Therefore, "cold vapor atomic absorption" can be used only for analysis of mercury.

Inductively coupled plasma mass spectrometry (ICP-MS) is not a spectrophotometric method but is a mass spectrometric method which is used for analysis of elements, especially trace elements which are found in very small quantities in biological specimens. This technique has much higher sensitivity than atomic absorption method capable of analyzing elements present in parts per billion. In addition, this method can analyze most elements (both metals and nonmetals) found in the Periodic Table. From a clinical laboratory perspective, a significant advantage of ICP-MS is in its capability for simultaneous measurement of multiple elements in a single analysis. In contrast, flame and graphite furnace atomic absorption method where the lamp is specific for a particular element, only one element can be measured in a single run. Coupled with short analysis time and simple sample preparation, ICP-MS offers the opportunity for very high sample throughput in the laboratory. However, due to high cost, such method is available in large medical centers and reference laboratories only.

There are six components in ICP-MS: the sample introduction system, inductively coupled plasma (ICP), interface, ion optics, mass analyzer, and detector. Liquid samples are first nebulized in the sample introduction system, creating a fine aerosol that is subsequently transferred to the argon plasma. The high-temperature plasma atomizes and ionizes the sample, generating ions which are then extracted through the interface region into a set of electrostatic lenses called the ion optics. The ion optics focuses and guides the ion beam into the mass analyzer. Usually quadrupole mass spectrometer is used in ICP-MS where only a singly charged ion can pass through the mass filter at a specific time. As a result, quadrupole mass analyzer separates ions according to their mass-charge ratio ($m/z$), and these ions are measured at the detector. ICP-MS technology is also capable of accurately measuring isotope of an element by using isotope dilution technique. Sometimes an additional separation method such as high-performance liquid chromatography (HPLC) can be coupled with ICP-MS [2].

## Enzymatic assays

Enzymatic assay often uses spectrophotometric detection of signal at a particular wavelength. For example, enzymatic assay of ethyl alcohol (alcohol) utilizes alcohol dehydrogenase enzyme to oxidize ethyl alcohol into acetaldehyde and in this process, co-factor NAD (nicotinamide adenine dinucleotide) is converted into NADH. While NAD does not absorb light at 340 nm, NADH absorbs light at 340 nm. Therefore, absorption of light is proportional to alcohol concentration in serum or plasma (see Chapter 18). Another example of enzymatic assay is determination of blood lactate. Lactate in the blood is converted into pyruvate by the enzyme lactate dehydrogenase and in this process, NAD is converted into NADH which is measured spectrophotometrically at 340 nm. Various enzymes, especially liver enzyme such as aminotransferases (AST and ALT), can be measured by coupled enzymatic reactions. For example, AST converts 2-oxoglutarate into L-glutamate and at the same time converts L-aspartate into oxaloacetate. Then oxaloacetate can be converted into L-malate by malate dehydrogenase and in this process NADH is converted into NAD. The disappearance of signal (NADH absorbs at 340 nm but NAD does not) is measured and can be correlated to AST concentration. In addition, enzyme activities can also be measured by utilizing its ability to convert its substrate to a product which has absorbance at visible or UV range. For example, gamma glutamyl transferase (GGT) activity can be measured by its ability to convert gamma-glutamyl *p*-nitroanilide into *p*-nitroaniline that absorbs at 405 nm. Enzymatic activity is expressed as U/L which is equivalent to IU/L (international unit/L).

Cholesterol, high-density lipoprotein cholesterol (HDL-C), and triglycerides are often measured using enzymatic assays where endpoint signals are measured using spectrophotometric principles of Beer's law. Cholesterol exists in blood mostly as cholesterol ester (approximately 85%). Therefore, it is important to convert cholesterol ester into free cholesterol prior to assay.

$$\text{Cholesterol esters} \xrightarrow{\text{Cholesterol ester hydrolase}} \text{Cholesterol + Fatty acids}$$

$$\text{Cholesterol + Oxygen} \xrightarrow{\text{Cholesterol Oxidase}} \text{Cholest} - 4 - \text{en} - 3 - \text{one} \\ + \text{Hydrogen peroxide}$$

Hydrogen peroxide ($H_2O_2$) then is measured in a peroxidase catalyzed reaction that forms a colored dye, absorption of which can be measured spectrophotometrically at the visible region and concentration of cholesterol can be calculated.

$$H_2O_2 + \text{Phenol} + 4 - \text{amainoantipyrine} \longrightarrow \text{Quinoneimine dye} + \text{water}$$

## Immunoassays

Immunoassays are based on principle of antigen antibody reaction and there are various formats of such immunoassay. However, in many immunoassays, the final signal generated (UV absorption, fluorescence, chemiluminescence, and turbidimetry) is measured using spectrophotometric principles using a suitable spectrophotometer. This topic is discussed in detail in Chapter 2.

## Turbidimetry and nephelometry: Methods based on light scattering

When particles are suspended in a solution in a cuvette, such particles make the solution turbid. Turbidity results in decrease in intensity of the light beam passing through a turbid solution due to scattering of light by particles, as well as absorption of some light by particles. However, a part of the light beam is also transmitted through the cuvette. In turbidimetry, intensity of transmitted light is measured by spectrophotometry to determine concentration of suspended particles in solution by calculating absorbance which is dependent on concentration of particles and particle size. Nephelometry, in contrast to turbidimetry, measures the scattered light rather than transmitted light using spectrophotometer which is placed at an angle (often 90 degrees) to the incident light path. One of the major advantages of turbidimetry and nephelometry is the ability to perform these measurements using spectrophotometry that are readily incorporated into high-throughput clinical analyzer systems. Turbidimetric method can be used for determination of concentration of total protein in urine or cerebrospinal fluid which contains small amounts of proteins using trichloroacetic acid, determination of lipase activity using triglycerides as substrate (lipase hydrolyzes fatty acids from an emulsion of oleic acid with simultaneous decrease in turbidity of the reaction mixture), etc. Nephelometry can be used for determination of immunoglobulins in serum. Since the amount of scattered light is higher than the transmitted light in a turbid solution, nephelometry has higher sensitivity than turbidimetry.

Both nephelometry and turbidimetry can be used as the detection method for immunoassays because when antigen-antibody complex is formed, such particles may cause turbidity. However, turbidimetric or nephelometric immunoassay systems must operate in the antibody excess zone where the concentration of antibody is held constant and the amount of antigen-antibody complex formed depends directly on the concentration of antigen in the mixture. This permits the formation of complexes of a constant size, providing a reproducible, stoichiometric relationship between the number of complexes formed at a given antigen concentration. One of the most important limitations of light-scattering methods, which are based upon the precipitin reaction, is the potential for a prozone effect where excess antigen can lead to a reduced signal.

## Ion selective electrodes

Ion selective electrodes selectively interact with a particular ion and measure its concentration by measuring the potential produced at the membrane-sample interface which is proportional to logarithm of the concentration (activity) of the ion. This is based on Nernst equation which is defined as

$$E = E_O - \frac{RT}{nF} \ln \frac{\text{Reduced ions}}{\text{Oxidized ions}}$$

where $E$ is measured electrode potential, $E_o$ is the electrode potential under standard condition (values are published), $R$ is the universal gas constant (8.3 J/K/mol), $n$ is number of electrons involved, and $F$ is Faraday's constant (96,485 C/mol). Putting these values, we can transform this equation into:

$$E = E_O - \frac{0.0592V}{n} \log \frac{\text{Reduced ions}}{\text{Oxidized ions}}$$

In ion selective electrodes, an interface or a specific membrane is used so that only ions of interest can filter through the membrane and can reach the electrode to create the membrane potential. Ion selective electrodes are used in clinical chemistry laboratories for measuring critical analytes such as sodium, potassium, chloride, calcium, magnesium, and lithium. Ion selective electrode methods can be classified as either indirect or direct. With indirect method, the specimen is diluted with diluent in ratios of 1:20 to 1:34, depending on specific method and then introduced into the measurement chamber. The use of large volume of diluent is advantageous because it adequately covers the entire surface of the electrode and significantly reduces interference from serum proteins. Indirect methods are commonly applied in large, high-throughput automated analyzers. In the direct method, undiluted specimen is introduced in the measuring chamber. Direct method is usually applied in point-of-care devices and blood gas analyzers. However, some laboratory analyzers such as Roche Integra system uses direct method. The main source of interference in ion selective electrode methodology is protein buildup on the membrane surface of the electrode which can be avoided by regular washing [3].

Polymer membrane electrodes are used to determine concentrations of electrolytes, such as sodium, potassium, chloride, calcium, lithium, magnesium, and bicarbonate ions. Glass membrane electrodes are used for measuring pH, sodium, and also a part of carbon dioxide sensor.

- Valinomycin could be used in liquid ion exchange membrane in potassium selective electrode.
- Sodium ion selective electrodes could be glass electrodes or polyvinyl chloride based membrane containing ionophore monensin or its derivatives.
- Membranes of chloride electrodes may contain a quaternary ammonium chloride as an ion exchanger.
- Calcium ion selective electrode may be polyvinyl chloride based with calcium ionophore for selective passage of calcium through membrane.
- Partial pressure of oxygen is measured in a blood gas machine using amperometry oxygen sensor.
- Optical oxygen sensors or enzymatic biosensors can also be used for measuring partial pressure of oxygen in blood.

## Basic principles of chromatography

Chromatography is a laboratory technique for the separation of a mixture into individual components where the mixture is dissolved in a fluid called mobile phase and is allowed to pass through a column packed with a stationary phase (for example, silica). Various components of the mixture are separated based on differential partitioning between mobile and stationary phase (partition coefficient). As a result, individual component elutes from the column at a regular time interval known as retention time. For example, if "A" (polar), "B" (intermediate polar) and "C" (nonpolar) are applied as a mixture to a silica (polar) column followed by passing hexane (nonpolar solvent) through the column. In general, polar compounds have highest affinity for polar compounds while nonpolar compounds have highest affinity for nonpolar compounds. Therefore, "A" being polar should have highest affinity for silica and least solubility in hexane. As a result, "A" should be retained in the column longest and would elute from the column last. In contrast, "C" should have little interaction with the silica but highest solubility in hexane and should elute first from the column while "B" should elute after "C" but before "A." The differential interaction of a component in the mixture with the solid phase and mobile phase (partition coefficient) is the basis of chromatographic analysis. A proper detector at the end of the column can be used for monitoring elution of these compounds from the column.

Chromatography was developed in early 1900 by Michael Tswett for separation of plant pigments such as chlorophyll, carotene, and xanthophyll. The creation of gas chromatography is credited to Archer Martin and his colleagues Richard Synge and Anthony James who won Nobel Prize in Chemistry in 1952 for their research. They, while conducting experiments with separation of amino acid mixtures in 1941, developed liquid partition chromatography and at the same work first predicted the use of a gas instead of a liquid as the mobile phase in a chromatographic process [4]. Later in 1952, James and Martin systemically separated volatile compounds (fatty acids) using gas chromatography (GC). Basis of this separation is differences in vapor pressure of the solutes and Raoult's law [5]. Widely used methods in clinical laboratories are gas chromatography also known as gas-liquid chromatography and high-performance liquid chromatography.

Thin-layer chromatography (TLC) is a typical chromatography method where different migrations of compounds on a specific absorbent (TLC plate) under specific developing solvent (mobile phase) are used for separation of individual components of a mixture. Typically, compounds are spotted on at the edge of the TLC plate and a solvent (or mixture of solvents) is allowed to migrate through the TLC plate as the mobile phase. Then, the plate is sprayed with a specific reagent to visualize spots representing individual compound. In TLC, retention factor ($R_f$) of each component is expressed by comparing the migration of the compound to the solvent front. TLC system consists of a TLC plate, preferably ready-made with a thin layer of stationary phase, for example, silica with fine particle size, and a TLC chamber for developing the plate. The chamber maintains a stable environment inside for proper development of spots. It also prevents the evaporation of solvents and keeps the process dust-free. A mobile phase containing high-purity solvent covering small part of the TLC plate is also used. Although TLC-based method such as TOXI-LAB (using paper as TLC plate) was popular in the past for drugs of abuse testing, today, this method is not used.

## Gas chromatography

The term gas chromatography (GC) indicates chromatographic techniques where an inert gas (such as helium, nitrogen, or argon) is used as the mobile phase which is passed through a GC column. A GC column is coated with a liquid stationary phases (this is the reason it is also known as gas-liquid chromatography; GLC). A compound can be analyzed by GC, if it is nonpolar with appropriate vapor pressure below 350–400°C. Originally, GC columns were wide-bore coiled columns packed with an inert support of high surface area, but later capillary columns were introduced for better resolution of compounds using GC. Such capillary columns are coated with liquid phases such as methyl, methyl-phenyl, propylnitrile, and other functional groups chemically bonded to the silica support. For example, a commonly used GC capillary column for drugs of abuse analysis marketed by Phenomenex is 30 m in length, 0.25 mm internal diameter, and coated with 1 μm layer of dimethylpolysiloxane (Zebron; ZB-1MS column). The effectiveness of the GC columns is based on number of theoretical plates ($n$) as defined by the equation,

$$n = 16(t_r/w_b)^2$$

$t_r$: retention time of the analyte and $w_b$: width of the peak at the baseline.

Major features of gas chromatography (GC) are:

- Gas chromatography can be used for separation of relatively volatile small molecules (approximately 10%–20% of all organic molecules) because GC separations are based on the differences in vapor pressures (boiling points), compounds with higher vapor pressures (low boiling points) will elute faster than the compounds with lower vapor pressures (high boiling points). It is also important to note that molecules should not degrade at high temperature such as 350–400°C.
- A gas chromatograph contains an injector port through which 5–50 μL of specimen dissolved in an appropriate solvent such as hexane or acetonitrile is injected into the capillary column. An oven where capillary column is kept usually in a coiled form to save space, is an essential component of GC. Oven temperature is programed in such a way that the initial oven temperature starting around 100–150°C is increased at a rate of 5–10°C/min to reach a final oven temperature around 300–350°C. As a result, more volatile compounds would elute from the column first

followed by less volatile compounds. Using temperature gradient, faster analysis time can be achieved.
- Generally, boiling point increases with increasing polarity.
- Compounds are typically identified by the retention time (RT) or travel time needed to pass through the GC column. Retention times depend on flow rate of gas (helium or an inert gas) through the column, nature of column, and boiling points of analytes.
- After separation by GC, compounds can be detected by flame-ionization detector (FID), electron-capture detector (ECD), nitrogen-phosphorus detector (NPD), or electrochemical detectors.
- Mass spectrometer (MS) is a specific detector for GC because mass spectral fragmentation patterns are specific for compounds except for optical isomers. Gas chromatography combined with mass spectrometry (GC-MS) is widely used in clinical laboratories for drugs of abuse and other analysis. Gas chromatography is used in toxicology laboratory for analysis of volatiles (methanol, ethanol, propanol, ethyl glycol, and propylene glycol), and selected drugs such as pentobarbital. In general, flame ionization detector is used for such analysis. A GC/MS system has an injector port where the specimen is introduced into the capillary column. Injector port temperature is usually high (100–300°C) to ensure that after injection compounds are easily volatilized. A carrier gas such as helium is typically used (although hydrogen or nitrogen can also be used as the carrier gas) and after elution of individual compound from the column, mass spectrometry is used as a director.
- Sometimes for GC analysis, a relatively nonvolatile compound, for example, a relatively polar drug metabolite can be converted into a nonpolar compound by chemically modifying a polar functional group into a nonpolar group. For example, a polar amino group ($-NH_2$) can be converted into a nonpolar group ($-NH-CO-CH_3$) by reaction with acetic acid and acetic anhydride. This process is called derivatization. GC/MS total ion chromatogram for benzodiazepines at 0.1 mg/L in blood after flash derivatization with *tert*-butyl-trimethylsilyl derivatization is shown in Fig. 1.

## High-performance liquid chromatography

One major limitation of gas chromatography is that only small molecules capable of existing in the vapor state without decomposition can be analyzed by this method. Therefore, polar molecules and molecules with higher molecular weight, for example, immunosuppressant cyclosporine, cannot be analyzed by gas chromatography. On the other hand, liquid chromatography can be used for analysis of both polar and nonpolar molecules. High-performance

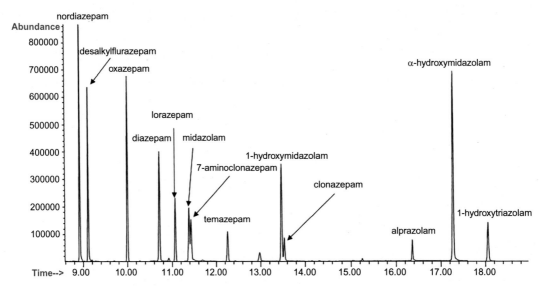

**FIG. 1**

GC/MS total ion chromatogram for benzodiazepines at 0.1 mg/L in blood after flash derivatization with *tert*-butyl-trimethylsilyl derivatization. *Adapted from Fig. 15.6 in Justin Holler and Barry Levine "Confirmation methods for SAMHSA drugs and other common abused drugs". Chapter 15 in "Critical issues in alcohol and drugs of abuse testing", 2nd edition (edited by Amitava Dasgupta), Elsevier 2019. Reprinted with permission.*

liquid chromatography (HPLC) also called high-pressure liquid chromatography is usually used in clinical laboratories in order to achieve better separation. An HPLC system contains following components:

- Solvent reservoir and pump: The mobile phase (one solvent or multiple solvents for solvent gradient) is stored in solvent reservoir which is usually glass container. The pump sucks mobile phase from the reservoir and forces the mobile phase through the column at high pressure (4000–6000 psi; 276–413 Bar). The operating pressure depends on particle size of stationary phase, flow rate, and composition of mobile phase.
- Injector: Injector allows the specimen to flow to the column at high pressure. Injector could be manual or automated where specimens can be analyzed in a batch one after another with a predetermined time interval.
- Column and mobile phase: HPLC columns are typically stainless steel column capable of withstanding high pressure. Columns are 50–300 mm in length with inner diameter of 2–5 mm. A column is packed with fine particles typically 3–5 μm in diameter but may vary with particular stationary phase. In order for the mobile phase to move through the column, a high pressure must be created which is achieved by using a high-performance pump. The elution of analytes from the column is

monitored by a detection method and computer can be used for data acquisition and analysis. For separation of polar compounds, a polar stationary phase such as silica is used and mobile phase should be nonpolar such as hexane, carbon tetrachloride, etc. This is called "normal phase chromatography." For separation of relatively nonpolar molecules a nonpolar stationary phase such as derivatized silica is used and mobile phase is a polar solvent such as methanol or acetonitrile. This is known as "reverse-phase chromatography." In clinical laboratories, reverse-phase chromatography is most commonly used. Commonly used derivatized silica in chromatographic columns includes C-18 (an 18 carbon fatty acid chain linked to the silica molecule), C-8, and C-6. Other types of chromatography such as ion-exchange chromatography (the stationary phase is having ionically charged surface opposite to charges of compounds of interest, and mobile phase is an aqueous buffer which controls the pH and ionic strength) and size-exclusion chromatography (molecules are separated based on molecular size using special stationary phase) are less commonly used in clinical laboratories.

- Detector: Elution of a compound from a liquid chromatography column can be monitored by ultraviolet-visible (UV-Vis) spectrophotometry, but UV detection is more common as many analytes absorb wavelength at UV region. Refractive index detection where change of refractive index of the mobile phase (solvent) due to elution of a peak from the column is measured is not used in clinical laboratory because this method is far less sensitive than UV detection. Fluorescence detection, a very sensitive technique more sensitive than UV, is also used in clinical laboratory as the detector of HPLC system. However, higher sensitivity and specificity can be achieved by using mass spectrometry detectors. High-performance liquid chromatography combined with tandem mass spectrometry (LC-MS/MS) is the most sensitive and robust method available in a clinical laboratory.
- Data collection: Signal from the detector must be stored in a computer which may produce a graph showing peaks eluting from the column in a specific time interval (chromatogram).

The time it takes for an analyte to elute from the column after injection is called "retention time" which depends on partition coefficient (differential interaction of the analyte with the stationary phase and the mobile phase). Retention time is usually expressed in minutes. When analytes of interest are separated from each other completely, it is called baseline separation. Basic principles of retention time of a compound include:

- Retention time of a compound can be reduced by increasing the flow rate. For example, if retention time of A is 5 min, retention time of B is 7 min, but retention time of C is 15 min and initial flow rate of the mobile phase through the column is 1 mL/min, then after elution of B at 7 min, the flow rate can be increased to 3 mL/min to shorten the retention tine of C in order to reduce the run time.

- If compounds A and B have the same or very similar partition coefficient for a particular stationary phase and mobile phase combination, then compounds A and B cannot be separated by chromatography using the same stationary-phase and mobile-phase composition. A different stationary phase, mobile phase or both stationary and mobile phase may be needed to separate compound A from B.
- Sometimes more than one solvent is used in composing the mobile phase by mixing predetermined amounts of two solvents. This is called "gradient" but if only one solvent is used in the mobile phase, it is termed as "isocratic condition." Using more than one solvent in mobile phase may improve the chromatographic separation.
- Sometime heating the column at 40–60°C may improve separation between peaks. This is often used for chromatographic analysis of immunosuppressants.

## Mass spectrometer as a detector

Mass spectrometry as mentioned earlier in this chapter is a very powerful detection method which can be coupled with a gas chromatography or a high-performance liquid chromatography analyzer. Mass spectrometric analysis takes place in a very low pressure except for atmospheric pressure chemical ionization mass spectrometry. During mass spectrometric analysis, analyte molecules in the gaseous phase are bombarded with high-energy electrons (electron ionization; commonly used in GC/MS) or a charged chemical compound with low molecular weight such as charged ammonia ions (chemical ionization). During collision, analyte molecules lose an ion to form a positively charged ion and may undergo further decomposition (fragmentation) into smaller charged ions. If the analyte molecule loses one electron and retain its identity, it forms a molecular ion ($m/z$) where $m$ is the molecular weight of the analyte and $z$ is the charge which is usually one. The fragmentation pattern depends on the molecular structure including the presence of various function groups in the molecule. Therefore, fragmentation pattern is like a finger print of the molecule and only optical isomers produce identical fragmentation pattern. The mass spectrometric detector can detect ions with various molecular mass and construct a chromatogram which is usually $m/z$ in the "$x$" axis, and intensity of the signal (ion strength) at the "$y$" axis. Although positive ions are more commonly produced during mass spectrometric fragmentation pattern, negative ions are also generated, especially during chemical ionization mass spectrometry. Therefore, negative ions can also be monitored, but it is done less often than positive ion mass spectrometry in clinical toxicology laboratories. Major features to remember in coupling a mass spectrometer with a gas chromatograph or HPLC system include:

- Because mass spectrometry occurs in vacuum, after elution of an analyte with the carrier gas from the gas chromatography column, the carrier gas must be removed quickly to have volatile analyte entering the mass spectrometer. This is achieved by high-performance turbo pump at the interface of gas chromatograph and mass spectrometer.

- Most commonly in GC/MS analysis, the mass spectrometer is operated in electron ionization (electron impact 70 eV) mode. However, gas chromatography combined with chemical ionization mass spectrometry is gaining more application in toxicology laboratories. For drugs of abuse analysis, drugs and metabolites are extracted from biological matrix (most commonly urine) using liquid-liquid or solid phase extraction. Sometimes polar conjugates such as glucuronic acid conjugate of a drug metabolite should be subjected to enzymatic or acid hydrolysis to break the conjugate prior to extraction because polar drug metabolite conjugated with glucuronic acid cannot be analyzed by GC/MS. If necessary, polar drug molecules are chemically modified to nonpolar compounds by a process known as "derivatization." In GC/MS system, in addition to chromatogram (showing retention time of each compound eluting from the capillary column), the computer in the mass spectrometer will also produce mass spectrum of individual compound. For drugs of abuse analysis, mass spectrometer is often operated in selected ion monitoring mode (where 2–3 characteristic ion of a particular compound is monitored) instead of full scan to increase sensitivity of the analysis.

- One advantage of chemical ionization mass spectrometry is that it is a soft ionization method and usually a strong molecular ion peak as adduct; ($M+H^+$, molecular ion adduct with hydrogen), or ($M+NH_4^+$, molecular ion adduct with ammonia) can be observed. In contrast, M + molecular ion peak in electron ionization method can be a very weak peak for certain analytes. However, chemical ionization is less common than electron impact in GC/MS analysis for drugs of abuse confirmation.

- Combining a high-performance liquid chromatography with a mass spectrometer is a big challenge because in GC/MS analysis, carrier gas can be easily pumped out by a turbo pump, but removing liquid mobile phase coming out of a column is more difficult. The interface between a liquid-phase technique (HPLC) with a continuously flowing eluent and a gas-phase technique carried out in a vacuum was difficult for a long time. However, with the advent of electrospray ionization, this problem was solved. Currently, the most common LC-MS interfaces are electrospray ionization (ESI), atmospheric pressure chemical ionization (APCI), and atmospheric pressure photoionization. ESI methods is most commonly used in clinical laboratories. ESI is a sensitive ionization technique for analytes that exist as ions in the eluent from HPLC column. In ESI, a solvent spray is formed by the application of a high voltage potential held between a stainless steel capillary and the instrument orifice, coupled with an axial flow of a nebulizing gas (typically nitrogen). Solvent droplets from the spray evaporate in the ion source of the mass spectrometer, releasing ions to the gas phase for analysis in the mass spectrometer. In some ESI sources, heat is used to increase the efficiency of removing solvent. While ESI is widely used, it is subject to matrix effects, particularly ion suppression, which must be taken into consideration during method development. A major advantage of liquid chromatography combines with mass spectrometry or tandem mass spectrometry over GC/MS is that both polar and nonpolar compounds can be analyzed using liquid chromatography. Therefore, for drugs of abuse analysis, no derivatization is needed for analysis using liquid chromatography combined with mass spectrometry of tandem mass spectrometry. However, for those molecules which are unstable, difficult to ionize or difficult to undergo fragmentation, derivatization could enhance quality of mass spectrum using LC-MS or LC-MS/MS.

- Liquid chromatography combined with tandem mass spectrometry (LC-MS/MS): In order to achieve high sensitivity and specificity, LC-MS/MS is preferred over a simple LC-MS system. It is sometimes called triple quad because the mass spectrometric detection is based on three subsequent steps. In the first step, compounds of interest are separated into individual compound from the mixture using HPLC column followed by ionization and analysis by the first mass spectrometer (first quadrupole) to produce fragment ions characteristic of that compound. In the second step, the selected fragment ion (precursor ion) enters a collision chamber containing inert gas molecules (second component of triple quad is sometimes erroneously called second quadrupole). In the collision chamber, the precursor ion undergoes further fragmentation producing daughter ions. Finally, all fragmented ions (precursor ions along with product ions, also known as daughter ions) are analyzed by a second quadrupole mass spectrometer to produce the final mass spectrum of the compounds. Fragmentation of precursor ions to product ions is referred as transition. This three-part strategy of triple quad makes the analytical technique very specific and sensitive. Multiple reaction monitoring (MRM, also known as selected reaction monitoring, or SRM) with careful selection of precursor ion $[M+H]^+$ molecular ion and subsequent product ion selection after collision-induced decomposition in the collision chamber creates increased specificity to identify a particular analyte. It is recommended to monitor more than one precursor/product ion (quantifier and qualifier) for increased sensitivity and specificity [6].

- For ease of quantitation, an internal standard (with known concentration) is added to the specimen prior to extraction in both GC and HPLC analysis. The internal standard should be structurally similar to the analyte of interest. For GC/MS or LC-MS/MS analysis, deuterated analog of the target drug or the analyte is most appropriate.

## Examples of application of chromatographic techniques in clinical toxicology laboratories

Chromatographic methods are used in toxicology laboratory for following reasons:

- For therapeutic drug monitoring of a drug where there is no commercially available immunoassay.
- Immunoassays are commercially available but have poor specificity. Good examples are immunoassays for immunosuppressants (cyclosporine, tacrolimus, sirolimus, everolimus, and mycophenolic acid) where metabolite cross-reactivity may produce 20%–50% positive bias in compared to a specific chromatographic method. For therapeutic drug monitoring of immunosuppressants, LC/MS or LC-MS/MS is the gold standard and preferred method of analysis.
- For legal blood alcohol determination, headspace GC is the gold standard (please see Chapter 18).
- GC/MS or LC-MS/MS is needed for confirmation of an abuse drug for legal drug testing.

Other than therapeutic drugs, and drugs of abuse, LC-MS/MS is also used in reference laboratories for analysis of vitamin D, hormones and diagnostic metabolites such as methylmalonic acid and free metanephrine.

Subramanian et al. described LC/MS analysis of nine anticonvulsants; zonisamide, lamotrigine, topiramate, phenobarbital, phenytoin, carbamazepine, carbamazepine, 10, 11-diol, 10-hydroxycarbamazepine, and carbamazepine 10, 11-epoxide. Sample preparation included solid-phase extraction for all anti-convulsants and HPLC separation was achieved by a reverse phase C-18 column (4.6 × 50 mm, 2.2 μm particle size) with a gradient mobile phase of acetate buffer, methanol, acetonitrile, and tetrahydrofuran. Four internal standards were used. Detection of peaks was achieved by atmospheric pressure chemical ionization mass spectrometry in selected ion monitoring mode with constant polarity switching [7].

LC-MS/MS is very useful in analysis of antifungals because no immunoassay is commercially available. Toussaint et al. described a liquid chromatography combined with tandem mass spectrometry method of simultaneous measurement of eight antifungals compounds: isavuconazole, voriconazole, posaconazole, fluconazole, caspofungin, flucytosine, itraconazole, and its metabolite hydroxy-itraconazole using only $50 \mu L$ of plasma [8]. Currently, no immunoassay is commercially available for analysis of antiretrovirals. Although HPLC combined with UV detection can be used for therapeutic drug monitoring of antiretrovirals, LC-MS/MS provides increased specificity for such analysis. Koal et al. described an online solid-phase extraction LC-MS/MS analysis of seven protease inhibitors and two nonnucleoside reverse transcriptase inhibitors in human plasma [9]. More recently, therapeutic drug monitoring using dried blood spot is gaining acceptance as an alternative specimen to blood. Immunoassays cannot analyze such low concentration of drugs present in dried blood spot. LC-MS/MS is the preferred standard for such analysis.

GC/MS is still widely used for drugs of abuse confirmation. The presence of benzoylecgonine, the inactive major metabolite of cocaine, must be confirmed by GC/MS in legal drug testing (such as preemployment drug testing) if initial immunoassay screen is positive. The carboxylic acid in benzoylecgonine must be derivatized prior to GC/MS analysis. A representative spectrum of the propyl ester of benzoylecgonine is shown in Fig. 2. Molecular ion and fragment ions

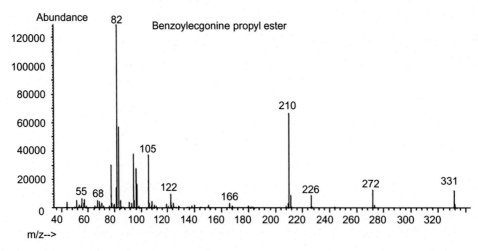

**FIG. 2**

Mass spectrum of benzoylecgonine propyl ester. *Figure courtesy of Dr. Buddha Dev Paul. Division of Forensic Toxicology (retired), Armed Forces Forensic Toxicology Laboratory, Rockville, Maryland.*

from the side chain are the major ions. Fragment ion $m/z$ 82 is unique to the core structure of the compound. The ion at $m/z$ 331 is the molecular ion.

More recently, LC-MS/MS is gaining popularity for drugs of abuse testing because many drugs can be analyzed in a single run. Kong et al. developed a liquid chromatography-tandem mass spectrometric (LC-MS/MS) method for analysis of 113 abuse drugs and their metabolites in human urine. A simple sample clean-up procedure using the "dilute and shoot" approach, followed by reversed phase separation using a C18 column, provided a fast and reliable method for routine analysis. Drugs were separated using a mobile phase gradient elution of 1 mM ammonium formate with 0.1% formic acid in water and acetonitrile. The total time for analysis was 32 min. The multiple reaction monitoring mode using two transitions (e.g., quantifier and qualifier) was optimized for both identification and determination. The authors successfully applied their method for the analysis of forensic urine samples obtained from 17 drug abusers [10].

## Automation in clinical laboratory

Automated analyzers are widely used in clinical laboratories for speed, ease of operation, and capability of a technologist to load a batch of samples for analysis, programming the instrument and then walk away when the analyzer automatically pipets small amount of specimen from the sample cup, mix it with reagent, and then recording the signal and finally producing the result. Therefore, automation follows the similar steps of analysis as a manual laboratory technique but each step is mechanized. Most common configuration of automated analyzers is "random access analyzers" where multiple specimens can be analyzed for different selection of tests. More recently modular analyzers have been introduced by manufacturers and to provide improved operational efficiency. Automated analyzers can be broadly classified under two categories:

- Open system where a technologist is capable of programing parameters for a test using reagents prepared in-house or obtained from a different vendor than the manufacturer.
- Closed system where the analyzer requires that the reagent should be in a unique container or format that is usually marketed by the manufacturer of the instrument or a vendor authorized by the manufacturer. Usually such proprietary reagents are more expensive than reagents available from multiple vendors which can be only be adapted to an open system analyzer.

Most automated analyzers have bar code readers so that the instrument can identify patient's specimen from the bar code. Moreover, many automated analyzers can be interfaced to the laboratory information system (LIS) so that after verification by the technologist and releasing the result, it is automatically transmitted to the patient record eliminating need for manual entry of the result in the computer. This is not only time efficient but also useful to prevent transcription error during manual entry of the result in the LIS.

More recently, total automation system is available for a laboratory where after receiving the specimen, the automated system can process the specimen including automated centrifugation, aliquoting, and delivering the aliquot to the analyzer. Invention of robotic arms makes this total automation in a clinical laboratory feasible.

## Electrophoresis including capillary electrophoresis

Electrophoresis is a technique that utilizes migration of charged solutes or analytes in a liquid medium under the influence of an applied electrical field. This is a very powerful technique for analysis of proteins in serum or urine and in analysis of various hemoglobin variants. Please see Chapter 24 for in-depth discussion on this topic.

## Summary/key points

- Major analytical methods used in clinical chemistry laboratory include spectrophotometry, ion selective electrodes, gas chromatography with various detectors, gas chromatography combined with mass spectrometry (GC/MS), high-performance liquid chromatography and liquid chromatography combined with mass spectrometry (LC-MS) or tandem mass spectrometry (LC-MS/MS).
- Spectrophotometric measurements are based on Beer's law or sometimes referred as Beer-Lambert Law. In spectrophotometry, often, transmittance is measured as absorption "$A$" because there is a linear relation between absorbance and concentration of the analyte in the solution.

$A = -\log T = -\log I_s/I_r = \log I_r/I_s$ where $I_r$ is the intensity of the light beam transmitted through the reference cell (containing only solvent) and $I_s$ is the intensity of the transmitted light through the cell containing the analyte of interest dissolved in the same solvent as reference cell. The scale of absorbance is from 0 to 2, where zero value is for no absorbance.

- Absorption of the light also depends on the concentration of the analyte in the solvent and length of the cell path. Therefore, $A = \log I_r/I_s = a \cdot b \cdot c$. where, "$a$" is proportionality constant termed as "absorptivity," "$b$" is the length of the cell path, and "$c$" is the concentration. If "$b$" is 1 cm and concentration of the analyte is expressed as mol/L, then "$a$" is termed as "molar absorptivity" and often designated as "$\varepsilon$" (epsilon). The value of "$\varepsilon$" is a constant for a particular compound for a particular wavelength under prescribed condition of pH, solvent, and temperature.

- In atomic absorption spectrophotometry (used for analysis of various elements including heavy metals), components of gaseous sample are converted into free atoms which can be achieved by a flame or by a flameless manner using a graphite chamber that can be heated after application of the sample. In atomic absorption spectrophotometry, a hollow cathode lamp (metal cathode contains the analyte of interest, for example, for copper analysis, cathode is made of copper) containing an inert gas like argon or neon at a very low pressure is used as a light source. Atoms in the ground state then absorb a part of the light emitted by the hollow cathode lamp to be in the excited state. Therefore, a part of the light beam is absorbed and resulting in a net decrease in the intensity of the beam arriving at the detector. Applying principles of Beer's law, concentration of the analyte of interest can be measured. Zimmerman correction is often applied in flameless atomic absorption spectrophotometry to correct for background noise to produce more accurate results. Mercury is vaporized at room temperature. Therefore, "cold vapor atomic absorption" can be used only for analysis of mercury.
- Inductively coupled plasma mass spectrometry (ICP-MS) is not a spectrophotometric method, but is a mass spectrometric method, which is used for analysis of elements, especially trace elements, which are found in very small quantities in biological specimens.
- When particles are suspended in a solution in a cuvette, such particles make the solution turbid. Turbidity results in decrease of intensity of the light beam passing through a turbid solution due to scattering of light by particles and absorption of some light by particles. However, a part of the light beam is also transmitted through the cuvette. In turbidimetry, intensity of transmitted light is measured by spectrophotometry to determine the concentration of suspended particles in solution by calculating absorbance, which is dependent on concentration of particles and particle size. Nephelometry, in contrast to turbidimetry, measures the scattered light rather than transmitted light using spectrophotometer which is placed at an angle (often 90 degrees) to the incident light path.
- Turbidimetry and nephelometry are light scattering-based methods and a potential problem is prozone effect where excess antigen can lead to a reduced signal and falsely low value.
- In ion selective electrodes, an interface or a specific membrane is used so that only ions of interest can filter through the membrane and can reach

the electrode to create the membrane potential. Ion selective electrodes are used in clinical chemistry laboratories for measuring critical analytes such as sodium, potassium, chloride, calcium, magnesium, and lithium. Ion selective electrode methods can be classified as either indirect or direct. With indirect method, the specimen is diluted with diluent in ratios of 1:20 to 1:34 depending on specific method and then introduced into the measurement chamber. Indirect methods are commonly applied in large, high-throughput automated analyzers. In the direct method, undiluted specimen is introduced in the measuring chamber. Direct method is usually applied in point-of-care devices and blood gas analyzers. However, some laboratory analyzers such as Roche Integra system use direct method.

- Valinomycin could be incorporated into a potassium selective electrode.
- Gas chromatography can be used for separation of relatively volatile small molecules where compounds with higher vapor pressures (low boiling points) will elute faster than the compounds with lower vapor pressures (high boiling points). Compounds are typically identified by the retention time (RT) or travel time needed to pass through the GC column. Retention times depend on flow rate of gas (helium or an inert gas) through the column, nature of column, and boiling points of analytes. After separation by GC, compounds can be detected by flame-ionization detector (FID), electron-capture detector (ECD), or nitrogen-phosphorus detector (NPD). However, mass spectrometer is the most specific detector for gas chromatography.
- Although gas chromatography can be applied only for analysis of relatively volatile compounds or compounds which can be converted into volatile compounds using chemical modification of the structure (derivatization), high-performance liquid chromatography (HPLC) is capable of analyzing both polar and nonpolar compounds. Common detectors used in HPLC system include ultraviolet detector (UV detector), fluorescence detector, or electrochemical detector. However, liquid chromatography combined with mass spectrometry is a superior technique and a very specific analytical tool. Electrospray ionization is commonly used in liquid chromatography combined with mass spectrometry or tandem mass spectrometry (MS/MS).
- Automated analyzers can be broadly classified under two categories: open system where a technologist is capable of programing parameters for a test using reagents prepared in house or obtained from a different vendor than the manufacturer of the analyzer and closed system where the analyzer requires that the reagent should be in a unique container or format that is usually marketed by the manufacturer of the instrument or a vendor authorized by the manufacturer.

# References

[1] Dasgupta A, Zaidi S, Johnson M, Chow L, Wells A. Use of fluorescence polarization immuno-assay for salicylate to avoid positive/negative interference by bilirubin in the Trinder salicylate assay. Ann Clin Biochem 2003;40:684–8.

[2] Wilschefski SC, Baxter MR. Inductively coupled plasma mass spectrometry: introduction to analytical aspects. Clin Biochem Rev 2019;40:115–33.

[3] Dimeski G, Badrick T, John AS. Ion selective electrodes (ISEs) and interferences—a review. Clin Chim Acta 2010;411:309–17.

[4] Kolomnikov IG, Efremov AM, Tikhomirova TI, Sorokina NM, Zolotov YA. Early stages in the history of gas chromatography. J Chromatogr A 2018;1537:109–17.

[5] James AT, Martin AJP. Gas-liquid partition chromatography: the separation and micro-estimation of volatile fatty acids from formic acid to dodecanoic acid. Biochem J 1952;50:679–90.

[6] van den Ouweland JM, Kema IP. The role of liquid chromatography-tandem mass spectrometry in the clinical laboratory. J Chromatogr B 2012;883–884:18–32.

[7] Subramanian M, Birnbaum AK, Remmel RP. High speed simultaneous determination of nine antiepileptic drugs using liquid chromatography-mass spectrometry. Ther Drug Monit 2008;30:347–56.

[8] Toussaint B, Lanternier F, Woloch C, Fournier D, et al. An ultra-performance liquid chromatography-tandem mass spectrometry method for the therapeutic drug monitoring of isavuconazole and seven other antifungal compounds in plasma samples. J Chromatogr B Anal Technol Biomed Life Sci 2017;1046:26–33.

[9] Koal T, Sibum M, Koster E, Resch K, et al. Direct and fast determination of antiretroviral drugs by automated online solid phase extraction-liquid chromatography-tandem mass spectrometry in human plasma. Clin Chem Lab Med 2006;44:299–305.

[10] Kong TY, Kim JH, Kim JY, In MK, et al. Rapid analysis of drugs of abuse and their metabolites in human urine using dilute and shoot liquid chromatography-tandem mass spectrometry. Arch Pharm Res 2017;40:180–96.

# CHAPTER 2

# Immunoassay design and issues of interferences

## Application of immunoassays for various analytes

Currently over one hundred immunoassays are commercially available and routinely used in clinical laboratory analysis. They help in the diagnosis and monitoring of various diseases, including allergy, anemia, autoimmune, cardiovascular, diabetes, metabolic, endocrine, and cancer. Commercially available immunoassays for various analytes are listed in Table 1. Most immunoassay methods use specimens without any pretreatment, are easy to use, and can run on fully automated, continuous, random-access systems. The assays use very small amounts of sample volumes (most <100 μL), reagents are stored in the analyzer, most have stored calibration curves on the system, often stable for 1–2 months, and report results in 10–30 min. Immunoassays offer fast throughput, automated rerun, relatively inexpensive tests of high sensitivity and specificity, and results are reported in a format that the laboratory wants and can be uploaded directly into laboratory information system (LIS). Moreover, software of automated analyzers is also capable of autoflagging (to alert for poor specimen quality such as hemolysis, high bilirubin, and lipemic specimen that may affect test result).

Immunoassays measure the analyte concentration in a specimen by forming a complex with a specific binding molecule, which in most cases is an analyte-specific antibody (or a pair of specific antibodies). The complex generates some kind of signal which is then converted into the analyte concentration via a "calibration curve." The immunoreaction is further utilized in various formats and labels, giving a whole series of immunoassay technologies, systems, and options. However, immunoassays also suffer from interferences from both endogenous and exogenous factors.

Clinical Chemistry, Immunology and Laboratory Quality Control. https://doi.org/10.1016/B978-0-12-815960-6.00007-8

Table 1 Examples of common analytes where immunoassays are commercially available.

| Type of analyte | Specific example |
|---|---|
| Allergy marker | Immunoglobulin E (IgE) |
| Anemia markers | Ferritin, folate, vitamin B12, etc. |
| Autoantibodies | Antinuclear antibody (ANA), rheumatoid factor, etc. |
| Cardiac markers | Troponin I, Troponin T, high-sensitivity troponin I, and T, B type-natriuretic peptide (BNP), NT-proBNP, CK-MB, etc. |
| Diabetes markers | Insulin, C-peptide, HbA1c, etc. |
| Drugs of Abuse | Amphetamine, Benzoylecgonine (cocaine metabolite), Opiates, Oxycodone, Hydrocodone, Methadone, Tramadol, Fentanyl, Marijuana metabolite, Phencyclidine, Benzodiazepines, Barbiturates, etc. |
| Hormones | ACTH, LH, FSH, prolactin, hCG, cortisol, parathyroid hormone (PTH), etc. |
| Infectious diseases | Antibodies to and antigens from infectious microorganisms such as hepatitis and HIV |
| Therapeutic drug monitoring | Digoxin, phenytoin, carbamazepine, valproic acid, phenobarbital, primidone, theophylline, Various aminoglycosides, vancomycin, immunosuppressants (cyclosporine, tacrolimus, sirolimus, everolimus, mycophenolic acid,) etc. |
| Thyroid diseases | T4, T3, TSH, thyroglobulin, etc. |
| Tumor markers | Alpha-fetoprotein, Beta-2 microglobulin, CA 15-3, CA 19-9, CA-125, calcitonin, carcinoembryonic antigen (CEA), gastrin, human chorionic gonadotropin (hCG), prostate-specific antigen (PSA), etc. |

## Immunoassay design and principle

Immunoassay design can be classified under two broad categories;

- Competition Immunoassay: This design uses only one antibody specific for the analyte molecule and is widely used for detecting small analyte molecules such as various therapeutic drugs and drug of abuse.
- Immunometric or non-competitive (Sandwich) Immunoassay: This design uses two analyte-specific antibodies recognizing different parts of the analyte molecule and is used for analysis of large molecules such as tumor markers, cardiac troponin I, and cardiac troponin T.

Depending on the need for separation between the bound labels (labeled antigen–antibody complex) versus free labels, the immunoassays may be further subclassified into homogeneous or heterogeneous format.

- Homogenous Immunoassay Format: After incubation, no separation between bound and free form is necessary.
- Heterogenous Immunoassay Format: Bound label must be separated from the free label before measuring the signal.

In competitive immunoassay, predetermined amounts of labeled antigen and antibody are added to the specimen followed by incubation. In the basic design of a competitive immunoassay, analyte molecules present in the specimen compete with analyte molecules attached with a tag (labeled analyte molecule) and added to the sample in a predetermined amount for limited number of binding sites in the antibody molecules (also added to the specimen in a predetermined amount). After incubation, signal is measured with separation (heterogenous format) or without separation (homogenous format) of antibody-bound labeled antigen from labeled antigen molecules which are free in reaction mixture. Let us take a hypothetical scenario presented in Fig. 1. In Scenario 1, four labeled antigen molecules and four antigen molecules present in the specimen are competing for four binding antibodies, while in scenario 2, more antigen molecules (analyte) are present. As expected, in the competitive assay format in scenario 1, more labeled antigen molecules would bind with the antibody than in scenario 2. If signal is produced when a labeled antigen is bound with an antibody molecule, then more signal will be generated in scenario 1 than scenario 2. Therefore, general conclusions are:

- If signal is generated when a labeled antigen bind with an antibody molecule, then signal is inversely proportional to analyte concentration in the specimen, for example, FPIA assay design.
- If signal is generated by an unbound labeled antigen, then assay signal is directly proportional to the analyte concentration, for example, EMIT assay (enzyme multiplied immunoassay technique).

In noncompetitive (sandwich) assay (Fig. 2), captured antibodies specific to the analyte are immobilized on a solid support (microparticle bead, microtiter

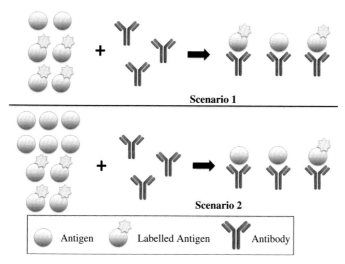

Scenario 1

Scenario 2

| Antigen | Labelled Antigen | Antibody |

**FIG. 1**
Competitive immunoassay. *(Figure courtesy of Stephen R. Master, MD, PhD, Children's Hospital of Philadelphia)*

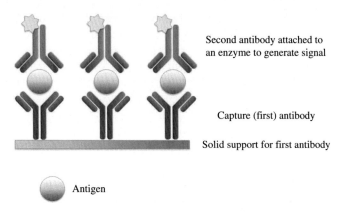

Second antibody attached to an enzyme to generate signal

Capture (first) antibody

Solid support for first antibody

Antigen

**FIG. 2**
Sandwich immunoassay. *(Figure courtesy of Stephen R. Master, MD, PhD, Children's Hospital of Philadelphia)*

plate, etc.). After specimen is added, a predetermined time is allowed for incubation of the analyte with antibody and then liquid reagent containing second antibody conjugated to a molecule for generating signal (label), for example, an enzyme, is added. Alternatively, after adding patient serum, liquid reagents may be added followed by single incubation. Then, a sandwich is formed. After incubation, excess antibody may be washed off by a washing step and a substrate for the label (for example, an enzyme) can be added for generating a signal which can be measured. Analyte concentration is directly proportional to the intensity of the signal.

The main reagent in immunoassays is the binding molecule, which is most commonly an analyte-specific antibody or its fragment. Several types of antibodies or their fragments are now used in immunoassays. There are polyclonal antibodies, which are raised in an animal when the analyte (as antigen) along with an adjuvant is injected into the animal. Small molecular weight analytes are most commonly injected as conjugates of a large protein, for example, albumin. Appearance of analyte-specific antibodies in the animal's sera is monitored, and when sufficient concentration of the antibody is reached, the animal is bled. The serum can directly be used as the analyte-specific binder in the immunoassay, but most commonly, the antibodies are purified from serum and used in the assay. Since there are many clones of the antibodies specific for the analyte, these antibodies are called polyclonal. In newer technologies, however, plasma cells of the animal can be selected as producing the optimum antibody, and then can be fused to immortal (usually myeloma) cells to form a hybridoma. The resulting tumor cell grows uncontrollably, producing only the single clone of desired antibody. Such antibodies, called monoclonal antibodies, now may be grown in live animals or cell culture. There are several benefits of the monoclonal antibodies over polyclonal antibodies:

- Binding properties of polyclonal antibodies are dependent on the animal producing the antibodies; if the source individual animal must be changed, the resultant antibody may be quite different.
- Since polyclonal antibodies constitute many antibody clones, they have less specificity than monoclonal antibodies.

Sometimes, instead of using the whole antibody, fragments of the antibody, generated by digestion of the antibody by peptidases, for example Fab, Fab', or their dimeric complexes, are also used as reagents.

The other main reagent component of immunoassays is the label. There are many different kinds of labels, generating different kinds of signals. For example, use of acridinium ester labels, when treated with peroxide, produces a chemiluminescent signal. As described earlier, an enzyme may be used as the label, which in its turn can generate different types of signals depending on the substrate used for the enzyme.

For ease of manufacturing, some manufactures use a solid phase with a common binding protein like streptavidin immobilized, and a common signal reagent. In such immunoassays the specific antibodies are conjugated with biotin. This assay architecture can be used with both competition and sandwich assays. While such design improves assay sensitivity and ease of manufacturing, such biotin-based assays may become vulnerable to interference from exogenous biotin interference if present in high concentration such as people taking high amount of biotin supplement. However, in most people who do not take biotin supplement in high dosage (5 mg or less per day) or people taking multivitamins containing much lower of biotin, there should be no biotin interference due to low amount of biotin present in sera. In competitive format, biotin-based assays show positive interference if biotin is present in high amount. In contrast, biotin interference is negative in sandwich immunoassay; this is discussed in Chapter 22.

## Various commercially available immunoassays

Many immunoassays are available commercially for analysis of a variety of analytes. These assays use different labels and different methods for generating and measuring signals, but basic principles are the same as described in the immunoassay design section. FPIA, EMIT, CEDIA, KIMS, and LOCI assays are examples of homogenous competitive immunoassay design. Design and formats of common commercially available immunoassays are summarized in Table 2.

**Table 2** Various types of common immunoassay kits.

| Immunoassay type | Immunoassay format | Example | Signal |
|---|---|---|---|
| Competition Analytes approximately 500–1000 Da | Homogenous | FPIA (TDx from Abbott) assays for therapeutic drugs and drugs of abuse | Fluorescence polarization, but Abbott Laboratories discontinued this format |
| Competition | Homogenous | EMIT (Siemens) on: ADVIA Chemistry, Dimension, or Viva SYNCHRON (Beckman-Coulter) | Absorbance at 340 nm (NAD is converted into NADH); enzyme modulation SYNCHRON uses particle-enhanced turbidimetric inhibition immunoassay method) |
| Competition | Homogenous | CEDIA (Microgenics) | Colorimetry (Enzyme Modulation); at system-dependent visible wavelengths, commonly using beta-galactosidase enzyme (fragments) |
| Competition | Homogenous | LOCI (Siemens) | chemiluminescence |
| Competition | Homogenous | KIMS (Roche) | Optical detection |
| Competition | Homogenous | PETINA | Optical detection |
| Sandwich (for large molecules such as proteins) | Homogenous | ADVIA Chemistry assays (Siemens), proteins, hormones | Turbidimetry latex |
| Sandwich | Heterogenous | Centaur assays (Siemens), Proteins, hormones | Chemiluminescence |
| Sandwich | Heteogenous | Elecsys assays (Roche) proteins, hormones | Electrochemiluminescence |
| Sandwich | Heterogenous | Architect assays (Abbott), proteins, hormones | Chemiluminescence |

- FPIA: In the fluorescent polarization immunoassay (FPIA), the free label (which is a relatively small molecule) attached to the analyte molecule has different Brownian motion than when label is complexed to a large antibody. FPIA is a homogenous competitive assay where after the incubation, fluorescence polarization signal is measured without separation of bound labels from free labels. If the labeled antigen is bound to the antibody molecule, then signal is generated and when the labeled antigen is free in the solution, no signal is produced. Therefore, the intensity of the signal is inversely proportional to the analyte concentration. Abbott Laboratories first introduced this assay design [1]. However, most FPIA assays have been discontinued by Abbott.
- Enzyme multiplied immunoassay technique (EMIT) was first introduced by the Syva Company and it is a homogenous competitive immunoassay. In this immunoassay design, antigen is labeled with glucose 6-phosphate dehydrogenase enzyme. The active enzyme reduces nicotinamide

adenine dinucleotide (NAD, no signal at 340 nm) to NADH (absorbs at 340 nm), and the absorbance is monitored at 340 nm. When labeled antigen binds with the antibody molecule, the enzyme becomes inactive. Therefore, signal is produced by free label and the intensity of the signal is proportional to the analyte concentration.

- CEDIA: The cloned enzyme donor immunoassay (CEDIA) method is based on recombinant DNA technology to produce a unique homogenous enzyme immunoassay system. The assay principle is based on the bacterial enzyme beta-galactosidase, which has been genetically engineered into two inactive fragments. The small fragment is termed as enzyme donor (ED) which can freely associate in the solution with the larger part called enzyme acceptor (EA) producing active enzyme which is capable of cleaving a substrate generating a color change in the medium that can be measured spectrophotometrically. In this assay, drug molecules in the specimen compete for limited antibody binding sites with drug molecules conjugated with ED fragment. If drug molecules are present in specimen, then they bind to antibody binding sites, leaving drug molecules conjugated with ED free to form active enzyme by binding with EA. Then activity of the intact enzyme generated can be monitored through hydrolysis of an appropriate substrate such as chlorophenol red–beta-D-galactopyranoside by the active enzyme. The intensity of the signal is directly proportional to the analyte concentration and is usually measured in the visible wavelength at 570 nm. Many therapeutic drug and drugs of abuse manufactured by Microgenics Corporation (Thermo Fisher parent company) use CEDIA format, although other commercial assays also use this format [2].
- Kinetic interaction of microparticle in solution (KIMS): In this assay, in the absence of antigen (analyte) molecules, free antibodies bind to drug microparticle conjugates, forming particle aggregates that result in an increase in absorption, which is optically measured at various visible wavelengths (500–650 nm). When antigen molecules are present in the specimen, antigen molecules bind with free antibody molecules and prevent formation of particle aggregates, resulting in diminished absorbance in proportion to the drug concentration. The On-Line Drugs of abuse testing immunoassays marketed by Roche Diagnostics (Indianapolis, IN) are based on the KIMS format.
- LOCI: Luminescent oxygen channeling immunoassays (LOCI) is a homogenous competitive immunoassay where reaction mixture is irradiated with light-generating singlet oxygen molecules resulting in the formation of chemiluminescent signal. This technology is used in the Siemens Dimension Vista automated assay system [3].

- Another format of homogeneous immunoassay is turbidimetric immunoassay (TIA) where an analog of the analyte is coupled to colloidal particles, for example, latex [4]. Such latex particles agglutinate in the presence of the antibody. However, in the presence of free analytes in the specimen, there is less agglutination. Using a spectrophotometer, the resulting turbidity can be monitored as an endpoint or as a rate. Measurement can also be performed using nephelometric scattered light at 90-degrees to the original light path is measured. Nephelometric immunoassays have several advantages over the turbidimetric assays: they are more sensitive and suffer from less endogenous interference. Homogeneous particle-enhanced turbidimetric inhibition immunoassay (PETINA) is also an example of turbidimetric immunoassay. In this format, antibody fragments and drug-latex particles will bind to form aggregates that increase the turbidity of the solution. However, the drug molecules present in the specimen will also compete with antibody for binding, thus decreasing the turbidity due to decreasing rate of aggregation. Both Siemens and Beckman utilize this assay format most commonly for therapeutic drug monitoring of certain drugs.

## Heterogenous immunoassays

In heterogeneous immunoassays the bound label is physically separated from the free (unbound) label prior to measuring the signal. The separation is often done magnetically using paramagnetic particles and after separation of bound label from free label using washing step, the bound label is reacted with other reagents to generate the signal. This is the mechanism in many chemiluminescent immunoassays (CLIA), where the label may be a small molecule that generates a chemiluminescent signal. Examples of immunoassay systems where the chemiluminescent labels generate signals by chemical reaction are the ADVIA Centaur from Siemens and the Architect from Abbott [5]. An example where the small label is activated electrochemically is the ELECSYS automated immunoassay system from Roche Diagnostics [6]. The label also may be an enzyme (enzyme-linked immunosorbent assay, or ELISA) that generates chemiluminescent, fluorometric, or colorimetric signal depending on the enzyme substrates used. Examples of commercial automated assay systems using the enzyme-linked immunosorbent assay (ELISA) technology and chemiluminescent labels are Immulite (Siemens) and ACCESS from Beckman-Coulter [7,8]. Another type of heterogeneous immunoassay uses polystyrene particles. If these particles are microsized, that type of assay is called microparticle enhanced immunoassay (MEIA) [9]. If the immunoassay format utilizes a radioactive label, such assay is called radio immunoassay (RIA) Today, RIA is rarely used due to safety and waste disposal issues involving radioactive materials.

## Calibration of an immunoassay

Like all quantitative assays, immunoassays also require calibration. Calibration is a process of analyzing samples containing analytes of known concentrations (calibrators) and then fitting the data in a calibration curve so that concentration of the analyte in an unknown specimen can be calculated by linking signal to a particular value on the calibration curve. For calibration purpose, known amounts of the analyte are added to a matrix similar to serum matrix to prepare a series of calibrators with varying concentrations from zero calibrator (contain no analyte) to a calibrator containing highest targeted concentration of the analyte which is also the upper limit of analytical measurement range (AMR). The minimum number of calibrator needed to calibrate an assay is two (one zero calibrator and another calibrator representing upper limit of AMR) and many immunoassays are based on two calibration system. However, in some immunoassays, five or six calibrators may be used with one zero calibrator, one representing upper end of AMR, and other calibrators in between concentrations.

The calibration curve can be straight line or a curved line fitting to a polynomial function or logit function. Regardless of curve fitting method, signal generated during analysis of an unknown patient sample can be extrapolated to determine the concentration of the analyte using the calibration curve. For example, LOCI myoglobin assay for application on the Dimension Vista analyzers (Siemens Diagnostics), a homogenous sandwich chemiluminescent immunoassay based on LOCI technology, uses six levels of calibrators for construction of the calibration curve. Level A (myoglobin concentration zero), Level B (110 ng/mL), and Level C (1100 ng/mL) calibrators are supplied by the manufacture and during calibration, the instrument autodilutes Level B and Level C calibrators to produce calibrators with intermediate myoglobin concentration. Chemiluminescence signal is measured at 612 nm and the intensity of the signal is proportional to the concentration of myoglobin in the specimen and calibration curve fits to a linear equation (Fig. 3).

## Various sources of interferences in immunoassays

Even though immunoassays are widely used in clinical laboratory, they suffer from the following types of interferences, rendering false-positive or false-negative results:

- Endogenous components, for example, bilirubin, hemoglobin, lipids, and paraproteins, may interfere with immunoassays.
- Interferences from the other endogenous and exogenous components.
- System or method-related errors, for example, pipetting probe contamination and carry-over. Most modern instruments have various

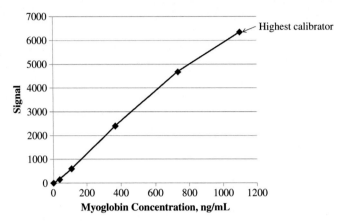

**FIG. 3**
Calibration curve of myoglobin using Vista 1500 analyzer (Siemens Diagnostics).

ways to eliminate carry-over issues most likely by using disposable probes or washing protocol between analyzes.

- Heterophilic interference is caused by endogenous human antibodies in the sample. Interferences from macro-analytes (endogenous conjugates of analyte and antibody), macro-enzymes, and rheumatoid factors.
- Prozone (or 'hook') effect: If very amount of analyte is present in the specimen, observed values may be much lower than the true analyte concentration (false negative result).

## Interferences from bilirubin, hemolysis, and high lipid content

Bilirubin is derived from hemoglobin of aged or damaged red blood cells. Bilirubin does not have iron, and is rather a derivative of the heme group. Some part of serum bilirubin is conjugated as glucuronides ("direct" bilirubin) and the unconjugated bilirubin referred as indirect bilirubin. In normal adults, total bilirubin concentrations in serum are from 0.3 to1.2 mg/dL. In different forms of jaundice, total bilirubin may increase to as high as 20 mg/dL. Major issues of bilirubin interference are as follows:

- Usually total bilirubin concentration below 20 mg/dL does not cause interferences but concentration over 20 mg/dL may cause problem.
- The interference of bilirubin in assays is mainly caused by bilirubin absorbance at 454 or 461 nm.
- Bilirubin may also interfere with an assay by chemically reacting with a component of the reagent.

Hemoglobin is mainly released from hemolysis of red blood cells (RBC). Hemolysis can occur in vivo, during venipuncture and blood collection or during processing of the sample. Hemoglobin interference depends on its concentration in the sample. Serum appears hemolyzed when the hemoglobin concentration exceeds 20 mg/dL. The absorbance maxima of the heme moiety in hemoglobin are at 540 to 580 nm wave lengths. However, hemoglobin begins to absorb around 340 nm and then absorbance increasing at 400–430 nm as well. Interference of hemoglobin (if the specimen is grossly hemolyzed) is due to interfering with optical detection system of the assay.

All lipids in plasma exist as complexed with proteins which are called lipoproteins and particle size varies from 10 nm to 1000 nm (the higher the percentage of the lipid, lower is the density of the resulting lipoprotein, and larger is the particle size). The lipoprotein particles with high lipid contents are micellar and are the main source of assay interference. Unlike bilirubin and hemoglobin, lipids normally do not participate in chemical reactions and mostly cause interference in assays due to their turbidity and capability of scattering light as in nephelometric assays.

## Interferences from the endogenous and exogenous components

Immunoassays are affected by variety of endogenous and exogenous compounds including heterophilic antibodies. The key points regarding immunoassay interferences include:

- Endogenous monoclonal (IgG, IgM, or IgA) and sometimes high concentrations of polyclonal immunoglobulins collectively known as paraproteins can interfere with many laboratory tests on different automated chemistry, nephelometry, turbidimetry, immunochemistry, and hematology platforms, causing both false-positive and false-negative results in many laboratory tests including total bilirubin, creatinine, inorganic phosphate, calcium, C-reactive protein, total albumin, glucose, iron, uric acid, high-density lipoprotein cholesterol, low-density lipoprotein cholesterol, insulin, total protein and therapeutic drugs, for example, vancomycin. Paraprotein may behave like heterophilic antibody or may precipitate in a nephelometric or turbidimetric assay causing interference. However, paraprotein interferences are method-dependent. For example, IgM causes negative interference with vancomycin PETINA immunoassay but has no effect on the EMIT assay. Paraprotein may also cause falsely low HDL-cholesterol level. In contrast paraprotein may falsely increase serum creatinine level.

- Structurally similar molecules are capable of cross-reacting with the antibody causing falsely elevated (positive interference) or falsely lower results (negative interference). Negative interference occurs less frequently than positive interference but may be clinically more dangerous. For example, if result of a therapeutic drug is falsely elevated compared to the previous measurement, clinician may question the result but if the value is falsely lower, the clinician may simply increase the dose without realizing that the value was falsely lower due to interference. That may cause drug toxicity in the patient.
- Interference from drug metabolite is the most common form of interference, although structurally similar other drugs may also be the cause of interference. See also Chapter 15.

## Interferences of heterophilic antibodies in immunoassays

Heterophilic antibodies are endogenous proteins that bind animal antigens. The heterophilic antibodies are polyclonal and heterogeneous in nature, consisting of the following types:

- Heterophilic antibodies which interact poorly and non-specifically with the assay antibodies. Anti-animal antibodies (HAAA) which interact strongly and specifically with the assay antibody; a common example is HAMA (human antimouse antibody).
- Endogenous human autoantibodies interfering with an assay.
- Therapeutic antibodies, where antibodies given therapeutically interfere with an assay.
- Rheumatoid factor can also be broadly classified as heterophilic antibody

Heterophilic antibodies may arise in a patient in response to exposure to certain animals or animal products, due to infection by bacterial or viral agents, or nonspecifically. Although many of the immunoglobulin clones in normal human serum may display antianimal antibody properties, only those antibodies with sufficient titer and affinity toward the reagent antibody used in assay may cause clinically significant interference. Prevalence of heterophilic antibody varied widely in various published reports. In one study, the prevalence of heterophilic antibody was 0.2%–3.7% [10]. Heterophilic antibodies are found more in sick and hospitalized patients with reported prevalence of 0.2%–15%. An individual can form heterophilic antibody at any time when exposed to foreign antigen, for example, during vaccination, antibody-targeted therapies (cancer, autoimmune disorder, etc.), and antibody-targeted imaging reagents. Exposure to animals may also lead to formation of heterophilic antibody. Moreover, blood transfusion, autoimmune disease may also result in formation of heterophilic antibody. It has been estimated that analytically important interferences due to the presence of heterophilic antibody in serum most likely occur in 0.5%–3% of specimens [11].

Heterophilic antibodies usually interfere with sandwich immunoassays but rarely with competitive immunoassays. In sandwich assay format, assay interference could happen when heterophilic antibodies bridge the capture and detection antibody in the assay design, resulting in elevated analyte concentration. However, rarely, heterophilic antibody may cause negative interference (falsely low analyte value) due to interfering with analyte binding to capture antibody. Heterophilic antibodies may affect a wide range of laboratory tests including false elevation of endocrine tests, tumor markers, cardiac markers, and very rarely therapeutic drug monitoring (digoxin, one case report, as well as cyclosporine and tacrolimus using ACMIA; antibody-conjugated magnetic immunoassay). However, heterophilic antibody may also cause falsely lower cortisol and thyroglobulin levels. Moreover, interferences of heterophilic antibody in immunoassay of various tumor markers are particularly important because it may cause false diagnosis of malignancy [12]. Falsely elevated D-dimer value may be due to heterophilic antibody interference [13]. A 58-year-old man without any familial risk for prostate cancer visited his primary care physician and his prostate-specific antigen (PSA) level was 83 ng/mL (normal 0–4 ng/mL). He was referred to a urologist and his digital rectal examination was normal. Despite such findings, he was treated for prostate cancer, but his PSA was still elevated to 122 ng/mL without any radiographic evidence of advanced cancer. At that point, his serum PSA was analyzed by a different assay (Immulite PSA, Cirrus Diagnostics, Los Angeles) and PSA level was <0.3 ng/mL. The treating physician therefore suspected a false-positive PSA by original Access Hybritech PSA assay (Hybritech, San Diego, CA) and interference of heterophilic antibody was established by treating specimens with heterophilic antibody blocking agent and re-analyzing high PSA specimen that showed a level below the detection limit. This patient received unnecessary therapy for his falsely elevated PSA level due to the interference of heterophilic antibody [14]. Analytes that are affected by heterophilic antibody interference are listed in Table 3.

Various approaches to overcome heterophilic antibody interference include:

- Heterophilic antibodies are absent in the urine. Therefore, if serum specimen is positive for an analyte, for example, human chorionic gonadotropin (hCG), but beta-hCG cannot be detected in the urine specimen, it indicates interference from heterophilic antibody in serum hCG measurement.

**Table 3** Interference of heterophilic antibody in various immunoassays.

| Test category | Example of individual tests | Comments |
|---|---|---|
| Cardiac markers | Troponin I, CK-MB, BNP | Usually falsely elevated levels |
| Thyroid function tests | TSH, FT3, FT4, thyroglobulin | In general, values are falsely elevated but both falsely elevated and falsely lower values have been reported with thyroglobulin testing |
| Hormones | ACTH, cortisol, calcitonin, FSH, LH, growth hormone, prolactin, estradiol, progesterone, testosterone, PTH | Usually falsely elevated values but negative interference in cortisol testing has been reported |
| Therapeutic Drugs | Digoxin, cyclosporine, tacrolimus | Falsely elevated values for all three drugs but only ACMIA cyclosporine and tacrolimus assays are affected by heterophilic antibody interference |
| Tumor markers | PSA, CEA, CA-19-9, CA-125, AFP, beta-hCG, calcitonin | Usually falsely elevated values which may cause wrong diagnosis of malignancy, a very serious issue of interference |
| Infectious disease | HIV testing | Weakly false-positive test result using point-of-care testing |
| Other analytes | C-reactive protein, D-dimer, Insulin, SHBG | Falsely elevated values |

*Abbreviations:* CK-MB, *creatine kinase MB isoenzyme;* BNP, *B-type natriuretic peptide or brain-type natriuretic peptide;* TSH, *thyroid-stimulating hormone;* FT3, *free thyroxine;* FT3, *free triiodothyronine;* FSH, *follicle-stimulating hormone;* LH, *luteinizing hormone;* ACTH, *adrenocorticotropic hormone;* PTH, *parathyroid hormone;* PSA, *prostate-specific antigen;* CEA, *carcinoembryonic antigen;* CA-19-9, *cancer antigen-19-9;* CA-125, *cancer antigen-125;* AFP, *alpha-fetoprotein;* beta-hCG, *beta-human chorionic gonadotropin;* SHBG, *sex hormone-binding globulin.*

*Adapted from Dasgupta A: Biotin and other interferences in immunoassays: a concise guide. Elsevier 2019. Reprinted with permission.*

- Another way to investigate heterophilic antibody interference is serial dilution of specimen. If serial dilution produces nonlinear result, it indicates interference in the assay.
- Interference from heterophilic antibodies may also be blocked by adding commercially available any heterophilic antibody-blocking agent in the specimen prior to analysis.
- For analytes which are also present in the protein-free ultrafiltrate (relatively small molecules), analysis of the analyte in the protein-free ultrafiltrate can eliminate interference from heterophilic antibodies because due to large molecular weights, heterophilic antibodies are absent in a protein-free ultrafiltrate. An example is elimination of heterophilic antibody interference in serum digoxin assay (rare) by monitoring free digoxin in the protein-free ultrafiltrate.

# Rheumatoid factor interferences

Rheumatoid Factors are IgM-type antibodies which interact with assay antibodies at the Fc area and present in serum from greater than 70% of patients with rheumatoid arthritis. Rheumatoid factors are also found in patients with other autoimmune diseases and titer may also increase during infection or inflammation. Interferences of rheumatoid factors are more commonly observed in sandwich immunoassays but may also occur in competitive immunoassay format. The mechanisms are similar to interferences caused by heterophilic antibodies. In competition-type immunoassays, rheumatoid factors bind to the assay antibody, preventing its reaction to the label reagent through steric-hindrance, thus generating false-positive results. If rheumatoid factors are suspected to cause interference, the patient's history needs to be examined. It is also advisable to measure the rheumatoid factor levels in serum by using commercially available immunoassays.

A 64-year-old male during a routine visit to his physician was diagnosed with hypothyroidism based on an elevated TSH (thyroid-stimulating hormone) and his clinician initiated therapy with levothyroxine (250 microgram per day). Despite therapy, there were increased level of TSH (33 mIU/L) and his FT4 level was also elevated. The endocrinologist at that point suspected that TSH levels measured by the Unicel Dxi analyzer (Beckman Coulter) were falsely elevated due to interference. Serial dilution of the specimen showed nonlinearity, an indication of interference. When the specimen was analyzed using a different TSH assay (immunoradiometric assay; IRMA also available from Beckman Coulter), the TSH value was 1.22 mIU/L, further confirming the interference with initial TSH measurement. The patient had high concentration of rheumatoid factor (2700 U/mL) and the authors speculated that his falsely elevated TSH was due to interference from rheumatoid factors [15].

Rheumatoid factor interference can cause false-positive result in the MEIA troponin I assay. The authors eliminated rheumatoid factor interference by incubating the sample with commercially available rheumatoid factor-blocking agent [16]. Astarita et al. described a case study of false-positive thyroglobulin in a woman with history of rheumatoid arthritis. The authors confirmed rheumatoid factor interference by (a) nonlinear dilution, (b) alternate thyroglobulin immunoassay employing different antibodies, and (c) precipitating out interfering RF and retesting [17].

# Interferences from autoantibodies

Autoantibody (immunoglobulin molecule) is formed by the immune system of an individual capable of recognizing an antigen on that person's own tissues.

Several mechanisms may trigger the production of autoantibodies, for example, an antigen, formed during fetal development and then sequestered may be released as a result of infection, chemical exposure, or trauma, as it occurs in autoimmune thyroiditis. The autoantibody may bind to the analyte-label conjugate in a competition-type immunoassay, producing false-positive or false-negative result. Circulating troponin I autoantibodies may be present in patient suffering from acute cardiac myocardial infarction where troponin I should be elevated. Unfortunately, circulating cardiac troponin I autoantibodies may falsely lower cardiac troponin I concentration (negative interference) using commercial immunoassays, thus complicating the diagnosis of acute myocardial infarction [18]. However, falsely elevated results due to the presence of autoantibodies are more common than false-negative results. Verhoye et al. found 3 patients with false-positive thyrotropin results that were caused by interference from an autoantibody against thyrotropin. The interfering substance in the affected specimens was identified as autoantibody by gel-filtration chromatography and poly-ethylene glycol precipitation [19].

## Macroenzyme interferences

Macroenzymes are high-molecular weight versions of specific serum enzymes formed either due to complex formation with a high-molecular weight serum component such as immunoglobulin or self-polymerization. Macroenzymes frequently interfere with enzyme assays, thus falsely increasing serum enzyme levels. In 1967, Berk et al. first reported the presence of macromolecular form of amylase in sera of patients with persistently elevated serum amylase levels. These patients had essentially normal renal functions [20]. Later reports of macromolecular forms of various enzymes appeared in the medical literature including macro-lactate dehydrogenase (macro-LDH), macro-alkaline phosphatase (macro-ALP), macro-aspartate aminotransferase (macro-AST), macro-alanine aminotransferase (macro-ALT), macro-creatine kinase (macro-CK), macro-gamma glutamyl transferase (macro-GGT), macro-acid phosphatase, macro-lipase, and macro-leucine aminopeptidase. Macroenzymes are usually discovered in patients who show persistent elevation of a particular serum enzyme but have no symptom [21]. Macroenzymes can be broadly classified under two categories: immunoglobulin-bound enzyme or nonimmunoglobulin-bound enzyme. Immunoglobulin-bound enzymes could be considered as a specific antigen–antibody complex. Macroenzymes are mostly benign in nature because many patients with macroenzymes are healthy. However, the presence of macroenzymes in many pathological conditions such as certain infection and parasitic disease, neoplasm, immune disorders, endocrine disorders, and others has been reported. Various methods including polyethylene glycol precipitation, gel filtration, and serum protein

**Table 4** Commonly encountered macroenzymes.

| Macroenzyme | Comments |
|---|---|
| Macro-Alkaline phosphatase | Complexed with IgG or lipid aggregates |
| Macro-Alanine aminotransferase | Complexed with IgG |
| Macro-Aspartate aminotransferase | Complexed with IgG |
| Macro-Gamma-glutamyl transferase | Complexed with IgG or lipid aggregates |
| Macro-Lactate dehydrogenase | Complexed with IgA |
| Macroamylase | Either complexed with IgA or exists as substrate-complex |
| Macrolipase | Mostly complexed with IgG |
| Macro-Creatine kinase | Mostly complexed with IgG (Macro-CK -1) or Macro-CK type 2 enzymes made up of mitochondrial-derived self CK polymerization |
| Macro-Trypsin | Protease-inhibitor complex |

*Adapted from Dasgupta A: Biotin and other interferences in immunoassays: a concise guide. Elsevier 2019. Reprinted with permission.*

electrophoresis can be used to detect the presence of macroenzyme in patient's sera and in eliminating such interferences. Commonly encountered macroenzymes are listed in Table 4.

Various macroenzymes are usually encountered in patients over 60 years of age. However, macro-AST is more frequent in subjects under the age of 60 years, particularly in females. Cases with this abnormality have also been reported in children, suggesting that macro-AST may start early in life but unlikely to be congenital. Macro-creatine kinase (macro-CK) is a neglected cause falsely elevated serum creatine kinase levels (CK), but such increases may not have any clinical significance. Macro-CKs are macroenzymes with high molecular weight and prolonged half-life. Macro-CK type 1 enzymes are complexes formed by one of the CK isoenzymes and an immunoglobulin and are also found in healthy individuals, as well as in a wide range of pathological conditions. Macro-CK type 2 enzymes are made up of mitochondrial-derived CK polymers and are associated with neoplasms. Both types of macro-CK have been observed but prevalence is very low. In one report, over a 10-year period, the authors observed only five cases related to macro-CK. Three patients had macro-CK type 1 and two patients had macro-CK type 2: a man with a neuroendocrine carcinoma and a woman with rheumatoid arthritis [22].

## Other macroanalytes

Macro-TSH, also known as macro-thyrotropin, interferes with TSH immunoassays. Another macro-analyte, macroprolactin, interferes with prolactin immunoassay. Macroprolactin is a large protein complex of 150 kDa or more formed due to binding of monomeric prolactin (molecular weight: 23 kDa) with immunoglobulin, predominately IgG. Macroprolactin has a slower clearance rate consistent with that of IgG, leading to accumulation in the circulation and causing falsely elevated prolactin levels using immunoassays. Macroprolactin is considered as benign [23].

## Prozone (or "hook") effect

Prozone or hook effect is observed when very high amount of an analyte is present in the sample but observed value is falsely lowered. This type of interference is observed more commonly in sandwich assays. The mechanism of this significant negative interference is the capability of high level of an analyte (antigen) to reduce the concentrations of 'sandwich' (antibody 1: antigen: antibody 2) complexes responsible for generating the signal by forming mostly single antibody: antigen complexes. Hook effect has been reported with assays of a variety of analytes such as $\beta$-hCG, prolactin, calcitonin, aldosterone, and cancer markers (CA 125, PSA). Nonlinearity during dilution is an indication of possible hook effect. In order to eliminate hook effect, 1:10, 1:100, or even 1:1000 dilution may be necessary so that the true analyte concentration falls within the analytical measurement range (AMR) of the assay.

## Summary/key points

- Immunoassays can be competitive or immunometric (non-competitive: also known as sandwich). In competitive immunoassays, only one antibody is used and this format is common for assays of small molecules such as a therapeutic drug or an abused drug. In sandwich format, two antibodies are used and this format is more commonly used for assay of relatively large molecules.
- Homogenous Immunoassay Format: After incubation, no separation between bound and free label is necessary.
- Heterogenous Immunoassay Format: Bound label must be separated from the free label before measuring the signal.
- Commercially available immunoassays use various formats including are FPIA, EMIT, CEDIA, KIMS, and LOCI. In the fluorescent polarization immunoassay (FPIA), the free label (a relatively small molecule) attached to the analyte (antigen) molecule has different Brownian motion than

when label is complexed to a large antibody (140,000 or more Daltons). FPIA is a homogenous competitive assay where after the incubation, fluorescence polarization signal is measured which is only produced if the labeled antigen is bound to the antibody molecule. Therefore, intensity of the signal is inversely proportional to the analyte concentration.

- Enzyme-multiplied immunoassay technique (EMIT) is a homogenous competitive immunoassay where antigen is labeled with glucose 6-phosphate dehydrogenase, an enzyme that reduces nicotinamide adenine dinucleotide (NAD, no signal at 340 nm) to NADH (absorbs at 340 nm), and the absorbance is monitored at 340 nm. When a labeled antigen binds with the antibody molecule, the enzyme label becomes inactive and no signal is generated. Therefore, intensity of signal is proportional to analyte concentration.

- Cloned enzyme donor immunoassay (CEDIA) method is based on recombinant DNA technology where bacterial enzyme beta-galactosidase is genetically engineered into two inactive fragments. When both fragments combine, signal is produced by adding a substrate of the active enzyme. The signal is proportional to analyte concentration and may be measured in visible wavelength.

- Kinetic interaction of microparticle in solution (KIMS) where in the absence of antigen molecules, free antibodies bind to drug microparticle conjugates, forming particle aggregates that result in an increase in absorption, which is optically measured at various visible wavelengths (500–650 nm).

- LOCI is the luminescent oxygen channeling immunoassay, where the immunoassay reaction is irradiated with light generating singlet oxygen molecules in microbeads ('Sensibead') coupled to the analyte. When bound to the respective antibody molecule, also coupled to another type of bead, it reacts with singlet oxygen and chemiluminescence signals are generated which are proportional to the concentration of the analyte-antibody complex.

- Usually total bilirubin concentration below 20 mg/dL does not cause interferences but concentration over 20 mg/dL may cause problem. The interference of bilirubin is mainly caused by its absorbance at 454 or 461 nm.

- Various structurally related drugs or drug metabolite can interfere with immunoassays.

- Heterophilic antibodies may arise in a patient in response to exposure to certain animals or animal products, due to infection by bacterial or viral agents, or use of murine monoclonal antibody products in therapy or imaging. Heterophilic antibodies interfere most commonly with sandwich assays used for measuring large molecules and rarely with competitive assays, causing mostly false-positive results.

• Heterophilic antibodies are absent in the urine. Therefore, if serum specimen is positive for an analyte, for example, if human chorionic gonadotropin (hCG) is detected in the serum but beta-hCG cannot be detected in the urine specimen, it indicates interference from heterophilic antibody because heterophilic antibody due to large molecular weight is absent in the urine. Another way to investigate heterophilic antibody interference is serial dilution of specimen. If serial dilution produces nonlinear result, it indicates interference in the assay. Interference from heterophilic antibodies may also be blocked by adding commercially available heterophilic antibody-blocking agents in the specimen prior to analysis.

• Autoantibody is formed by the immune system of the person that recognizes an antigen on that person's own tissues and may interfere with an immunoassay, producing false-positive results and less frequently false-negative results. Often the endogenous analyte of interest may conjugate with immunoglobin or other antibodies to generate macro-analytes, which may falsely elevate a result. For example, macroamylasemia and macroprolactinemia may produce falsely elevated results in amylase and prolactin assays, respectively. Such interference may be removed by polyethylene glycol precipitation.

• Prozone (or "hook") effect: Very high levels of antigen may reduce the concentrations of sandwich' (antibody 1: antigen: antibody 2) complexes responsible for generating the signal by forming mostly single antibody: antigen complexes. This effect known as prozone or hook effect (excess antigen) mostly cause negative interference (falsely lower result). Best way to eliminate hook effect is serial dilution.

# References

[1] Jolley ME, Stroupe SD, Schwenzer KS, et al. Fluorescence polarization immunoassay III. An automated system for therapeutic drug determination. Clin Chem 1981;27:1575–9.

[2] Jeon SI, Yang X, Andrade JD. Modeling of homogeneous cloned enzyme donor immunoassay. Anal Biochem 2004;333:136–47.

[3] Snyder JT, Benson CM, Briggs C, et al. Development of NT-proBNP, troponin, TSH, and FT4 LOCI(R) assays on the new dimension (R) EXL with LM clinical chemistry system. Clin Chem 2008;54:A92 [Abstract #B135].

[4] Datta P, Dasgupta A. A new turbidimetric digoxin immunoassay on the ADVIA 1650 analyzer is free from interference by spironolactone, potassium canrenoate, and their common metabolite canrenone. Ther Drug Monit 2003;25:478–82.

[5] Dai JL, Sokoll LJ, Chan DW. Automated chemiluminescent immunoassay analyzers. J Clin Ligand Assay 1998;21:377–85.

[6] Forest J-C, Masse J, Lane A. Evaluation of the analytical performance of the Boehringer Mann-heim Elecsys® 2010 immunoanalyzer. Clin Biochem 1998;31:81–8.

[7] Babson AL, Olsen DR, Palmieri T, et al. The IMMULITE assay tube: a new approach to hetero-geneous ligand assay. Clin Chem 1991;37:1521–2.

[8] Christenson RH, Apple FS, Morgan DL. Cardiac troponin I measurement with the ACCESS® immunoassay system: analytical and clinical performance characteristics. Clin Chem 1998;44:52–60.

[9] Montagne P, Varcin P, Cuilliere ML, Duheille J. Microparticle-enhanced nephelometric immu-noassay with microsphere-antigen conjugate. Bioconjug Chem 1992;3:187–93.

[10] Preissner CM, Dodge LA, O'Kane DJ, Singh RJ, et al. Prevalence of heterophilic antibody inter-ference in eight automated tumor marker immunoassays. Clin Chem 2005;51:208–10.

[11] Preissner CM, O'Kane DJ, Singh RJ, Morris JC, et al. Phantoms in the assay tube: heterophile antibody interferences in serum thyroglobulin assays. J Clin Endocrinol Metab 2003;88:3069–74.

[12] Morton A. When lab tests lie: heterophilic antibody. Aust Fam Physician 2014;43:391–3.

[13] Wu Y, Xiao YX, Huang TY, Zhang XY, Zhou HB, et al. What makes D-dimer assays suspicious–heterophilic antibodies? J Clin Lab Anal 2019;33:e22687.

[14] Henry N, Sebe P, Cussenot O. Inappropraite treatment of prostate cancer caused by hetero-philic antibody interference. Nat Clin Pract Urol 2009;6:164–7.

[15] Georges A, Charrie A, Raynaud S, Lombard C, et al. Thyroxin overdose due to rheumatoid fac-tor interferences in thyroid-stimulating hormone assays. Clin Chem Lab Med 2011;49:873–5.

[16] Dasgupta A, Banerjee SK, Datta P. False positive troponin I in the MEIA due to the presence of rheumatoid factors in serum. Am J Clin Pathol 1999;112:753–6.

[17] Astarita G, Gutiérrez S, Kogovsek N, et al. False positive in the measurement of thyroglobulin induced by rheumatoid factors. Clin Chim Acta 2015;447:43–6.

[18] Tang G, Wu Y, Zhao W, Shen Q. Multiple immunoassays systems are negatively interfered by circulating cardiac troponin I autoantibodies. Clin Exp Med 2012;12:47–53.

[19] Verhoye E, AVD B, Delanghe JR, et al. Spuriously high thyrotropin values due to anti-thyrotropin antibody in adult patients. Clin Chem Lab Med 2009;47:604–6.

[20] Berk JE, Kizu H, Wilding P. Macroamylasemia: a newly recognized cause for elevated serum amylase activity. N Engl J Med 1967;277:941–6.

[21] Moriyama T, Tamura S, Nakano K, Otsuka K, et al. Laboratory and clinical features of abnor-mal macroenzymes found in human sera. Biochim Biophys Acta 2015;1854:658–67.

[22] Aljuani F, Tournadre A, Cecchetti S, Soubrier M, et al. Macro-creatine kinase: a neglected cause of elevated creatine kinase. Intern Med J 2015;45:457–9.

[23] Vilar L, Vilar CF, Lyra R, Freitas MDC. Pitfalls in the diagnostic evaluation of hyperprolactine-mia. Neuroendocrinology 2019;109:7–19.

# CHAPTER 3

# Preanalytical variables

## Laboratory errors

Accurate clinical laboratory test results are important for proper diagnosis and treatment of patients. Factors that are important to obtain accurate laboratory test results include:

- The right patient is identified prior to specimen collection by matching at least two criteria.
- The right technique and right blood collection tube are used for collecting the sample avoiding tissue damage, prolonged venous stasis, or hemolysis.
- After collection, the specimen is labeled properly with correct patient information because specimen misidentification is a major source of preanalytical error.
- Proper centrifugation (in case of analyzing serum or plasma specimen) and proper transportation of specimen to the laboratory.
- Maintaining proper storage of specimen prior to analysis in order to avoid artifactual changes in analyte, for example, storing blood gas specimen in ice if analysis cannot be completed within 30 min of specimen collection.
- Proper analytical steps to obtain correct result avoiding interferences.
- Correctly reporting the result to the laboratory information system (LIS) if the analyzer is not interfaced with the LIS.
- The report reaching the clinician contains the right result, together with interpretative information, such as a reference range and other comments, aiding clinicians in decision-making process.

Failure at any of these steps can result in an erroneous or misleading laboratory result, sometimes with adverse outcomes. The analytical part of the analysis involves measurement of concentration of the analyte corresponding to its "true" level (as compared to a "gold standard" measurement) within a clinically acceptable margin of error (the total acceptable analytical error, TAAE). Errors

**47**

| Table 1 Commonly occurring laboratory errors. |
| --- |
| **Type of error** |
| **Preanalytical errors** |
| Tube filling error<br>Patient identification error<br>Inappropriate container<br>Insufficient specimen collected<br>Order not entered in laboratory information system<br>Specimen collected wrongly from an infusion line<br>Specimen stored improperly<br>Contamination of culture tube |
| **Analytical errors** |
| Inaccurate result due to interference<br>Random error caused by the instrument<br>*Postanalytical errors*<br>Result communication error<br>Excessive turnaround time due to instrument downtime |

can occur at any stage of analysis (preanalytical, analytical, and postanalytical). It has been estimated that preanalytical errors account for more than two-thirds of all laboratory errors while errors in the analytical phase and postanalytical phase account for only one-third of all laboratory errors. Carraro and Plebani reported that among 51,746 clinical laboratory analysis performed in a 3-month period in the author's laboratory (7615 laboratory orders, 17,514 blood collection tubes), clinicians contacted the laboratory regarding 393 questionable results, out of which 160 results were confirmed due to laboratory errors. Of 160 confirmed laboratory errors, 61.9% were determined to be preanalytical errors, 15% were analytical errors while 23.1% were postanalytical errors [1]. Types of laboratory errors (preanalytical, analytical, and postanalytical) are summarized in Table 1.

To avoid preanalytical errors, several approaches can be taken including:

- Handheld devices connected to the LIS that can objectively identify the patient by scanning a patient attached barcode, typically a wrist band.
- Current laboratory orders can be retrieved from the LIS.
- Barcoded labels are printed at the patient's side, minimizing the possibility of misplacing the labels on the wrong patient samples.

When classifying sources of error, it is important to distinguish between *cognitive errors*, or mistakes, which are due to poor knowledge or judgment, from *noncognitive errors*, commonly known as slips and lapses, due to interruptions in a process during even routine analysis involving automated analyzers. Cognitive errors can be prevented by increased training, competency evaluation, and process aids, such as checklists while noncognitive errors can be reduced by improving work environment which is re-engineered to minimize distractions and fatigue. The vast majority of errors are noncognitive slips and lapses performed by the personnel directly involved in the process which can be avoided.

The worst preanalytical error is incorrect patient identification where a physician may act on test results from the wrong patient. Another common error is blood collection from intravenous line which may falsely increase the test results of glucose, electrolytes, or a therapeutic drug due to contamination with infusion fluid. A 59-year-old woman was admitted to the hospital due to transient ischemic heart attack. During hospitalization, she had seizure and was treated with phenytoin administered intravenously. On day 5, phenytoin concentration was $17.0\,\mu g/mL$ and on day 7, phenytoin concentration was $13.4\,\mu g/mL$. Surprisingly on day 8, phenytoin concentration was in life-threatening level of $80.7\,\mu g/mL$ although the patient did not show any symptom of phenytoin toxicity. Another sample drawn 7 h later showed a phenytoin level of $12.4\,\mu g/mL$. It was suspected that falsely elevate serum phenytoin level was due to drawing of the specimen from the same line through which the intravenous phenytoin was administered [2]. This is an example of serious preanalytical error.

## Order of draw of blood collection tubes

The correct order of draw for blood specimens are:

- Microbiological blood culture tubes (yellow top)
- Royal Blue tube (no additive); trace metal analysis if desired
- Citrate tube (light blue)
- Serum tube (red top) or tube with gel separator/clot activator (gold top or tiger top)
- Heparin tube (green top)
- EDTA tube (ethylenediamine tetraacetic acid; purple/lavender top)
- Oxalate-Fluoride tube (gray top)

Tubes with additives must be thoroughly mixed by gentle inversion as per manufacturer recommended protocols. Erroneous test results may be obtained when the blood is not thoroughly mixed with the additive. When trace metal testing on serum is ordered it is advisable to use trace element tubes. Royal-Blue Monoject Trace Element Blood Collection Tubes are available for this purpose. These tubes are free from trace and heavy metals.

## Errors in patient preparation

There are certain important issues regarding patient preparation for obtaining meaningful clinical laboratory test results. For some analytes, overnight fast is needed. Increases in serum glucose, triglycerides, bilirubin, and aspartate aminotransferase are commonly observed after meal consumption. On the other hand, fasting will increase fat metabolism and increase the formation of acetone, $\beta$-hydroxybutyric acid, and acetoacetate both in serum and in urine. However, overnight fasting is required for glucose and triglyceride measurement because both values are significantly elevated after meals. In contrast cholesterol levels are less affected after meals. Other well-known examples of analytes showing variation with fasting interval include serum bilirubin and serum iron.

Physiologically, blood distribution differs significantly in relation to body posture. Gravity pulls the blood into various parts of the body when recumbent and the blood moves back into the circulation, away from tissues, when standing or ambulatory. Blood volume of an adult in upright position is 600–700 mL less than when the person is lying on a bed and this shift directly affects certain analytes due to dilution effects. Therefore, concentrations of proteins, enzymes and protein-bound analytes (thyroid stimulating hormone; TSH, cholesterol, T4, and medications like warfarin) are affected by the posture but most affected are factors directly influencing hemostasis including renin, aldosterone, and catecholamines. It is vital for laboratory requisitions to specify the need for supine samples when these analytes are requested.

## Biological rhythm and time for specimen collection

Predictable patterns in the temporal variation of certain analytes, reflecting patterns in human needs, constitute biological rhythms. Different analytes have different rhythms, ranging from a few hours to monthly changes. These changes can be divided into circadian, ultradian, and infradian rhythms according to the time interval of their completion. During a 24-h period of human metabolic activity, programming of metabolic needs may cause certain laboratory tests to fluctuate between a maximum and a minimum value. This rhythm is known as circadian rhythm. A good example is cortisol which shows much higher concentration in the morning compared to evening. Therefore, the time of specimen collection may affect the test result. Common analytes that show circadian rhythm are listed in Table 2.

Patterns of biological variation occurring on cycles less than 24 h are known as ultradian rhythms. Therefore, an ultradian rhythm is a recurrent period or cycle repeated throughout a 24-h day. In contrast, circadian rhythms (diurnal rhythm) complete one cycle daily. Analytes that are secreted in a pulsatile

**Table 2** Common analytes that show circadian cycle.

| Analyte | Comment |
|---------|---------|
| Cortisol | Much higher concentration in the morning than afternoon |
| Renin | Maximum activity early morning, minimum in the afternoon |
| Iron | Higher levels in the morning than afternoon |
| TSH | Maximum level 2 a.m.–4 a.m. while minimum level 6 p.m.–10 p.m. |
| Insulin | Higher in the morning than later part of the day |
| Phosphate | Lowest in the morning, highest in early afternoon |
| ALT | Higher level in the afternoon than morning |

TSH, *thyroid stimulating hormone*; ALT, *alanine aminotransferase*.

manner throughout the day show this pattern. Testosterone, which usually peaks between 10:00 a.m. and 5 p.m., is an example of an analyte showing this pattern. However, many physiological process such as sleep cycle and appetite also follow ultradian rhythm. The final pattern of biological variation is infradian. This involves cycles greater than 24 h. The example most commonly cited is the monthly menstrual cycle, which takes approximately 28–32 days to complete. Constituents such as pituitary gonadotropin, ovarian hormones, and prostaglandins are significantly affected by this cycle.

## Errors with patient identification and related errors

Accurate patient and specimen identification is required for providing ordering clinicians with correct results. Regulatory agencies like Joint Commission on Accreditation of Healthcare Organization (JCAHO) have made it a top priority to ensure patient safety. Patient and specimen misidentification occurs mostly during the preanalytical phase.

- Accurate identification of a patient requires verification of at least two unique identifiers from the patient and ensuring that those match the patient's prior records.
- If a patient is unable to provide identifiers (i.e., neonate or a critically ill patient) a family member or nurse should verify the identity of the patient.
- Information on laboratory requisitions or electronic orders must also match patient information in their chart or electronic medical record. Specimens should not be collected unless all identification discrepancies have been resolved.

The specimens should be collected and labeled in front of the patient and then sent to the laboratory with the test request. For nonbarcoded specimens, specimens should be accessioned, labeled with a barcode (or re-labeled, if necessary), processed (either manually or on an automated line), and sent for analysis. Identification of the specimen should be carefully maintained during centrifugation, aliquoting, and analysis. Most laboratories use barcoded labeling systems to preserve sample identification. Patient misidentification may have serious adverse outcome on a patient especially if wrong blood is transfused to a patient due to misidentification of blood specimen sent to the laboratory for cross-matching. In this case, a patient may die from receiving wrong blood group.

Although errors in patient identification occur mostly in preanalytical phase, errors may also occur during analytical and even postanalytical phase. Results from automated analyzers are electronically transferred to the LIS through an interface but if direct transfer of result from a particular instrument is not available, error may occur during manual transfer of results. Dunn and Morga reported that out of 182 specimen misidentification they studied, 132 misidentifications occurred in preanalytical stage. These misidentifications were due to wrist bands labeled for wrong patient, laboratory tests were ordered for the wrong patient, selection of wrong medical record from a menu of similar names and social security numbers, specimen mislabeling during collection associated with batching of specimens and printed labels, misinformation from manual entry of laboratory form, failure of two source patient identification for clinical laboratory specimens, and failure of two person verification of patient identify for blood bank specimens. Out of 37 misidentification, errors during analytical phase were associated with mislabeled specimen containers, tissue cassettes, or microscopic slide. Only 13 events of misidentification occurred in postanalytical stage which was due to the reporting of results into the wrong medical record and incompatible blood transfusions due to failure of two persons' verification of blood products [3].

Mislabeled specimen for ABO typing or receiving wrong blood belonging to another patient for transfusion has serious consequences. Transfusion of blood that is typed on specimens that are mislabeled can result in acute hemolytic transfusion reactions resulting in serious morbidity and death. To reduce the risk of error, American Association of Blood Banks standards require that patient blood sample tubes have affixed to them labels bearing at least two unique patient identifiers and the dates on which specimens were collected. Further, several studies have demonstrated that barcoding patient specimens reduces specimen misidentification. To assess the rates of blood bank ABO typing specimens that are mislabeled and/or contain blood belonging to another patient, the College of American Pathologists conducted a Q-probe study where a total of 30 institutions submitted data on 41,333 ABO blood typing specimens. There were only 306 mislabeled specimens submitted for ABO typing (0.74%) and 10 instances of wrong blood in the tube for transfusion out of 23,234 specimens (0.043%). Mislabeling rates were lower in institutions requiring that specimens be labeled with patients' birth dates than those that did not [4].

Delta checks are a simple way to detect mislabels. A delta check is a process of comparing a patient's result to his or her previous result for any one analyte over a specified period of time. The difference or "delta," if outside preestablished rules, may indicate a specimen mislabel or other preanalytical error.

## Error of collecting blood in wrong tubes: Effect of anticoagulants

Blood specimen must be collected in right tube to get accurate test result. It is important to have correct anticoagulant in the tube (different anticoagulant tubes have different colored tops). Anticoagulants are used to prevent coagulation of blood or blood proteins to obtain plasma or whole blood specimens. The most routinely used anticoagulants are ethylenediaminetetraacetic acid (EDTA), heparin (sodium, ammonium, or lithium salts), and citrates (trisodium and acid citrate dextrose). The optimal anticoagulant: blood ratio is essential to preserve analytes and prevent clot or fibrin formation via various differing mechanisms. Proper anticoagulants for various tests are as follows:

- Potassium EDTA (ethylenediamine tetraacetic acid; purple top tube) is the anticoagulant of choice for the complete blood count (CBC).
- EDTA is also used for blood bank pretransfusion testing, flow cytometry, hemoglobin A1C, and most common immunosuppressive drugs, such as cyclosporine, tacrolimus, sirolimus and everolimus, although another immunosuppressant mycophenolic acid is measured in serum or plasma instead of whole blood.
- Heparin (green top tube) is the only anticoagulant recommended for the determination of pH blood gases, electrolytes, and ionized calcium. Lithium heparin is commonly used instead of sodium heparin for general chemistry tests. Heparin is not recommended for protein electrophoresis and cryoglobulin testing because the presence of fibrinogen, which comigrates with beta-2 monoclonal proteins.
- For coagulation testing, citrate (light blue top) is the appropriate anticoagulant.
- Potassium oxalate is used in combination with sodium fluoride, and sodium iodoacetate to inhibit enzymes involved in the glycolytic pathway. Therefore, oxalate/fluoride (gray top) tube should be used for collecting specimen for measuring glucose level.

Although lithium heparin tubes are widely used for blood collection for the analysis of many analytes in the chemistry section of a clinical laboratory, a common mistake is to collect specimen for lithium analysis in a lithium heparin tube. This may cause clinically significant falsely elevated lithium that may confuse the ordering physician.

## Blood collection from catheter and intravenous line

Although most blood collections are through venipuncture (except collection of arterial blood gas), blood may also be collected from catheter or intravenous (IV) line. For single phlebotomy, it is generally better to avoid an area near an IV line. However, for patients periodically requiring numerous specimens, collecting blood through IV lines and indwelling catheters, including central venous lines or arterial catheters, offers the advantage of collecting blood without venipuncture. Unfortunately, this creates an inherent risk for improper specimen collection. To avoid contamination and dilution with IV fluid; it is recommended that the valve be closed for at least 3 min prior to specimen collection. In order to clear the IV fluid from the line, approximately 6–10 mL should be withdrawn and discarded. Because heparin is often in a line to maintain patency, larger volumes may need to be discarded for coagulation studies. Blood drawn from these lines should not be cultured due to the high risk of contamination from bacteria growing in the line. Some drugs such as cyclosporine adhere so tightly to these lines that specimens for drug monitoring should always be obtained from a site unrelated to drug administration.

## Issues with collecting urine specimens

Urinalysis remains one of the key diagnostic tests in the modern clinical laboratory and as such proper timing and collection techniques are important. Urine is essentially an ultrafiltrate of blood. Examination of urine may take several forms: microscopic, chemical (including immunochemical), and electrophoresis. Three different timings of collection are commonly encountered. The most common is the random or "spot" urine collection. However, if it would not unduly delay diagnosis, the first voided urine in the morning is generally the best sample. This is because the first voided urine is generally the most concentrated and contains the highest concentration of sediment. The third timing of collection is the 12 or 24 h collection. This is the preferred technique for quantitative measurements such as creatinine, electrolytes, steroids and total protein. The usefulness of these collections is limited, however, by poor patient compliance.

For most urine testing, a clean catch specimen is optimal with a goal to collect a "midstream" sample for testing. In situations where the patient cannot provide a clean catch specimen, catheterization represents another option but this must be performed only by trained personnel. Urine collection from infants and young children prior to toilet training can be facilitated through the use of disposable plastic bags with adhesive surrounding the opening.

For point-of-care urinalysis (e.g., urine dipstick and pregnancy testing) any clean and dry container is acceptable. Disposable sterile plastic cups and even clean waxed paper cups are often employed. If the sample is to be sent for culture, the specimen should be collected in a sterile container. For routine urinalysis and urine culture, the containers should not contain preservative. For specific analyses some preservatives are acceptable. The exception to this if for timed collections where hydrochloric acid, boric acid, or glacial acetic acid is used as a preservative.

Storage of the urine specimens at room temperature is generally acceptable for up to 2 h. After this time, the degradation of cellular and some chemical elements becomes a concern. Likewise, bacterial overgrowth of both pathologic as well as contaminating bacteria may occur with prolonged storage at room temperature. Therefore, if greater than 2 h will elapse between collections and testing of the urine specimen it must be refrigerated. Storage in refrigeration for up to 12 h is acceptable for urine samples destined for bacterial culture. Again, proper patient identification and specimen labeling is important to avoid errors in reported results.

## Issues with specimen processing and transportation

Specimens after collection require transportation to the clinical laboratory. If specimens are collected in the outpatient clinic of the hospital and analyzed in the hospital laboratory, transportation time may not be a factor. However, if specimens are transported to the clinical laboratory or a reference laboratory, care must be taken in shipping specimens especially by putting ice pack or cold pack to preserve specimens at lower temperature because analytes are more stable at lower temperature. Turbulence during transportation such as transporting specimens in a van to the main laboratory even may affect concentrations of certain analytes.

Many clinical laboratory tests are performed by using serum or plasma. Due to instability of certain analytes in unprocessed serum or plasma, separation of serum or plasma from blood components must be performed a soon as possible, but definitely within 2 h of collection. Appropriate preparation of specimens prior to centrifugation is required to ensure accurate laboratory results. Serum specimens must be allowed ample time to clot prior to centrifugation. Tubes with clot activators require sufficient mixing and at least 30 min of clotting time. Plasma specimens must be mixed gently according to manufacturer's instruction to ensure efficient release of additive/anticoagulant.

## Special issues: Blood gas and ionized calcium analysis

Specimens collected for blood gas determinations require special care, as the analytes are very sensitive to time, temperature, and handling. In standing whole blood samples, pH falls at a rate of $0.04–0.08\,h^{-1}$ at 37°C,

$0.02–0.03\,h^{-1}$ at 22°C, and $<0.01\,h^{-1}$ at 4°C. This drop in pH is concordant with decreased glucose and increased lactate. In addition, $pCO_2$ increases around 5.0 mmHg/h at 37°C, 1.0 mmHg/h at 22°C, and 0.5 mmHg/h at 4°C. At 37°C, $pO_2$ decreases by 5–10 mmHg/h, but only 2 mmHg/h at 22°C. Ideally, all blood gas specimens should be measured immediately and never stored. A plastic syringe, transported at room temperature is recommended if analysis will occur within 30 min of collection but a glass syringe if more than 30 min needed prior to analysis and specimens must be stored in ice. Bubbles must be completely expelled from the specimen prior to transport, as the $pO_2$ will be significantly increased and $pCO_2$ decreased within 2 min [5].

Blood gas analyzers re-heat samples to 37°C for analysis to recapitulate physiological temperature. However, for patients with abnormal body temperature, either hyperthermia due to fever, or induced hypothermia in patients undergoing cardiopulmonary bypass, a temperature correction should be made to determine accurate pH, $pO_2$, and $pCO_2$ results.

Ionized calcium is often measured in by ion-sensitive electrodes in blood gas analyzers. Ionized calcium is inversely related to pH: decreasing pH decreases albumin-binding to calcium, thereby increasing free, ionized calcium. Therefore, specimens sent to the lab for ionized calcium determinations should be handled with the same caution as other blood gas samples, since preanalytical errors in pH will impact ionized calcium results [6].

## Summary/key points

- Errors in clinical laboratory can occur in preanalytical, analytical, or postanalytical step. Most errors occur in preanalytical steps.
- During specimen collection, a patient must be identified by matching at least two criteria. Blood should be collected in correct tube following correct order of draw.
- Correct order of drawing blood: First microbiological blood culture tubes (yellow top), second royal blue tube (no additive) if trace metal analysis if desired, third citrate tube (light blue), fourth serum tube (red top) or tube with gel separator/clot activator (gold top or tiger top), fifth heparin tube (green top), sixth EDTA tube (purple/lavender top), and last oxalate-fluoride tube (gray top).
- Proper centrifugation (in case of analyzing serum or plasma specimen) and proper transportation of specimen to the laboratory as well as maintaining proper storage of specimen prior to analysis is necessary to avoid artifactual changes in the analyte.
- EDTA (purple top tube) is the anticoagulant of choice for the complete blood count (CBC). EDTA tube is also used for Blood Bank

pretransfusion testing, flow cytometry, hemoglobin A1C and most common immunosuppressive drugs, such as cyclosporine, tacrolimus, sirolimus, and everolimus although another immunosuppressant mycophenolic acid is measured in serum or plasma instead of whole blood.

- Heparin (green top tube) is the only anticoagulant recommended for the determination of pH blood gases, electrolytes, and ionized calcium. Lithium heparin is commonly used instead of sodium heparin for general chemistry tests. Heparin is not recommended for protein electrophoresis and cryoglobulin testing because the presence of fibrinogen, which comigrates with beta-2 monoclonal proteins.
- For coagulation testing, citrate (light blue top) is the appropriate anticoagulant.
- Potassium oxalate is used in combination with sodium fluoride, and sodium iodoacetate to inhibit enzymes involved in the glycolytic pathway. Therefore oxalate/fluoride (gray top) tube should be used for collecting specimen for measuring glucose level.
- Ideally, all blood gas specimens should be measured immediately and never stored. A plastic syringe, transported at room temperature is recommended if analysis will occur within 30 min of collection. Otherwise, a specimen must be stored in ice. Glass syringes are recommended for delayed analysis because glass does not allow the diffusion of oxygen or carbon dioxide. Bubbles must be completely expelled from the specimen prior to transport, as the $pO_2$ will be significantly increased and $pCO_2$ decreased within 2 min.

## References

[1] Carraro P, Plebani M. Errors in STAT laboratory; types and frequency 10 years later. Clin Chem 2007;53:1338–42.

[2] Murphy JE, Ward ES. Elevated phenytoin concentration caused by sampling through the drug-administered line. Pharmacotherapy 1991;11:340.

[3] Dunn EJ, Morga PJ. Patient misidentification in laboratory medicine: a qualitative analysis of 227 root cause analysis reports in the Veteran Administration. Arch Pathol Lab Med 2010;134:244–55.

[4] Novis DA, Lindholm PF, Ramsey G, Alcorn KW, et al. Blood bank specimen mislabeling: a College of American Pathologists Q-Probes Study of 41 333 blood bank specimens in 30 institutions. Arch Pathol Lab Med 2017;141:255–9.

[5] Knowles TP, Mullin RA, Hunter JA, Douce FH. Effects of syringe material, sample storage time, and temperature on blood gases and oxygen saturation in arterialized human blood samples. Respir Care 2006;51:732–6.

[6] Toffaletti J, Blosser N, Kirvan K. Effects of storage temperature and time before centrifugation on ionized calcium in blood collected in plain vacutainer tubes and silicone-separator (SST) tubes. Clin Chem 1984;30:553–6.

# CHAPTER 4

# Laboratory statistics and quality control

## Mean, standard deviation, and coefficient of variation

In an ideal situation, when measuring value of the analyte in a specimen, the same value should be produced over and over again. However, in reality, the same value is not reproduced by the instrument but a similar value is observed. Therefore, most basic statistical operation is to calculate the mean, standard deviation, and then to determine coefficient of variation (CV). Mean value is defined as:

$$\text{Mean}\,(\overline{X}) = \frac{X_1 + X_2 + X_3 + \ldots + X_n}{n}$$

where $X_1$, $X_2$, $X_3$, etc., are individual values and "$n$" is the number of values.

After calculation of the mean value, standard deviation (SD) of sample can be easily determined by using the following formula:

$$SD = \sqrt{\frac{\sum (x_1 - \overline{x})^2}{n - 1}}$$

where $X_1$ is the individual values from the sample and $n$ is again the number of observations.

Standard deviation represents the average deviation of an individual value from the mean value. Smaller is the standard deviation, better is the precision of the measurement. Standard deviation is the square root of variance. Variance indicates deviation of a sample observation from the mean of all values and is expressed as sigma. Therefore,

$$\sigma = \sqrt{SD}$$

Coefficient of variation is also a very important parameter because CV can be easily expressed as a percent value, and lower the CV, better is the precision for the measurement. Advantage of CV is that one number can be used to express

**59**

Clinical Chemistry, Immunology and Laboratory Quality Control. https://doi.org/10.1016/B978-0-12-815960-6.00028-5

precision instead of stating both mean value and standard deviation. CV can be easily calculated from the following formula:

$$CV = (SD/Mean) \times 100$$

Sometimes standard error of mean is also calculated.

Standard error of mean $= SD/\sqrt{n}$, where $n$ is the number of data point in the set.

## Precision and accuracy

Precision is a measure of how reproducible values are in a series of measurements of a specimen for a particular analyte, while accuracy indicates how close a determined value is to the target values. CV is a good estimate of precision, for example, if CV is 2% for a glucose test, it implies that if one glucose value is 100 mg/dL, then if the specimen is reanalyzed, values will be between 98 and 102 mg/dL indicating good precision. However, CV does not provide any information on accuracy because an assay may be very precise but not very accurate. Accuracy can be determined for a particular analyte by analysis of an assayed control where the target value is known (provided by the manufacturer or made in-house by measuring accurately a predetermined amount of analyte and then dissolving it in a predetermined amount of a solvent matrix where matrix is similar to plasma). An ideal assay has both excellent precision and accuracy, but good precision of an assay may not always guarantee good accuracy.

## Gaussian distribution and reference range

Gaussian distribution, also known as normal distribution, is a bell-shaped curve and it is assumed that during any measurement values will follow a normal distribution with equal number of measurements above the mean value and below the mean value. In order to understand normal distribution, it is important to know definitions of "mean," "median," and "mode." The "mean" is calculated average of all values. "Median" is the value at the center point (midpoint) of the distribution, while "mode" is the value that was observed most frequently during the measurement. If a distribution is normal, then value of mean, median, and mode is the same. However, the value of mean, median, and mode may be different if the distribution is skewed (not Gaussian distribution). Other characteristics of Gaussian distributions are as follows:

- Mean $\pm 1$ SD contain 68.2% of all values.
- Mean $\pm 2$ SD contain 95.5% of all values.
- Mean $\pm 3$ SD contain 99.7% of all values in the distribution.

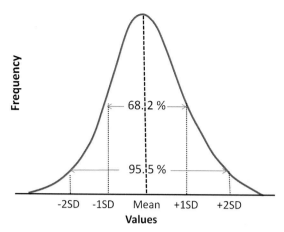

**FIG. 1**

A Gaussian distribution showing percentage of values within certain standard deviation from the mean. *Courtesy: Andres Quesda, MD.*

A Gaussian distribution is shown in Fig. 1. Usually reference range is determined by measuring value of an analyte in a large number of normal subjects (at least 100 normal healthy people, but preferably 200–300 healthy individuals). Then mean and standard deviations are determined.

Reference range : mean value − 2 SD to mean value + 2 SD

This incorporates 95% of all values. The rationale for reference range to be mean ± 2 SD is based on the fact that lower end of abnormal values and upper end of normal values may often overlap. Therefore, mean ± 2 SD is a conservative estimate of reference range based on measurement of the analytes in a healthy population. Important points of reference range include:

- Reference range may be same between male and female for many analytes, but reference range may differ significantly between male and female for certain analytes such as sex hormones
- Reference range of an analyte in an adult population may be different from infants or elderly patients
- Although less common, reference range of certain analytes may be different between different ethnic populations
- For certain analytes such as glucose, cholesterol, triglycerides, and high-density and low-density cholesterol, there is no reference range, but there are desirable ranges which are based on study of a large population and risk factors associated with certain values of analytes, for example, desirable value to cholesterol is below 200 mg/dL.

## Sensitivity, specificity, and predictive value

An assay cannot be 100% sensitive or specific because there is some overlap between values of a particular biochemical parameter observed in normal individuals and patients with a particular disease (Fig. 2). Therefore, during measurement of any analyte, there is a gray area where few abnormal values are generated from analysis of specimens from healthy people (false positive) and few normal results are generated from patients (false negative).

- The gray area depends on width of normal distribution and reference range of the analyte.
- False-positive results may mislead the clinician leading to unnecessary investigation and diagnostics tests, as well as increased anxiety of the patient.
- False-negative result is more dangerous than a false-positive result because diagnosis of a disease may be missed or delayed causing serious problem.
- For a test, as clinical sensitivity increases, specificity decreases. For calculating clinical sensitivity, specificity, and predictive value of a test, following formulas can be used.

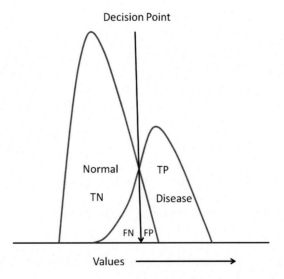

**FIG. 2**
Distribution of values in normal and diseased state, where *TN*, true negative values; *TP*, true positive values; *FN*, false negative values; *FP*, false positive values. *Courtesy: Andres Quesda, MD.*

$$\text{Sensitivity (individuals with disease who show positive test results)} = \frac{TP}{TP + FN} \times 100$$

$$\text{Specificity (individuals without disease who show negative test results)} = \frac{TN}{TN + FP} \times 100$$

$$\text{Positive predictive value} = \frac{TP}{TP + FP} \times 100$$

TP = True positive, a result correctly identifies a disease.

FP = False positive, result falsely identifies a disease.

TN = True negative, result correctly excludes a disease when the disease is not present in an individual.

FN = False negative, result incorrectly excludes a disease when the disease is present in an individual.

Therefore, when assay results are positive, results are combination of TP and FP and when assay results are negative, results are combination of TN and FN.

Positive predictive value is proportion of individuals with disease who showed positive value compared to all individual tested. Let us consider an example where a particular analyte was measured in 100 normal individuals and 100 individuals with disease. Then following observations were made:

$TP = 95, FP = 5, TN = 95, FN = 5$

Therefore, sensitivity $= (95/95 + 5) \times 100 = 95\%$.

## Random and systematic errors in measurements

Random errors and systematic errors are important issues in laboratory quality control process. Random errors are unavoidable and occur due to imprecision of any analytical method. On the other hand, systematic errors have certain characteristics and are often due to errors in measurement using a particular assay. Because random errors cannot be eliminated or controlled, goal of quality control in a clinical laboratory is to avoid or minimize systematic errors. Usually recalibration of the assay is the first step taken by a clinical laboratory technologist to correct systematic error, but more serious problem such as instrument malfunction may also be responsible for systematic errors. In this case, laboratory should immediately contact technical specialist from the diagnostic company to fix the problem.

## Laboratory quality control: Internal and external

Good quality control is the heart of a good laboratory operation. Because value of an analyte in a patient's specimen is unknown, clinical laboratory professionals rely on producing accurate result using controls for an assay. Controls can be purchased from a commercial source or can be made in house. A control is defined as a material that contains the analyte of interest with a known concentration. It is important that the control material has similar matrix to serum or plasma. Different types of controls used in clinical laboratories are listed below:

- Assayed control: The value of the analyte is predetermined. Most commercially available controls have predetermined values of various analytes. Target value must be verified before use.
- Unassayed control: Target value is not predetermined. This control must be fully validated (at last run 20 times in a single run to determine within-run precision and then run once a day for 20 consecutive days for establishing between-run precision).
- Homemade control: If assayed control material is not easily commercially available, for example, for an esoteric test, control material may be prepared by the laboratory staff by dissolving correctly weighted pure material in an aqueous based solvent or in serum or whole blood (for an analyte not present in human, for example, a drug).

Commercially available control materials may be obtained as a ready-to-use liquid control or as a lyophilized powder. If control material is available in the form of lyophilized powder, it must be reconstituted prior to use strictly by following manufacturer's recommended protocol. Control materials must be stored in a refrigerator following manufacturer's recommendation and expiration date of the control must be clearly visible so that expired control is not used by mistake. Expired controls must also be discarded. Usually low, medium, and high controls of an analyte are used indicating analyte concentrations in both normal physiological state and diseased state. At least two controls must be used for each analyte (high and low controls). Control materials may be run along with patient samples, but more commonly controls are analyzed once at the beginning of each shift, after an instrument is serviced, when reagent lots are changed, after calibration, and when validity of patient result is questioned by a clinician.

## CAP proficiency testing

Quality control in the laboratory may be both internal and external. Internal quality control is essential and results are plotted in a Levey-Jennings chart as discussed later in this chapter.

The most common example of external quality control is analysis of CAP (College of American Pathologists) proficiency samples for most tests offered by a clinical laboratory. Proficiency samples may not be available for few esoteric tests. CLIA 88 (Clinical Laboratory Improvement Amendments) requires all clinical laboratories to register with the government and to disclose all tests these laboratories offer. The test may be "waived tests" or "nonwaived tests."

- "Waived tests," where laboratories can perform such tests as long as they follow manufacturer's protocol. Enrolling in an external proficiency testing program, such as CAP survey, is not required for waived tests. However, accuracy of waived tests must be documented twice a year.
- "Nonwaived tests" are moderately complex or complex tests and laboratories performing such tests are subjected to all CLIA regulations and must be inspected by CLIA inspectors every 2 years or by inspectors from nongovernment organizations such as CAP or Joint Commission on Accreditation of Healthcare Organization (JCAHO). In addition, a laboratory must participate in external proficiency such as CAP proficiency surveys and must successfully pass proficiency testing in order to operate legally. A laboratory must produce correct results for four of five external proficiency specimens for each analyte and must have at least 80% score except for certain blood bank related testing where passing score is 100%.

After April 2003, clinical laboratory must perform method validation for each new test even if such test already has FDA approval. Currently most common external proficiency testing samples are offered by CAP and there are proficiency specimens for over 580 analytes. The major features of CAP external proficiency testing include:

- CAP proficiency samples are mailed to participating laboratories three times a year and there are at least five samples for each analytes during this period. Results must be reported to CAP on or before due date. This is very important because if results are even 1 day late, it is considered as proficiency testing failure. After results are reported, proficiency samples must be stored frozen because in case of proficiency testing failure such specimen can be retrieved and retested to investigate source of error in the initial testing of failed analyte.
- CAP proficiency samples must be treated as a regular patient specimen and must be analyzed by a technologist performing those tests for routine analysis of patient specimens. CAP proficiency specimen cannot be assigned to a special person in the laboratory, for example, laboratory supervisor. In addition, CAP proficiency samples cannot be run in duplicate and must not been sent to a different laboratory or a reference laboratory for analysis. Such violations have severe consequences.
- CAP proficiency test results are reported in several columns with the purpose of comparing results obtained by the laboratory with the peer group (using same assay and analyzer as your laboratory). From left to right of the column, results reported by your laboratory, peer group mean result, peer group standard deviation (SD), the number of peer group reporting results, your SDI (standard deviation index), upper limit of

acceptable result, lower limit of acceptable result, and your grade (acceptable grade means the laboratory passed that test). However, for certain analytes, results could be positive or negative. In this case, at the left column response of the laboratory is recorded and in the next column the intended response is given followed by grade. If intended response is positive and the laboratory reported positive response, then as excepted the grade is "acceptable."

SDI = Your laboratory mean − Peer group mean/Peer group standard deviation

- Every year the laboratory must renew subscription of CAP proficiency samples and pay appropriate fee because proficiency samples are shipped on a regular basis for 1 year only and such subscription is not automatically renewed.
- When CAP proficiency test results arrive at the laboratory for review by laboratory professionals, appropriate action must be taken if any analyte failed the survey. Action must then be documented and laboratory director or designee must sign results of CAP survey approving corrective action taken by the laboratory staff in case of a failed analyte.
- The best way to evaluate CAP proficiency testing result of an individual clinical laboratory is to use e-lab solution available from the CAP for downloading.
- If CAP proficiency testing is not available, then the laboratory must validate the test every 6 months by comparing test results of the analyte obtained by the laboratory with test results obtained by another laboratory offering the test, using split samples (at least 10 or more specimens). Alternatively, if proficiency samples are available from another source, for example, AACC (American Association for Clinical Chemistry), passing such proficiency testing is also acceptable.
- A laboratory may participate in addition to CAP external proficiency testing program, other proficiency testing program. However, for laboratory accreditation by CAP, it is required that the laboratory must participate in CAP proficiency survey provided that proficiency specimen is available from the CAP.

There are number of publications that indicate that participating in external proficiency survey such as offered by CAP is useful in improving quality of a clinical laboratory operation [1–3].

## Passing CAP proficiency testing

There are different criteria for acceptable results in a CAP proficiency survey for different analytes. A target value of the analyte must be selected for grading results by CAP. For most analytes peer group mean is selected as the target value, although for few analytes other approaches may be taken. When a target value is set then performance of a laboratory result is evaluated by comparing value reported by the laboratory with the target value. For some analytes, fixed range criteria are used; for example, a fixed range of $\pm 4$ mmol/L of peer mean value is used as acceptable range for sodium. Therefore, if peer mean is 140 mmol/L, then acceptable range would be 136–144 mmol/L. Another approach is using percentage of peer mean, for example, $\pm 10\%$ of peer mean for albumin. Therefore, if peer mean is 4.0 g/dL, then acceptable range should be $\pm 10\%$ of 4.0 g/dL, which is 3.6–4.4 g/dL. Variable range approach may also be applied to determine acceptable range. For example, evaluation criterion for bilirubin is $\pm 0.4$ mg/dL or 20% of peer mean, whichever is greater. Therefore, if peer mean is 5.0 mg/dL, then using 20% criteria, acceptable range should be 4.0–6.0 mg/dL, and using 0.4 mg/dL, acceptable range should be 4.6–5.4 mg/dL. In this case 20% is greater, so acceptable range is 4–6 mg/dL and this should be used for setting acceptable limit. However, for most analytes $\pm 3$ SD from peer mean value is used to determine acceptability of a result.

Sometimes an analyte in the proficiency testing specimens may not be graded. For example, if there is a problem with one or more challenge samples, perhaps due to contamination, then all participants will pass the event with a 100% score. In other instances, individual laboratories may temporarily be excused from participating in CAP proficiency testing, such as a case in which an analyzer is being repaired and not being used for patient testing or a natural disaster has occurred. In this type of extenuating circumstance, the laboratory can avoid an unsatisfactory score provided the laboratory notifies its accreditation organization when factors beyond its control prevent normal performance of CAP proficiency testing.

Laboratories must demonstrate successful performance in CAP proficiency testing in order to remain in good standing. In most cases, there are five challenges per analyte per event and there are three such events per year for most analytes. In order to pass CAP proficiency survey successfully, laboratories must get at least four of the five challenges correct (80%) for each analyte, or the event will be considered as unsatisfactory. However, for certain critical analytes such as ABO typing, Rh(D) typing, compatibility testing, etc., 100% performance must be achieved. There are no sanctions if the laboratory has an occasional, isolated unsatisfactory proficiency testing event for a specific analyte. However, if a laboratory has two unsatisfactory events out of three consecutive events, it will be cited with a deficiency for unsuccessful performance.

The mechanism for scoring challenges depends upon whether the analyte is qualitative or quantitative in nature. Many times, qualitative analytes have only two possible answers, for example, reactive/nonreactive or positive/negative. In addition, in the specialty of hematology, the presence or absence of a certain cell type or the correct identification of various cells or cell inclusions are considered qualitative results. For passing proficiency testing, laboratory result must agree with intended response.

## What action to take if a laboratory failed CAP survey for certain analytes?

CLIA mandates proficiency testing for a subset of moderate and high complexity testing analytes termed as "regulated analytes" but not for waived tests. Therefore, if a laboratory fails in a proficiency testing for one or more analytes, corrective action must be taken and should be properly documented. Moreover, only one failed proficiency testing is allowed each year for a particular analyte except for ABO typing and Rh(D) typing and compatibility testing where passing score is perfect 100%. If failed twice in a year that test must be discontinued and specimen should be sent to a different laboratory before CAP allows the laboratory to do that test again after documentation of extensive corrective action. For waived testing analytes, laboratory must verify accuracy of the test twice annually.

## Levey-Jennings chart and Westgard rules

In addition to participating in CAP program, clinical laboratories must run control specimens every shift, at least three times in a 24-h cycle. In addition, instruments must be calibrated as needed in order to maintain good laboratory practice. Calibration is needed for all assays that a clinical laboratory offers. Calibration of immunoassays is discussed in Chapter 2. However, other assays are calibrated using calibrators which are either commercially available or homemade.

- Calibrators are defined as materials that contain known amounts of the analyte of interest. For a single assay at least two calibrators are needed for calibration, zero calibrator (contains no analyte) and a high calibrator containing amount of the analyte representing upper end of analytical measurement range. However, commonly five to six calibrators are used for calibration. One calibrator must be zero calibrator and the highest calibrator must contain concentration of the analyte at the upper end of the analytical measurement range. Other calibrators usually have concentrations in between zero calibrator and the highest calibrator, representing normal values of the analyte and values expected in diseased state (for drugs, values below therapeutic range, between therapeutic range, and then toxic range).

- Controls are materials that contain known amount of the analyte. Matrix of the control must be similar to matrix of the patient's sample, for example, matrix of the control must resemble serum for assays conducted in serum or plasma.

Levey-Jennings chart is commonly used for recording observed values of controls during daily operation of a clinical laboratory. Levey-Jennings chart is a graphical representation of all controls values for an assay during an extended period of laboratory operation. During this graphical representation, values are plotted with respect to the calculated mean and standard deviation and for if all controls are within the mean and $\pm 2$ SD, then all controls values were within acceptable limit and all runs during that period have acceptable performance (Fig. 3). In this figure, all glucose low controls were within acceptable limit for the entire month. Levey-Jennings chart must be constructed for each control (low and high control or low, medium, and high control) for each assay the laboratory offers. For example, if the laboratory runs two controls (low and high) for each test and offer 100 tests, then there will be $100 \times 2 = 200$ Levey-Jennings charts each month. Usually Levey-Jennings chart is constructed for one control for 1 month. The laboratory director or designee must review all Levey-Jennings charts each month and signs them for compliance with accrediting agency. However, technologists review result of the control during run and accept the run if the value of the control is within acceptable range established by the laboratory (usually the range is mean $\pm 2$ SD) and then laboratory supervisor reviews all control data daily and manager or supervisor also reviews all control data weekly.

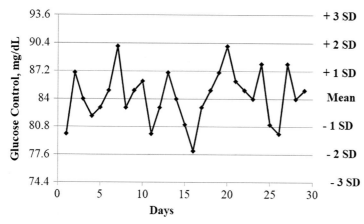

**FIG. 3**
Levey-Jennings chart with no violation.

**Table 1** Westgard rules.

| Violation | Comments | Accept/reject run | Error type |
|---|---|---|---|
| $1_{2s}$ | One control value is outside ±2 SD limit, but other control within ±2 SD limit | Accept run | Random |
| $1_{3s}$ | One control exceeds ±3 SD | Reject run | Random |
| $2_{2s}$ | Both controls outside ±2 SD limit or two consecutive control outside limit | Reject run | Systematic |
| $R_{4s}$ | One control +2 SD and another −2 SD | Reject run | Random |
| $4_{1S}$ | Four consecutive control exceeding +1 SD or −1 SD | Reject run[a] | Systematic |
| $10_x$ | Ten consecutive control values falling on one side of the mean | Reject run[a] | Systematic |

*[a]Although these are rejection rules, a laboratory may consider these violations as warning and may accept the run but takes step to correct such systematic errors.*

Usually Westgard rules are used for interpreting Levey-Jennings chart and for certain violation, run must be rejected and problem must be resolved before resuming testing of patient's samples. Various errors can occur in Levey-Jennings chart including shift, trend, and other violations (Table 1). The basic principle is that control values must fall within ±2 SD of the mean, but there are some situations when violation of Westgard rules occurs despite control values are within the ±2 SD limits of the men. Usually $1_{2s}$ is a warning rule and occurs due to random error (Fig. 4), other rules (Figs. 4 and 5) may be warning rules or

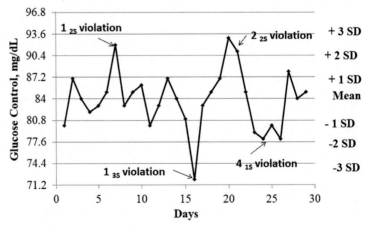

**FIG. 4**
Levey-Jennings chart showing certain violations ($1_{2S}$, $1_{3S}$, $2_{2S}$, and $4_{1S}$).

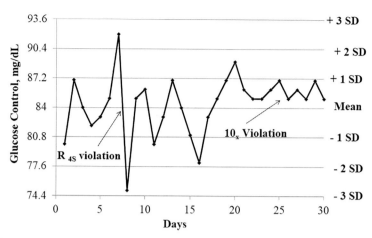

**FIG. 5**

Levey-Jennings chart showing certain violations ($R_{4S}$ and $10_x$).

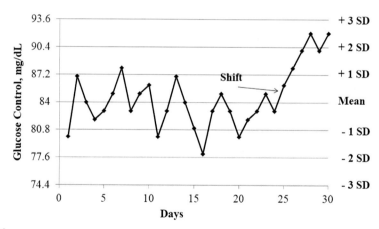

**FIG. 6**

Levey-Jennings chart showing shift of control values.

rejection rules (see Table 1). In addition, shift (Fig. 6) and trend (Fig. 7) may be observed in Levey-Jennings chart indicating systematic errors where corrective actions must be taken. When 10 or more consecutive control values are falling on one side of the mean (Fig. 5), a shift is observed (10X rule). Moreover, when a 10X violation is observed, it may also indicate trend when control values also indicate upward trend.

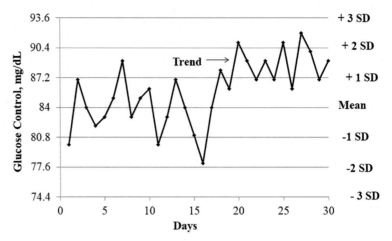

**FIG. 7**
Levey-Jennings chart showing trend.

## Delta checks

Delta check is an additional quality control measure adopted by the computer of an automated analyzer or the laboratory information system (LIS) where a value is flagged if the value deviates more than predetermined limit from the previous value in the same patient. The limit of deviation for each analyte is set by the laboratory professionals. The basis of the delta check is that value of an analyte in a patient which should not deviate significantly from the previous value unless certain intervention is done, for example, a high glucose value may decrease significantly following administration of insulin. If a value is flagged as failed delta check, then further investigation should be made. A phone call to the nurse may address issues such as erroneous result due to collection of a specimen from IV line or collection of wrong specimen. Quality control of the assay must also be addressed to ensure that the erroneous result is not due to instrument malfunction.

Value of delta check is usually based on one of the following criteria:

- Delta difference: current value—previous value should be within a predetermined limit
- Delta percent change: delta difference/current value
- Rate difference: delta difference/delta interval × 100
- Rate percent change: delta percentage change/delta interval

## Method validation for a new test using existing analyzer

After April 2003, clinical laboratory must perform method validation for each new test implemented in the laboratory despite such tests is FDA approved. In this section, adding a new test to an existing analyzer is discussed. Following steps are followed for method validation and implementation of a new method in the clinical laboratory.

- Within-run precision of the assay must be validated by running low, medium, and high controls or low and high controls 20 times each in a single run. Mean, standard deviation, and CV must then be calculated individually for low, medium, and high control.
- Between-run precision of the assay must be established by running low, medium, and high control or low and high control once daily for 20 consecutive days. Mean, standard deviation, and CV must then be calculated.
- Although linearity of the assay is provided by the manufacturer, linearity must be validated in the clinical laboratory prior to running patient specimens. Linearity is essentially the calibration range of the assay, which is also termed as "analytical measurement range." In order to validate the linearity, a high-end calibrator or standard can be selected and then it can be diluted to produce at least four to five dilutions that cover the entire analytical measurement range. If the observed value matches with the expected value, then the assay should be considered linear over the stated range.
- Detection limit should be traditionally determined by running zero calibrator or blank specimen for 20 times and then determining the mean and standard deviation. The detection limit (also called lower limit of detection) is considered as: mean + 2 SD value. However, guidelines of Clinical and Laboratory Standards Institute (CLSI; E17 protocol) advise running a specimen with no analyte (blank specimen) should be run and then limit of blank (LoB) = mean + 1.654 SD. This should be established by running blank specimens 60 times, but if a company already established a guideline, then 20 runs are enough. Limit of quantification is usually defined as a concentration where CV is 20% or less [4].
- Comparison of new method with an existing method is a very important step in method validation. Ideally, at least 100 patient specimens must be run by the existing method in the laboratory and at the same time by the new method. However, for validation of an FDA-approved test, at least 20 patient specimens must be analyzed using the new added test to a reference method such as the same test performed by another laboratory using the same analyzer and the same reagents, or if not available, results of the same analyte using a different analyzer.

It is advisable to batch patient samples and then run these specimens by both methods on the same day, if possible at the same time (by splitting specimens). Results obtained by the existing method should then be plotted in the $x$ axis (reference method) and corresponding values obtained by the new method should be plotted in the $y$ axis. Linear regression is the simplest way of comparing results obtained by the existing method in the laboratory and the new method. Linear regression equation is the line of best fit with all data points and computer can produce the linear regression line, as well as an equation called linear regression equation which is the equation representing a straight line (regression line):

$$y = mx + b$$

Here "$m$" is called the slope of the line and "$b$" is the intercept. The computer calculates the equation of the regression line using least square approach. The software also calculates "$r$," the correlation coefficient.

## How to interpret regression equation?

The regression equation, $y = mx + b$, provides many important information regarding how well the new method ($y$) compares with the reference method ($x$). Interpretations of a linear regression equation include:

- Ideal value; $m = 1$, $b = 0$. therefore $y = x$. In reality this never happens.
- If value of $m$ is less than 1.0, then the method shows negative bias compared to the reference method. Bias can be calculated as $1 - m$, for example, if the value of "$m$" is 0.95, then negative bias is $1 - 0.95 = 0.05$ or $0.05 \times 100 = 5\%$.
- If value of $m$ is over 1.0, it indicates positive bias in the new method. For example, if $m$ is 1.07, then positive bias in the new method is $1.07 - 1 = 0.07$ or $0.07 \times 100 = 7\%$.
- The intercept "$b$" can be a positive or negative value and must be relatively small number.
- Ideal value of "$r$" (correlation coefficient) is 1, but any value above 0.95 is considered good and a value of 0.99 is considered excellent. The correlation coefficient indicates how well the new method can be compared with the existing method, but cannot tell anything about any inherent bias in new method. Therefore, slope must be considered to determine bias.

In our laboratory, we evaluated a new immunoassay for mycophenolic acid, an immunosuppressant with a HPLC-UV method, the current method in our laboratory, using specimens from 60 transplant recipients after deidentifying specimens [5]. The regression equation was:

$y = 1.1204x + 0.0881 \ (r = 0.98)$

This equation indicated that there was an average 12.04% positive bias with the new immunoassay method compared to the reference HPLC-UV method in determining mycophenolic acid concentration. This was most likely due to cross-reactivity of mycophenolic acid acyl glucuronide with the mycophenolic acid assay antibody because metabolite does not interfere with mycophenolic acid determination using HPLC-UV. However, correlation coefficient of 0.98 indicated good agreement between both methods.

## Bland-Altman plot

Although linear regression analysis is useful for method comparison, such analysis is affected by extreme values (where one or a series of "$x$" values differ widely from the corresponding "$y$" values) because equal weights are given to all points. Bland-Altman plot compares two methods by plotting the difference between the two measurements on the $y$ axis, and the average of the two measurements on the $x$ axis. Difference between two methods can be expressed as a percentage difference between two methods or a fixed difference such as 1 SD or 2 SD or a fixed number. It is easier to see bias between two methods using Bland-Altman plot.

## How to implement a new analyzer in the laboratory?

Usually hospital administration obtains analyzers by lease agreement, most commonly for 5 years. However, analyzers can also be purchased outright. After 5 years administration may consider another set of analyzers from a different diagnostic company for a variety of reasons including more advanced technology, more automation, cost savings, etc. For this purpose, laboratory professions must undertake a thorough evaluation of new analyzers to ensure proper functioning of such analyzers and laboratory tests results to ensure results produced by the new analyzers are comparable to results produced by old methods, or if necessary, slight modification of reference ranges.

All essential analyzers such as automated chemistry analyzer, hematology analyzers, etc., are typically acquired at least in duplicate (same manufacturer and preferably same mode; large medical centers and reference laboratories may acquire multiple analyzers due to large volume) because one analyzer may function as a backup just in case another analyzer breaks down. However, both analyzers must be used at the same time for proper functioning of such analyzers. CLIA regulation requires that two instruments must be compared against each other every 6 months. This can be easily achieved by running 20 specimens for each analyte using both analyzers and then comparing results. Usually correlations are good. Any result that deviates from each other by 20% is considered as discordant specimen and is unacceptable. Differences should be within ±10%.

When new analyzers come to the laboratory to replace old analyzer, service representatives from the company will come to the laboratory to set up new analyzers, connecting them to hospital LIS and also to perform initial validation of tests including calibration of each assay, within- and between-run precision, linearity, and patient correlation. The patient correlation is an important part of validating new instruments. It is usually done by analyzing 20 or more patient specimens using the old analyzer and the new analyzer. Regression equation is then used to estimate the bias. A slope of 1.00 is ideal, but most commonly slope is between 0.95 and 1.05. However, if the slope is 0.9 or lower and 1.10 or higher, the analytical bias is considered significant and adjustment of reference range may be needed for that particular analyte as per discretion of the medical director. For correlation coefficient a value of 0.95 ($r = 0.95$) or higher is needed, desirably 0.98 or 0.99 (ideal correlation coefficient of 1.0 cannot be achieved).

## Analytical measurement range, reference range, and critical values

In general, analytical measurement range (the range of analyte concentration that can be measured accurately by the assay without dilution) is given in the package insert of each analyte provided by the company. The reportable range is the range of the analyte concentrations that can be accurately reported by the laboratory for a particular analyte. For example, in a digoxin assay the analytical measurement range is 0.2–4.0 ng/mL. If one patient specimen showed initial value of above the range, the specimen can be diluted 1:1 by saline or a diluent provided by the diagnostic company and the observed result was 2.2; then that can be multiplied by 2 to provide a digoxin value of 4.4 and that result can be reported as it is within reportable range. However, with higher dilution, matrix issue may arise. Some instruments have capability of autodilute a specimen and rerun the specimen if original value is above the reportable range. However, for very high values, for example, a highly elevated cancer marker, dilution may not provide an accurate value and in these cases values should be reported higher than reportable range.

Reference ranges are included in the patient report to help clinicians to interpret laboratory test results. Many text books in clinical chemistry and laboratory medicine list reference ranges of most analytes. In addition, diagnostic companies also provide reference range for each analyte in the package insert. CLIA

Regulation mandates that manufacturer's reference range (normal values) must be verified by each laboratory to ensure that such range is appropriate for the laboratory's patient population. This can be done by analyzing 20 specimens from normal volunteers (such as healthy laboratory personnel) to ensure values are within reference range of the manufacturer

Critical values are life-threatening laboratory test results (for example, glucose value of 35 mg/dL). Critical values are given along with reference values in clinical chemistry or laboratory medicine test books and also by the manufacturer. The laboratory director may also set critical values based on local population which is approved by the medical staff council. The CLIA regulation requires that when a critical value is observed in the laboratory it must be reported immediately (preferably within 15 min) to the provider by phone with a read back by the provider and a documentation of the call. The laboratory policy and procedure must have a list of critical values but not all analyte has a critical value, for example, prostate-specific antigen. However, a provider may not answer the phone call or a page in a reasonable amount of time. The critical value report policy must have a list of alternative professional to call in case the physician is not responding or a nurse in care of the patient is not available. One approach is to call the "on-call physician" to report the value. However, this situation must also be documented.

## Quality assurance indicators in the laboratory

Quality control (QC) looks at daily performance of the assays, while quality assurance (QA) investigates a broader picture of total quality of the laboratory including at processes that occur before (preanalytical), during (analytical), and after testing (postanalytical). In general, approximately 15% errors occur at the analytical step with approximately 65% errors occur in the preanalytical step. CLIA regulations require that a clinical laboratory must have protocols for both QC and QA. However, CLIA regulation does not have any guidelines on which quality assurance indicators should be monitored [6]. QA indicators could be a collection of preanalytical, analytical, postanalytical, physician satisfaction survey, etc. Common QA indicators include:

- Test specimens are labeled properly with patient name, patient identification, etc., all the time.
- Specimen rejection rate due to various causes should be less than 5%.
- Critical values are reported to the provider appropriately; goal 100% within 15 min.
- Proper documentation of reporting critical value to the provider; goal 100%.
- Significant findings and actionable results must be properly transmitted and documented; goal 100%.

- Tests ordered STAT (including STAT emergency room tests ordered) should be reported within an hour. If 90% or more STAT orders are reported within 1 h it is a good QA indicator.
- For critical analytes such as troponin I, goal is to report result within 35 min.
- For blood gas ordered STAT, goal is to report result within 15 min in 95% or more specimens.
- STAT tests ordered for in-patients goal is to report the result within 90 min from the time of blood drawn.
- Routine tests are reported within 24 h except for special tests which are sent to a reference laboratory; goal 100%.

Detail discussion on QA indicators are beyond the scope of this book.

## Receiver operating characteristic curve

Receiver operating characteristic (ROC) curve is often used to make an optimal decision level for a test. ROC plots the true positive rate of a test (sensitivity) either as a scale of 0–1 (1 highest sensitivity) or as percent on the $y$ axis versus false-positive rate (1 specificity). As sensitivity increases, the specificity decreases. In Fig. 8, a hypothetical ROC curve is given. If decision point 1 is selected for the test value, then sensitivity of the test is 0.6 or 60% but specificity is very high (99%, in the scale 1 specificity, 0.01). On the other hand, if a higher value of the test is selected for a decision point (decision point 3), the sensitivity is increased to 90% but specificity is decreased to 42% (in the scale 1 specificity, 0.58). Therefore, a decision point can be made which can be used for making a clinical decision. In general, closer the decision point to the $y$ axis, better is the specificity.

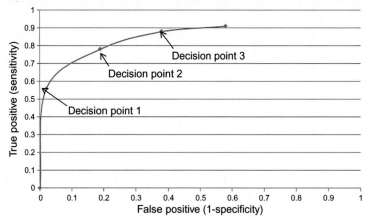

**FIG. 8**

Receiver operating characteristic (ROC) curve showing various decision points.

## What is six sigma?

Six sigma originally originated from Motorola Corporation's approach for total quality management during manufacturing with an objective to reduce defects in manufacturing. Although six sigma was originally developed in manufacturing process, the principles can be applied in total quality improvement of any operation including clinical laboratory operation. The goal of six sigma is to achieve an error rate of 3.4 of the 1 million processes or error rate is only 0.00034%. An error rate of 0.001% is considered as 5.8 sigma. The goal of clinical laboratory operation is to reduce error rate to at least 0.1% (4.6 sigma), but preferably 0.01% (5.2 sigma) or higher. Improvement can be made during any process of the laboratory operations (preanalytical, analytical, or postanalytical) with an overall goal of reducing laboratory errors.

## Errors associated with reference range

Reference ranges are given with patient's values to help clinicians in interpreting laboratory test results. However, most reference ranges include values in the range of mean $\pm 2$ SD as observed with normal population. Therefore, reference range only accounts for 95% values observed in healthy individuals for the particular tests and statistically 5% values of normal population should fall outside reference range. If more than one test is used then greater percentage of values should fall outside the reference range. Likelihood on "$n$" test results falling within the reference range can be calculated from the formula

%Results falling within normal range $= 0.95^n \times 100$

Therefore, %result falling outside reference range in normal people $= (1 - 0.95^n) \times 100$

For example, if five tests are ordered for health screening of a healthy person, then

%Results falling outside normal range $= (1 - 0.95^5) \times 100 = (1 - 0.773) \times 100 = 22.7\%$

Examples of number of tests falling within reference range and falling outside reference range are given in Table 2.

## Basic statistical analysis: Student's *t*-test and related tests

A new method can be validated against an existing method by using regression analysis as stated earlier in this chapter. Bias can be calculated based on the

**Table 2** For multiple tests ordered in a healthy subject, chances of number of tests falling within reference ranges and number of tests falling outside reference range.

| Number of tests | Results within reference range (%) | Outside reference range (%) |
|---|---|---|
| 1 | 95 | 5 |
| 2 | 90 | 10 |
| 3 | 85.7 | 14.3 |
| 4 | 81.4 | 18.6 |
| 5 | 77.3 | 22.7 |
| 6 | 73.5 | 26.5 |
| 10 | 59.8 | 40.2 |

analysis of the slope or Bland-Altman plot. However, in some instances, bias between the two methods can be significant and in this case a laboratory professional need to know if values on an analyte determined by the reference method is significantly different than the values determined by the new method. This can be calculated by the mean of two sets of values and standard deviation using Student's t-test.

- Student's t-test is useful to determine if one set of values are different from another set of values based on the difference between mean values and standard deviations. This statistical test is also useful in clinical research to see if values of an analyte in the normal state are significantly different from the values observed in a diseased state.
- Student's t-test is only applicable if both distributions of values are normal (Gaussian).
- If the "$t$" value is significant based on the degree of freedom ($n_1 + n_2 - 1$), where $n_1$ and $n_2$ represent number of values in set 1 distribution and set 2 distribution, then null hypothesis (there is no difference between two sets of values) is rejected and it is assumed that values in set 1 distribution are statistically different from values in set 2 distribution. Value of $t$ can be easily obtained from published tables.
- F test is a measure of differences in variances and can also be used to see if one set of data is different from another set of data. F test can be used for the analysis of multiple set of data when it is called ANOVA (analysis of variance).
- If the distribution of data is non-Gaussian, then neither t-test nor F test can be used. For this purpose, Wilcoxon's rank sum test (also known as Mann-Whitney $U$ test) should be used.

Formula for *t*-test and Mann-Whitney *U* test can be found in any text book on statistics. However, detail discussion on these statistical methods is beyond the scope of this book.

## Summary/key points

- Formula for coefficient of variation $(CV) = SD/Mean \times 100$.
- Standard error of mean $= SD/\sqrt{n}$, where $n$ is the number of data point in the set.
- If a distribution is normal, value of mean, median, and mode is the same. However, the value of mean, median, and mode may be different if the distribution is skewed (not Gaussian distribution).
- In a Gaussian distribution, mean $\pm 1$ SD contains 68.2% of all values, mean $\pm 2$ SD contains 95.5% of all values, and mean $\pm 3$ SD contains 99.7% of all values in the distribution.
- Reference range when determined by measuring an analyte in at least 100 healthy people and distribution of values in a normal Gaussian distribution is calculated as: Mean $\pm 2$ SD
- For calculating sensitivity, specificity, and predictive value of a test, following formulas can be used (where TP = true positive, FP = false positive, TN = true negative, and FN = false negative):

$$\text{Sensitivity (individuals with disease who show positive test results)} = \frac{TP}{TP + FN} \times 100$$
$$\text{Specificity (individuals without disease who show negative test results)} = \frac{TN}{TN + FP} \times 100$$
$$\text{Positive predictive value} = \frac{TP}{TP + FP} \times 100$$

- In a clinical laboratory, three types of control materials are used: assayed control where value of the analyte is predetermined, unassayed control where target value is not predetermined, and homemade control where control material is not easily commercially available, for example, for an esoteric test.
- Quality control in the laboratory may be both internal and external. Internal quality control is essential and results are plotted in a Levey-Jennings chart, while most common example of external quality control is analysis of CAP (College of American Pathologists) proficiency samples.
- "Waived tests" are not complex and laboratories can perform such tests as long as they follow manufacturer's protocol. Enrolling in an external proficiency testing program such as CAP survey is not required for waived tests.

- "Nonwaived tests" are moderately complex or complex tests and laboratories performing such tests are subjected to all CLIA regulations and must be inspected by CLIA inspectors every 2 years or by inspectors from nongovernment organizations such as CAP or Joint Commission on Accreditation of Healthcare Organization (JCAHO). In addition, for all nonwaived tests, laboratories must participate in external proficiency program, most commonly CAP proficiency surveys, and must successfully pass proficiency testing in order to operate legally. A laboratory must produce correct results for four of five external proficiency specimens for each analyte and must have at least 80% score for three consecutive challenges except for certain blood bank related testing where passing score is 100%.
- After April 2003, clinical laboratory must perform method validation for each new test even if such test already has FDA approval.
- Levey-Jennings chart is a graphical representation of all controls values for an assay during an extended period of laboratory operation. During this graphical representation, values are plotted with respect to the calculated mean and standard deviation. If all controls are within the mean and $\pm 2$ SD, then all controls values were within acceptable limit and all runs during that period have acceptable performance. Levey-Jennings chart must be constructed for each control (low and high control or low, medium, and high control) for each assay the laboratory offers. The laboratory director or designee must review all Levey-Jennings charts each month and signs them for compliance with accrediting agency.
- Usually Westgard rules are used for interpreting Levey-Jennings chart and for certain violation run must be rejected and problem must be resolved before resuming testing of patient's samples. Various errors can occur in Levey-Jennings chart including shift, trend, and other violations. Usually $1_{2s}$ is a warning rule and occurs due to random error, other rules are rejection rule (see Table 1).
- Delta check is important to identify laboratory errors and can be based on any of the criteria including delta difference, delta percent change (delta difference/current value), rate difference (delta difference/delta interval $\times$ 100), or rate percent change (delta percentage change/delta interval).
- Usually within- and between-run precision are expressed as CV. Then linearity of the assay is revalidated. Detection limit should be determined by running zero calibrator or blank specimen for 20 times and then determining the mean and standard deviation. The detection limit (also called lower limit of detection) is considered as: mean + 2 SD value, but more sophisticated method of calculating limit of detection has also been described.
- Comparison of new method with an existing method is a very important step in method validation. For this purpose, at least 100 patient

specimens must be run by the existing method in the laboratory and at the same time by the new method, but for FDA-approved test comparing 20 patient specimens could be sufficient. Then values are plotted and linear regression equation is the line of best fit expressed by the equation: $y = mx + b$, where "$m$" is called the slope of the line and "$b$" is the intercept. The computer calculates the equation of the regression line using least square approach. The software also calculates "$r$" the correlation coefficient by using a complicated formula. Ideal value of $m$ is 1, while ideal value of $b$ is zero. In reality, if slope is less than 1.0, it indicates negative bias with the new method compared to old method and if slope is over 1.0, it indicates positive bias.

- Usually hospital administration obtains analyzers by lease agreement most commonly for 5 years. All essential analyzers such as automated chemistry analyzer, hematology analyzers, etc., are typically acquired at least in duplicate (same manufacturer and preferably same mode; large medical centers and reference laboratories may acquire multiple analyzers due to large volume) because one analyzer may function as a backup just in case another analyzer breaks down. However, both analyzers must be used at the same time for proper functioning of such analyzers.
- CLIA regulation requires that two instruments must be compared against each other every 6 months. This can be easily achieved by running 20 specimens for each analyte using both analyzers and then comparing results.
- When new analyzers come to the laboratory to replace old analyzer, service representatives from the company will come to the laboratory to set up new analyzers, connecting them to hospital LIS, and also to perform initial validation of tests including calibration of each assay, within- and between-run precision, linearity, and patient correlation. The patient correlation is an important part of validating new instruments. It is usually done by analyzing 20 or more patient specimens using the old analyzer and the new analyzer.
- In general, analytical measurement range (the range of analyte concentration that can be measured accurately by the assay without dilution) is given in the package insert of each analyte provided by the company. The reportable range is the range of the analyte concentrations that can be accurately reported by the laboratory for a particular analyte. Some instruments have capability of autodilute a specimen and rerun the specimen if original value is above the reportable range. However, for very high values, dilution may not provide an accurate value and in these cases values should be reported higher than reportable range.
- Reference ranges are included in the patient report to help clinicians to interpret laboratory test results. Many text books in clinical chemistry and

laboratory medicine list reference ranges of most analytes. In addition, diagnostic companies also provide reference range for each analyte in the package insert. CLIA Regulation mandates that manufacturer's reference range (normal values) must be verified by each laboratory to ensure that such range is appropriate for the laboratory's patient population. This can be done by analyzing 20 specimens from normal volunteers (such as healthy laboratory personnel) to ensure values are within reference range of the manufacturer

- Critical values are life-threatening laboratory test results (for example, glucose value of 35 mg/dL). The CLIA regulation requires that when a critical value is observed in the laboratory it must be reported immediately to the provider by phone with a read back by the provider and a documentation of the call.
- Quality control (QC) looks at daily performance of the assays, while quality assurance (QA) investigates a broader picture of total quality of the laboratory including at processes that occur before (preanalytical), during (analytical), and after testing (postanalytical). CLIA regulations require that a clinical laboratory must have protocols for both QC and QA. However, CLIA regulation does not have any guidelines on which quality assurance indicators should be monitored. QA indicators could be a collection of preanalytical, analytical, postanalytical, physician satisfaction survey, etc.
- Receiver operating characteristic (ROC) curve is often used to make an optimal decision level for a test. ROC plots the true positive rate of a test (sensitivity) either as a scale of 0–1 (1 highest sensitivity) or as percent on the $y$ axis versus false positive rate (1 specificity).
- Six sigma goal is achieved if error rate is only 3.4 out of 1 million processes or error rate is only 0.00034%.
- Likelihood on "$n$" test results falling within the reference range can be calculated from the formula:

$$\%\text{Results falling within normal range} = 0.95^n \times 100$$

Therefore % result falling outside reference range in normal people $= (1 - 0.95^n) \times 100$.

- Student's $t$-test is useful to determine if one set of values are different from another set of values based on the difference between mean values and standard deviations. This statistical test is also useful in clinical research to see if values of an analyte in the normal state are significantly different from the values observed in a diseased state.

# References

[1] Jenny RW, Jackson KY. Proficiency test performance as a predictor of accuracy of routine patient testing for theophylline. Clin Chem 1993;39:76–81.

[2] Theolen D, Lawson NS, Cohen T, Gilmore B. Proficiency test performance and experience with College of American Pathologist's programs. Arch Pathol Lab Med 1995;119:307–11.

[3] Boone DJ. Literature review of research related to the Clinical Laboratory Improvement Amendments of 1988. Arch Pathol Lab Med 1992;116:681–93.

[4] Armbuster DA, Pry T. Limit of blank, limit of detection and limit of quantification. Clin Biochem Rev 2008;29(suppl 1):S49–51.

[5] Dasgupta A, Tso G, Chow L. Comparison of mycophenolic acid concentrations determined by a new PETINIA assay on the dimension EXL analyzer and a HPLC-UV method. Clin Biochem 2013;46:685–7.

[6] How To Be A Lab Director 2020 edition by Dr. Philip A. Dauterman; Independently published.

# CHAPTER 5

# Water homeostasis electrolytes and acid-base balance

## Distribution of water and electrolytes in human body

Water is a major constituent of human body representing approximately 60% of body weight in male and 55% body weight in women. Two-third of water in human body is associated with intracellular fluid and one-third is found in extracellular fluid. Extracellular fluid is composed mostly of plasma (containing 92% water) and interstitial fluid. Major extracellular electrolyte is sodium. Human body contains approximately 4000 mmol of sodium out of which 70% is present in exchangeable form and rest is found in bone. The intracellular concentration of sodium is 4–10 mmol/L. Normal sodium level in human serum is 135–145 mmol/L. Potassium is the major intracellular electrolyte with intracellular concentration of approximately150 mmol/L. The normal potassium level in serum is usually considered as 3.5–5.1 mmol/L. The balance between extracellular and extracellular electrolyte is maintained by sodium, potassium ATPase pump present in cell membranes.

Along with sodium and potassium other major electrolytes of human body are chloride and bicarbonate. Electrolytes are classified either as positively charged ions known as cations (sodium, potassium, calcium, magnesium, etc.) or as negatively charged ions known as anions (chloride, bicarbonate, phosphate, sulfate, etc.). Four major electrolytes of human body (sodium, potassium, chloride, and bicarbonate) play important roles in human physiology, including

- Maintaining water homeostasis of the body
- Maintaining proper pH of the body (7.35 to 7.45)
- Maintaining optimal function of the heart
- Participating in various physiological reactions
- Cofactors for some enzymes

It is important to drink plenty of water and adequate salt intake on a daily basis to maintain proper health. Healthy adults (age 19–50) should consume 1.5 g of sodium and 2.3 g of chloride each day or 3.8 g salt each day to replace salt loss each day. Tolerable upper limit of salt is 5.8 g (5800 mg) each day, but many

Clinical Chemistry, Immunology and Laboratory Quality Control. https://doi.org/10.1016/B978-0-12-815960-6.00024-8

Americans exceed this limit because average daily sodium intake is 3.5–6 g (3500–6000 mg) per day. Processed foods contain high amount of sodium because sodium is added by the manufacturer for food preservation. For example, a can of tomato juice may contain up to 1000 mg of sodium. Adults should consume 4.7 g of potassium each day but many Americans do not meet this recommended potassium requirement. Potassium rich foods include banana, mushroom, spinach, almonds, and a variety of other fruits and vegetables. High sodium intake may cause hypertension. Dietary Approaches to Stopping Hypertension (DASH) eating plan recommends not more than daily intake of 1600 mg (1.6 g) of sodium. In general, high sodium intake increases blood pressure and replacing high sodium diet with a diet low in sodium and high in potassium can decrease blood pressure. Sodium and potassium are freely absorbed from the gastrointestinal tract and excess sodium is excreted by the kidney. Potassium filtered through glomerular filtration in kidneys is almost completely reabsorbed in the proximal tubule and is secreted in the distal tubules in exchange of sodium under the influence of aldosterone. Interestingly, African Americans excrete less urinary potassium than Caucasians even while consuming similar diets in the DASH trial. However, consuming diet low in sodium may reduce this difference [1].

## Plasma and urine osmolality

Plasma osmolality is a way to measure electrolyte balance of the body. Osmolality (measured by an osmometer in a clinical laboratory) is technically different from osmolarity, which can be calculated based on the measured sodium, urea, and glucose concentration of the plasma. Osmolality is a measure of osmoles of solutes per kilogram of a solution where osmolarity is a measure of osmoles per liter of solvent. Because one kilogram of plasma is almost one liter in volume, osmolality and osmolarity of plasma can be considered as the same for all practical purposes. Normal plasma osmolality is 275–300 milliosmoles/kg (mOsm/kg) of water while urine osmolality is 50–1200 mOsm/kg of water. Although plasma and urine osmolality can be measured by using an osmometer, it is also calculated by the following formula.

Plasma osmolality $= 2 \times$ Sodium $+$ Glucose $+$ Urea (all concentrations in mmol/L)

Although sodium value is expressed as mmol/L, in clinical laboratories concentrations of glucose and urea are expressed as mg/dL. Therefore, the formula should be modified as follows to calculate osmolality:

Plasma osmolality $= 2 \times$ Sodium in mmol/L $+$ Glucose mg/dL/18 $+$ BUN, mg/dL/2.8

BUN stands for blood urea nitrogen.

Although this formula is commonly used, a stricter approach to calculate plasma osmolality also accounts for other osmotically active substance in plasma such as potassium, calcium, and proteins by adding 9 mOsm/kg to the equation

$$\text{Plasma osmolality} = 1.86\,\text{Sodium in mmol/L} + \text{Glucose mg/dL}/18 \\ + \text{BUN, mg/dL}/2.8 + 9$$

Plasma osmolality increases with dehydration and decreases with over hydration. Plasma osmolality regulates secretion of antidiuretic hormone (ADH). Another important laboratory parameter is osmolar gap defined as:

$$\text{Osmolar gap} = \text{Observed osmolality} - \text{Calculated osmolality}$$

If the measured osmolality is higher than the calculated osmolality, then this is referred to as osmolar gap and can be due to the presence of abnormal osmotically active substance such as overdose with ethanol, methanol, and ethylene glycol or if fractional water content of plasma is reduced due to hyperlipidemia or paraproteinemia. Although normal urine osmolality measured using random urine is relatively low, fluid restriction can raise urine osmolality to 850 mOsm/kg or higher (although within normal range of urine osmolality). However, greater than normal urine osmolality may be seen when:

- Reduced renal perfusion, e.g., dehydration, shock, renal artery stenosis.
- Excessive water retention without renal hypoperfusion, e.g., Syndrome of Inappropriate Antidiuretic Hormone Secretion (SIADH).
- Osmotically active substances in urine, e.g., glycosuria.

## Hormones involved in water and electrolyte balance

Antidiuretic hormone (ADH) and aldosterone play important roles in water and electrolyte balance of human body. ADH along with oxytocin are produced in the supra optic and paraventricular nuclei of the hypothalamus. These hormones are stored in the posterior pituitary and released in response to appropriate stimuli. For ADH, secretion is regulated by plasma osmolality. If plasma osmolality increases, it stimulates secretion of ADH, which acts at the collecting duct of the nephron where it causes reabsorption of only water and producing concentrated urine. In this process, water is conserved in the body and as a result plasma osmolality should be reduced. A low serum osmolality on the other hand reduces secretion of ADH, and more water is excreted as urine (diluted urine) and plasma osmolality is corrected. In addition, ADH at high concentrations causes vasoconstriction, thus raising blood pressure. Increased water retention due to ADH can result in following conditions:

- Concentrated urine
- Increased plasma volume
- Reduced plasma osmolality

Therefore, it is logical to assume that ADH secretion is stimulated by low plasma volume and increased plasma osmolality. In human, urine produced during sleep is more concentrated than urine produced during waking hours. Usually urine in the morning (first void) is most concentrated. This may be partly due to less or no fluid intake during sleeping hours and plasma ADH concentration is also higher during night than day time. It has been postulated that REM (rapid eye movement) sleep or dreaming sleep induces ADH secretion [2].

## Renin-angiotensin-aldosterone system

With low circulating blood volume, the juxtaglomerular apparatus of the kidney secretes renin, a peptide hormone into the blood stream. Renin converts angiotensinogen released by the liver into angiotensin I, which is then converted into angiotensin II in the lungs by angiotensin converting enzyme (ACE). Angiotensin II is a vasoconstrictor and also stimulates release of aldosterone from adrenal cortex. This is defined as "Renin-Angiotensin-Aldosterone" system. Aldosterone is a mineralocorticoid secreted from the zona glomerulosa of the adrenal cortex. It acts on the distal tubules and collecting ducts of the nephron and causes:

- Retention of water
- Retention of sodium
- Loss of potassium and hydrogen ions

Retention of water and sodium results in increased plasma volume and blood pressure. Increase in plasma potassium is a strong stimulus for aldosterone synthesis and release.

Atrial natriuretic peptide (ANP) and brain natriuretic peptide (BNP) are secreted by the right atrium and ventricles, respectively. The main stimulus for secretion of these peptides is volume overload. With volume overload, ANP and BNP secretion results in natriuresis.

## Diabetes insipidus

Diabetes insipidus is an uncommon condition that occurs when kidney is unable to concentration urine properly. As a result, diluted urine is produced, which results in increased plasma osmolality. This will result in increased water intake (polydipsia). The cause of diabetes insipidus is lack of secretion of ADH (cranial diabetes insipidus, also known as central diabetes insipidus) or due to inability of ADH to work at the collecting duct of the kidney (nephrogenic diabetes insipidus). Cranial diabetes insipidus is due to hypothalamic damage or pituitary damage and major causes of such damage include the following conditions:

- Head injury
- Stroke
- Tumor
- Infections affecting central nervous system
- Sarcoidosis
- Surgery involving hypothalamus or pituitary

Diabetes insipidus due to viral infection is rarely reported, but one report illustrates diabetes insipidus due to type A (sub-type: H1N1, swine flu) influenza virus infection in a 22-year-old man who produced up to 9 liter of urine per day [3]. Neuroendocrine complication following meningitis in neonates may also cause diabetes insipidus [4]. Pituitary abscess is a rare life-threatening condition that may also cause central diabetes insipidus. Autoimmune diabetes insipidus is an inflammatory noninfectious form of diabetes insipidus, which is rare but is presented with antibody to ADH secreting cells. Adipsic diabetes insipidus (ADI) is also a rare disorder consisting of central diabetes insipidus and a deficient or absent thirst response to hyperosmolality. Patients with ADI experience marked morbidity and mortality. Diagnosis and management of these patients may be very challenging [5].

Nephrogenic diabetes insipidus is due to inability of kidney to concentrate urine in the presence of ADH and major causes of nephrogenic diabetes include:

- Chronic renal failure
- Polycystic kidney disease
- Hypercalcemia, hypokalemia
- Drugs such as amphotericin B, demeclocycline, lithium

In both type of diabetic insipidus, patients usually present with diluted urine with low osmolality, but plasma osmolality should be higher than normal. These patients also experience excessive thirst and drink lots of fluid. Even if a patient is not allowed to drink fluid, urine still remains diluted with a possibility of dehydration. In contrast in a normal healthy individual, fluid deprivation results in concentrated urine. This observation is the basis of the water deprivation test to establish the presence of diabetes insipidus in a patient. In order to differentiate cranial diabetes insipidus from nephrogenic diabetes insipidus, intranasal vasopressin is administered. If urine osmolality increases, then the diagnosis is cranial diabetes insipidus but if urine is still dilute with no change in urine osmolality, then the diagnosis is nephrogenic diabetes insipidus. Congenital form of nephrogenic diabetes is a rare disease and most commonly inherited in an X-linked manner with mutations of the arginine vasopressin receptor type 2 (AVPR2) [6].

## The syndrome of inappropriate antidiuretic hormone secretion (SIADH)

The syndrome of inappropriate antidiuretic hormone secretion (SIADH also known as Schwartz-Bartter syndrome) is due to excessive and inappropriate release of antidiuretic hormone (ADH). Usually reduction of plasma osmolality causes reduction of ADH secretion but in SIADH, reduced plasma osmolality does not inhibit ADH release from the pituitary gland causing water overload. The main clinical features of SIADH include:

- Hyponatremia (plasma sodium <131 mmol/L)
- Decreased plasma osmolality (<275 mOsm/kg)
- Urine osmolality >100 mOsm/kg) and high urinary sodium (>20 mmol/L)
- No edema

Various causes of SIADH are listed in Table 1. Peri et al. reviewed differential diagnosis and clinical management of SIADH patients [7].

**Table 1** Causes of syndrome of inappropriate antidiuretic hormone secretion SIADH.

| Type of disease | Specific disease/comments |
| --- | --- |
| Pulmonary diseases | Pneumonia, pneumothorax, acute respiratory failure, bronchial asthma, atelectasis, tuberculosis |
| Neurological | Meningitis, encephalitis, stroke, brain tumor. Infection |
| Malignancies | Lung cancer, especially small cell carcinoma, head and neck cancer, pancreatic cancer |
| Hereditary | Two genetic variant, one affecting renal vasopressin receptor and another affecting osmolality sensing in hypothalamus have been reported |
| Hormone therapy | Use of desmopressin or oxytocin can cause SIADH |
| Drugs | Cyclophosphamide, carbamazepine, valproic acid, amitriptyline, SSRI, monoamine oxidase inhibitors and certain Chemotherapeutic agents may also cause SIADH |

# Hyponatremia, sick cell syndrome, and hypernatremia

Hyponatremia can be either absolute hyponatremia or dilutional hyponatremia, although in clinical setting, dilutional hyponatremia is encountered more commonly than absolute hyponatremia. In absolute hyponatremia, total sodium content of the body is low. Patient is hypovolemic, which results in activation of rennin angiotensin system causing secondary hyperaldosteronism and also increased levels of ADH. In dilutional hyponatremia, total body sodium is not low, rather total body sodium may be increased. Patient is volume overload with resultant dilution of sodium levels. Examples of such conditions include congestive heart failure, renal failure, nephrotic syndrome, and cirrhosis of liver. Although hyponatremia is defined as any sodium value less than reference range (135 mEq/L), usually clinical features such as confusion, restlessness leading to drowsiness, myoclonic jerks, convulsions, and coma are observed at much lower sodium levels. Hyponatremia is common among hospitalized patients affecting up to 30% patients [8]. However, sodium level below 120 mEq/L is associated with poor prognosis and even fatal outcome [9]. Major types of hyponatremia include:

- Absolute hyponatremia (patient is hypovolemic) related to loss of sodium through gastrointestinal tract or loss through kidney due to kidney diseases (pyelonephritis, polycystic disease, interstitial disease) or through kidney due to glycosuria or therapy with diuretics or less retention of sodium by the kidney due to adrenocortical insufficiency
- Dilutional hyponatremia (patient hypervolemic): This condition is related to SIADH or conditions like congestive heart failure, renal failure, nephrotic syndrome, and cirrhosis of liver
- Pseudohyponatremia as seen in patients with hyperlipidemia and hypergammaglobulinemia (also known as factitious hyponatremia)

In critical illness, cell membranes may leak allowing solutes normally found in intracellular fluid to escape in extracellular fluid. This is known as "sick cell syndrome." The solutes that escape are also osmotically active causing water to move from intracellular fluid to extracellular fluid leading to dilution of plasma sodium and consequently hyponatremia. Sick cell hyponatremia is usually associated with a positive osmolar gap (the difference between measured and calculated osmolality) because escaped molecules contribute to measured plasma osmolality. These patients also produce high levels of ADH, which causes water retention causing hyponatremia.

Hypernatremia is due to elevated serum sodium level above 150 mEq/L. Symptom of hypernatremia is usually neurological due to intraneuronal loss of water to extracellular fluid. Patients exhibit features of lethargy, drowsiness, and eventually become comatose. Hypernatremia may be hypovolemic or hypervolemic. The most common cause of hypovolemic hypernatremia is dehydration, which may be due to decreased water intake or excessive water loss through the skin (heavy sweating) or kidney or gastrointestinal tract (diarrhea). Patients usually presents with concentrated urine (osmolality over 800 mOsm/kg) and low urinary sodium (<20 mmol/L). Hypervolemic hypernatremia may be observed in hospitalized patients receiving sodium bicarbonate or hypertonic saline. Hyperaldosteronism, Cushing's syndrome, and Conn's disease may also cause hypervolemic hypernatremia.

## Hypokalemia and hyperkalemia

Hypokalemia is defined as serum potassium concentration <3.5 mEq/L, which may be caused by loss of potassium or redistribution of extracellular potassium into the intracellular compartment. Hypokalemia may occur due to:

- Loss of potassium from the gastrointestinal tract due to vomiting, diarrhea, active secretion of potassium from villous adenoma of rectum.
- Loss of potassium from the kidneys due to diuretic therapy, glucocorticoid, and mineralocorticoid excess. Increased levels of lysozyme (seen in monocytic leukemia) may also cause renal loss of potassium. Bartter's, Liddle, and Gitelman syndromes are rare inherited disorders due to mutations in the ion transport proteins of the renal tubules that may cause hypokalemia.
- Intracellular shift due to drug therapy with beta-2 agonists (salbutamol), which drives potassium into the cell, or due to alkalosis (hydrogen ions move out of the cell in exchange of potassium), or insulin therapy or familial periodic paralysis and hypothermia.

Clinically patients with hypokalemia presents with muscle weakness, areflexia, paralytic ileus, and cardiac arrhythmias. Electrocardiogram findings include prolonged PR interval, flat T, and tall U. Most of potassium of the body resides intracellularly. Hyperkalemia is elevated serum or potassium plasma level and a common cause is hemolysis of blood where potassium leaks from red blood cells into serum, thus artificially increasing potassium level. Causes of hyperkalemia include:

- Lysis of cells: in vivo hemolysis, rhabdomyolysis, and tumor lysis.
- Intracellular shift: in acidosis intracellular potassium is exchanged with extracellular hydrogen ions, causing hyperkalemia. Therefore, hyperkalemia typically accompanies metabolic acidosis. An exception is renal tubular acidosis (RTA) type I and II where acidosis without hyperkalemia is observed.
- Acute digitalis toxicity (therapy with digoxin or digitoxin) may cause hyperkalemia. Hyperkalemia is a result of the inhibition of Na-K-ATPase

and the subsequent increase in extracellular potassium. However, in chronic toxicity, hypokalemia is more worrisome. Hypokalemia, hypomagnesemia, and hypercalcemia have all been demonstrated to augment the toxicity of digoxin [10].
- Renal failure.
- Pseudohyperkalemia: Although pseudohyperkalemia or artificial hyperkalemia is most commonly seen in secondary to red cell hemolysis, it is also seen in patients with thrombocytosis and rarely in patients with familial pseudohypokalemia. Patients with highly elevated white blood cell count such as patients with chronic lymphocytic leukemia (CLL) may also show pseudohyperkalemia. Diagnosis of pseudohyperkalemia can be made from observation of higher serum potassium than plasma potassium (serum potassium exceeds plasma potassium by 0.4 mEq/L provide both specimens are collected carefully and analyzed within 1 h) or measuring potassium in whole blood (using blood gas machine) where whole blood potassium is within normal range.

Clinical features of hyperkalemia include muscle weakness, cardiac arrhythmias, and cardiac arrest. EKG findings include flattened P, prolonged PR interval, wide QRS complex, and tall T waves. Drugs that may cause hyperkalemia are listed in Table 2.

## Introduction to acid-base balance

In general, acid is defined as a compound that can donate hydrogen ion, and base is a compound that can accept hydrogen ion. In order to determine if a solution is acidic or basic, pH scale is used, which is the abbreviation of power of hydrogen ion, and pH is equal to negative log of hydrogen ion concentration in solution. The neutral pH is 7.0, and if a solution is acidic, pH is below 7.0 but

**Table 2** Drugs that may cause hyperkalemia.

Potassium supplement and salt substitute
Beta blockers
Digoxin and digitoxin (acute intoxication)
Potassium sparing diuretics (spironolactone and related drugs)
NSAIDs (nonsteroidal antiinflammatory drugs)
ACE inhibitors
Angiotensin II-blockers
Trimethoprim/sulfamethoxazole combination (Bactrim)
Immunosuppressants (cyclosporine and tacrolimus)
Heparin

basic if pH is above 7.0. Therefore, physiological pH of 7.4 is slightly basic. The concentrations of hydrogen ion, which are present in both the extracellular and intracellular compartments of human body, are tightly controlled. Although normal human diet is almost at a neutral pH containing very low amount of acid, human body produces about 50–100 mEq of acid in a day principally from the cellular metabolism of proteins, carbohydrates and fats, generating sulfuric acid, phosphoric acid, and other acids. Although excess base is excreted in feces, excess acid generated in the body must be neutralized or excreted in order to tightly control near normal pH of the blood (arterial blood 7.35–7.45 and venous blood 7.32–7.48). Carbonic acid ($H_2CO_3$) is generated in human body due to dissolution of carbon dioxide in water present in the blood:

$$CO_2 + H_2O = H_2CO_3 = H^+ + HCO_3^-$$

The hydrogen ion concentration of human blood can be calculated from the Henderson-Hasselbalch equation:

$$pH = Pka + \log [Salt]/[acid]$$

Where salt is concentration of bicarbonate [$HCO_3^-$] and concentration of acid is concentration of carbonic acid, which can be calculated from the measured partial pressure of carbon dioxide. The value of $p$Ka is 6.1, which is dissociation constant of carbonic acid at physiological temperature. The concentration of carbonic acid can be calculated by multiplying partial pressure of carbon dioxide by 0.03. Therefore, Henderson-Hasselbalch equation can be expressed as:

$$pH = 6.1 + \log \frac{[HCO_3^-]}{0.03 \times PCO_2}$$

Body has three mechanisms to maintain acid-base homeostasis:

- Physiological buffer present in the body, which consists of bicarbonate-carbonic acid buffer system, phosphate in the bone, and intracellular proteins.
- Respiratory compensation where lung can excrete more carbon dioxide or less depending on acid-base status of the body.
- Kidney can also correct acid-base balance of the human body if other mechanisms are ineffective.

Respiratory compensation to correct acid-base balance is the first compensatory mechanism, which is effective immediately, but it may take longer time for initiation of renal compensatory mechanism where at the collecting duct, sodium is retained in exchange of either potassium or hydrogen ions. As a result, if excess acid is present, more hydrogen ions should be excreted by the kidney to balance acid-base homeostasis. In the presence of excess acid (acidosis), kidneys excrete hydrogen ion and retain bicarbonate while during alkalosis; kidneys excrete bicarbonate and retain hydrogen ions. However, when there is an excess acid, hydrogen ions may also move into the cells in exchange of potassium moving out of the cell. As a result, metabolic acidosis usually causes hyperkalemia. Concurrently, bicarbonate concentration is reduced because hydrogen ions react with bicarbonate ions to produce carbonic acid. The kidneys need to reabsorb more of the filtered bicarbonate, which takes place at the proximal tubule.

## Diagnostic approach to acid-base disturbance

Major acid-base disturbances can be divided into four categories: metabolic acidosis, respiratory acidosis, metabolic alkalosis, and respiratory alkalosis. In general, metabolic acidosis or alkalosis is related to abnormalities in regulation of bicarbonate and other buffers in blood while abnormal removal of carbon dioxide may cause respiratory acidosis or alkalosis. However, both states may also coexist. However, it is important to know normal values of certain parameters measured in blood for diagnosis of acid-base disturbances.

- Normal pH of arterial blood: 7.35–7.45
- Normal $pCO_2$ is 35–45 mmHg
- Normal bicarbonate level is 23–25 mmol/L
- Normal chloride level is 95–105 mmol/L

The first question is whether the pH value is higher or lower than normal? If the pH is lower than normal, then it is acidosis, and if the pH is higher than normal, then diagnosis of alkalosis can be made. If the diagnosis is acidosis, then the next question to ask is whether the acidosis is metabolic or respiratory in nature. Similarly, if the pH is above normal, the question is whether the alkalosis is metabolic or respiratory in nature. In general, if the direction of change from normal for pH is the same direction for change of $pCO_2$ and bicarbonate, then the disturbance is metabolic in nature and if the direction of change of normal pH is in the opposite direction for change of $pCO_2$ and bicarbonate, then the disturbance is respiratory. Therefore, four different scenarios are possible:

- Metabolic Acidosis: The value of pH is decreased along with decreases in the values of $pCO_2$ and bicarbonate (both values below normal range).
- Respiratory Acidosis: The value of pH is decreased, but values of both $pCO_2$ and bicarbonate are increased from normal values.
- Metabolic Alkalosis: The value of pH is increased along with values of both $pCO_2$ and bicarbonate (both values above reference range).
- Respiratory Alkalosis: The value of pH is increased, but values of both $pCO_2$ and bicarbonate are decreased.

## Metabolic acidosis

Metabolic acidosis may occur with increased anion gap (high) or normal anion gap. Anion gap is defined as the difference between measured cations (sodium and potassium) and anions (chloride and bicarbonate) in serum. However, sometimes concentration of potassium is omitted because it is low compared to sodium ion concentration in serum.

$$\text{Anion gap} = [\text{sodium}] - ([\text{chloride}] + [\text{bicarbonate}])$$

Normal value: 8–12 mmol/L (mEq/L)

In metabolic acidosis, bicarbonate should decrease resulting in increased anion gap metabolic acidosis. If the chloride level increases, then even with decline in bicarbonate the anion gap may remain normal. This is normal anion gap metabolic acidosis. Thus, normal anion gap metabolic acidosis is also referred to as hyperchloremic metabolic acidosis. Causes of normal anion gap metabolic acidosis include loss of bicarbonate buffer from the gastrointestinal tract (chronic diarrhea, pancreatic fistula, and sigmoidostomy) or renal loss of bicarbonate due to kidney disorder such as renal tubular acidosis and renal failure. Causes of increased anion gap metabolic acidosis is remembered by the mnemonic MUDPILES (M for methanol, U for uremia, D for diabetic ketoacidosis, P for paraldehyde, I for isopropanol, L for lactic acidosis, E for ethylene glycol, and S for salicylate). In addition, alcohol abuse and other toxins such as formaldehyde toluene and certain drug overdose may also cause metabolic acidosis with increased anion gap. In general, if any other metabolic disturbance coexists with increased anion gap metabolic acidosis, this can be diagnosed from corrected bicarbonate level.

Corrected bicarbonate = measured value of bicarbonate + (anion gap-12).

If corrected bicarbonate is <24 mmol/L, then there exists additional metabolic acidosis and if corrected bicarbonate is >24 mmol/L, then there exists additional metabolic alkalosis.

Winter's formula is used to assess whether there exits adequate respiratory compensation with metabolic disturbance:

$$\text{Winter's formula}: \text{expected } pCO_2 = [1.5 \times \text{Bicarbonate}] + (8 \pm 2)$$

If $pCO_2$ is as expected by Winter's formula, then there is adequate respiratory compensation, but if $pCO_2$ is less than expected, then additional respiratory alkalosis may be present. However, if $pCO_2$ is more than expected, then there is additional respiratory acidosis

## Metabolic alkalosis

Metabolic alkalosis is related to loss of hydrogen ion or due to gain of bicarbonate or alkali:

- Loss of acid from gastrointestinal track, e.g., vomiting, diarrhea
- Loss of acid from kidneys, e.g., glucocorticoid or mineralocorticoid excess, diuretics
- Gain of alkali, e.g., "milk-alkali syndrome"

Milk-alkali syndrome (MAS) consists of hypercalcemia, various degrees of renal failure, and metabolic alkalosis as a result of ingestion of large amounts of milk and absorbable alkali (sodium carbonate or a combination), which was the standard treatment of peptic ulcer in 1930s. Based on the chronicity of symptoms and prognosis, MAS can be further classified as chronic (Burnett syndrome or acute (Cope syndrome). Although MAS became less common after development of proton pump inhibitors for treating peptic ulcer, more recently it is coming back because people take calcium supplements for preventing osteoporosis (especially in postmenopausal women) [11]. Interestingly, there is a more recent case report where a 74-year-old woman who developed MAS due to excessive intake of calcium carbonate (Tums) for her worsening heartburn. Her hypercalcemia and alkalosis recovered completely with aggressive hydration along with improvement in her renal function [12].

In general, body attempts to compensate metabolic acidosis by using respiratory compensation mechanism where enhanced carbon dioxide elimination can be achieved by hyperventilation (Kussmaul respiration), but this process may lead to respiratory alkalosis. In the case of metabolic alkalosis, depression of respiratory mechanism causes retention of carbon dioxide to compensate for metabolic alkalosis. However, respiratory response to metabolic alkalosis may be erratic. In addition, during metabolic alkalosis, kidneys try to compensate increased pH by decreasing excretion of hydrogen ion and sodium ions. When adequate compensation mechanism is absent, mixed acidosis may occur.

## Respiratory acidosis

Respiratory acidosis is due to carbon dioxide retention due to type II respiratory failure. Causes include:

- CNS disorders that damage or suppress the respiratory center, e.g., stroke, tumor, drugs, alcohol
- Neuropathy or myopathy affecting muscles of ventilation, e.g., Guillain-Barré syndrome, myasthenia gravis
- Reduced movement of chest wall, e.g., flail chest, severe obesity (Pickwickian syndrome)
- Airway obstruction, e.g., severe acute asthma

### Respiratory alkalosis

Major causes of respiratory alkalosis include:

- CNS stimulation, e.g., drugs such as aspirin, ketamine
- Hysteria
- Bronchial asthma (early stage)

If the acid-base disturbance is related to respiratory disturbance, then it is important to establish whether such disturbance is acute or chronic. In acute respiratory disturbance for any 10 mmHg $pCO_2$ change (assuming a normal value 40 mmHg), the change in pH is 0.08 units. In chronic respiratory disturbance for any 10 mmHg $pCO_2$ change, the change in pH is 0.03 units.

## Short cases: Acid-base disturbances

**Case 1:** A patient is overdosed with aspirin in an attempted suicide and brought to the ER. Her arterial blood pH was 7.57, $pCO_2$ was 20 mmHg, and bicarbonate was 22 mmol/L. Because the pH was above the normal range, the patient presented with alkalosis. In addition, both $pCO_2$ and bicarbonate were also decreased, but these two values were opposite in direction of pH (which was increased). Therefore, the patient had respiratory alkalosis. In addition to establishing the diagnosis of respiratory alkalosis, it is also important to establish if this is acute or chronic respiratory disturbance. The decrease of $pCO_2$ was 20 (normal value 40 mmHg). Multiplying 20 with 0.08 (in acute respiratory disturbance for any 10 mmHg pCO2 change, the change in pH is 0.08 units) indicates a value of 0.16. The increase of pH was 0.18 (assuming normal value of pH as 7.4), which was comparable to 0.16, and the patient showed acute respiratory alkalosis as expected with acute aspirin overdose.

**Case 2:** A patient with admitted myasthenia gravis admitted to the hospital showed arterial blood gas pH of pH 7.13, $pCO_2$ 80 mmHg, and bicarbonate of 26 mmol/L. Because blood pH was below the reference range, the patients suffered from acidosis. Moreover, both $pCO_2$ and bicarbonate were increased but of pH (decrease) (change in opposite direction), indicating that the patient had respiratory acidosis. Moreover, $pCO_2$ was 80 and assuming (for purposes of calculation) 40 is normal, the change was 40, which when multiplied by 0.08 was equal 0.32. The patient's pH was 7.13, which was lower by 0.27 from a normal value of 7.4. Therefore, the patient had acute respiratory disturbance (respiratory acidosis).

**Case 3:** An adult pregnant female with persistent vomiting was brought to the ER and her arterial blood pH was 7.62, $pCO_2$ was 47 mmHg, and bicarbonate was 38 mmol/L. Because pH was increased from the normal value the patient presented with alkalosis. In addition, both $pCO_2$ and bicarbonate were increased along with pH (all changes in same direction), the patient had metabolic alkalosis. Using Winter's formula, expected $pCO_2$ should be $1.5 \times$ bicarbonate$) + 8 \pm 2$ or $65 \pm 2$, that is, between 63 and 67. If pCO2 was as expected by Winter's formula, then adequate respiratory compensation was present, but this patient showed a $pCO_2$ of 47 indicating that in addition to metabolic alkalosis, additional respiratory alkalosis was also present.

**Case 4:** An adult male with insulin-dependent diabetes mellitus (IDDM) was admitted with altered mental status and had the following values: pH 7.22, $pCO_2$ 25 mmHg, bicarbonate 10 mmol/L, sodium 130 mmol/L, and chloride 80 mmol/L. Because pH was lower than normal, he had acidosis. In addition, all three parameters (pH, $pCO_2$ 25, and bicarbonate) were decreased (changed in same direction), establishing the diagnosis as metabolic acidosis. The anion gap of the patient was 40 elevated. Therefore, the patient presented with metabolic acidosis with increased anion gap. The corrected bicarbonate of the patient as 38 using the formula:

Corrected bicarbonate = measured value of bicarbonate + (anion gap − 12)

Because the corrected anion gap was higher than 24, the patient had additional metabolic alkalosis (corrected anion gap >24 mmol/L, additional metabolic acidosis present, corrected anion gap >24 mmol/L, additional metabolic alkalosis present). Moreover, using Winter's formula; expected $pCO_2 = [1.5 \times$ Bicarbonate$] + (8 \pm 2)$, expected $pCO_2$ should be between 21 and 25. Because measured $pCO_2$ 25, adequate respiratory compensation was present in the patient. In summary, this patient had increased anion gap metabolic acidosis with additional metabolic alkalosis but adequate respiratory compensation.

## Key points

- Plasma osmolality $= 2 \times$ Sodium in mmol/L + Glucose mg/dL,/18 + BUN, mg/dL/2.8 (BUN: blood urea nitrogen. Osmolar gap = Observed osmolality − Calculated osmolality.
- Higher osmolar gap can be due to the presence of abnormal osmotically active substance such as ethanol, methanol, and ethylene glycol (overdosed patients) or if fractional water content of plasma is reduced due to hyperlipidemia or paraproteinemia.
- Diabetes insipidus is due to lack of secretion of ADH (cranial diabetes insipidus, also known as central diabetes insipidus) or due to inability of ADH to work at the collecting duct of the kidney (nephrogenic diabetes insipidus).

- The main clinical features of SIADH (syndrome of inappropriate antidiuretic hormone secretion) include hyponatremia (plasma sodium <131 mmol/L), decreased plasma osmolality (<275 mOsm/kg), urine osmolality >100 mOsm/kg), and high urinary sodium (>20 mmol/L) with no edema.
- Major categories of hyponatremia include absolute hyponatremia (patient is hypovolemic) due to loss of sodium through gastrointestinal tract and kidneys, and dilutional hyponatremia (patient hypervolemic) related to SIADH, volume overload state, and pseudohyponatremia.
- Hypokalemia may occur due to loss of potassium from the gastrointestinal tract and intracellular shift.
- Causes of hyperkalemia include lysis of cells, intracellular shift, renal failure, or pseudohyperkalemia.
- Metabolic Acidosis: The value of pH is decreased along with decreases in the values of $pCO_2$ and bicarbonate (both values below normal range). May be normal anion gap or increased anion gap where anion gap = [sodium] − ([chloride] + [bicarbonate]) (normal value: 8–12 mmol/L (mEq/L). Causes of normal anion gap metabolic acidosis include loss of bicarbonate buffer from the gastrointestinal tract (chronic diarrhea, pancreatic fistula, and sigmoidostomy) or renal loss of bicarbonate due to kidney disorder such as renal tubular acidosis and renal failure. Causes of increased anion gap metabolic acidosis is remembered by the mnemonic MUDPILES (M for methanol, U for uremia, D for diabetic ketoacidosis, P for paraldehyde, I for isopropanol, L for lactic acidosis, E for ethylene glycol, and S for salicylate).
- Metabolic Alkalosis: The value of pH is increased along with values of both $pCO_2$ and bicarbonate (both values above reference range). Metabolic alkalosis is related to loss of hydrogen ion or due to gain of bicarbonate or alkali for any of the following reasons: loss of acid, from gastrointestinal tract (vomiting, diarrhea), loss of acid from kidneys (glucocorticoid or mineralocorticoid excess, diuretics), or gain of alkali, e.g., "milk-alkali syndrome" (also called Burnett's syndrome caused by excess intake of milk and alkali leading to hypercalcemia).
- Winters formula is used to assess whether there exits adequate respiratory compensation with metabolic disturbance: Winter's formula: expected $pCO_2 = [1.5 \times \text{Bicarbonate}] + (8 \pm 2)$. If $pCO_2$ is as expected by Winter's formula, then there is adequate respiratory compensation, but if $pCO_2$ is less than expected, then additional respiratory alkalosis may be present. However, if $pCO_2$ is more than expected, then there is additional respiratory acidosis.
- Respiratory Acidosis: The value of pH is decreased, but values of both $pCO_2$ and bicarbonate are increased from normal values. Respiratory acidosis is due to carbon dioxide retention due to type II respiratory

failure. Causes include CNS disorders that damage or suppress the respiratory center, e.g., stroke, tumor, drugs, alcohol, neuropathy, or myopathy affecting muscles of ventilation, e.g., Guillain-Barré syndrome, myasthenia gravis, reduced movement of chest wall, e.g., flail chest, severe obesity (Pickwickian syndrome), or airway obstruction, e.g., severe acute asthma.

- Respiratory Alkalosis: The value of pH is increased, but values of both $pCO_2$ and bicarbonate are decreased. Major causes of respiratory alkalosis include CNS stimulation due to drugs such as aspirin, ketamine, hysteria, or bronchial asthma (early stage). If the acid-base disturbance is related to respiratory disturbance, then it is important to establish whether such disturbance is acute or chronic. In acute respiratory disturbance for any 10 mmHg $pCO_2$ change (assuming a normal value 40 mmHg), the change in pH is 0.08 units. In chronic respiratory disturbance for any 10 mmHg $pCO_2$ change, the change in pH is 0.03 units.

# References

[1] Turban S, Thompson CB, Parekh RS, Appel LJ. Effects of sodium intake and diet on racial difference in urinary potassium excretion: results from the Dietary Approaches to Stop Hypertension (DASH) sodium trail. Am J Kidney Dis 2013;61:88–95.

[2] Steiger A. Sleep and the hypothalamo-pituitary-adrenocortical system. Sleep Med Rev 2002;6:125–38.

[3] Kobayashi T, Miwa T, Odawara M. A case of central diabetes insipidus following probable type A/H1N1 influenza infection. Endocr J 2011;58:913–8.

[4] Cohen C, Rice EN, Thomas DE, Carpenter TO. Diabetes insipidus as a hallmark neuroendocrine complication of neonatal meningitis. Curr Opin Pediatr 1998;10:449–52.

[5] Eisenberg Y, Frohman LA. Adipsic diabetes insipidus: a review. Endocr Pract 2016;22:76–83.

[6] Bockenhauer D, Bichet DG. Nephrogenic diabetes insipidus. Curr Opin Pediatr 2017;29:199–205.

[7] Peri A, Grohé C, Berardi R, Runkle I. SIADH: differential diagnosis and clinical management. Endocrine 2017;55:311–9.

[8] Pillai B, Unnikrishnan AG, Pavithran P. Syndrome of inappropriate antidiuretic hormone secretion: revisiting a classical endocrine disorder. Indian J Endocrinol 2011;15(Supply3): S208–15.

[9] Gill GV, Osypiw JC, Shearer ES, English PJ, et al. Critical illness with hyponatremia and impaired cell membrane integrity-the sick cell syndrome revisited. Clin Biochem 2005;38:1045–8.

[10] Kanji S, MacLean RD. Cardiac glycoside toxicity: more than 200 years and counting. Crit Care Clin 2012;28(4):527–35.

[11] Medarov BI. Milk-alkali syndrome. Mayo Clin Proc 2009;84(3):261–7.

[12] Timilsina B, Tachamo N, Parajuli PR, Gabriely I. Acute milk-alkali syndrome. Endocrinol Diabetes Metab Case Rep 2018;2018:18–0075.

# CHAPTER 6

# Lipid metabolism and disorders

## Lipids and lipoproteins

Along with proteins, carbohydrates, and nucleic acids, various lipids are also vital building blocks of life. In contrast to some proteins, carbohydrates, and nucleic acids, all lipids are insoluble in water. This is essential because lipids are integral structural part of cell membranes in animals and human. Because lipids are insoluble in water, when transported in blood these molecules must combine with water soluble proteins forming lipoproteins. Carbohydrates and lipids (especially fatty acids) are major sources of energy. Steroids are also lipids and many steroids act as hormones. Major lipids are listed below:

- Triglycerides: Formed when a glycerol molecule, which has three hydroxy groups, is esterified with three fatty acid molecules. There are two main sources of triglycerides, exogenous and endogenous. Exogenous triglycerides are dietary triglycerides.
- Fatty acids: These molecules are integral part of triglyceride molecule but a small amount may exist in the circulation as free fatty acids. Metabolism of fatty acids is a major energy source of the body.
- Phospholipids: Integral building block of cell membrane where two hydroxyl groups are esterified with a fatty acid but the third hydroxyl group is esterified with a phosphorus containing ester. Phospholipids are integral part of cell membranes.
- Cholesterol: Cholesterol is an integral component of cell membrane and also acts as a precursor of steroid hormone. In contrast to structures of triglycerides and phospholipids, cholesterol has a four membered ring structure with a side chain containing a hydroxyl group. In circulation, most cholesterol molecules exist as cholesterol ester (60%–80%) where the hydroxyl group is esterified with a fatty acid.

In addition to these lipids, there are some specific lipids known as sphingolipids found in cell membranes, especially in central nervous system and gray matter of the brain. Sphingolipids are formed when amino alcohol sphingosine

**105**

Clinical Chemistry, Immunology and Laboratory Quality Control. https://doi.org/10.1016/B978-0-12-815960-6.00010-8

is esterified with fatty acids. When sphingosine is bound to one fatty acid molecule containing 18 or more carbon, it is called ceramide. When a ceramide binds with a phosphocholine it forms sphingomyelin, which is also found in cell membranes. Sphingolipids are complex molecules, which play important role in communication between cells. These molecules may accumulate in certain lipid disorders.

When lipoproteins are devoid of lipids, they are called apolipoprotein. Apolipoproteins can be classified under several groups:

- Apolipoprotein A (Apo A) consisting mostly of Apo AI and Apo AII
- Apolipoprotein B (Apo B): Most abundant is large Apo B known as Apo B-100 while the less abundant is a smaller particle known as Apo B-48
- Apolipoprotein C (Apo C): Three forms are found: Apo CI, Apo CII, and Apo CIII
- Apolipoprotein D (Apo D)
- Apolipoprotein E (Apo E)

Characteristics of various apolipoproteins are summarized in Table 1.

## Various classes of lipoproteins

Lipoproteins are classified based on their density following ultracentrifugation of serum preferably overnight. In general lipids are lighter than water. Therefore, as the protein content of the lipoprotein increases, the particle become denser. The gold standard for separation and analysis of plasma lipoprotein fractions is "ultracentrifugation" [1]. Then after isolation and quantification

**Table 1** Characteristics of various apolipoproteins.

| Apolipoprotein | Characteristics |
| --- | --- |
| A-1 | Activates LCAT, found in HDL and chylomicron |
| Apo-AII | Found only in HDL |
| Apo-AIV | Activates LCAT and found in HDL |
| Apo B-100 | LDL receptors recognize apo B-100 and remove cholesterol from circulation, found in LDL, IDL, and VLDL |
| Apo B-48 | Smaller particle than Apo B-100, mostly found in chylomicron but may also be associated with LDL |
| Apo CI | Activated LCAT and found in chylomicron, VLDL, and HDL |
| Apo CII | Cofactor for lipoprotein lipase and found in chylomicrons, VLDL, and HDL |
| Apo CIII | Inhibits activation of lipoprotein lipase (opposite action of Apo II) and found in chylomicrons, VLDL, and HDL |
| Apo E | Facilitates uptake of chylomicrons remnant and IDL, found in chylomicrons, VLDL, and HDL |

LCAT, *lecithin cholesterol acyltransferase.*

of individual fractions, plasma cholesterol, triglycerides, and apolipoprotein can be measured. Usually low protein containing lipoproteins such as VLDL (very-low-density lipoprotein) stays at the top of the specimen while other denser lipoproteins are found in various other fractions. In reality, lipid ultra-centrifugation test is offered in relatively few reference laboratories. Major lipo-proteins found in plasma are:

- Chylomicrons: Lightest fraction containing approximately 2% lipoprotein and mostly lipids, especially triglycerides. This fraction is absent in fasting specimens unless the patients are suffering from a lipid disorder.
- Very-low density lipoprotein (VLDL): Denser than chylomicron but lighter than LDL, this fraction contains 4%–10% proteins and the rest lipids, most notably triglycerides (45%–60%).
- Intermediate-density lipoprotein (IDL): This fraction is lighter than LDL and contains approximately 15% proteins and the rest lipids. This fraction is usually a transient fraction and is absent in fasting specimen except for certain lipid disorders.
- Low-density lipoprotein (LDL): This fraction is denser than VLDL but lighter then HDL and contains approximately 25% proteins and the rest lipids, most commonly esterified cholesterol (approximately 50%).
- High-density lipoprotein (HDL): Most dense fraction containing approximately 50% protein and 50% lipids.

Characteristics of various lipoprotein fractions are summarized in Table 2.

**Table 2** Characteristics of various lipoproteins.

| Lipoproteins | Lipid/ protein ratio | Major lipids |
| --- | --- | --- |
| Chylomicron (lowest density) | 99:1 | Triglycerides (86%), coming from diet |
| VLDL (higher than chylomicron) | 90:10 | Triglycerides (55%), endogenously synthesized |
| IDL (higher density than VLDL) | 85:15 | Contain less triglycerides and more cholesterol ester than VLDL, a transitory particle between VLDL and HDL |
| LDL (higher density than IDL) | 80:20 | Cholesterol-rich lipoprotein and higher levels are associated with higher risk for cardiovascular diseases |
| HDL | 50:50 | Highest content of phospholipid among all lipoprotein particles. HDL removes cellular cholesterol from peripheral cells to liver for excretion by reverse cholesterol transport pathway |

## Lipid metabolism

There are two sources of lipids in human body: exogenous lipids from diet and endogenous lipids, which are synthesized mostly in the liver. Dietary triglycerides are broken down into fatty acids, glycerol, and monoglycerides in the small intestine. In the intestinal epithelial cell, triglycerides are resynthesized, which are then incorporated into chylomicrons and finally enter into the systematic circulation. Chylomicron contains very small amount of lipoproteins. In the circulation, lipoprotein lipase breaks down the triglycerides present in chylomicron into glycerol and free fatty acids. Lipoprotein lipase is found in the capillary endothelium of adipose tissue, skeletal, and cardiac muscle. Apo CII, which is present in chylomicron, plays an important role in activating lipoprotein lipase. The resultant particle is chylomicron remnant which is quickly removed by hepatic lysosomes. Free fatty acids generated during catabolism of chylomicrons are taken up by cells for oxidation to produce energy or can be utilized for resynthesis of triglycerides for storage. In this process, chylomicron particles are converted into chylomicron remnants. Therefore, chylomicron is absent in a fasting specimen unless in case of a specific lipid disorder. If chylomicron is present, they are found to float at the top of serum of plasma as a creamy layer.

Chylomicron is the major transport form of exogenous triglycerides. Liver is the site of endogenous triglycerides synthesis, but endogenously produced triglycerides are not incorporated into chylomicrons. Instead they are incorporated into VLDL, which is the major transport form of endogenous triglycerides. Triglycerides present in VLDL are also hydrolyzed by lipoprotein lipase and as a result IDL is formed which is a transitory particle because IDL is eventually converted into LDL. VLDL, IDL, and LDL all have the apoprotein B-100. LDL is removed from the circulation by the liver and other tissues. Uptake of LDL is receptor dependent, and Apo B-100 interacts with the LDL receptor present in the liver.

Cholesterol is present in the diet and is also synthesized in the liver. The rate-limiting step is catalyzed by 3-hydroxy-3-methylglutaryl-CoA reductase (HMG-CoA reductase). Cholesterol is an integral part of cell membranes and is a precursor for steroid hormones and bile acids. Fatty acids are derived from triglycerides. Cholesterol after synthesis is released into circulation as lipoprotein and approximately 60%–80% of cholesterol is esterified because esterified cholesterol can be more readily transported by lipoproteins. Formation of cholesterol ester also reduces accumulation of free cholesterol in cells [2]. Essential fatty acids cannot be synthesized by the human body and must be obtained from diet. Free fatty acids are transported by albumin in the plasma. There are two essential fatty acids: linoleic acid (18 carbon chain with two double bonds; omega-6 fatty acid) and linolenic acid (18 carbon chain with three double bonds; omega-3 fatty acid). Other common fatty acids found in human tissue include lauric acid (12 carbon chain), myristic (14 carbon chain), palmitic (16 carbon chain), stearic (18 carbon chain), oleic (18 carbon chain with one double bond), arachidic (20 carbon chain), and arachidonic acid (20 carbon chain with four double bonds). It is formed by the synthesis from dietary linoleic acid and is a precursor in the biosynthesis of prostaglandins, thromboxanes, and leukotrienes.

## LDL metabolism

Of all the lipoproteins, LDL has the highest amount of cholesterol. LDL is taken up by tissue with LDL receptors. Apo B-100 interacts with the LDL receptors present mostly in the liver. Lysosomal degradation of LDL releases free cholesterol. Cholesterol released from LDL then inhibits HMG-CoA reductase (3-hydroxy-3-methylglutaryl-CoA reductase), thus preventing endogenous synthesis of cholesterol by the liver. If the Apo B-100 protein is defective, uptake of LDL by LDL receptor is impaired. Patients with familial hypercholesteremia have a defect in the gene that codes for LDL receptor. As a result, LDL receptor may be absent or deficient causing elevated plasma cholesterol level. These patients are very susceptible to coronary atherosclerosis artery disease at a very early age.

## HDL metabolism

HDL is principally produced in the liver. HDL particles can be classified under subclass, $HDL_2$ and $HDL_3$. Nascent HDL acquires free cholesterol from tissues, chylomicrons, and VLDL. Major role of HDL is to remove cholesterol from peripheral cells and then return it to liver for excretion, a pathway called reverse cholesterol transport. Cholesterol transfer from cell membranes to HDL is stimulated by ATP binding cassette protein A1 (ABCA1). The free cholesterol is converted to cholesteryl esters by the enzyme lecithin cholesterol acyltransferase (LCAT). This enzyme is present in nascent HDL. Apo A-1 also present in HDL activates this enzyme. The cholesteryl esters are then transferred to chylomicron remnants and IDL. Cholesteryl ester transport proteins (CETP) are involved in this transfer. Chylomicron remnants and IDL are removed from the circulation by the liver.

## Lipid profile and risk of cardiovascular disease

Relation between plasma cholesterol and risk for arthrosclerosis has been extensively investigated by the Framingham Heart Study. This study was initiated in 1948 in Framingham, Massachusetts, by enrolling 5209 men and women to study risk factors for heart diseases. The study was under the direction of National Heart Institute now known as National Heart, Lung and Blood Institute. Many important guidelines regarding risk of cardiovascular diseases emerged from Framingham Heart Study, which led to publication of many scientific papers in leading medical journals.

Cardiovascular diseases including myocardial infarction are leading causes of morbidity and mortality throughout the world. Numerous studies have demonstrated the link between elevated cholesterol levels and risk of cardiovascular diseases. In the large international INTERHEART study, lipid disorders and smoking were shown to be the two most important risk factors for cardiovascular diseases. Other important risk factors are hypertension, obesity, and diabetes [3]. Risk factors for cardiovascular diseases are listed below:

## Unmodifiable risk factors

- Male sex
- Advanced age (male >45 years, female >55 years)
- Postmenopausal
- Family history (myocardial infarction of sudden death below 55 years of age in father or other male first degree relative or below 65 years of age in mother or other first degree female relative)
- Genetic factors (African Americans, Mexican Americans, Native Indians, people from Indian subcontinent have higher risk of heart diseases than Caucasians)

## Modifiable risk factors

- Abnormal lipid profile (also can be genetic)
- Hypertension
- Diabetes
- Smoking
- Obesity (more than 20% of ideal body weight)
- Physical inactivity
- Excessive use of alcohol (moderate drinking protects against cardiovascular diseases and stroke)
- Poor diet (no fruits and vegetables and high in carbohydrate)
- Excessive stress

According to World Health Organization (WHO), majority of cardiovascular diseases can be prevented by risk factor modification and change in lifestyle. Unfortunately, approximately 70% of Americans are overweight and fewer than 15% children and adults exercise sufficiently. Among American adults, 11%–13% have diabetes and 34% have hypertension, indicating the depth of the problem and risk of cardiovascular diseases in Americans [4]. Initial results of the Framingham study established the link between high total cholesterol and low HDL-cholesterol and risk for cardiovascular diseases, but elevated triglyceride was thought to play little role in elevating risks of cardiovascular diseases [5]. However, later reports indicated the link between elevated triglycerides as risk of heart diseases. Currently, the basis of treatment of lipid disorders is the third report of the expert panel of the National Cholesterol Education program and desirable total cholesterol level of less than 200 mg/dL has been universally accepted. Desirable and elevated lipid parameters related to risk of cardiovascular diseases are listed in Table 3. Unfortunately, in United States, approximately 16.3% population suffer from high cholesterol levels of 240 mg/dL or higher. This population has a cardiovascular risk factor twice as high as people with optimal cholesterol level of 200 mg/dL or lower [6].

## High LDL and risk for cardiovascular disease

Although desirable total cholesterol level of less than 200 mg/dL is universally accepted, guidelines for desirable LDL levels have changed significantly over time. Older guidelines of desirable LDL-cholesterol level of less than 130 mg/dL has been lowered to less than 100 mg/dL in National Cholesterol Education Program expert recommendation for Adult Treatment panel III (ATP III). Even if LDL-cholesterol level is borderline high (100–129 mg/dL),

**Table 3** Lipid profile and risk of cardiovascular diseases.

| Analyte | Value | Comment |
|---|---|---|
| Total cholesterol | <200 mg/dL | Desirable |
| | 200–239 mg/dL | Borderline high |
| | >240 | High |
| Low-density lipoprotein cholesterol (LDL-cholesterol) | <100 mg/dL | Optimal |
| | 100–129 mg/dL | Near optimal |
| | 130–159 | Borderline high |
| | >160 mg/dL | Highly elevated |
| High-density lipoprotein cholesterol (HDL-cholesterol) | <40 mg/mL | Low |
| | ≥60 mg/dL[a] | High (desirable) |
| Triglycerides | <150 mg/dL | Desirable |
| | 150–199 | Borderline high |
| | 200–499 | High |
| | >500 mg/dL | Very high |

[a]High HDL-cholesterol is not a risk factor for cardiovascular disease because only low HDL-cholesterol (<40 mg/dL) is a risk factor for both men and women.

some atherogenesis occurs and at level above 130 mg/dL the process is accelerated. Therefore, in a high risk patient drug therapy may be initiated with LDL-cholesterol level over100 mg/dL. However, some investigators reported that desirable LDL-cholesterol in high risk patients is around 70 mg/dL because LDL-cholesterol level is most important in predicting risk of cardiovascular diseases and especially oxidized LDL infiltrates the intima where is stimulates inflammation, endothelial dysfunction and finally atherosclerosis. Atherosclerosis has been observed in individuals with relatively lower LDL-cholesterol levels (90–130 mg/dL) [7]. Therefore, it has been recommended that for very high risk patients targeting LDL-cholesterol level below 70 mg/dL is a valid therapeutic option [8]. Multiple statin trials and meta-analyses support a treatment target of LDL-cholesterol less than 70 mg/dL in very high risk patients [9]. Because LDL level is significantly associated with cardiovascular disease, the primary goal of lipid lowering therapy using statins target lowering LDL level.

## High triglycerides and risk for cardiovascular disease

The extent to which high triglycerides directly promote cardiovascular diseases has been debated over three decades. Accepted guidelines consider serum triglyceride concentration of 150 mg/dL or less as optimal, but approximately 31% adult US population has triglyceride levels over 150 mg/dL [10]. There are speculations that some triglyceride-rich lipoproteins may be atherogenic, especially remnant lipoproteins such as remnant VLDL and IDL. In addition, when triglyceride levels are above 200 mg/dL, the increased concentration of triglyceride-rich remnant lipoproteins may further increase the risk of cardiovascular disease. The major causes of elevated triglycerides are:

- Overweight and obesity
- Excessive alcohol use
- Very high carbohydrate diet
- Diseases such as Type 2 diabetes, nephrotic syndrome
- Certain drug therapy
- Genetic factor

Hypertriglyceridemia is also a risk factor for acute pancreatitis.

## HDL cholesterol and lower risk of cardiovascular disease

Many epidemiological studies have shown correlation between low HDL-cholesterol and higher risk of cardiovascular disease while high HDL-cholesterol is associated with lower risk. HDL can remove cholesterol from atherosclerotic plaque by reverse cholesterol transport. In addition, antioxidant and antiinflammatory properties of HDL also protect against atherogenesis. Although the level is set at <40 mg/dL for low HDL-cholesterol, women typically have higher HDL-cholesterol than men. However, if HDL-cholesterol is below 50 mg/dL in a woman, it may be a marginal risk factor requiring lifestyle change in order to elevate HDL-cholesterol level. Sometimes low HDL-cholesterol is encountered in individuals with high triglycerides. Causes of low HDL-cholesterol include:

- High serum triglycerides
- Obesity and physical inactivity
- Cigarette smoking
- Type 2 diabetes
- Certain drug therapy such as therapy with beta-blockers
- Genetic factors

## Non-HDL cholesterol, Lp(a) and risk of cardiovascular disease

More recently, the role of non-HDL cholesterol in risk stratification for coronary artery diseases has been investigated extensively.

$$Non - HDL\,cholesterol = VLDL\,cholesterol + LDL\,cholesterol$$

However, in clinical laboratories non-HDL cholesterol is calculated as:

$$Non - HDL\,cholesterol = Total\,cholesterol - HDL\,cholesterol$$

Therefore, non-HDL cholesterol includes all lipoproteins that contain apo B. The amount of non-HDL cholesterol is important in patients with elevated triglyceride levels. In individuals with triglyceride levels between 200 and 499 mg/dL, most cholesterol found in VLDL fraction is associated with smaller (remnant) VLDL particles making such particle atherogenic. Some studies have indicated better correlation between non-HDL cholesterol and cardiovascular mortality than LDL cholesterol in patients with high triglycerides [11]. In addition, non-HDL cholesterol is also highly correlated with apo B, the major atherogenic lipoprotein. In patients not requiring therapy, target for non-HDL cholesterol is <130 mg/dL (apo B target <90 mg/dL). However, during statin therapy it may be necessary to reduce non-HDL cholesterol <100 mg/dL to get optimal benefit [12].

Total cholesterol to HDL cholesterol ratio is also used for calculating risk factor for cardiovascular disease. Usually if the ratio is above 5, it is considered as high. Similarly, apo B/Apo AI ratio have also been used for calculating risk of cardiovascular disease. Goswami et al. reported that mean total cholesterol to HDL cholesterol ratio 5.15 in 100 patients who suffered from myocardial infarction but 100 controls, the ratio was 3.45 indicating that high total cholesterol to HDL cholesterol ratio increases risk of myocardial infarction. The

Apo B to Apo AI ratio was also higher in patients (0.96 in patients vs 0.71 in controls) [13].

Lipoprotein(a), also known as Lp(a), was discovered in 1963 and is synthesized by the liver. Lp(a) has two different components. The first one is an LDL-like particle containing an apo B-100. However, the LDL-like particle in Lp(a) is larger in size and higher in lipid content, with a density marginally lower than the LDL particle isolated from the same individual. The second component is a hydrophilic glycoprotein called apo(a) that shares homology with plasminogen, giving the particle atherogenic properties. In Lp(a) particle, the apo B present in the outer part of LDL-like particle is covalently connected by a disulfide bond with apo(a) [14]. The serum level of Lp(a) is controlled genetic makeup and is an independent risk factor for cardiovascular disease. Usually normal range of Lp(a) range is up to 30 mg/dL [15]. In general, Lp(a) level between 30 and 50 mg/dL is considered as borderline high while value above 50 mg/dL is associated with increased risk for cardiovascular disease. In addition to increasing risk of cardiovascular disease, Lp(a) also has proinflammatory and potentially prothrombotic properties. Current guidelines recommend Lp(a) measurement for patients with an intermediate-to-high risk of cardiovascular disease, familial hypercholesterolemia, or a family history of early cardiovascular disease [16].

## Various types of hyperlipidemia

Hyperlipidemia is also termed as hyperlipoproteinemia can be primary or secondary in origin. Various primary hyperlipidemia includes:

- Familial hypercholesterolemia: This disease is transmitted as an autosomal dominant disorder. Mutations affect LDL receptor synthesis or its proper function as well as mutation of Apo B-100 gene which results in decreased binding of LDL with Apo B-100. Total cholesterol and LDL-cholesterol are highly elevated in these individuals making them susceptible to myocardial infarction at a young age.
- Polygenic hypercholesterolemia: In these individuals, both genetics and environmental factors play important roles in producing high cholesterol levels.
- Familial hypertriglyceridemia: This disease is also transmitted as an autosomal dominant disorder characterized by an increased production of VLDL by the liver.
- Familial hyperchylomicronemia: This disease is transmitted as autosomal recessive and caused by deficiency of the enzyme lipoprotein lipase or apo CII. Triglycerides levels are high in these individuals.

- Familial dysbetalipoproteinemia: In these individuals, there is an increased level of IDL and chylomicron remnants. Both cholesterol and triglycerides are subsequently increased. The apoprotein E exhibits polymorphism showing three isoforms: apo E2, apo E3, and apo E4. Common phenotype is E3/E3. Individuals with familial dysbetalipoproteinemia tend to have E2/E2 phenotype. This phenotype results in impaired hepatic uptake of chylomicron remnants and IDL by the liver.
- Familial combined hyperlipidemia: In these individuals, either cholesterol or triglyceride or both are elevated. It is possibly transmitted as autosomal dominant.

Secondary hyperlipidemia is common and causes include diabetes mellitus, hypothyroidism, nephrotic syndrome, cholestasis, and alcohol.

Lipid analysis can also be performed using electrophoresis where chylomicron is observed at the point of application, and second band above point of application is VLDL (prebeta band) followed by LDL band and the band furthest from the point of application is HDL. IDL band is usually not observed during lipid electrophoresis, but if present it can be found between VLDL and LDL band. Lipid disorders are also classified according to the Frederickson classification, which is an older classification. In this classification, there are five types of hyperlipidemia:

- Type I: In these individuals elevated chylomicrons are found due to lipoprotein lipase or apo CII deficiency causing elevated levels of triglycerides. This is actually familial hyperchylomicronemia.
- Type IIa: In these individuals, elevated LDL cholesterol and total cholesterol are observed due to familial hypercholesterolemia, polygenic hypercholesterolemia, familial combined hyperlipidemia, as well as nephritic syndrome and hypothyroidism.
- Type IIb: In these individuals, elevated LDL and VLDL are seen as observed in individuals with familial combined hyperlipidemia. Both cholesterol and triglycerides levels may be elevated.
- Type III: These individuals have elevated IDL, and this disorder is actually dysbetalipoproteinemia due to apo E/apo E2 profile. Both cholesterol and triglycerides levels may be elevated. In type III lipid disorder, VLDL/triglyceride ratio is usually above 0.3, while normal value is around 2.0.
- Type IV: These individuals have elevated VLDL as seen in familial hypertriglyceridemia, or familial combined hyperlipidemia. As a result, triglyceride level is elevated. Type IV disorder also may be due to secondary cause such as diabetes and nephrotic syndrome.
- Type V: These individuals have elevated VLDL and chylomicrons causing elevated triglycerides.

It is important to note that Type IIa, IIb, and also type III are associated with significantly increased risk for cardiovascular diseases.

## Various types of hypolipidemia

Hypolipidemias also termed as hypolipoproteinemia are also classified as primary and secondary. Secondary hypolipidemias can be seen in severe liver diseases, malabsorption of protein energy malnutrition states. However, primary hypolipidemia are rare diseases. Primary hypolipidemias include:

- Tangier disease
- Abetalipoproteinemia
- Familial hypobetalipoproteinemia
- Chylomicron retention disease

In Tangier disease, there is a loss of function of ABCA1 protein as a result of mutation of ABCA1 gene causing very low level of HDL in serum. This disease is inherited in an autosomal recessive pattern. Normally, ABCA1 protein helps in the uptake of cholesterol by HDL. Therefore, cholesteryl esters cannot be removed from peripheral cells by reverse cholesterol transport mechanism. The tonsils appear hyperplastic and orange in color. Individuals with Tangier disease have higher risk for cardiovascular disease. Low HDL level in the absence of orange tonsil may also indicate use of anabolic steroids. Anabolic steroids not only decrease HDL cholesterol level but also increase LDL cholesterol level.

Abetalipoproteinemia is a rare inherited disease with approximately 100 cases reported worldwide. In this disease, there is a total absence of apo B-100, thus concentrations of triglycerides and cholesterol carrying lipoprotein concentrations (chylomicrons, VLDL, IDL, and LDL) are highly reduced. This causes malabsorption of dietary fats, cholesterol, and fat soluble vitamins such as A, D, E, and K. The signs and symptoms of this disease appear within first few months of life including failure to growth and gaining weight, steatorrhea, and red cell acanthocytosis. Eventually, people develop poor muscle balance and ataxia.

Familial hypobetalipoproteinemia is inherited as a recessive form as a result of two mutations of MTTP gene. The MTTP gene provides instructions for producing microsomal triglyceride transfer protein, which is involved in production of beta-lipoproteins. This disease is more common than abetalipoproteinemia affecting one in 1000 individuals. Due to partial absence of apo B-100 lipoproteins, reduced levels of chylomicron, VLDL, IDL, and LDL are observed causing malabsorption of fats and fat soluble vitamins. The progress of this disease is severe if manifested in early childhood. Welty reviewed various types of hypobetalipoproteinemia and abetalipoproteinemia [17].

Chylomicron retention disease is also an inherited disease (autosomal recessive pattern), affecting absorption of dietary fats and cholesterol and fat soluble vitamins. Chylomicrons are essential for transport of dietary fats. In this disease, which is very rare (approximately 40 cases worldwide), mutation of SAR1B gene causes impaired release of chylomicrons in the blood.

## Newer lipid parameters and other factors related to risk for cardiovascular disease

In addition to traditional lipid parameters such as cholesterol, triglycerides, HDL-cholesterol, and Lp(a), there are also other lipid markers and nonlipid markers that can be used for assessing risk of cardiovascular disease in individuals. These markers include:

- Lipoprotein associated phospholipase $A_2$ (Lp-PLA2) (lipid parameter)
- LDL particle size (lipid parameters)
- C-reactive protein (nonlipid parameter)
- Homocysteine (nonlipid parameter)
- Myeloperoxidase

Lipoprotein-associated phospholipase $A_2$ (Lp-PLA2) is a monomeric enzyme that catalyzes oxidized phospholipid found in LDL into lysophosphatidylcholine and oxidized fatty acid, both of which are atherogenic. Therefore, Lp-PLA2 is an inflammatory biomarker like C-reactive protein. Lp-PLA2 is mostly associated with LDL (particularly small dense LDL) but some may be also associated with HDL. Lp-PLA2 levels are elevated in patients with elevated cholesterol, especially LDL-C. However, measurement of this parameter is not recommended for routine screening of patients for assessing risk of cardiovascular disease.

Two different phenotype of LDL particle has been described. Pattern B with mostly small dense LDL particles (peak diameter $<25.5$ nm) and pattern A with a higher proportion of large more buoyant LDL particles (peak diameter $>25.5$ nm). Small LDL particles contain more apo B and tend to coexist with elevated triglycerides, HDL cholesterol, and apo AI concentration (atherogenic dyslipidemia) and are heritable. Women tend to have less small LDL particles than men. Small LDL particle size is associated with several other cardiovascular risk factors, including metabolic syndrome, type 2 diabetes mellitus, and postprandial hypertriglyceridemia [18].

C-reactive protein is an inflammation marker and is also a predictor for risk of cardiovascular disease. C-reactive protein is found in low level (1 mg/L) in normal individuals and may increase 100 fold in response to acute phase. Levels usually return to normal in 8–10 days. Traditional assays are not sensitive to measure C-reactive protein for risk assessment and highly sensitive C-reactive protein assay (analytical measurement range: 0.1–100 mg/L) is used for this purpose. A C-reactive protein level of <1 mg/L is associated with low risk, 1–3 mg/L with moderate risk, and >3 mg/L with high risk for cardiovascular disease.

Homocysteine is a thiol-containing amino acid intermediate formed during methionine metabolism. McGully in 1992 reported the presence of atherosclerosis in children and young adults with inborn error of homocysteine metabolism such as cystathionine-beta-synthase deficiency. The disorders are associated with markedly elevated plasma homocysteine levels (>100 μmol/L). McGully's work raised the possibility that mild to moderate elevation in homocysteine concentrations could contribute to atherosclerotic vascular disease. Such increases in homocysteine levels can occur with aging, menopause, hypothyroidism, low plasma level of vitamin cofactors ($B_6$, $B_{12}$, and folate), and chronic renal failure [19]. Genetic variation in enzymes involved in the metabolism of homocysteine may contribute to difference in homocysteine levels in different individual. One such polymorphism in methylene tetrahydrofolate reductase may lead to mild or moderate elevation in homocysteine levels and about 15% of Caucasians may have that genetic polymorphism [20]. A homocysteine level over 15 μmol/L is associated with and increased risk for cardiovascular disease and each 5 μmol/L increase is equivalent to an increase in 20 mg/dL in cholesterol concentration. Before treatment of elevated homocysteine level is considered, vitamin $B_{12}$ status should be evaluated to ensure that there is no such deficiency. Treatment with 1 mg/day of folic acid is effective in reducing mild to moderately elevated homocysteine level. Supplementing folic acid therapy with $B_{12}$ and $B_6$ is also practiced. However, patients with renal failure may need a much higher dose of folic acid (up to 20 mg/day) for effective reduction of their elevated homocysteine level.

Myeloperoxidase, an enzyme released by activated neutrophils, has prooxidant and proinflammatory properties. It is stored in azurophilic granules of polymorphonuclear neutrophils and macrophages. Because myeloperoxidase is involved in oxidative stress and inflammatory process, it is a biomarker for inflammation in ischemic heart disease and acute coronary syndromes. Various laboratory parameters used for assessing risk factor for cardiovascular diseases are listed in Table 4.

## Laboratory measurements of various lipids

Lipid profile that consists of total cholesterol, triglyceride, LDL-cholesterol (LDL-C) and HDL-cholesterol (HDL-C) are measured routinely. Traditionally, blood specimens for analysis of lipid parameters are collected after an overnight

**Table 4** Various laboratory parameters used for assessing risk factor for cardiovascular diseases.

| Parameter | Comments |
| --- | --- |
| **Lipid parameters** | |
| Total cholesterol | Values over 240 mg/dL indicates high risk |
| LDL-cholesterol | Optimal value is <100 mg/dL, values >160 mg/dL indicates high risk |
| HDL-cholesterol | Value <40 mg/dL indicates risk, while value >60 mg/dL is good as it reduces the risk for cardiovascular disease (negative risk) |
| Cholesterol/HDL-C ratio | High ratios especially value of 5 or more is associated with increased risk |
| High non-HDL-C | Optimal is <130 mg/dL, but in high-risk patient optimal level is <100 mg/dL. Higher levels are associated with increased risk |
| Lp(a) | Value over 30 mg/dL indicates increased risk |
| **Nonlipid parameters** | |
| C-reactive protein | Desirable level is <1 mg/L, while level >3 mg/L indicates high risk |
| Homocysteine | Value >15 μmol/L indicates high risk |

fast of 10–12 h to ensure chylomicrons are cleared from plasma. In fact, before 2009 essentially all societies, guidelines, and statements required fasting before measuring a lipid profile for cardiovascular risk prediction. However, lipids and lipoproteins only change minimally in response to normal food intake: in four large prospective studies, maximal mean change was observed with triglycerides only. Other lipid parameters were less affected. Further, using 108,602 individuals in the Copenhagen General Population Study, the authors observed good correlations between lipid parameters obtained using random nonfasting specimens and cardiovascular risk factors. Many countries are currently changing their guidelines toward a consensus on measuring a lipid profile for cardiovascular risk prediction in the nonfasting state, simplifying blood sampling for patients, laboratories, and clinicians worldwide [21].

In serum, majority of cholesterol exists as cholesterol ester. Therefore, in the first step cholesterol ester is hydrolyzed by cholesterol ester hydrolase enzyme. Then cholesterol is oxidized by cholesterol oxidase generating cholest-4-en-3-one and hydrogen peroxide. Hydrogen peroxide generated is proportional to serum cholesterol concentration and is measured by its reaction with a suitable compound, for example, 4-aminoantipyrene (reaction catalyzed by peroxidase), forming a colored dye. HDL is usually measured as HDL-cholesterol after precipitating out other lipoprotein fractions using polyanions such as dextran sulfate-magnesium chloride or phospho-tungstate-magnesium chloride or heparin sulfate-manganese chloride.

For serum triglyceride measurement, lipase enzyme is used, which converts triglyceride into glycerol and free fatty acid. Then glycerol is oxidized by glycerol kinase into glycerophosphate. Glycerophosphate is then measured by either its reaction with nicotinamide adenine dinucleotide (NAD, no absorption at 340 nm) to form NADH (absorbs at 340 nm) or its oxidation by glycerophosphate oxidase enzyme generating dihydroxyacetone and hydrogen peroxide.

Plasma LDL values are typically calculated from the formula (Friedewald formula):

LDL cholesterol = Total cholesterol − HDL cholesterol − Triglyceride/5

where all measurements are in mg/dL.

This formula is invalid if triglyceride values are above 400 mg/dL. In such situations, direct measurement of LDL is indicated. In addition, this is only applicable for calculating LDL-cholesterol in an overnight fasting specimen. For certain patients, calculated LDL-cholesterol may not reflect true LDL-cholesterol level. If there is a discrepancy between measured LDL-cholesterol level and calculated LDL-cholesterol level, it indicates that there is a modification of lipoprotein metabolism.

Elevated chylomicrons cause the plasma to appear as milky, and when plasma is allowed to stand, a creamy layer at the top is visible. Elevated triglycerides cause the entire plasma to appear turbid. The various lipoproteins have distinct electrophoretic patterns as seen on serum protein electrophoresis. Chylomicron has minimum protein and is found at the origin. The HDL fraction is seen in the alpha-1 region. LDL migrates in the beta region and VLDL is present in the pre beta region. Apolipoproteins are measured by using appropriate immunoassays.

## Drugs for treating lipid disorders

Statins are the first-line treatment for hypercholesterolemia. Statins decrease hepatic cholesterol synthesis by inhibiting HMG-CoA reductase, which is the rate-limiting step in cholesterol biosynthesis, resulting in increased LDL receptor activity with increased clearance of LDL particles and precursors. Statins are capable of reducing LDL-C concentration by up to 60%. In addition, statins also lower triglycerides and modestly increase HDL-C. However, statin dose dependently may cause abnormal liver function tests in about 0.5%–3% patients due to elevated aminotransferase levels. Muscle complaints represent the most frequent adverse reports among patients treated with statins [22]. Commonly used statins include atorvastatin, fluvastatin, lovastatin, pravastatin, rosuvastatin, simvastatin, and pitavastatin.

Cholesterol absorption inhibitors such as ezetimibe modestly lower LDL-cholesterol and can be used in combination with statins. Fibrates are primarily used in treating patients with hypertriglyceridemia, but these agents may also increase HDL-cholesterol level. Nicotinic acid (niacin) raises HDL-cholesterol very effectively with modest reduction of LDL-cholesterol. Nicotinic acid may also lower Lp(a) level.

However, additional therapies are necessary for patients who cannot reach the target LDL-C level when taking the maximum-tolerated dose of a statin. More recently, FDA approved use of PCSK9 inhibitors (human monoclonal antibody against PCSK9) for significantly lowering LDL-cholesterol. PCSK9 (proprotein convertase subtilisin/kexin type 9) is an enzyme that binds to low-density lipoprotein receptors (LDL receptors) and promotes its degradation. As a result, LDL-cholesterol is not effectively removed by LDL-receptors. Discovery that human loss-of-function mutations in PCSK9 gene are associated with low levels of LDL-cholesterol and decreased risk of cardiovascular disease lead to development of monoclonal antibodies against PCSK9 to be used as therapeutic agents to lower LDL-cholesterol. In randomized trials, the use of monoclonal antibodies that inhibits PCSK9 significantly reduced LDL cholesterol levels and reduced the incidence of cardiovascular events [22].

## Summary/key points

- Apolipoprotein A (Apo A) consists of Apo AI and Apo AII.
- Apolipoprotein B (Apo B): Most abundant is large Apo B known as Apo B-100, while the less abundant is a smaller particle known as Apo B-48. Apo B-48 is more atherogenic than Apo B-100.
- Apolipoprotein C (Apo C): Exists in three forms: Apo CI, Apo CII, and Apo CIII.
- Apolipoprotein D (Apo D).
- Apolipoprotein E (Apo E).
- Major lipoproteins found in plasma are chylomicrons, VLDL, LDL, and HDL.
- Chylomicrons: Lightest fraction containing approximately 2% lipoprotein and mostly lipids, especially triglycerides. This fraction is absent in fasting specimens (preferred specimen for lipid analysis) unless the patients is suffering from a lipid disorder.
- Very-low-density lipoprotein (VLDL): Denser than chylomicron but lighter than LDL, this fraction contains 4%–10% proteins and the rest lipids, most notably triglycerides (45%–60%).
- Intermediate-density lipoprotein (IDL): This fraction is lighter than LDL and contains approximately 15% proteins and the rest lipids. This fraction is usually a transient fraction and is absent in fasting specimen except for certain lipid disorders.
- Low-density-lipoprotein (LDL): This fraction is denser than VLDL but lighter then HDL and contains approximately 25% proteins and the rest lipids most commonly esterified cholesterol (approximately 50%).

- High-density-lipoprotein (HDL): Most dense fraction containing approximately 50% protein and 50% lipids.
- In the circulation, lipoprotein lipase found in the capillary endothelium of adipose tissue, skeletal, and cardiac muscle breaks down triglycerides present in chylomicrons into glycerol and free fatty acids. Apo CII present in chylomicrons activates lipoprotein lipase. Chylomicrons if present are found to float at the top of serum of plasma as a creamy layer.
- Chylomicron is the major transport form of exogenous triglycerides. Liver is the site of endogenous triglycerides synthesis but endogenously produced triglycerides are not incorporated into chylomicron. Instead they are incorporated into VLDL, which is the major transport form of endogenous triglycerides. Triglycerides present in VLDL are also hydrolyzed by lipoprotein lipase and as a result IDL is formed, which is a transitory particle because IDL is eventually converted into LDL. VLDL, IDL, and LDL all have the apoprotein B-100. LDL is removed from the circulation by the liver and other tissues. Uptake of LDL is receptor dependent and Apo B-100 interacts with the LDL receptor present in the liver.
- Cholesterol is present in the diet and is also synthesized in the liver. The rate-limiting step is catalyzed by 3-hydroxy-3-methylglutaryl-CoA reductase (HMG-CoA reductase). Cholesterol is an integral part of cell membranes and is a precursor for steroid hormones and bile acids.
- Of all the lipoproteins, LDL has the highest amount of cholesterol. LDL is taken up by tissue with LDL receptors. Apo B-100 interacts with the LDL receptors present mostly in the liver. Lysosomal degradation of LDL releases free cholesterol. Cholesterol released from LDL then inhibits HMG-CoA reductase, thus preventing endogenous synthesis of cholesterol by the liver. If the apo B-100 protein is defective, uptake of LDL by LDL receptor is impaired. In addition, patients with familial hypercholesteremia have a defect in the gene that codes for LDL receptor.
- Plasma LDL values are typically calculated using Friedewald formula: LDL cholesterol = Total cholesterol − HDL cholesterol − Triglyceride/5, where all measurements are in mg/dL. This formula is invalid if triglyceride values are above 400 mg/dL. In such situations, direct measurement of LDL is indicated.
- Major role of HDL is to remove cholesterol from peripheral cells and then return it to liver for excretion, a pathway called reverse cholesterol transport. Cholesterol transfer from cell membranes to HDL is stimulated by ATP binding cassette protein A1 (ABCA1). The free cholesterol is converted to cholesteryl esters by the enzyme lecithin cholesterol acyltransferase (LCAT). This enzyme is present in nascent HDL. ApoA-1 also present in HDL activates this enzyme. The cholesteryl esters are then

transferred to chylomicron remnants and IDL. Cholesteryl ester transport proteins (CETP) are involved in this transfer.

- Currently desirable levels of total cholesterol, LDL-cholesterol, and triglycerides are <200 mg/dL, <100 mg/dL (<70 mg/dL in high-risk patients), and <150 mg/dL, respectively. Hypertriglyceridemia is also a risk factor for acute pancreatitis.

- Many epidemiological studies have shown correlation between low HDL-cholesterol and higher risk of cardiovascular disease. The desired level of HDL is set at >40 mg/dL in male and >50 mg/dL in female.

- Non-HDL cholesterol = Total cholesterol − HDL cholesterol (this is the way it is measured in a laboratory). Some studies have indicated better correlation between non-HDL cholesterol and cardiovascular mortality than LDL cholesterol in patients with high triglycerides. In patients not requiring therapy, target for non-HDL cholesterol is <130 mg/dL. However, during statin therapy it may be necessary to reduce non-HDL cholesterol <100 mg/dL to get optimal benefit.

- Total cholesterol to HDL cholesterol ratio is also used for calculating risk factor for cardiovascular disease. Usually if the ratio is above 5, it is considered as high.

- Hyperlipidemia is also termed as hyperlipoproteinemia can be primary or secondary in origin.

- Familial hypercholesterolemia: This disease is transmitted as an autosomal dominant disorder. Mutations affect LDL receptor synthesis or its proper function, as well as mutation of Apo B-100 gene, which results in decreased binding of LDL with Apo B-100. Total cholesterol and LDL-cholesterol are highly elevated in these individuals making them susceptible to myocardial infarction at a young age.

- Polygenic hypercholesterolemia: In these individuals, both genetics and environmental factors play important roles in producing high cholesterol levels.

- Familial hypertriglyceridemia: This disease is also transmitted as an autosomal dominant disorder where there is an increased production of VLDL by the liver.

- Familial hyperchylomicronemia: This disease is transmitted as autosomal recessive where there is a deficiency of the enzyme lipoprotein lipase or apo CII. Triglycerides levels are high in these individuals.

- Familial dysbetalipoproteinemia: In these individuals, increased level of IDL and chylomicron remnants are observed in circulation. Both cholesterol and triglycerides are also subsequently increased. The apoprotein E exhibits polymorphism showing three isoforms: apo E2, apo E3, and apo E4. Common phenotype is E3/E3. Individuals with familial dysbetalipoproteinemia tend to have E2/E2 phenotype. This phenotype results in impaired hepatic uptake of chylomicron remnants and IDL by the liver.

- Familial combined hyperlipidemia: In these individuals, either cholesterol or triglyceride or both are elevated. It is possibly transmitted as autosomal dominant.
- Secondary hyperlipidemia is common and causes include diabetes mellitus, hypothyroidism, nephrotic syndrome, cholestasis, and alcohol.
- Lipid analysis can also be performed using electrophoresis where chylomicron is observed at the point of application, and second band above point of application is VLDL (prebeta band), followed by LDL band and the band furthest from the point of application is HDL.
- Lipid disorders are also classified according to the Frederickson classification, which is an older classification. In this classification, there are five types of hyperlipidemia:
  - Type I: In these individuals, elevated chylomicrons are found due to lipoprotein lipase or apo CII deficiency causing elevated levels of triglycerides. This is actually familial hyperchylomicronemia.
  - Type IIa: In these individuals, elevated LDL cholesterol and total cholesterol are observed due to familial hypercholesterolemia, polygenic hypercholesterolemia, familial combined hyperlipidemia, as well as nephritic syndrome and hypothyroidism.
  - Type IIb: In these individuals, elevated LDL and VLDL are seen as observed in individuals with familial combined hyperlipidemia. Both cholesterol and triglycerides levels may be elevated.
  - Type III: These individuals have elevated IDL, and this disorder is actually dysbetalipoproteinemia due to apo E/apo E2 profile. Both cholesterol and triglycerides levels may be elevated. In type III lipid disorder, VLDL/triglyceride ratio is usually close to 0.3, while normal value is around 2.0.
  - Type IV: These individuals have elevated VLDL as seen in familial hypertriglyceridemia, or familial combined hyperlipidemia. As a result, triglyceride level is elevated. Type IV disorder also may be due to secondary cause such as diabetes and nephrotic syndrome.
  - Type V: These individuals have elevated VLDL and chylomicrons causing elevated triglycerides.
- Type IIa, IIb, and also type III are associated with significantly increased risk for cardiovascular diseases.
- In Tangier disease, there is a loss of function of ABCA1 protein due to mutation of ABCA1 gene causing very low level of HDL in serum. This disease is inherited in an autosomal recessive pattern. The tonsils appear hyperplastic and orange in color. Individuals with Tangier disease have higher risk for cardiovascular disease.

- Abetalipoproteinemia is a rare inherited disease with approximately 100 cases reported worldwide. In this disease, there is a total absence of apo B-100, thus concentrations of triglycerides and cholesterol carrying lipoprotein concentrations (chylomicrons, VLDL, IDL, and LDL) are highly reduced.
- In addition to traditional lipid parameters such as cholesterol, triglycerides, HDL-cholesterol, and Lp(a) (reference range up to 30 mg/dL), there are also other lipid markers and nonlipid markers that can be used for assessing risk of cardiovascular disease in individuals. These markers include lipoprotein associated phospholipase $A_2$ (Lp-PLA2) (lipid parameter), LDL particle size (lipid parameters), C-reactive protein (nonlipid parameter), homocysteine (nonlipid parameter), and myeloperoxidase.
- Statins are first line of lipid-lowering drugs. Cholesterol absorption inhibitors such as ezetimibe modestly lower LDL-cholesterol and can be used in combination with statins. Fibrates are primarily used in treating patients with hypertriglyceridemia. Nicotinic acid may also lower Lp(a) level. More recently, FDA approved use of PCSK9 inhibitors (human monoclonal antibody against PCSK9) for significantly lowering LDL-cholesterol.

# References

[1] Sawle A, Higgins MK, Olivant MP, Higgins JA. A rapid single step centrifugation method for determination of HDL, LDL and VLDL cholesterol and TG and identification of predominant LDL subclass. J Lipid Res 2002;43:335–43.

[2] Luo J, Yang H, Song BL. Mechanisms and regulation of cholesterol homeostasis. Nat Rev Mol Cell Biol 2020;21:225–45.

[3] Yusuf S, Hawken S, Ounpuu S, Dans T, et al. Effect of potentially modifiable risk factors associated with myocardial infarction in 52 countries (the INTERHEART study): case control study. Lancet 2004;364:937–52.

[4] Kones R. Peimary prevention of coronary heart disease: integration of new data, evolving views, revised goals, and role of rosuvastatin in management: a comprehensive survey. Drug Des Devel Ther 2011;5:325–80.

[5] Gordon T, Kannel WB, Castelli WP, Dawber TR. Lipoproteins, cardiovascular disease and death. The Framingham study. Arch Intern Med 1981;141:1128–31.

[6] Vijayakrishnan R, Kalyatanda G, Srinivasan I, Abraham GM. Compliance with the adult treatment panel III guidelines for hyperlipidemia in a resident run ambulatory clinic: a retrospective study. J Clin Lipidol 2013;7:43–7.

[7] Law MR, Wald NJ. Risk factor thresholds: their existence under scrutiny. BMJ 2003;327:518.

[8] Grundy SM, Cleeman JI, Merz CN, Brewer HB, et al. Implications of recent clinical trials for National Cholesterol Education Program Adult Treatment Panel III guidelines. J Am Coll Cardiol 2004;44:720–32.

[9] Martin SS, Blumenthal RS, Miller M. LDL cholesterol: lower the better. Med Clin North Am 2012;96:13–26.

[10] Miller M, Stone NJ, Ballantyne C, Bittner V, et al. Triglycerides and cardiovascular disease: a scientific statement from American Heart Association. Circulation 2011;123:2292–333.

[11] Liu J, Sempos CT, Donahue RP, Dorn J, et al. Non high density lipoprotein and very low density lipoprotein cholesterol and their risk predictive values in coronary heart disease. Am J Cardiol 2006;98:1363–8.

[12] Ballantyne CM, Raichlen JS, Cain VA. Statin therapy alters the relationship between apolipoprotein B and low density lipoprotein cholesterol and non-high density lipoprotein cholesterol targets in high risk patients: the MERCURY II (measuring effective reductions in cholesterol using rosuvastatin) trail. J Am Coll Cardiol 2008;52:626–32.

[13] Goswami B, Rajappa M, Malika V, Kumar S, et al. Apo B/apo AI ratio, a better discriminator for coronary artery disease risk than other conventional lipid ratios in Indian patients with acute myocardial infarction. Acta Cardiol 2008;63:749–55.

[14] Tada H, Takamura M, Kawashiri MA. Lipoprotein(a) as an old and new causal risk factor of atherosclerotic cardiovascular disease. J Atheroscler Thromb 2019;26:583–91.

[15] Warner C, Rader D, Bartens W, Kramer J, et al. Elevated plasma lipoprotein (a) in patients with nephrotic syndrome. Ann Intern Med 1993;119:263–9.

[16] Wu MF, Xu KZ, Guo YG, Yu J, et al. Lipoprotein(a) and atherosclerotic cardiovascular disease: current understanding and future perspectives. Cardiovasc Drugs Ther 2019;33:739–48.

[17] Welty FK. Hypobetalipoproteinemia and abetalipoproteinemia: liver disease and cardiovascular disease. Curr Opin Lipidol 2020;31:49–55.

[18] Roheim PS, Asztalos BF. Clinical significance of lipoprotein size and risk for coronary atherosclerosis. Clin Chem 1995;41:147–52.

[19] McGully KS. Homocystinuria, arteriosclerosis, methylmalonic aciduria and methyltransferase deficiency: a key case revisited. Nutr Rev 1992;50:7–12.

[20] Jacques PF, Bostom AG, Williams RR, et al. Relation between folate status a common mutation in methylene tetrahydrofolate reductase and plasma homocysteine concentration. Circulation 1996;93:7–9.

[21] Langsted A, Nordestgaard BG. Nonfasting versus fasting lipid profile for cardiovascular risk prediction. Pathology 2019;51:131–41.

[22] Xu RX, Wu YJ. Lipid-modifying drugs: pharmacology and perspectives. Adv Exp Med Biol 2020;1177:133–48.

# CHAPTER 7

# Carbohydrate metabolism, diabetes, and hypoglycemia

## Carbohydrates: An introduction

Carbohydrates including sugar and starch are important in human physiology because glucose provides more than half of the total energy requirement of the human body. Glucose is a breakdown product of dietary carbohydrates. In addition, glucose is also produced by glycogenolysis and gluconeogenesis.

Carbohydrates contain only carbon, hydrogen, and oxygen and the ratio of hydrogen to oxygen is 2:1, the same proportion present in water. Carbohydrates are often referred to as saccharides (derived from a Greek word meaning sugar) and can be subclassified into four categories:

- Monosaccharides: Simplest carbohydrates, which cannot be further hydrolyzed. The common monosaccharides are glucose, fructose, and galactose, and all contain six carbons. Ribose, a monosaccharide containing five carbons, is an integral part of RNA and several cofactors such as ATP, NAD, etc.
- Disaccharides: When two monosaccharide molecules are joined together, a disaccharide is formed, which can be hydrolyzed to monosaccharides. Most abundant disaccharide in blood is sucrose, which can be hydrolyzed to glucose and fructose. Lactose, the disaccharide present in milk, is composed of galactose and glucose. Maltose is another common disaccharide, which breaks down into two glucose molecules.
- Oligosaccharides: These molecules are less complex than polysaccharides and usually contain 10 or less monosaccharide molecules in one oligosaccharide molecule. Oligosaccharides are often found in glycoprotein molecules.
- Polysaccharides: Complex carbohydrates containing many monosaccharides (200–2500). Polysaccharides may serve as energy storage (starch, glycogen, etc.), and/or may act as structural components, for example, cellulose, which is an integral component of cell membranes of plants.

**127**

Clinical Chemistry, Immunology and Laboratory Quality Control. https://doi.org/10.1016/B978-0-12-815960-6.00009-1

## Regulation of blood glucose concentration

The blood glucose concentration is tightly controlled and increased blood glucose concentration is encountered in patients suffering from diabetes. After eating a meal, starch and glycogen present in food are partially digested by salivary amylase and then further digested by pancreatic amylase and disaccharidase present in intestinal mucosa into monosaccharides (glucose, fructose, and galactose). These monosaccharides are absorbed into circulation by active carrier-mediated transfer process. After absorption into portal vein, these monosaccharides are transported into the liver and depending on the physiological need, glucose can be metabolized completely into carbon dioxide and water to provide immediate energy or stored in the liver as glycogen. Major biochemical processes involved in regulation of blood glucose level includes:

- Glycolysis: Metabolism of glucose through a series of biochemical reactions into lactate or pyruvate is glycolysis. In this process, two molecules of ATP (adenosine triphosphate) are formed, thus providing energy. Fructose and galactose can also enter glycolysis process after phosphorylation. Oxidation of glucose into carbon dioxide and water can also take place through hexose monophosphate shunt pathways.
- Glycogenesis: Conversion of glucose into glycogen for storage in liver. Glycogen can also be found in skeletal muscle.
- Glycogenolysis: A process by which glycogen breaks down into glucose during fasting period when needed for energy.
- Gluconeogenesis: During extended fasting period, this biochemical pathway is used for production of glucose from noncarbohydrate sources such as lactate, glycerol, pyruvate, glucogenic amino acids, and odd chain fatty acids.

Major hormones involved in blood glucose regulation are insulin and glucagon. Insulin is a 51-amino-acid polypeptide, secreted by the beta cells of the islets of Langerhans in the pancreas. The molecule proinsulin is cleaved to form insulin and C-peptide. Whenever insulin is secreted, C-peptide molecule is also found in blood. The ratio if C-peptide to insulin in a normal individual is 5:1. In Type 1 and Type 2 diabetes, there is an inappropriate increase in release of proinsulin. Proinsulin has 10%–15% of biological activity of insulin.

Insulin induces:

- Cellular uptake of glucose
- Glycogen and protein synthesis
- Fatty acid and triglyceride synthesis

Insulin inhibits:

- Glycogenolysis
- Gluconeogenesis
- Proteolysis and lipolysis

Cell membranes are not permeable to glucose. Specialized glucose transporter (GLUT) proteins transport glucose through the cell membranes. Insulin-mediated cellular uptake of glucose is regulated through GLUT-4 proteins. The insulin receptor consists of alpha and beta subunits. The insulin molecule binds with the alpha subunit. The beta subunits traverse the cell membrane and conformational changes take place in the beta subunit when insulin binds with the alpha subunit. This results in various intracellular responses including translocation of vesicles carrying GLUT-4 proteins to the cell membrane.

Glucagon is a 29-amino-acid polypeptide secreted by the alpha cells of the islets of Langerhans in the pancreas. It has the opposite action to that of insulin. Thus, it increases:

- Glycogenolysis
- Gluconeogenesis
- Lipolysis

In addition, to insulin and glucagon, other hormones are also involved in regulation of blood glucose concentration. Somatostatin, a polypeptide found in several organs but mostly in hypothalamus and in delta cells of pancreatic islet, can regulate secretion of insulin and glucagon, thus modulating actions of these two important hormones. In general, insulin decreases blood glucose and glucagon increases blood glucose, thus counteracting the effect of insulin. Other hormones that can increase blood glucose include epinephrine, cortisol, and growth hormone.

## Diabetes mellitus: Basic concepts

Diabetes mellitus is a syndrome, characterized by hyperglycemia due to relative insulin deficiency or insulin resistance. It is important to note that diabetes insipidus, which is also characterized by polyuria, is different from diabetes mellitus because diabetes insipidus is not related to insulin secretion or insulin resistance but is an uncommon condition that occurs when kidney is unable to concentrate urine properly. As a result, diluted urine is produced affecting plasma osmolality. The cause of diabetes insipidus is lack of secretion of antidiuretic hormone (ADH), which results in cranial diabetes insipidus (also known as central diabetes insipidus). Another cause of diabetes insipidus is the inability of ADH to work at the collecting duct of the kidney (nephrogenic diabetes insipidus). Please see Chapter 5 for more detail.

Diabetes mellitus can be primary or secondary in nature. Primary diabetes mellitus can be monogenic or polygenic. Monogenic diabetes mellitus covers a heterogeneous group of diabetes caused by a single gene mutation and characterized by impaired insulin secretion by beta cells of the pancreas. Maturity onset diabetes of the young (MODY), mitochondrial diabetes and neonatal diabetes are examples of monogenic diabetes mellitus. The diagnosis of monogenic diabetes mellitus and differentiating this type of diabetes from Type 1 and Type 2 diabetes mellitus (which are polygenic) is essential. Monogenic diabetes accounts for 2%–5% of all diabetes and is less common than Type 1 and Type 2 diabetes mellitus which are the most common forms of diabetes mellitus encountered in clinical practice [1].

Polygenic diabetes mellitus can be either Type 1 or Type 2. Type 1 diabetes mellitus (formerly called insulin-dependent diabetes) is characterized by an absolute deficiency of insulin due to islet cell destruction and usually presents in younger people with an acute onset. Type 2 diabetes mellitus (formerly called noninsulin-dependent diabetes mellitus) is characterized by insulin resistance and beta cell dysfunction in the face of insulin resistance and hyperglycemia. There is also secondary diabetes, which may be drug related or due to various diseases.

## Monogenic diabetes mellitus

All cases of MODY and most cases of neonatal diabetes are due to defects in insulin secretion. MODY is the most common form of monogenic diabetes. The clinical pattern of MODY is characterized by young age onset of diabetes (10–45 years) but most likely before age of 25 years, marked family history of diabetes in every generation due to autosomal dominant inheritance, absence of obesity and insulin resistance, negative autoantibody against pancreatic beta cells, and mild hyperglycemia. Usually, patients respond to sulfonylurea therapy if needed. Many genetic mutations have been reported in patients with MODY but most commonly encountered mutations are due to mutation in genes encoding the enzyme glucokinase (GCK) and mutation of genes encoding nuclear transcription factors of the hepatocytes nuclear factors (HNF). At present sequencing of common genes causing MODY such as GCK, HNF1A (transcription factor-1), and HNF1 B (transcription factor-2) are available in reference laboratories for confirming diagnosis of MODY. However, new gene mutations related to MODY are regularly described in the medical literature [2]. MODY can be further subclassified as MODY 1, 2, and 3, while MODY 3 is the most commonly encountered form, which is caused by mutation of HNF 1A. It is important to note that patients with MODY require lower doses of sulfonylurea than other groups of patients.

Neonatal diabetes mellitus, a rare disease, may occur up to the age of 6 months due to mutation of different genes involved in organogenesis, formation of beta cells, and insulin synthesis. Depending on genetic mutation, neonatal diabetes can be transient or permanent. Neonatal diabetes diagnosed before 6 months of age is frequently due to mutation of genes that encode Kir6.2 (ATP-sensitive inward rectifier potassium channel) or sulfonylurea receptor 1 subunit of ATP-sensitive potassium channel, and these patients respond to high-dose sulfonylurea therapy rather than insulin therapy [3]. Mitochondrial diabetes mellitus is due to mutation of mitochondrial DNA and the disease may be manifested as early as 8 years of age but mean onset is 35 years. Because diabetes develops due to failure on insulin secretion, most patients will eventually require insulin therapy [4].

## Type 1 diabetes mellitus

Type 1 diabetes, formerly known as insulin-dependent diabetes or juvenile onset diabetes encountered in 5%–10% of all patients with diabetes mellitus and is characterized by polyuria, polydipsia, and rapid weight loss. Type 1 diabetes is due to autoimmune destruction of pancreatic beta cells by T lymphocytes. Markers of immune destruction of beta cells in these patients include islet cell autoantibodies as well as autoantibodies to insulin, glutamic acid decarboxylase (GAD) and tyrosine phosphatases. Usually 85%–90% patients with Type 1 diabetes mellitus have one or more autoantibodies. Individuals with antibodies are classified as Type 1A and other individuals with Type 1 diabetes but without any evidence of autoimmunity or any known cause of islet cell destruction are classified as Type 1B. This form of Type 1 diabetes mellitus is less common and is often referred as idiopathic diabetes.

Both genetic susceptibility and environmental factors play important roles in the pathogenesis of Type 1 diabetes. Genetic susceptibility is polygenic with the greatest contribution coming from the HLA region but no single gene responsible for Type 1 diabetes has been characterized. More than 90% of patients with Type 1 diabetes mellitus carry HLA-DR3-DQ2, HLA-DR4-DQ8, or both. Multiple other loci also regulate specific immune responses and modify the vulnerability of β-cells to inflammatory mediators. Compared with the genetic factors, environmental factors that affect the development of Type 1 diabetes are less well characterized but contact with particular microorganisms is emerging as an important factor. Environmental factors that have been implicated include dietary constituents, Coxsackie viruses, and vaccinations. Rate of destruction of beta cells in patients with Type 1 diabetes mellitus is variable with more rapid destruction usually observed in infants and children compared to adults. Children and adolescents may first present with ketoacidosis as the manifestation of the disease. Patients are usually dependent on insulin as very little or no insulin is produced [5].

## Type 2 diabetes mellitus

Type 2 diabetes mellitus is the most common form of diabetes mellitus accounting for over 90% of all cases and was formerly referred as noninsulin dependent diabetes mellitus. Type 2 diabetes mellitus has adult onset and is characterized by insulin resistance and also may be accompanied by beta cell dysfunction causing insulin deficiency. Many patients with Type 2 diabetes mellitus are obese because obesity itself can cause insulin resistance. However, these patients may not need insulin initially after diagnosis or maybe through-out their life. Timely diagnosis of Type 2 diabetes mellitus is important because early intervention can prevent many complications of Type 2 diabetes mellitus, including neuropathy, nephropathy, and retinopathy, but such diagnosis is often difficult because hyperglycemia develops gradually and at early stage a patient may not notice any classical symptoms of diabetes. Ketoacidosis seldom occurs in Type 2 diabetes and when seen it is usually associated with a stress factor such as infection.

Insulin resistance in Type 2 diabetes mellitus includes downregulation of the insulin receptor, abnormalities in the signaling pathway, and impairment of fusion of GLUT-4 (glucose transporter Type 4) containing vesicles with the cell membrane. Initially with insulin resistance hyperinsulinemia is observed, which attempts to compensate for the insulin resistance. With time beta cell dysfunction may be encountered (both quantitative and qualitative) causing hyperglycemia. Various features of Type 1 diabetes mellitus, Type 2 diabetes mellitus and MODY are summarized in Table 1.

**Table 1** Major features of Type 1, Type 2 diabetes mellitus and MODY.

| Clinical feature | Type 1 diabetes mellitus | Type 2 diabetes mellitus | MODY |
|---|---|---|---|
| Typical age of diagnosis | <25 years | >25 years | <25 years |
| Body weight | Usually not obese | Overweight to obese | No obesity |
| Autoantibodies | Present (90%) | Absent | Absent |
| Insulin dependence | Yes | No | No |
| Family history | Infrequent | Frequent | Yes, in multiple generations |
| Diabetic keto acidosis | High risk | Low risk | Low risk |

MODY, *maturity onset diabetes of young.*

## Metabolic syndrome or syndrome X

Metabolic syndrome or syndrome X was first described in 1988 by Gerald Reaven where he proposed the existence of a new syndrome or syndrome X, which is characterized by insulin resistance, hyperinsulinemia, hyperglycemia, dyslipidemia, and arterial hypertension. Metabolic syndrome is currently defined by WHO as a pathologic condition characterized by abdominal obesity, insulin resistance, hypertension, and hyperlipidemia. Metabolic syndrome is a major public health concern because individuals with this syndrome have higher risk of coronary artery disease, stroke, and Type 2 diabetes. American Heart Association/National Heart, Lung and Blood Institute (AHA/NHLBI) criteria for metabolic syndrome include three or more of following risk factors:

- Central obesity (waist circumference: 40 in. or more in men, 35 in. or more in women).
- Insulin resistance: Fasting glucose over 100 mg/dL.
- Elevated triglyceride (>150 mg/dL).
- Reduced high density lipoprotein cholesterol (HDL-cholesterol: <40 mg/dL for men, <50 mg/dL for women).
- Elevated blood pressure (≥130 mm of Hg for systolic or ≥85 mm of Hg for diastolic) or drug treatment for hypertension.

Other risk factors for metabolic syndrome include genetic makeup, advanced age, lack of exercise, and hormonal changes. Weight control, daily exercise, and healthy food habit are the primary goals of therapy. Drug therapy may be initiated depending on the clinical judgment of the physician [6, 7]. Smokers with metabolic syndrome are advised to quit smoking.

## Complications of diabetes

Diabetic complications can be divided into two broad categories: acute complications and chronic complications. Acute complications include diabetic ketoacidosis (DKA), hyperosmolar non ketosis and lactic acidosis. Chronic complications could be either macrovascular (stroke, myocardial infarction, gangrene, etc.) or microvascular such as diabetic retinopathy, diabetic eye diseases and diabetic neuropathy.

Diabetic ketoacidosis may be the presenting feature of Type 1 diabetes mellitus or it may occur in a diabetic individual managed on insulin who does not take insulin or insulin requirement has been increased due to infection, myocardial infection or due to other causes. Diabetic ketoacidosis is a medical emergency and if not treated on time may be fatal. In typical diabetic ketoacidosis absolute insulin deficiency, along with increased secretion of glucagon and other counter regulatory hormones results in decreased uptake of glucose into cells with increased glycogenolysis and gluconeogenesis producing more sugar, which are released in the circulation. As a result, marked hyperglycemia, glycosuria and osmotic diuresis may result causing water and salt loss through the kidney. Reduction in plasma volume may cause renal hypoperfusion and eventually acute renal failure. Absence of insulin also leads to release of fatty acids mostly from adipose tissue (lipolysis) causing generation of excess fatty acids that result in formation of ketone bodies, including acetoacetic acid, beta hydroxybutyric acid, and acetone. Acetoacetic acid and beta-hydroxybutyric acid contribute to the acidosis and body attempts to neutralize such excess acid by bicarbonate compensatory mechanism. As bicarbonate is depleted, body attempts other mechanism of compensation such as hyperventilation and in some patients, extreme form of hyperventilation (Kussmaul respiration) is observed. Acetone is volatile and responsible for the typical ketone odor present in patients with diabetic ketoacidosis. Typically, in patient with diabetic ketoacidosis, blood glucose is above 250 mg/dL, arterial blood pH < 7.3, and bicarbonate between 15 and 18 mmol/L but may be lower than 10 mmol/L in severe cases. In addition, ketone bodies are present in urine. Diabetic ketoacidosis is more commonly encountered in patients with Type 1 diabetes although under certain circumstances such as trauma, surgery, infection or severe stress, stress diabetic ketoacidosis may also be observed in patients with Type 2 diabetes. Other than diabetes, ketoacidosis may be observed in patients with alcoholism (alcoholic ketoacidosis), starvation, and also may be drug induced, for example, salicylate poisoning.

Ketosis-prone diabetes (KPD) is a heterogeneous condition characterized by patients who present with diabetic ketoacidosis but lack the phenotype of autoimmune Type 1 diabetes. KPD is also known as Flatbush diabetes (cases observed in East Flatbush, Brooklyn). KPD comes in four forms depending upon the presence or absence of β-cell autoantibodies (A+ or A−) and β-cell functional reserve (β+ or β−). KPD is highly prevalent in African populations but also observed in Hispanic and some Asian populations, as well as Caucasians. KPD patients are usually middle aged, overweight or mildly obese, and more likely to be male. After intensive initial insulin therapy, many patients become insulin independent and can be well controlled on diet alone or diet plus oral medications. Therefore, clinical course of KPD is having more resemblance with Type 2 diabetic patients than Type 1 diabetic patients [8].

Hyperosmolar nonketosis is typically seen in patients with Type 2 diabetes mellitus. Insulin deficiency is not absolute and as a result ketosis is not significant. There is also minimal acidosis. Characteristic clinical features include significant hyperglycemia with high plasma osmolality and dehydration. Lactic acidosis is an uncommon situation with diabetics. It may be seen in patients on biguanide therapy (phenformin was withdrawn from US market due to lactic acidosis) with liver or renal impairment. Currently metformin is the only biguanide available in United States but metformin-associated lactic acidosis is a rare condition, with an estimated prevalence of 0.03–0.06 per 1000 patient-years [9].

Macrovascular complication of diabetes mellitus is related to atherosclerosis and diabetes is a major risk factor for cardiovascular diseases. The reason for this is multifactorial. Diabetic individuals have abnormal lipid metabolism with increased low-density-lipoprotein cholesterol (LDL-C) and decreased high-density-lipoprotein cholesterol (HDL-C). Triglyceride level is typically increased in patients with diabetes mellitus. In addition, glycation of lipoproteins may lead to altered functions of these lipoproteins.

Microvascular complications are related to following mechanisms:

- Nonenzymatic glycosylation of proteins
- Activation of protein kinase C
- Disturbance in the polyol pathway

The degree of nonenzymatic glycosylation is related to the blood glucose level and measurement of glycosylated hemoglobin, and less frequently fructosamine (glycated proteins) level in blood is measured on a regular basis in patients with diabetes. Hyperglycemia induced activation of protein kinase C results in production of proangiogenic molecules, which can cause neovascularization and also formation of profibrogenic molecules, which lead to deposition of extracellular matrix and basement membrane material. Increase in intracellular glucose leads to increased production of sorbitol by the enzyme aldose reductase. Sorbitol is a polyol, which is converted to fructose. Increased accumulation of sorbitol and fructose can cause cellular injury.

## Secondary causes of diabetes mellitus

Gestational diabetes mellitus is typically seen during the second or third trimester of pregnancy most likely due to increased levels of hormones such as estrogen, progesterone, cortisol, etc., which counteract action of insulin. Although gestational diabetes may resolve after delivery, these women may have a higher risk of developing Type 2 diabetes mellitus later. Pancreatic disease, endocrine diseases, and various drugs may also cause diabetes, and these are all considered as secondary diabetes. Various drugs may also cause diabetes. Various causes of secondary diabetes are summarized in Table 2.

## Diagnostic criteria for diabetes

The classic clinical presentation of diabetes mellitus patients includes:

| Table 2 Various causes of secondary diabetes. | |
| --- | --- |
| **Causes** | **Specific examples** |
| Pancreatic diseases | Cystic fibrosis, chronic pancreatitis, hemochromatosis |
| Endocrine diseases | Cushing's syndrome, acromegaly, pheochromocytoma |
| Drugs | Thiazide diuretics, glucocorticoids, thyroid hormones |

- Polyuria
- Polydipsia
- Weight loss

American Diabetic Association recommends screening for diabetes mellitus for any individual over 45 years of age. Fasting blood glucose and glycosylated hemoglobin (HbA1c) are best criteria for diagnosis of diabetes mellitus. Guidelines of Expert Committee on Diagnosis and Classification of Diabetes Mellitus indicate that normal fasting glucose level should be 70–99 mg/dL (3.9–5.5 mmol/L). Individuals with fasting glucose levels between 100 (5.6 mmol/L) and 125 mg/dL (6.9 mmol/L) are classified as having impaired fasting glucose. Impaired glucose tolerance and impaired fasting glucose are both considered to be prediabetic condition as individuals are at higher risk of developing diabetes. Impaired glucose tolerance is present in an individual when in glucose tolerance test 2 h glucose value is in the range of 140 mg/dL (7.8 mmol/L) to 199 mg/dL (11.0 mmol/L). These conditions are regarded as prediabetic condition as individuals are at higher risk of developing diabetes later in life. Glycated hemoglobin also known as glycosylated hemoglobin (HbA1c) is formed when a nonenzymatic ketoamine reaction occurs between glucose and the N-terminal amino acid of the β chain of hemoglobin. The amount of HbA1c generated is proportional to the average blood glucose during the 8–10 weeks before the test. Therefore, HbA1c is a very useful biomarker of long term glucose control. The HbA1c values between 5.7% and 6.4% may also be considered as prediabetic in adults.

The criteria for diagnosis of diabetes mellitus are fasting plasma glucose 126 mg/dL (7.0 mmol/L) or higher on more than one occasion with no calorie intake in last 8 h or in a patient with classic symptoms of hyperglycemia, a random plasma glucose 200 mg/dL (11.1 mmol/L) or higher is an indication of diabetes mellitus. In addition, in glucose tolerance test after an oral dose of 75 g of glucose, 2-h plasma glucose of 200 mg/dL (11.1 mmol/L) or higher indicates diabetes mellitus. However, oral glucose tolerance tests to establish diagnosis of diabetes mellitus is only recommended in pregnant woman to establish the diagnosis of gestation diabetes. An International Expert Committee recommended use of HbA1c test for diagnosis of diabetes mellitus with a

**Table 3** Various laboratory based criteria for diagnosis of diabetes mellitus and gestational diabetes mellitus.

| Laboratory test | Value | Comment |
|---|---|---|
| **Diabetes mellitus** | | |
| Fasting blood glucose[a] | 70–99 mg/dL | Normal value |
| | 100–125 | Impaired fasting glucose |
| | ≥126 mg/dL | Determined in at least two occasions indicative of diabetes |
| Random blood glucose | ≥200 mg/dL | Indicative of diabetes in a patient with suspected diabetes mellitus |
| Glucose tolerance test (2 h)[b] | <140 mg/dL | Normal |
| | 140–199 mg/dL | Impaired glucose tolerance |
| | ≥200 mg/dL | Indicative of diabetes |
| Hemoglobin A1c | 5.7%–6.4% | Increased risk of diabetes |
| | ≥6.5% | Indicative of diabetes |
| **Gestational diabetes** | | |
| Glucose tolerance test | ≥92 mg/dL (fasting) | Any of the criteria (fasting glucose or 1 h or 2 h glucose in glucose tolerance test is equal or exceeding the limit |
| | ≥180 mg/dL (1 h) | |
| | ≥153 mg/dL (2 h) | |

[a]Fasting glucose means no calorie intake for at least 8 h.

[b]Glucose tolerance test is typically performed using 75 g of glucose given orally. The test is preferably performed in the morning in ambulatory patients after overnight fasting.

cut-off value of 6.5% [10]. Criteria for diagnosis of diabetes mellitus are same in both adults and children.

For diagnosis of gestational diabetes, typically glucose tolerance test is performed using 75 g of oral anhydrous glucose during 24–28 weeks of gestation in pregnant women. The glucose tolerance test is typically performed in the morning after 8 h of overnight fasting. Various laboratory-based criteria for diagnosis of diabetes mellitus and gestational diabetes are summarized in Table 3. If a woman has gestational diabetes mellitus, then risk of developing diabetes later in life is higher for both mother and the child.

## Hypoglycemia

Hypoglycemia is arbitrarily defined as blood glucose below 50 mg/dL (2.8 mmol/L), but some authors favor to consider the cut-off value as 55 or 60 mg/dL. Epinephrine is mostly responsible for symptoms of hypoglycemia

| Table 4 Common causes of hypoglycemia. |
| --- |
| *Fasting hypoglycemia* |
| Endocrine diseases, for example, hypoadrenalism, hypopituitarism, etc.<br>Hepatic failure<br>Renal failure<br>Neoplasms<br>Inborn errors of metabolism, for example, glycogen storage disease Type 1<br>Insulinoma |
| *Reactive hypoglycemia* |
| Post prandial due to gastric surgery or idiopathic<br>Inborn errors of metabolism, for example, galactosemia, fructose 1, 6-diphosphate deficiency<br>Drug induced but most likely due to insulin therapy or oral hypoglycemic agents such as sulfonylureas<br>Alcohol abuse |

including trembling, sweating, lightheadedness, hunger, and possibly epigastric discomfort. Brain is dependent on glucose for energy. In neonates, blood glucose is usually lower than adults and blood glucose level of 30 mg/dL may be encountered in a neonate without any symptoms of hypoglycemia. Hypoglycemia may be related to prolong fasting, but other disease conditions may precipitate such condition. Postprandial or reactive hypoglycemia may be related to insulin therapy, inborn errors of metabolism, or other causes. Various causes of fasting and reactive hypoglycemia are summarized in Table 4. Hypoglycemia may occur in both Type 1 and Type 2 diabetes but more common in Type 1 diabetic patients receiving insulin. Some Type 1 diabetic patients may experience hypoglycemia even once or twice a week. Insulinomas are tumors of the insulin-secreting beta cells of the islets of Langerhans. Insulin levels are high along with high levels of C-peptide. Individuals with insulinomas usually have characteristic features of hypoglycemia and high insulin, as well as C-peptide levels. Imaging techniques are used to detect the tumor.

A diagnosis of hypoglycemia is not based only on symptoms. Clinicians look for the presence of Whipple's Triad, named after Allen O. Whipple, MD (low plasma glucose concentration, clinical signs/symptoms consistent with hypoglycemia, and resolution of signs or symptoms when the plasma glucose concentration increases). Whipple's Triad should be documented in patients presenting with hypoglycemia prior to initiating an evaluation. Medications should be reviewed. Critical illnesses, malnutrition, hormone deficiencies especially adrenal insufficiency, and nonislet-cell tumors secreting insulin-like growth factor-II should be considered in those critically ill patients. Hypoglycemia can also follow bariatric surgery [11].

## Laboratory methods

In a clinical laboratory glucose concentration is usually measured in serum or plasma. However, occasionally glucose concentration may be measured in whole blood. It is important to note that fasting whole blood glucose concentration is approximately 10%–12% lower than corresponding plasma or serum glucose concentration. Glycolysis reduces blood sugar level in an uncentrifuged specimen by 5–10 mg/dL per hour. Therefore, timely centrifugation of specimen is required. However, the best practice is to collect blood in tubes containing sodium fluoride and potassium oxalate (fluoride/oxalate tube; Gray top) because sodium fluoride inhibits glycolysis and glucose level is stable up to 3 days at room temperature. Less commonly used preservative for glucose collection tube is sodium iodoacetate.

Measurement of glucose can be done using either hexokinase method or glucose oxidase method. Hexokinase method is considered as the reference method.

Hexokinase method is based on following reactions:

$$\text{Glucose} + \text{ATP} \xrightarrow{\text{Hexokinase}} \text{Glucose 6} - \text{phosphate} + \text{ADP}$$

$$\text{Glucose 6-phosphate} \xrightarrow[\text{NAD} \quad \text{NADH}]{\text{Glucose 6-phosphate dehydrogenase}} \text{6-Phosphaogluconate}$$

In the second reaction, NAD, which has no absorption at 340 nm, is converted into NADH, which absorbs at 340 nm, and the absorbance is proportional to glucose concentration.

In glucose oxidase method, following reaction takes place:

$$\text{Glucose} \xrightarrow{\text{Glucose oxidase}} \text{Gluconic acid} + \text{Hydrogen peroxide}$$

In this enzymatic reaction, hydrogen peroxide is generated and concentration of hydrogen peroxide is measured to determine glucose concentration. Concentration of hydrogen peroxide can be measured by addition of peroxidase enzyme and oxygen receptor such as $o$-dianisidine, which when oxidized forms a colored complex that can be measured spectrophotometrically.

In glucose dehydrogenase method, glucose is oxidized to gluconolactone and in this reaction, NAD is converted into NADH.

$$\text{Glucose} + \text{NAD} \xrightarrow{\text{Glucose dehydrogenase}} \text{Gluconolactone} + \text{NADH}$$

Mutarotase is added to shorten the time needed to reach the end point. The amount of NADH formed (signal at 340 nm) is proportional to glucose concentration. Although NAD is a cofactor required for catalytic reaction for glucose oxidase, other forms of glucose oxidase can use pyrroloquinoline quinone (PQQ) as a cofactor or flavin dinucleotide (FAD) as a cofactor. These methods are used in point-of-care glucose meters.

## Glucose meters

Since 1980s portable glucose meters (glucometer) are available for blood glucose monitoring both in point-of-care testing sites and home monitoring of glucose especially patients receiving insulin or patients with Type 2 diabetes who have difficulty in maintaining good glucose control. In order to perform a measurement, a sample of blood (usually a fingerstick) is placed on the test strip, and then this test strip is inserted into the meter or test strip may already be inserted in the meter, and there is a point for application of a drop of blood. After a short period, a digital reading is observed and some more recently introduced meters also have memory where a value can be stored for a period of time. Glucose meters utilize glucose oxidase, hexokinase, or glucose dehydrogenase with PQQ (pyrroloquinoline quinone, cofactor) or glucose oxidase combined with NAD (nicotinamide adenine dinucleotide) for glucose measurement and final reading could be reflectance photometry or electrochemistry measurement methods. Accuracy of glucose meter is of major concern. Following criteria must be met by a glucose home monitoring method:

- Current FDA (Food and Drug Administration) criteria for acceptability of a glucometer is that 95% of all values should fall within ±15 mg/dL of the glucose value obtained by a clinical laboratory-based reference method at glucose concentration <75 mg/dL. For glucose values over 75 mg/dL, 95% individual values must fall within ±20% of glucose valued determined by a reference method.
- Glucose reading obtained by a glucose meter should not be used for diagnosis of diabetes.
- American Diabetes Association criterion suggests that the glucose value measured by a glucose meter should be within ±5% of value obtained by a laboratory-based glucose assay.

Major limitations of glucose meters are inaccuracy of measurement compared to a reference glucose method, as well as various interferences. During peritoneal dialysis, infuse may contain icodextrin, which is converted into maltose by human body. Maltose falsely increases glucose value if glucose meter is based on glucose oxidase method. In 2009, 13 deaths of patients who were on peritoneal dialysis were reported to FDA due to falsely elevated glucose reading using glucose meters based on glucose dehydrogenase-PQQ method. Falsely elevated glucose value may result in insulin overdose where a clinician may think that a patient was severely hyperglycemic where in reality glucose value may be within acceptable limits. Insulin overdose may cause death. Even failure to dry hands, after hand washing prior to prick for glucose monitoring may falsely decrease glucose reading due to hemodilution [12]. Major interferences include:

- Elevated concentration of ascorbic acid (vitamin C) can falsely elevate a reading in glucose-dehydrogenase-based methods, but small changes in value in both directions (positive/negative) may occur with glucose meters using glucose oxidase method. However, a recent study where patients with sepsis received high dose vitamin C, The Bland-Altman analysis showed a mean difference between point of care and laboratory blood glucose levels of 8.9 mg/dL, which may not have resulted in a change in clinical action [13].
- Maltose, xylose, or galactose can falsely increase glucose value in glucose meter using glucose dehydrogenase-PQQ method but not glucose-dehydrogenase-NAD-based method.
- Hematocrit affects the reading and regardless of methods, anemia can falsely elevate glucose reading.
- Hypoxia or increased altitude can falsely elevate readings in glucose meters using glucose oxidase method.
- Diabetic ketoacidosis may falsely decrease glucose reading regardless of method.

Currently, in addition to traditional self-monitoring of blood glucose using a glucometer, there are new strategies to measure glucose levels, including the detection of interstitial glucose through continuous glucose monitoring or Flash Glucose Monitoring (FGM). In September 2017, the FDA approved a Freestyle Libre, an FGM device for personal use by patients. The system's disposable sensor is applied to the back of a patient's arm and can be worn for 10 days with the personal device and up to 14 days with the Pro version. The device is a size of coin with a 4-mm wire, which must be inserted subcutaneously. Freestyle Libre Pro and Freestyle Libre use wired enzyme glucose sensing technology (wired glucose oxidase enzyme co-immobilized on an electrochemical sensor) and automatically measure glucose every minute, and readings are stored in 15-min intervals. The Libre Reader is held near the sensor when a glucose reading is needed. This technology is factory-calibrated, and patients do not have to calibrate with blood sample glucose meter readings [14].

Glucose and galactose are major reducing monosaccharides present in urine. Most common cause of galactose in urine is galactosemia, a rare inborn error of metabolism. Usually Clinitest is used to detect the presence of reducing sugars in urine. This test utilizes ability of reducing sugar to reduce copper sulfate to cuprous oxide in the presence of sodium hydroxide and the characteristic orange color of cuprous oxide (copper sulfate is blue) qualitatively indicate the presence of reducing sugar in urine. If Clinitest is positive and urine dipstick indicates negative glucose, then most likely galactose is present and enzymatic test should be performed to establish the diagnosis of galactosemia.

Glycated hemoglobin (HbA1c) in whole blood can be determined by various methods including high-performance liquid chromatography, ion exchange microcolumn, affinity chromatography, capillary electrophoresis, and immunoassays. However, liquid chromatography combined with mass spectrometry is usually considered as the reference method although most clinical laboratories do not use this method. Patients do not need to fast and specimen can be collected in EDTA tubes or oxalate and fluoride tube. However, glycosylated hemoglobin value may be falsely lowered due to autoimmune hemolysis where life spans of erythrocytes are shorter than normal. Ribavirin and other drugs that are associated with anemia may also cause false decline in glycosylated hemoglobin value [15]. In contrast, iron, vitamin B12, or folate deficiency anemia can all increase HbA1c levels, irrespective of blood glucose levels [16]. Uremia and alcoholism may also cause elevated HbA1c levels.

Since HbA1c is a product of the glycosylation of hemoglobin A, its use in patients with hemoglobinopathies may be less accurate for predicting disorders of glucose metabolism, especially in patients with homozygous hemoglobinopathies. HbA1c test results are falsely low and may not correlate with blood glucose results in patients with sickle cell disease or sickle cell trait. Lacy et al. based on retrospective cohort study using data collected from 7938 participants in two community-based cohorts concluded that African Americans with sickle cell trait had lower levels of HbA1c at any given concentration of fasting or 2-h glucose compared with participants without sickle cell trait. Therefore, HbA1c may systematically underestimate past glycemia in black patients with sickle cell trait [17]. HbA1c results may not be accurate in patients with other hemoglobinopathies such as hemoglobin C and E. Common causes of falsely lower or falsely elevated HbA1c values are summarized in Table 5.

**Table 5** False decline or elevated HbA1c levels in various conditions.

| False decline in HbA1c value | Falsely elevated HbA1c value |
| --- | --- |
| • Hemoglobinopathies most commonly sickle cell trait and sickle cell disease are associated with falsely lower HbA1c levels<br>• Autoimmune hemolysis where life spans of erythrocytes are shorter than normal<br>• Drug-induced anemia from Ribavirin and other drugs | • Iron, vitamin B12, or folate deficiency anemia can all increase HbA1c levels, irrespective of blood glucose levels<br>• Uremia/renal failure<br>• Alcoholism |

For selected group of patients where HbA1c value cannot be used for diagnosis of diabetes, fructosamine may be used for evaluating glucose control. Fructosamine is a generic name for plasma protein ketoamines (nonenzymatic attachment of glucose to amino groups of proteins). In principle, glycated albumin is the major fructosamine present in serum because albumin is the most abundant extracellular plasma protein accounting for 60%–70% of total serum proteins with an approximate half-life of 20 days. Fructosamine provides status of blood glucose control for the previous 2–3 weeks, whereas glycosylated hemoglobin can indicate glucose control over past 2–3 months. Normal level of fructosamine in adults is 160–240 µmol/L [18]. However, normal reference range of 200–285 µmol/L has also been reported. A number of methods have been developed for the measurement of fructosamine in serum and plasma. Colorimetric-based assays are indeed the most widely used and those better standardized, and typically exploit the unique property of fructosamine to reduce nitroblue tetrazolium into formazan in an alkaline solution. The formation of formazan (measured at 546 nm) is directly proportional to fructosamine concentrations.

In addition to fructosamine, glycated albumin may also be used for assessment of glucose control and diagnosis of diabetes mellitus. Glycated albumin level is not influenced by the concentration of other serum proteins and reflects short-term glycemia due to the half life time of the albumin, which is approximately 3 weeks. Compared to HbA1C, glycated albumin is not affected by the presence of hemolytic processes, hemoglobinopathies, as well as other conditions such as anemia, pregnancy, and postprandial hyperglycemia. Like HbA1c, glycated albumin can be measured in nonfasting specimen. Glycated albumin is measured as the percent of total albumin using the formula:

Percent glycated hemoglobin = glycated hemoglobin/total hemoglobin × 100

In 2006, the Japan Diabetes Society (JDS) established a reference interval for glycated albumin from 12.3% to 16.9%, but more recent studies indicated lower cut-off of 14.9%. Glycated albumin can be measured in serum or plasma by ion-exchange high-performance liquid chromatography (HPLC), boronate affinity chromatography, as well as immunoassays [19]. Enzymatic assay utilizing ketoamine oxidase and albumin specific protease is also available to measure glycated albumin where total albumin is measured using bromocresol purple. Then percent glycated albumin value is calculated.

## Summary/key points

- Common monosaccharides are glucose, fructose, and galactose. Lactose, the disaccharide present in milk is composed of galactose and glucose. Maltose is composed of two glucose molecules.

- Insulin is a 51-amino-acid polypeptide, secreted by the beta cells of the islets of Langerhans in the pancreas. The molecule proinsulin is cleaved to form insulin and C-peptide. Insulin induces cellular uptake of glucose, glycogen and protein synthesis, fatty acid and triglyceride synthesis. Insulin inhibits glycogenolysis, gluconeogenesis, proteolysis, and lipolysis.
- Diabetes mellitus is a syndrome, characterized by hyperglycemia due to relative insulin deficiency or insulin resistance. Diabetes mellitus can be primary or secondary in nature. Primary diabetes mellitus can be monogenic or polygenic. Monogenic diabetes mellitus covers a heterogenous group of diabetes caused by a single gene mutation and characterized by impaired insulin secretion by beta cells of the pancreas. Maturity onset diabetes of the young (MODY), mitochondrial diabetes, and neonatal diabetes are examples of monogenic diabetes mellitus. Polygenic diabetes mellitus can be either Type 1 or Type 2.
- Type 1 diabetes mellitus (formerly called insulin dependent diabetes) is characterized by an absolute deficiency of insulin due to islet cell destruction and usually presents in the younger people with an acute onset.
- Type 1 diabetes is due to autoimmune destruction of pancreatic beta cells by T lymphocytes. Markers of immune destruction of beta cells in these patients include islet cell autoantibodies, as well as autoantibodies to insulin, glutamic acid decarboxylase (GAD), and tyrosine phosphatases. Usually 85%–90% patients with Type 1 diabetes mellitus have one or more autoantibodies.
- Type 2 diabetes mellitus (formerly called noninsulin-dependent diabetes mellitus) is characterized by insulin resistance and beta-cell dysfunction in the face of insulin resistance and hyperglycemia.
- Type 2 diabetes mellitus has adult onset and is characterized by insulin resistance and also may be accompanied by beta-cell dysfunction causing insulin deficiency. Many patients with Type 2 diabetes mellitus are obese because obesity itself can cause insulin resistance.
- Diabetic complications can be divided into two broad categories: acute complications and chronic complications. Acute complications include diabetic ketoacidosis (DKA), hyperosmolar nonketosis, and lactic acidosis. Chronic complications could be either macrovascular (stroke, myocardial infarction, gangrene, etc.) or microvascular such as diabetic retinopathy, diabetic eye diseases, and diabetic neuropathy.
- Microvascular complications are related to nonenzymatic glycosylation of protein, activation of protein kinase C, and disturbance in the polyol pathway. The degree of nonenzymatic glycosylation is related to the blood glucose level and measurement of glycosylated hemoglobin

(HbA1c) and less frequently fructosamine (glycosylated serum proteins) level in blood is measured on a regular basis in patients with diabetes.

- American Diabetic Association recommends screening for diabetes mellitus for any individual over 45 years of age. Fasting blood glucose and HbA1c are best criteria for diagnosis of diabetes mellitus. Guidelines of Expert Committee on Diagnosis and Classification of Diabetes Mellitus indicate that normal fasting glucose level should be 70–99 mg/dL (3.9–5.5 mmol/L). Individuals with fasting glucose levels between 100 (5.6 mmol/L) and 125 mg/dL (6.9 mmol/L) are classified as having impaired fasting glucose. The criteria for diagnosis of diabetes mellitus are fasting plasma glucose 126 mg/dL (7.0 mmol/L) or higher on more than one occasion with no calorie intake in last 8 h or in a patient with classic symptoms of hyperglycemia, a random plasma glucose 200 mg/dL (11.1 mmol/L) or higher is an indication of diabetes mellitus. In addition, in glucose tolerance test after an oral dose of 75 g of glucose, 2-h plasma glucose of 200 mg/dL (11.1 mmol/L) or higher indicates diabetes mellitus. However, oral glucose tolerance tests to establish diagnosis of diabetes mellitus is only recommended in pregnant a woman to establish the diagnosis of gestation diabetes. An International Expert Committee recommended use of HbA1c test to for diagnosis of diabetes mellitus with a cut-off value of 6.5%. Criteria for diagnosis of diabetes mellitus are same in both adults and children.
- It is important to note that HbA1c correlates with average glucose concentration in blood.
- For diagnosis of gestational diabetes typically glucose tolerance test is performed using 75 g of oral anhydrous glucose during 24–28 week of gestation in pregnant women.
- In a clinical laboratory, glucose concentration is usually measured in serum or plasma. However, occasionally glucose concentration may be measured in whole blood. Fasting whole blood glucose concentration is approximately 10%–12% lower than corresponding plasma or serum glucose concentration. Glycolysis reduces blood sugar level in an uncentrifuged specimen by 5–10 mg/dL per hour. The best practice is to collect blood in tubes containing sodium fluoride and potassium oxalate (fluoride/oxalate tube; Gray top) because sodium fluoride inhibits glycolysis and glucose level is stable up to 3 days at room temperature.
- Measurement of glucose can be done using either hexokinase method or glucose oxidase method. Hexokinase method is considered as the reference method.
- Glucose meters also utilize glucose oxidase, hexokinase, glucose dehydrogenase with PQQ (pyrroloquinoline quinone) or glucose oxidase combined with NAD (nicotinamide adenine dinucleotide) for

glucose measurement. Glucose reading obtained by a glucose meter should not be used for diagnosis of diabetes.

- Major interferences while using glucose meters can occur from vitamin C, acetaminophen, hematocrit, hypoxia, maltose, xylose, galactose, and diabetic ketoacidosis.
- Glucose monitoring in urine is usually performed using urine dip sticks, which has a pad for detection of glucose. Many such test strips use glucose oxidase method. However, glucose test in urine lacks specificity because false positive result may be encountered if hydrogen peroxide or a strong oxidizing agent is present and false negative result may occur if reducing substances such as ascorbic acid, ketones, or salicylate is present in urine.
- Glucose and galactose are major reducing monosaccharides present in urine. Most common cause of galactose in urine is galactosemia, a rare inborn error of metabolism. Usually Clinitest is used to detect the presence of reducing sugars in urine. If Clinitest is positive and urine dipstick indicates negative glucose, then most likely galactose is present.
- Tests for detecting ketone bodies in urine usually detect acetoacetate but not beta-hydroxybutyric acid, the major ketone in urine. However, enzyme assays are available for estimation of beta-hydroxybutyric acid.
- Measuring fructosamine levels or percent glycated albumin levels are useful in some patients where HbA1c value may not be reliable. One common example is falsely lower HbA1c in patients with sickle cell trait or sickle cell disease.

## References

[1] Fajans SS, Graeme IB, Polonksy KS. Molecular mechanisms and clinical pathophysiology of maturity-onset diabetes in the young. N Engl J Med 2001;345:971–80.

[2] Johansson S, Irgens H, Chudssama KK, Molnes J, et al. Exon sequencing and genetic testing for MODY. PLoS One 2012;7, e38050.

[3] Murphy R, Ellard S, Hattersley AT. Clinical implications of a molecular genetics classification of monogenic beta-cell diabetes. Nat Clin Pract Endocrinol Metab 2008;4:200–13.

[4] Henzen C. Monogenic diabetes mellitus due to defect in insulin secretion. Swiss Med Wkly 2012;142:w13690.

[5] Ilonen J, Lempainen J, Veijola R. The heterogeneous pathogenesis of type 1 diabetes mellitus. Nat Rev Endocrinol 2019;15:635–50.

[6] Balkau B, Valensi P, Eschwege E, Slama G. A review of metabolic syndrome. Diabetes Metab 2007;33:405–13.

[7] Saklayen MG. The global epidemic of the metabolic syndrome. Curr Hypertens Rep 2018;20:12.

[8] Levovitz HE, Banerjii MA. Ketosis-prone diabetes (Flatbush diabetes): an emerging worldwide clinically important entity. Curr Diab Rep 2018;18:120.

[9] DeFronzo R, Fleming GA, Chen K, Bicsak TA. Metformin-associated lactic acidosis: current perspectives on causes and risk. Metabolism 2016;65:20–9.

[10] International Expert Committee. International Expert Committee report on the role of the A1C assay in the diagnosis of diabetes mellitus. Diabetes Care 2009;32:1327–34.

[11] Douillard C, Jannin A, Vantyghem MC. Rare causes of hypoglycemia in adults. Ann Endocrinol (Paris) 2020;81:110–7.

[12] Hellman R. Glucose meter inaccuracy and the impact on the care of patients. Diabetes Metab Res Rev 2012;28:207–9.

[13] Smith KE, Brown CS, Manning BM, May T, et al. Accuracy of point-of-care blood glucose level measurements in critically ill patients with sepsis receiving high-dose intravenous vitamin C. Pharmacotherapy 2018;38:1155–61.

[14] Blum A. Freestyle libre glucose monitoring system. Clin Diabetes 2018;36:203–4.

[15] Trask L, Abbott D, Lee HK. Low HbA1c: good diabetic control? Clin Chem 2012;58:648–9.

[16] Shanthi B, Revathy C, Manjula Devi AJ, Subhashree. Effect of iron deficiency on glycation of haemoglobin. J Clin Diagn Res 2013;7:15–7.

[17] Lacy ME, Wellenius GA, Sumner AE, Correa A, et al. Association of sickle cell trait with HbA1c in African Americans. JAMA 2017;317:507–15.

[18] Chen HS, Wu TE, Lin HD, Jap TS, et al. Hemoglobin A (1c) and fructosamine for assessing glycemic control in diabetic patients with CKD stage 3 and 4. Am J Kidney Dis 2010;55:867–74.

[19] Freitas PAC, Ehlert LR, Camargo JL. Glycated albumin: a potential biomarker in diabetes. Arch Endocrinol Metab 2017;61:296–304.

# CHAPTER 8

# Cardiac markers

## Myocardial infarction

Coronary artery diseases account for approximately 37% of all deaths in the United States making it the number one cause of mortality. Acute myocardial infarction (AMI), one of the most serious coronary artery diseases, is associated with increased morbidity and mortality as well as impaired quality of life. Approximately, one-sixth of patients diagnosed with AMI would have unplanned readmission within 30 days of hospital discharge, with estimated direct costs of approximately 1 billion dollars per year Medicare expenditures in the United States [1].

Acute coronary syndrome is a broad term that covers unstable angina, non-ST-segment-elevated myocardial infarction (NSTEMI) and ST-segment-elevated myocardial infarction (STEMI). Although every year several million individuals report to the emergency department with symptoms suggestive of acute coronary syndrome, estimated 1.4 million individuals are admitted to the hospital with acute coronary syndrome every year in US hospitals and almost 70% of these patients suffer from unstable angina or NSTEMI. In patients with NSTEMI, assessment of clinical symptoms and measurement of cardiac biomarkers are critical for proper diagnosis [2]. In addition, myocardial infarction (MI) is responsible for approximately 500,000 deaths per year in the United States. The classical guideline for the diagnosis of MI recommends two of the following three criteria for diagnosis:

- Typical symptoms
- Characteristic rise-and-fall pattern of a cardiac marker
- A typical electrocardiogram (ECG) pattern involving the development of Q waves

However, the classical guidelines have been updated with 2018 publication of fourth universal definition of myocardial infraction being the latest where cardiac troponins are considered preferred biomarkers for the diagnosis of MI or myocardial injury.

**149**

Clinical Chemistry, Immunology and Laboratory Quality Control. https://doi.org/10.1016/B978-0-12-815960-6.00008-X

## Overview of cardiac markers

In broad sense, cardiac markers are endogenous substances that are released in the circulation when the heart is damaged or stressed. Accurate coronary syndrome is caused by rupture of a plaque formed due to atherosclerosis causing thrombus formation in the damaged coronary artery that results in sudden decrease in the amount of blood and oxygen reaching the heart. Angina is due to reduced blood supply in the heart and when such blood flow is interrupted for 30–60 min, it can cause necrosis of heart muscle resulting in MI. Cardiac biomarkers are released in the circulation due to damage or death of cardiac myocytes and measuring these biomarkers in serum or plasma is useful in the diagnosis of MI. There are four established biomarkers for myocardial necrosis:

- Myoglobin
- Creatine kinase isoenzyme (CK-MB)
- Cardiac troponin I
- Cardiac troponin T

Creatine kinase (CK) is an enzyme (and also a protein) which is often referred to as a cardiac enzyme. However, it is important to note that troponin I and troponin T are also proteins but not enzymes. Out of all these biomarkers, troponin I is the most specific for myocardial necrosis. Currently, troponin I and troponin T (both conventional and high sensitivity) are most frequently used for the diagnosis of AMI. Creatine kinase isoenzyme (CK-MB) was used more frequently in the past but are less frequently used these days. Myoglobin is rarely used as a marker of AMI. Characteristics of established cardiac markers are listed in Table 1.

**Table 1** Various cardiac markers.

| Cardiac biomarker | Increases | Peak | Return to baseline | Comments |
|---|---|---|---|---|
| Myoglobin | 1–4h | 4–12h | 24–36h | Nonspecific old cardiac biomarker, no longer used |
| CK-MB | 4–9h | 24h | 48–72h | Widely used biomarker before introduction of cardiac troponins. Now its application is limited |
| Cardiac troponin I | 4–9h, conventional troponin I assay 3–4h, high-sensitivity troponin I assay | 12–24h | 10–14 days | Cardiac troponin I is the most specific marker of myocardial injury |
| Cardiac troponin T | 4–9h, conventional troponin T assay 3–4h, high-sensitivity troponin T assay | 12–24h | 10–14 days | Troponin T is slightly less specific than troponin I because troponin T is also elevated in chronic renal failure |

There are also biomarkers of myocardial ischemia. These biomarkers include:

- Ischemic modified albumin
- Heart-type fatty acid-binding protein
- Glycogen phosphorylase BB

Biomarkers of hemodynamic stress of the heart include:

- BNP (B-type natriuretic peptide) and NT-proBNP (N-terminal proBNP)
- Atrial natriuretic peptide (ANP)

Inflammatory and prognostic markers include:

- C-reactive protein
- Homocysteine
- Myeloperoxidase
- sCD40L

Potential biomarker of atherosclerotic plaque instability:

- Pregnancy-associated plasma protein A

Moreover, lipid parameters are useful in determining risk factors for cardiovascular diseases. Please see Chapter 6 for in-depth discussion on this topic.

Historically, aspartate transaminase (AST) was the first cardiac biomarker extensively used in 1960s. However, AST is not specific for cardiac muscle. Therefore, in 1970s, two biomarkers, lactate dehydrogenase (LDH) and CK, were introduced. Although neither biomarkers are specific for cardiac muscle, CK is more specific than LDH in the diagnosis of AMI. Eventually, advances in electrophoresis methods, cardiac-specific isoenzymes of CK known as CK-MB, as well as cardiac-specific isoenzymes of LDH (LDH 1 and 2) gained acceptance and were included as one of the diagnostic criteria to rule out AMI by the World Health Organization (WHO). Myoglobin test in serum was developed in 1978 and myoglobin became a useful biomarker for the early detection of AMI. However, myoglobin test is rarely recommended in current practice due to the availability of more specific markers such as troponin I and troponin T [3].

Currently, cardiac markers are used in the diagnosis and risk stratification of patients with chest pain and suspected acute coronary syndrome. In this context, cardiac troponin especially troponin I is particularly useful in the diagnosis of myocardial infection due to its superior specificity over other cardiac biomarkers [4]. Cardiac markers are not necessary for the diagnosis of patients who present with ischemic chest pain and diagnostic ECGs with ST-segment elevation. These patients may be candidates for thrombolytic therapy or primary angioplasty. Treatment should not be delayed to wait for cardiac marker

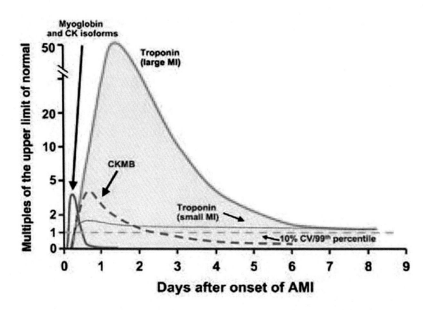

**FIG. 1**

Timing of release of various cardiac biomarkers after myocardial injury. *Adapted from Anderson JL, Adams CD, Antman EM, Bridges CR, et al. ACC/AHA 2007 guidelines for the management of patients with unstable angina/non ST-elevation myocardial infarction: a report of the American College of Cardiology/American Heart Association Task Force on Practice Guidelines. Circulation 2007;116:e148–e304. ©American Heart Association. Reprinted with permission.*

results, especially since the sensitivity is low to moderate in the first 3–6 h after onset of initial symptoms. Timing of release of various cardiac markers in circulation after MI is shown in Fig. 1.

## Myoglobin

Myoglobin is a heme protein found in both skeletal and cardiac muscle. Myoglobin is typically released in the circulation as early as 1 h after MI, with gradual increase reaching a peak at 4–12 h, and returns to normal within 24–36 h. Rapid release of myoglobin probably reflects its low molecular (17 KDa) weight and cytoplasmic location. Although myoglobin was extensively used in the past, currently myoglobin is rarely used for the diagnosis of AMI because of its poor clinical specificity (60%–90%) due to the presence of large quantities of myoglobin in skeletal muscle. Myoglobin being a small protein is excreted in urine. In the absence of AMI, myoglobin concentration is also elevated due to renal failure as well rhabdomyolysis [5]. Woo et al. reported that in a group of 42 patients, 22 patients were later diagnosed to have myocardial infection, but in 11 patients who did not have MI, myoglobin concentrations were falsely elevated and these patients had various degree of muscular trauma [6]. Myoglobin is no longer mentioned in the European guidelines for the management of acute coronary syndromes in patients presenting without persistent ST-segment elevation. As a result, some clinical laboratories no longer offer this test [7].

## CK isoenzyme: CK-MB

Prior to the introduction of cardiac troponin T or cardiac troponin I, the biochemical marker of choice for the diagnosis of AMI was CK isoenzyme. CK is a dimeric enzyme that catalyzes reversible transformation of creatine and ATP into creatine phosphate and ADP. Therefore CK is also called creatine phosphokinase. CK is found in many tissues that consume large amounts of energy, including myocardium and skeletal muscle. CK has two subunits: M-type (for muscle) and B-type (for brain). The CK-MM isoenzyme is dominant in adult skeletal muscle (98% of total muscle CK) while CK-BB is found in the brain, kidney, and gastrointestinal tract. In myocardium, CK-MB is present in a relatively high concentration (20%–30% of the total myocardial CK) while the rest is CK-MM. However, in addition to myocardium, small amounts of CK-MB are found in skeletal muscle, small intestine, diaphragm, uterus, tongue, and prostate. Molecular weight of CK-MB is roughly 86,000 and during myocardial injury, CK-MB first rises, then reach a peak level and returns to normal usually according to the following pattern [8].

- CK-MB concentration gradually rises 4–9 h after onset of chest pain.
- Peak concentration is reached in approximately 24 h after MI.
- Concentration of CK-MB begins to decline reaching baseline level in 48–72 h.
- However, due to high molecular weight, CK-MB cannot detect minor myocardial damage.

The criterion most commonly used for the diagnosis of AMI using CK-MB is two serial elevations above the diagnostic cut-off level or a single result more than twice the upper limit of normal. Although CK-MB is more concentrated in the myocardium, it is also present in skeletal muscle, and false-positive elevations occur in a number of clinical settings, including trauma, heavy exertion, and myopathy. Elevation of the total CK level is not cardiac-specific and may be observed in patients with skeletal muscle injury and other disorders.

Because CK-MB remains elevated for longer period of time following MI, it is useful to detect reinfarction using serial CK-MB measurement. Following myocardial or skeletal muscle injury, both the total CK and CK-MB levels gradually increase. To differentiate from cardiac and skeletal muscle as the source of elevation of CK-MB, calculation of relative index can be used:

$$CK-MB\,index = \frac{CK-MB\,(ng/mL)}{Total\,CK\,(U/L)} \times 100$$

It is important to note that in calculation of index, CK-MB concentration is expressed in ng/mL while the total CK level is expressed in U/L, and in strict mathematical sense, ratio can only be calculated when both numerator and denominator is expressed in the same unit. However, when this index was originally proposed, CK-MB was measured as activity (U/L) just like the total CK but due to interferences in CK-MB activity assay, more specific assays for CK-MB were developed using two different antibodies and such specific assays measure CK-MB mass. Nevertheless, this approach is useful and an index of greater than 2.5 is indicative of some myocardial damage, while an index greater than 5% is suggestive of AMI. An exception may occur with patients with chronic myopathic disorders in which the skeletal muscle CK-MB content may be increased. Chronic myopathic disorder usually produces a persistent elevation in the CK-MB level as opposed to the typical rise-and-fall temporal pattern seen in patients with AMI.

The CK-MB level also may be elevated in patients with nonischemic cardiac injury, including myocarditis and in noncardiac diseases such as seizure, pulmonary embolism, and skeletal muscle trauma. Marathon and long distance runners usually have elevated CK-MB. In addition, alcohol and various drug abuse including opiates may increase CK-MB level. Various common causes of elevated CK-MB other than due to MI are summarized in Table 2. False-negative CK-MB is rarely encountered in clinical situation and mostly likely due to time when specimen was collected (too soon after infarction) or when the infarction was small causing only intranormal bump [9].

After passing into blood, CK-MB is divided into two groups; CK-MB1 and CK-MB2. Laboratory determination of CK-MB actually represents the simple sum of the two isoforms. Macro-CKs are high molecular weight enzyme-macromolecule complex with prolonged half-life that results in falsely increased CK levels in serum. There are two types of macro-CK: macro-CK type 1 where one of the CK isoenzyme is complexed with immunoglobulin and macro-CK type 2 which are made up of mitochondrial-derived CK polymers. While macro-CK type 1 may be found also in normal population and may not have any pathological significance, macro-CK type 2 may be associated with neoplasm and liver cirrhosis. Lee et al. reported that prevalence of macro-CK type 1 is only 0.43% and while prevalence of macro-CK-type 2 may be up to 1.2% and usually associated with autoimmune disease or

Table 2 Various common causes of elevated CK-MB other than myocardial infarction.

| Disease/cause | Comments |
| --- | --- |
| Myocarditis | During active inflammation, CK-MB may be significantly elevated but not always |
| Myositis | Elevated CK-MB, may have cardiac involvement |
| Cardiac surgery | May increase both total CK and CK-MB concentration |
| Muscular dystrophy | Elevated CK-MB may be found |
| Muscle trauma | Both total CK and CK-MB elevated |
| Pulmonary embolism | Elevated CK-MB usually observed |
| Rhabdomyolysis | Increases both total CK and CK-MB |
| Hypothermia | May cause elevation of CK-MB level due to possible myocardial damage |
| Hypothyroidism | Elevated level due to reduced clearance |
| Seizure | Elevated CK-MB may be related to cardiac involvement |
| Long distance runner | Elevated CK-MB probably due to muscle damage |
| Renal failure | Mechanism of elevated CK-MB is not clear |
| Alcohol overdose | Toxic effect of excess alcohol to muscle and myocardium |
| Opiate overdose | Morphine and other opiate overdose may increase both total CK and CK-MB |

malignancy [10]. Aljuani et al. reported five cases over a 10-year period where macro-CK was identified. Three patients had macro-CK type 1: one patient with fibromyalgia, another patient with elevated CK-MB but cardiologist ruled out cardiomyopathy and the third patient had osteoarthritis. In addition, two patients had macro-CK type 2: a man with a neuroendocrine carcinoma and a woman with rheumatoid arthritis. The authors used CK electrophoresis to identify macro-CK. Macro-CK type 1 are complexes of most commonly CK-BB with immunoglobulins (most common IgG). The authors also commented that due to the presence of macro-CK, total CK activity is usually lower than 500 U/L. CK-MB activity levels can be falsely elevated by the presence of macro-CK especially if immune-inhibition assays are used in the measurement. When CK-MB level is higher than total CK, it may be an indication of the presence of macro-CK [11]. Macro-CK interference can be resolved by CK-electrophoresis. In addition, polyethylene glycol precipitation may be helpful in resolving macro-CK issue, but it may not work for all specimens.

## Troponin I and troponin T

Troponin, a three-piece protein complex, is a component of the contractile apparatus within skeletal and cardiac myocytes. Troponin proteins along with calcium ions regulate and facilitate the interaction between actin and myosin filaments for muscle contraction. Both cardiac troponin T and troponin I are derived from genes specific to myocardium, but troponin C is present in both cardiac and skeletal muscle with no specific isoform for myocardium. The three subunits are designated as follow:

- Troponin C (the calcium-binding component; molecular weight 18.4 KDa)
- Troponin T (the tropomyosin-binding component; molecular weight 36 KDa)
- Troponin I (inhibitory unit blocking actin-myosin interaction; molecular weight 23.8 KDa)

Although troponin C is synthesized in both cardiac and skeletal muscle, both troponin I and T are highly specific to cardiac myocytes, and as a result are also called cardiac troponins. Majority of both cardiac troponin T and I (92%–95%) are attached to the actin thin filament in the cardiac sarcomere and a small amount (5%–8%) is found in the myocyte cytoplasm (free troponins). Following myocardial damage, cytosolic-free (unbound) troponin is released first and as further damage occurs, complexes of cardiac troponin T, I, and C (T-I-C or ternary complex) and complexes of cardiac troponin I and C (I-C or binary complex) present in sarcomere are released into circulation, making troponin a both early and late marker of AMI. Kinetics of release of both troponin T and troponin I after myocardial damage are as follows:

- Levels of both troponin T and I start increasing 3–4 h after AMI if measured by 5th generation high-sensitivity troponin assays or 4–9 h if measured by conventional (4th generation) assays.
- Peak at 12–24 h.
- May remain elevated 10–14 days.

The half-life of cardiac troponins is approximately 2 h in serum. Cardiac troponin T and I have many advantages over CK-MB. Small elevation of troponin but normal CK-MB level may indicate microscopic zone of myocardial narcosis (microinfarction). Although cardiac troponin I is not elevated due to injury of noncardiac tissues, the situation is more complex for cardiac troponin T because injured skeletal muscle expresses proteins that are detected by the cardiac troponin T assay, leading to some situations where elevations of cardiac troponin T could originate from skeletal muscle injury. Moreover, in humans, cardiac troponin T isoform expression has been reported in patients with muscular dystrophy, polymyositis, dermatomyositis, and end-stage renal disease. In general, cardiac troponin T levels may be elevated in patients with chronic renal failure without any acute myocardial injury, which is a major limitation of troponin T [12].

The European Society of Cardiology, the American College of Cardiology, the American Heart Association, and the World Heart Federation have issued the fourth universal definition of MI. It is now a standard practice to first categorize MI based on ECG findings as STEMI versus NSTEMI, and then further distinguish MI into five groups. According to these guidelines, cardiac troponin I and cardiac troponin T are the preferred biomarkers for the diagnosis of myocardial injury, and high-sensitivity cardiac troponin I or T assays are recommended for routine clinical use. Other biomarkers, such as CK-MB isoform, are less sensitive and less specific. Myocardial injury is defined as being present when blood levels of cardiac troponins are increased above the 99th percentile upper reference limit. The injury may be acute, as evidenced by a newly detected dynamic rising and/or falling pattern of cardiac troponin values above the 99th percentile upper reference limit, or chronic in the setting of persistently elevated cardiac troponin levels, which clearly separates MI from myocardial injury. The five groups of MI include:

- Type 1 MI is the classical type due to acute coronary atherothrombotic myocardial injury with plaque rupture or erosion or dissection leading to intraluminal thrombus in one or more artery causing ischemia and resultant myocyte necrosis. Most STEMI and NSTEMI patients fit into this category. Detection of a rise and/or fall of cardiac troponin values with at least one value above the 99th percentile upper reference limit is one criterion of diagnosis. Any other symptoms including evidence of myocardial ischemia, development of pathological Q waves or new ischemia-related ECG changes or imaging studies or identification of a coronary thrombus by angiography including intracoronary imaging or during autopsy may confirm such diagnosis.
- Type 2 MI includes patients who do not have acute coronary atherothrombotic injury but instead have oxygen supply-demand imbalance most likely due to coronary artery disease. Most patients present with NSTEMI. Detection of a rise and/or fall of cardiac troponin values with at least one value above the 99th percentile upper reference limit along with other symptoms may confirm this diagnosis.
- Type 3 MI is secondary to sudden cardiac death where cardiac biomarker values are unavailable. Patients can manifest a typical presentation of myocardial ischemia/infarction, including presumed new ischemic ECG changes or ventricular fibrillation, and die before it is possible to obtain blood for cardiac biomarker determination or the patient may succumb soon after the onset of symptoms before an elevation of biomarker values has occurred.
- Type 4 MI are related to coronary procedural events such as percutaneous coronary intervention (type 4a), stent thrombosis (type 4b MI), and

postprocedural restenosis (type 4c MI). Diagnosis is based on the same criteria for type 1 MI diagnosis.

- Type 5 MI is related to coronary artery bypass grafting. It is arbitrarily defined as elevation of cardiac troponin values >10 times the 99th percentile upper reference limit in patients with normal baseline cardiac troponin values. In patients with elevated preprocedure cardiac troponin values, the postprocedure cardiac troponin values must rise by >20%. However, the absolute postprocedural value still must be >10 times the 99th percentile reference limit. In addition, development of new pathological Q waves or other criteria must be present.

The fourth Universal Definition of Myocardial Infarction document also has a new section on takotsubo syndrome (type of nonischemic cardiomyopathy in which there is a sudden temporary weakening of the muscular portion of the heart) is also included. Takotsubo syndrome can mimic MI and is found in ~1%–2% of patients presenting with suspected STEMI. The onset of takotsubo syndrome is often triggered by intense emotional or physical stresses such as bereavement. Over 90% of patients are postmenopausal women [13].

## High-sensitivity troponin assays

Detection of increased or decrease of at least one value of troponin above 99th percentile of the upper reference limit is an essential component of diagnosis of AMI.

Many conventional 4th generation troponin immunoassays do not achieve desirable less than 10% CV at the 99th percentile upper reference limit due to poor sensitivity and usually cannot detect troponin I at lower than 0.02 ng/mL concentrations. However, high-sensitivity troponin assays are capable of detecting 10- to 100-fold lower concentrations of troponin. For example, Abbott high-sensitivity troponin I assay has a lower limit of detection of 1.9 ng/L (0.0019 ng/mL). The high-sensitivity troponin assays should be able to detect troponin levels below the 99th percentile cut-off values in at least 50% of normal population. Moreover, CVs are less than 10% at 99th percentile cut-off for all high-sensitivity troponin assays. Currently, only one high-sensitivity troponin T assay is available from Roche Diagnostics (cut off: 5.0 ng/L or 0.005 ng/mL) and several high-sensitivity troponin I assays are commercially available (cut off: 2.5 ng/L; Beckman Coulter, 0.8 ng/L, Siemens etc.), it is important to note that 99th percentile upper reference limit is higher in males than females, for example, the limit is 15.6 pg/mL for female and 34.2 pg/ml for male using Abbott's high sensitivity troponin I assay.

## Clinical application of high-sensitivity troponin assays

Because of the higher sensitivity, it may be possible to early rule in or rule out AMI in patients complaining about chest pain. Reichlin et al. developed the 2-h algorithm based on high-sensitive troponin T values and other criteria where 60% of patients could be ruled out and 16% patients could be ruled in for MI while 24% patients were placed in the "observational-zone" [14]. In another study, the authors developed a simple algorithm incorporating high-sensitivity troponin I baseline values and absolute 2-h changes allowing a triage toward safe rule-out or accurate rule-in of AMI in the majority of patients [15].

## Elevated troponin levels in conditions other than myocardial infarction

Clinical conditions other than myocardial infraction may also increase serum or plasma levels of troponin. These conditions include sepsis, cerebrovascular accidents, cardiotoxic medications, renal failure, acute respiratory failure, severe burn (over 30% of body's surface area), carbon monoxide exposure, cardiomyopathy, heart failure, myocarditis, myocardial contusion, rhabdomyolysis, pulmonary embolism, and sequelae of malignancies. Among these conditions, most studied are sepsis, neurogenic diseases such as stroke, subarachnoid hemorrhage, and cardiotoxic medicines such as anthracyclines and trastuzumab [16]. Typical angina pectoris and high-sensitive troponin I are independently associated with coronary artery disease and future cardiovascular events. However, in clinically stable angina pectoris patients without known cardiovascular disease, a thorough chest pain history in combination with high-sensitivity troponin I testing can identify a significant low-risk group. A number of studies have demonstrated that cardiac troponin T can be used for risk stratification of patients with chronic renal failure without ischemia [17]. However, significant rise and then fall of troponin T in patients with chronic renal disease may still can be used for the diagnosis of MI. Dialysis does not affect cardiac troponin T or cardiac troponin I levels. Recently, elevated cardiac troponin levels are reported in patients with COVID-19 (severe acute respiratory syndrome coronavirus-2, SARS-CoV-2) infection. In the setting of COVID-19, myocardial injury, defined by an increased troponin level, occurs especially due to nonischemic myocardial processes, including severe respiratory infection with hypoxia, sepsis, systemic inflammation, pulmonary thrombosis and embolism, cardiac adrenergic hyperstimulation during cytokine storm syndrome, and myocarditis [18]. Elevation of cardiac troponin in conditions other than AMI are listed in Table 3.

**Table 3** Various common causes of elevated troponin other than myocardial infarction.

| Cardiac causes | Other causes |
|---|---|
| • Angina pectoris<br>• Cardiomyopathy<br>• Cardiac contusion<br>• Congestive heart failure<br>• Cardiac surgery and cardiac procedures<br>• COVID-19 infection (may damage heart)<br>• Heart block<br>• Heart failure (both acute and chronic)<br>• Myocarditis<br>• Myocardial trauma<br>• Tachycardia/tachyarrhythmia | • Acute pulmonary edema<br>• Acute pulmonary embolism<br>• Amyloidosis<br>• Carbon monoxide exposure<br>• Chronic obstructive pulmonary disease (patients with cardiac troponin I or T may have higher mortality)<br>• Cardiotoxic medicines<br>• Chronic renal failure (most commonly affect troponin T)<br>• Pulmonary embolism<br>• Pulmonary hypertension<br>• Rhabdomyolysis<br>• Severe burn (>30% body's surface area)<br>• Sepsis<br>• Stroke, subarachnoid hemorrhage |

## Issues of interferences

In addition, many interferences have been reported in troponin I assays including endogenous substances such as hemolysis and high bilirubin. However, more commonly, heterophilic antibodies, rheumatoid factors, and macrotroponin are known to interfere with cardiac troponin immunoassays (Table 4). Macrotroponin I, a troponin-immunoglobulin (IgG) complex, should be considered when troponin I values are inconsistent with the clinical picture and fail to demonstrate a rise and/or fall pattern in suspected cases of AMI. Macrotroponin T, a complex with troponin T and immunoglobulin causing falsely elevated troponin T value has also been reported. In one study, the authors identified macrotroponin I in 5% of patients with elevated high-sensitivity troponin I values with the potential to be clinically misleading [19]. Falsely elevated and misleading troponin T values due to the presence of macrotroponin T has been reported. Heterophilic antibody interference in high-sensitivity troponin T has also been recently reported [20]. Biotin if present in high concentrations may cause negative interference with troponin measured in immunoassays that are biotin-based. In one study, the authors reported that daily biotin ingestion (10 mg) resulted in negative interference in 5th generation high-sensitivity troponin T assay yielding nondetected values [21]. The FDA reported that one patient taking high levels of biotin died from falsely low troponin results when a troponin test known to have biotin interference was used. However, the FDA did not specify which diagnostic company marketed the troponin assay. Please see Chapter 22 for more detail on biotin interferences in various immunoassays including troponin assays.

| Table 4 Issues of interferences in troponin assays. | |
|---|---|
| **Interfering agent** | **Comments** |
| Heterophilic antibody including human antimouse antibody (HAMA) | Falsely increased troponin I or troponin T values depending on the assay design. May also produce discordant results between different assays. High-sensitivity troponin assays are also affected |
| Rheumatoid factor | Falsely increases both troponin I and T values and magnitude of interference is assay-dependent |
| Autoantibodies | Cardiac troponin I autoantibodies to the stable portion of troponin I near the carboxyterminal end of the molecule may cause **false-negative troponin I result** |
| Macrotroponin | Macrotroponin is either a troponin I complexing with immunoglobulin or troponin T complexing with immunoglobulin which falsely increases troponin value |
| Interference from high bilirubin, gross hemolysis | **False-negative results** have been attributed to bilirubin, hemoglobin in serum due to gross hemolysis |
| Fibrin clot | May cause false-positive test result with troponin I |
| Biotin | Biotin if present in high concentration **falsely lower troponin T or troponin I** level if assay design utilizes biotinylated antibody and streptavidin |

## LDH: An old cardiac marker

LDH isoenzymes was used widely in the past for the diagnosis of MI but more recently due to availability of troponin immunoassays LDH isoenzyme assay has been mostly discontinued in clinical setting for the diagnosis of MI but may be used in evaluating certain hepatic disorder. Briefly, LDH exists in five isoenzyme forms:

- LDH-1: Present primarily in cardiac myocytes and erythrocytes
- LDH-2: Present mostly in white blood cells
- LDH-3: Present in highest quantity in lung tissue
- LDH-4: Highest amounts found in pancreas, kidney, and placenta
- LDH-5: Highest amounts found in liver and skeletal muscle

Usually, LDH isoenzymes increases 24–72 h following MI reaching a peak concentration in 3–4 days and remains elevated for 8–14 days making it a late marker for MI. Normally, concentration of LDH-1 is lower than LDH 2 but after MI, LDH-1 concentration becomes elevated exceeding concentration of LDH-2 and this phenomenon is called flipped LDH pattern. However, hemolysis (LDH is present in erythrocytes in similar concentration) produces this characteristic flip and it is important to ensure that specimen is not hemolyzed prior to analysis. Moreover, LDH is a nonspecific marker for MI and its concentration can be elevated in hemolytic anemia, stroke, pancreatitis, ischemic cardiomyopathy, and variety of other diseases.

## Biomarkers of myocardial ischemia

Biomarkers of myocardial ischemia include ischemic-modified albumin, heart-type fatty acid-binding protein (H-FABP) and glycogen phosphorylase.

Early diagnosis of impaired myocardial perfusion before the occurrence of necrosis in patients presenting to the emergency department with acute chest pain is difficult. Ischemia-modified albumin, a form of albumin displaying reduced cobalt-binding affinity, is significantly elevated within minutes after myocardial ischemia and returns to baseline in 6 h after restoring perfusion. The ischemic-modified albumin level in serum can be measured by the commercially available and FDA-approved albumin-cobalt-binding (ACB) assay where cobalt(II) chloride (approximately 1.5 mol equivalents per albumin molecule) is added to a serum sample to allow ACB. Dithiothreitol (DTT), a metal chelator that forms a colored complex with cobalt, is then added. The resulting brown DTT-cobalt product is measured by absorption spectro-photometry at 470 nm and compared to a serum-cobalt blank without DTT present. The reduced cobalt-binding capacity of ischemic-modified albumin leaves more unbound cobalt ion to complex with DTT, resulting in higher absorbance readings. The ACB test has an excellent negative predictive value, i.e., low ischemic-modified albumin readings correspond well to the absence of myocardial ischemia. However, a severe shortcoming is the high incidence of false-positives results in many disease states including chronic liver and kidney diseases, infectious diseases such as malaria, and pregnancy-related conditions such as preeclampsia. In addition, elevated ischemic-modified albumin levels have been measured in metabolic syndrome, diabetes, and obesity [22].

H-FABP is a type of intracardiac protein that plays an essential part in the metabolism of fatty acid inside cardiomyocytes. H-FABP is a biomarker for immediate myocardial injury and even for relatively long-term postischemic prognosis such as ischemic reperfusion injury. Although fatty acid can bind to albumin, there are also fatty acid-binding proteins that are responsible for the transportation of fatty acids and lipophilic materials into or out of cells. There are several types of tissue-specific fatty acid-binding proteins such as liver-type, intestinal-type, heart-type, adipocyte-type, epidermal-type, ileal-type, brain-type, myelin-type, and testis-type fatty acid-binding proteins. As expected H-FABP is mainly found in the heart and has a molecular weight of 15 kDa. Following ischemic injury to the myocardium, H-FABP is rapidly released into the circulatory system and can be detected within 1 h after injury. H-FABP is eliminated by the kidney [23]. Immunoassays are available for the measurement of H-FABP in clinical laboratories. One such assay, the Biocheck ELISA (Biocheck, Foster City, CA) assay uses a monoclonal anti-H-FABP antibody. It provides a minimum detection limit of 0.25 ng/mL.

Glycogen phosphorylase is an essential enzyme in the regulation of glycogen metabolism where this enzyme converts glycogen into glucose 1-phosphate in the first step of glycogenolysis. Three different isoenzymes have been identified in humans, glycogen phosphorylase BB (brain and heart muscle), glycogen phosphorylase LL (liver), and glycogen phosphorylase MM (muscle). Glycogen phosphorylase BB is released in circulation 2–4 h after the onset of cardiac ischemia and returns to baseline 1–2 days after AMI, making it an early marker. However, more research is needed to fully evaluate this marker for early diagnosis of cardiac ischemia.

## Biomarkers of hemodynamic stress of the heart

The natriuretic peptide family consists of three biologically active peptides: ANP, brain (or B-type) natriuretic peptide (BNP), and C-type natriuretic peptide. Among these, ANP and BNP are secreted by the heart and act as cardiac hormones. Both ANP and BNP preferentially bind to natriuretic peptide receptor-A (NPR-A or guanylyl cyclase-A) and exert similar effects through increases in intracellular cyclic guanosine monophosphate within target tissues. Human ANP has three molecular forms, α-ANP, β-ANP, and proANP (or γ-ANP) with proANP predominating in healthy atrial tissue. However, after secretion, proANP is proteolytically cleaved into bioactive α-ANP which is the major circulating fragment. ProANP and β-ANP are minor forms in the circulation but are increased in patients with heart failure. BNP and NT-proBNP are the most commonly used biomarkers for evaluating hemodynamic stress of the heart [24].

BNP is a hormone secreted primarily by the ventricular myocardium in response to wall stress such as volume expansion and pressure overload. BNP has shown promises as a diagnostic marker of congestive heart failure. In addition, multiple studies have demonstrated that BNP may also be a useful prognostic indicator for myocardial stress and correlated with long-term cardiovascular mortality in patients with AMI. Studies have shown that the BNP level predicted cardiac mortality and other adverse cardiac events across the entire spectrum of acute coronary syndrome. The mortality rate is nearly doubled when both cardiac troponin I and BNP levels were elevated. In addition, BNP level is also a good predictor of left ventricular ejection fraction and heart failure in these patients.

BNP is initially synthesized as prepro BNP which contains 134 amino acids but it is cleaved into proBNP containing 108 amino acids. On secretion, it splits into biologically active BNP (amino acids 77–108) and the remaining N-terminal proBNP (NT-pro BNP: 1–76 amino acids) which is biologically inactive. BNP is a smaller molecule than NT-proBNP and it is cleared from circulation earlier than NT-proBNP. Therefore, concentration of NT-proBNP in serum or plasma is higher than BNP. In addition, NT-proBNP is more stable in serum or plasma than BNP. Although studies have demonstrated that both BNP and NT-proBNP have similar effectiveness as biomarkers, some authors tend to favor NT-proBNP as a slightly superior biomarker than BNP. Assays are available (both laboratory-based assay and point of care device) for the analysis of both BNP and NT-proBNP and in general both BNP and NT-proBNP follow similar pattern after heart failure but quantitative BNP value does not match with quantitative proBNP value. Major points to remember regarding BNP and NT-proBNP include:

- Major application of both BNP and proBNP testing is evaluating patients with congestive heart failure. Although a single determination may be helpful in diagnosis, multiple determinations may provide more useful information. If heart failure responds to therapy, concentration of BNP and NT-proBNP should decline indicating progress of therapy. If a patient does not respond, values may be increased gradually.
- In general, NT-proBNP is more stable (up to 7 days at room temperature and up to 4 months if stored at −20°C) than BNP which is not stable even for a day even if specimen is stored in a refrigerator. Therefore BNP analysis must be performed as soon as possible after collecting specimen.
- Cut-off level of BNP and NT-proBNP depends on age as values tend to increase with advancing age. In general, heart failure is unlikely if BNP value is less than 100 pg/mL and heart failure is very likely if the value is over 500 pg/mL. For NT-proBNP, normal value for a person 50 years or younger is usually 125 pg/mL but heart failure is unlikely if NT-proBNP value is <300 pg/mL. However, heart failure is likely if the value is >450 pg/mL (>900 pg/mL in a patient of age 50 and above) [25].
- Patients with end-stage renal disease and dialysis patients usually show higher BNP and NT-proBNP in serum than normal individuals.

Interestingly, the blood levels of BNP and NT-proBNP also reflect the severity of congestive heart failure in children. In particular, NT-proBNP is a useful biomarker for evaluating congestive heart failure in children [26].

## Inflammatory and prognostic biomarkers

Commonly monitored inflammatory and prognostic biomarkers are C-reactive protein (CRP), homocysteine, and myeloperoxidase (MPO).

CRP, a nonspecific marker of inflammation, is considered to be directly involved in coronary plaque atherogenesis. Studies show that an elevated

CRP level independently predicted adverse cardiac events at the primary and secondary prevention levels. Data indicate that CRP is a useful prognostic indicator in patients with acute coronary syndrome, as elevated CRP levels are independent predictors of cardiac death, AMI as well as congestive heart failure. In combination with cardiac troponin I and BNP, CRP may be a useful adjunct, but its nonspecific nature limits its use as a diagnostic cardiac marker for acute coronary syndrome in patients presenting at emergency department. However, for assessing cardiovascular risk factor high-sensitivity CRP assay (analytical measurement range 0.1–100 mg/L) must be used instead of conventional assays which are used for measuring CRP as a marker of inflammation with much lower sensitivity.

Homocysteine is a thiol-containing amino acid intermediate formed during methionine metabolism. Markedly elevated homocysteine levels are associated with increased risk for cardiovascular disease. Please see Chapter 6 for more detail on CRP and homocysteine.

MPO is an enzyme stored in azurophilic granules of polymorphonuclear neutrophils and macrophages and released into extracellular fluid in the setting of inflammatory process. Initial studies showed significantly increased MPO levels in patients with angiographically documented coronary artery disease. In patients presenting to the emergency department with chest pain, elevated MPO levels independently predicted increased risk for major adverse cardiac events, including MI, reinfarction, need for revascularization, or death at 30 days and at 6 months. Among the patients who presented to the emergency department with chest pain but who were ultimately ruled out for MI, an elevated MPO level at presentation predicted subsequent major adverse cardiovascular outcomes. MPO may be a useful early marker in the emergency department based on its ability to detect plaque vulnerability that precedes acute coronary syndrome [27]. An enzymatic assay is available for measuring MPO levels in serum or plasma.

CD40 ligand (CD40L) and its soluble form (sCD40L) are proteins of the tumor necrosis factor superfamily that exhibit prothrombotic and proinflammatory properties when bind to CD40. CD40L is expressed mainly by platelets and activated T-lymphocytes but also expressed in variety of cells. CD40L could be cleaved and soluble form (sCD40L) with lower molecular weight is released in blood circulation. High circulating levels of sCD40L have been associated with poor prognosis in patients with acute coronary syndrome, stroke, and sepsis. However, elevated levels of sCD40L have also reported in chronic hepatitis C infection, nonalcoholic fatty liver disease, liver cirrhosis, and hepatocellular carcinoma. Enzyme-linked immunosorbent assay (ELISA) assay is available for measuring serum level of sCD40L.

## Pregnancy-associated plasma protein A

Pregnancy-associated plasma protein A (PAPP-A) was first isolated in 1974 from the plasma of a pregnant woman. It has been shown that women with low blood levels of PAPP-A at 8–14 weeks of gestation have an increased risk of intrauterine growth restriction, Down syndrome, premature delivery, pre-eclampsia, and stillbirth. However, PAPP-A has also been shown to have abundant expression in unstable but not stable atherosclerotic plaques, and elevated serum PAPP-A levels are observed in patients with both MI and unstable angina. This may mean that PAPP-A helps to identify soft, inflammatory plaques that are vulnerable to rupture or erosion and is a potential biomarker of atherosclerotic plaque instability. However, PAPP-A serum levels do not correlate with those of troponins, which may mean that PAPP-A helps to identify patients with unstable coronary disease without infarction. Therefore PAPP-A may be insufficiently sensitive for the detection of MI at presentation to the emergency department. However, it is a promising marker for the detection of unstable coronary disease and may have tremendous potential as a prognostic marker, especially if incorporated into a multimarker strategy. ELISA is available for its measurement [28].

## Summary/key points

- There are four established biomarkers for myocardial necrosis: myoglobin, creatine kinase isoenzymes, cardiac troponin I, and cardiac troponin T. However, myoglobin, an old marker is rarely measured today in clinical laboratories.
- Myoglobin is a heme protein found in both skeletal and cardiac muscle. Myoglobin is typically released in the circulation as early as 1 h after myocardial infarction, with gradual increase reaching a peak at 4–12 h, and returns to normal within 24–36 h. Rapid release of myoglobin probably reflects its low molecular (17 KDa) weight and its cytoplasmic location. Myoglobin is an early marker of acute myocardial infarction and exhibits a high negative predictive value. Myoglobin has poor clinical specificity (60%–90%) because it is also found in large quantities in skeletal muscle. Myoglobin being a small protein is excreted in urine and high level of myoglobin is encountered in patient with acute renal failure or uremic syndrome.
- Creatine kinase is an enzyme and is it is often called cardiac enzyme. Creatine kinase has two subunits including M-type (for muscle) and B-type (for brain). The CK-MM isoenzyme is dominant in adult skeletal muscle (98% of total muscle CK) while CK-BB is found mostly in central nervous system. In myocardium, CK-MB is present in a relatively high

concentration (15%–20% of the total myocardial CK) while about 85% is CK-MM). Therefore, CK-MM is the most abundant isoenzyme of creatine kinase.

- CK-MB concentration gradually rises from 4 to 9 after the onset of chest pain with a peak concentration reaching in approximately 24 h. Concentration of CK-MB begins to decline reaching baseline level in 48–72 h. Because CK-MB remains elevated for longer period of time following myocardial infarction, it is useful to detect reinfarction using serial CK-MB measurement.

- Although CK-MB is more concentrated in the myocardium, it also present in skeletal muscle and false-positive elevations occur in a number of clinical settings, including trauma, heavy exertion, and myopathy.

- There are two types of macro-CK: macro-CK type 1 where one of the CK isoenzyme commonly CK-BB is complexed with immunoglobulin, commonly IgG and macro-CK type 2 which are made up of mitochondrial-derived CK polymers. While macro-CK type 1 may be found also in normal population and may not have any pathological significance, macro-CK type 2 may be associated with neoplasm and liver cirrhosis. Macro-CK can persist in serum for a long time and may falsely increase CK-MB level if a mass assay is used, causing confusion in diagnosis of myocardial infarction.

- Troponin, a three-piece protein complex is a component of the contractile apparatus within skeletal and cardiac myocytes. The three subunits are designated as troponin C (the calcium-binding component; molecular weight 18.4 KDa), troponin T (the tropomyosin-binding component; molecular weight 36 KDa) and troponin I (the inhibitory component; molecular weight 23.8 KDa).

- Although troponin C is synthesized in both cardiac and skeletal muscle, both troponin I and T are highly specific to cardiac myocytes and as a result are also called cardiac troponins. Majority of both cardiac troponin T and I (92%–95%) are attached to the actin thin filament in the cardiac sarcomere and a small amount (5%–8%) is found in the myocyte cytoplasm (free troponins). Following myocardial damage, cytosolic-free (unbound) troponin is released first and as further damage occurs, complexes of cardiac troponin T, I, and C (T-I-C or ternary complex) and complexes of cardiac troponin I and C (I-C or binary complex) present in sarcomere are released into circulation, making troponin a both early and late marker of acute myocardial infarction.

- Levels of both troponin T and I start increasing 3–4 h after acute myocardial infarction if measured by 5th generation high-sensitivity troponin assays or 4–9 h if measured by conventional (4th generation) assays, peak at 12–24 h and may remain elevated 10–14 days.

- Cardiac troponin T and I have many advantages over CK-MB as cardiac markers. First, levels of troponin in normal individuals are very low or none-detected. Therefore, significant elevation of troponin indicates myocardial injury. Small elevation of troponin but normal CK-MB level may indicate microscopic zone of myocardial narcosis (microinfarction). Troponin I is very specific for myocardium because only one isoform of cardiac troponin I has been identified which is found exclusively in cardiac myocytes. Cardiac troponin I is not expressed in skeletal muscle. Although cardiac troponin T has a different amino acid sequence when compared to other troponins, small amounts of cardiac troponin T have been identified in skeletal muscle. In humans, cardiac troponin T isoform expression has been reported in patients with muscular dystrophy, polymyositis, dermatomyositis, and end-stage renal disease.
- The European Society of Cardiology, the American College of Cardiology, the American Heart Association, and the World Heart Federation have issued the fourth universal definition of myocardial infarction. It is now a standard practice to first categorize myocardial infarction (MI) based on ECG findings as ST-segment elevation myocardial infarction (STEMI) versus non-ST segment elevation myocardial infraction (NSTEMI) and then further distinguish MI into five groups. According to these guidelines, cardiac troponin I and cardiac troponin T are the preferred biomarkers for the diagnosis of myocardial injury, and high-sensitivity cardiac troponin I or T assays are recommended for routine clinical use. Other biomarkers, such as creatine kinase MB isoform (CK-MB), are less sensitive and less specific. Myocardial injury is present when blood levels of cardiac troponins are increased above the 99th percentile upper reference limit. The injury may be acute, as evidenced by a newly detected dynamic rising and/or falling pattern of cardiac troponin values above the 99th percentile upper reference limit, or chronic, in the setting of persistently elevated cardiac troponin levels, which clearly separates myocardial infarction from myocardial injury. According to these guideline, there are five types of myocardial infarction.
- Studies have revealed a high prevalence of elevated cardiac troponin levels especially troponin T in patients with chronic renal failure. A single elevated cardiac troponin T level in patients with chronic renal failure is nondiagnostic for acute myocardial infarction in the absence of other findings.
- The fourth Universal Definition of Myocardial Infarction document also has a new section on Takotsubo syndrome (type of nonischemic cardiomyopathy in which there is a sudden temporary weakening of the muscular portion of the heart) is also included. Takotsubo syndrome can

mimic MI and is found in ~ 1%–2% of patients presenting with suspected STEMI. The onset of Takotsubo syndrome is often triggered by intense emotional or physical stresses such as bereavement. Over 90% of patients are postmenopausal women.

- In general, conventional 4th generation troponin immunoassays do not achieve desirable less than 10% CV at the 99th percentile upper reference limit due to poor sensitivity and usually cannot detect troponin I at lower than 0.02 ng/mL concentrations. However, high-sensitivity troponin assays are capable of detecting 10- to 100-fold lower concentrations of troponin. For example, Abbott high-sensitivity troponin I assay has a lower limit of detection of 1.9 ng/L (0.0019 ng/mL). The high-sensitivity troponin assays should be able to detect troponin levels below the 99th percentile cut-off values in at least 50% of normal population. Moreover, CVs are less than 10% at 99th percentile cut-off for all high-sensitivity troponin assays. Currently, only one high-sensitivity troponin T assay is available from Roche Diagnostics but several high-sensitivity troponin I assays are commercially available. it is important to note that 99th percentile upper reference limit is higher in males than females, for example, the limit is 15.6 pg/ml for females and 34.2 pg/mL in males using Abbott's high sensitivity troponin I assay.

- Many interferences have been reported in troponin I assays including endogenous substances such as hemolysis and high bilirubin. However, more commonly, heterophilic antibodies, rheumatoid factors, and macrotroponin are known to interfere with cardiac troponin immunoassays. Macrotroponin I, a troponin-immunoglobulin (IgG) complex, should be considered when troponin I values are inconsistent with the clinical picture, and fail to demonstrate a rise and/or fall pattern in suspected cases of acute myocardial infarction. Macrotroponin T, a complex with troponin T and immunoglobulin causing falsely elevated troponin T value has also been reported.

- LDH isoenzymes, an old cardiac biomarker are no longer used.

- Biomarkers of myocardial ischemia include ischemic modified albumin (reduced capacity of binding cobalt), heart-type fatty acid-binding protein and glycogen phosphorylase.

- BNP and NT-proBNP are most commonly used biomarkers for evaluating hemodynamic stress of the heart. Major application of both BNP and proBNP testing is evaluating patients with heart failure including congestive heart failure. BNP is initially synthesized as prepro BNP which contains 134 amino acids but it is cleaved into proBNP containing 108 amino acids. On secretion, it splits into biologically active BNP (amino acids 77–108) and the remaining N-terminal proBNP (NT-pro BNP: 1–76 amino acids) which is biologically inactive. BNP is a smaller molecule than NT-proBNP and it is cleared from circulation earlier than NT-proBNP. Therefore, concentration of NT-proBNP in serum or plasma is higher than BNP. In addition, NT-proBNP is more stable in serum or plasma than BNP.

- Commonly monitored inflammatory and prognostic biomarkers are C-reactive protein, homocysteine, and myeloperoxidase.
- Potential biomarker of atherosclerotic plaque instability is pregnancy-associated plasma protein A.

# References

[1] Wang H, Zhao T, Wei X, Lu H, Lin X. The prevalence of 30-day readmission after acute myocardial infarction: a systematic review and meta-analysis. Clin Cardiol 2019;42:889–98.

[2] Movahed MR, John J, Hashemzadeh M, Hashemzadeh M. Mortality trends in non-ST-segment elevation myocardial infarction (NSTEMI) in the United States from 1998 to 2004. Clin Cardiol 2011;34:689–92.

[3] Garg P, Morris P, Fazlanie AL, Vijayan S, et al. Cardiac biomarkers of acute coronary syndrome: from history to high sensitivity cardiac troponin. Intern Emerg Med 2017;12:147–55.

[4] Wright RS, Anderson JL, Adams CD, Bridges CR, et al. ACCF/AHA focused update of the guidelines for the management of patients with unstable angina/non-ST-elevation myocardial infarction (updating 2007 guidelines). J Am Coll Cardiol 2011;57:1920–59.

[5] Lackner KJ. Laboratory diagnostics of myocardial infarction-troponins and beyond. Clin Chem Lab Med 2013;51:83–9.

[6] Woo J, Lacbawan FL, Sunheimer R, LeFever D, et al. Is myoglobin useful in the diagnosis of acute myocardial infarction in the emergency department setting? Am J Clin Pathol 1995;103:725–9.

[7] Servonnet A, Dubost C, Martin G, Lefrère B, et al. Myoglobin: still a useful biomarker in 2017? Ann Biol Clin (Paris) 2018;76:137–41.

[8] Aydin S, Ugur K, Aydin S, Sahin İ, Yardim M. Biomarkers in acute myocardial infarction: current perspectives. Vasc Health Risk Manage 2019;15:1–10.

[9] Lee T, Goldman L. Serum enzymes in the diagnosis of acute myocardial infarction. Ann Int Med 1986;105:221–33.

[10] Lee KN, Casko G, Bernhardt P, Elin R. Relevance of macro creatine kinase type 1 and type 2 isoenzymes to laboratory and clinical data. Clin Chem 1994;40:1278–83.

[11] Aljuani F, Tournadre A, Cecchetti S, Soubrier M, Dubost JJ. Macro-creatine kinase: a neglected cause of elevated creatine kinase. Intern Med J 2015;45:457–9.

[12] deFilippi C, Seliger S. The cardiac troponin renal disease diagnostic conundrum: past, present, and future. Circulation 2018;137:452–4.

[13] Thygesen K, Alpert JS, Jaffe AS, Chaitman BR, et al. Fourth universal definition of myocardial infarction (2018). J Am Coll Cardiol 2018;72:2231–64.

[14] Reichlin T, Cullen L, Parsonage WA, Greenslade J, et al. Two-hour algorithm for triage toward rule-out and rule-in of acute myocardial infarction using high-sensitivity cardiac troponin T. Am J Med 2015;128:369–79.

[15] Boeddinghaus J, Reichlin T, Cullen L, Greenslade JH, et al. Two-hour algorithm for triage toward rule-out and rule-in of acute myocardial infarction by use of high-sensitivity cardiac troponin I. Clin Chem 2016;62:494–504.

[16] Sternberg M, Pasini E, Chen-Scarabelli C, Corsetti G, et al. Elevated cardiac troponin in clinical scenarios beyond obstructive coronary artery disease. Med Sci Monit 2019;25:7115–25.

[17] Ferguson JL, Beckett GJ, Stoddart M, Fox KAA. Myocardial infarction redefined: the new ACC/ESC definition, based on cardiac troponin, increases the apparent incidence of infarction. Heart 2002;88:343–7.

[18] Imazio M, Klingel K, Kindermann I, Brucato A, et al. COVID-19 pandemic and troponin: indirect myocardial injury, myocardial inflammation or myocarditis? Heart 2020;106:1127–31.

[19] Warner JV, Marshall GA. High incidence of macrotroponin I with a high-sensitivity troponin I assay. Clin Chem Lab Med 2016;54:1821–9.

[20] Hedley J, Menon V, Cho L, McShane AJ. Fifth generation troponin T assay is subject to antibody interference. Clin Chim Acta 2020;505:98–9.

[21] Frame IJ, Joshi PH, Mwangi C, Gunsolus I, et al. Susceptibility of cardiac troponin assays to biotin interference. Am J Clin Pathol 2019;151:486–93.

[22] Coverdale JPC, Katundu KGH, Sobczak AIS, Arya S, et al. Ischemia-modified albumin: crosstalk between fatty acid and cobalt binding. Prostaglandins Leukot Essent Fatty Acids 2018;135:147–57.

[23] Ye XD, He Y, Wang S, Wong GT, et al. Heart-type fatty acid binding protein (H-FABP) as a biomarker for acute myocardial injury and long-term post-ischemic prognosis. Acta Pharmacol Sin 2018;39:1155–63.

[24] Nakagawa Y, Nishikimi T, Kuwahara K. Atrial and brain natriuretic peptides: hormones secreted from the heart. Peptides 2019;111:18–25.

[25] Weber M, Hamm C. Role of B-type natriuretic peptide (BNP) and NT-proBNP in clinical routine. Heart 2006;92:843–9.

[26] Sugimoto M, Manabe H, Nakau K, Furuya A, Okushima K, et al. The role of N-terminal pro-B-type natriuretic peptide in the diagnosis of congestive heart failure in children. – Correlation with the heart failure score and comparison with B-type natriuretic peptide. Circ J 2010;74:998–1005.

[27] Kolodziej AR, Abo-Aly M, Elsawalhy E, Campbell C, et al. Prognostic role of elevated myeloperoxidase in patients with acute coronary syndrome: a systemic review and meta-analysis. Mediators Inflamm 2019;2019:2872607.

[28] Body R, Ferguson C. Pregnancy-associated plasma protein A: a novel cardiac marker with promise. Emerg Med J 2006;23:875–7.

# CHAPTER 9

# Endocrinology

## Introduction to various endocrine glands

Homeostasis is maintained by both the nervous system and the endocrine system in the human body. Endocrine activity can be classified as autocrine, paracrine, and classical endocrine activity. In autocrine activity, chemicals produced by a cell act on the cell itself. In paracrine activity, chemicals produced by a cell act locally. However, in classical endocrine activity, chemicals produced by an endocrine gland act at a distant site after their release in the circulation and these chemicals are termed as hormones. Major endocrine glands include pituitary, thyroid, parathyroid, adrenals, gonads (testis in male and ovary in female), and pancreas. However, the pineal gland secretes melatonin (N-acetyl-5-methoxytryptamine), which may contribute to the regulation of biological rhythm and may induce sleep because melatonin synthesis is stimulated by darkness [1]. Chemical structures of hormones vary widely, and hormones may be polypeptides, glycoproteins, steroids, or amines.

Most classical hormones are secreted into the systemic circulation. However, hypothalamic hormones are secreted into the pituitary portal system. Hormones may be bound to certain proteins in blood, and such binding proteins include thyroxine-binding globulin (TBG), sex hormone-binding globulin (SHBG), cortisol-binding globulin (CBG), and insulin-like growth factor (IGF)-binding proteins (IGF-BP3). However, albumin can also bind certain hormones. In addition, prealbumin can also act as a binding protein for a hormone.

Hormones usually act by binding to receptors. Receptors for hormones may be:

- Cell surface or membrane receptors or
- Nuclear receptors

Cell surface or membrane receptors may be G-protein-coupled receptors or dimeric transmembrane receptors. G-protein-coupled receptors bind hormones in the extracellular domain which activate membrane G protein complex. The activated G protein complex is then responsible for generating

Clinical Chemistry, Immunology and Laboratory Quality Control. https://doi.org/10.1016/B978-0-12-815960-6.00003-0

secondary messengers. Most peptide hormones act via this mechanism. Dimeric transmembrane receptors bind hormones in their extracellular component and the intracellular component is responsible for phosphorylation of intracellular messengers, which leads to the activation of various messengers. Growth hormone (GH) and insulin-like growth factor-I (IGF-I) act by this mechanism.

Steroid and thyroid hormones act via nuclear receptors. These hormones pass through the cell membrane and bind with the receptors in the cytoplasm, and the complex is translocated to the nucleus causing an increased transcription of genes.

Secretion of hormones from endocrine glands may be under positive feedback and negative feedback. For example, thyrotropin-releasing hormone (TRH) secreted by the hypothalamus stimulates the release of thyroid-stimulating hormone (TSH) by the anterior pituitary which in turn causes the thyroid gland to release thyroxine (T4) and triiodothyronine (T3). T3 and T4 once released cause negative feedback on the secretion of TSH and TRH.

Hormone secretion may be continuous or intermittent. Thyroid hormone secretion is continuous. As a result, levels may be measured at any time to assess hormonal status. Secretion of follicle-stimulating hormone (FSH), luteinizing hormone (LH), and GH is pulsatile. Thus, a single measurement may not reflect hormonal status. Some hormones exhibit biological rhythms. Cortisol exhibits circadian rhythm where levels are highest in the morning and lowest during late night. The menstrual cycle is an example of a longer biological rhythm, where different level of a hormone is observed during a specific part of the cycle. During normal menstrual cycle, there is interplay of feedback between hypothalamus, anterior pituitary, and ovaries. In follicular phase, low estrogen level stimulates secretion of LH and FSH by negative feedback mechanism. Level of progesterone is also low during follicular phase. At the end of follicular phase, both estradiol and estrogen levels are high triggering release of gonadotropin-releasing hormone (GnRH) from the hypothalamus that stimulates secretion of LH from anterior pituitary. Level of LH is highest in midcycle during ovulation and levels of estrogen. LH surge during midcycle is a good indication of ovulation. After midcycle, progesterone level starts increasing in luteal phase reaching its highest level during 8 days after ovulation. In luteal phase, levels of FSH and LH also decline gradually (Fig. 1).

Certain hormone levels are elevated during stress. These include:

- Adrenocorticotropic hormone (ACTH) and cortisol
- Growth hormone (GH)
- Prolactin
- Adrenaline and noradrenaline

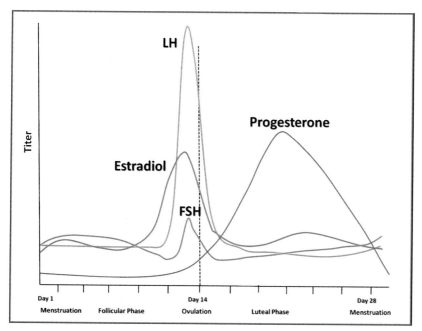

**FIG. 1**
Titers of various hormones during menstrual cycle. *Figure courtesy of Andres Quesada, MD.*

Therefore, it is important for a phlebotomist to wait for some after initial venipuncture prior to collect blood specimen for measuring blood levels of such hormones. Certain hormone levels are increased during sleep such as growth hormone (GH) and prolactin.

This chapter is an overview of activities of the hypothalamus, pituitary, thyroid, adrenal, gonads, and pancreas in relation to their endocrine activity as well as endocrine disorders related to these organs. Diabetes mellitus is the most prevalent endocrine disorder in the United States, and diabetes mellitus is discussed in detail in Chapter 7. Most common endocrine disorders are listed in Table 1. With hormone replacement and other therapies, endocrine disorders can be treated well, and severe consequence of endocrine disorder is observed rarely today.

## Hypothalamus

The hypothalamus produces thyrotropin-releasing hormone (TRH), corticotrophin hormone (CRH), gonadotropin-releasing hormone (GnRH), growth hormone-releasing hormone (GRH), and somatostatin (growth hormone inhibitory hormone). These hormones act on the anterior pituitary resulting

**Table 1** Most common endocrine disorders.

| Endocrine disorder | Cause |
|---|---|
| Diabetes mellitus | Pancreatic beta-cell dysfunction that produces insulin |
| Acromegaly | Overproduction of growth hormone |
| Addison's disease | Decreased production of hormones by adrenal glands |
| Cushing syndrome | High cortisol produced by adrenal glands |
| Grave's disease | Hyperthyroidism due to excess production of thyroid hormones |
| Hashimoto's thyroiditis | Autoimmune disease related to hypothyroidism |
| Hyperthyroidism | Excess production of thyroid hormones |
| Hypothyroidism | Underproduction of thyroid hormones |
| Prolactinoma | Overproduction of prolactin by pituitary gland |
| Polycystic ovary syndrome | Excessive production of androgenic hormone |

in release of various other hormones, including thyroid-stimulating hormone (TSH), adrenocorticotropin (ACTH), FSH, LH, and GH. Somatostatin inhibits the release of GH. Dopamine (also known as prolactin inhibitory hormone), a neurotransmitter, is also produced by the hypothalamus. Dopamine can inhibit GH secretion.

The supra optic and paraventricular nuclei of the hypothalamus produce anti-diuretic hormone (ADH; also known as vasopressin) and oxytocin. These hormones are stored in the posterior pituitary and act on certain body parts rather than acting on pituitary like other tropic hormones. ADH acts on the collecting ducts of the renal tubules and causes absorption of water. ADH secretion is linked to serum osmolality, and increased serum osmolality results in increased secretion of ADH. Lesions of the hypothalamus may result in inadequate ADH secretion, also known as cranial diabetes insipidus. Failure of ADH to act on the collecting ducts results in nephrogenic diabetes insipidus. Causes of nephrogenic diabetes insipidus include hypercalcemia, hypokalemia, and lithium therapy. In both types of diabetes insipidus, polyuria with low osmolality is common symptoms. Please see Chapter 5 for more detail.

Oxytocin is a nonapeptide hormone (nine amino acids) primarily synthesized in the magnocellular neurons of paraventricular and supraoptic nuclei of hypothalamus, and most of oxytocin produced is transported to the posterior pituitary where it is released to regulate parturition and lactation. In addition, oxytocin plays an important role in the development of the capacity to form social bonds, the mediation of the positive aspects of early-life nurturing on adult bonding capacity, and the maintenance of social bonding [2]. Hormones released by hypothalamus and their characteristics are listed in Table 2.

**Table 2** Characteristics of hormones released by hypothalamus.

| Hormone | Composition | Action |
|---|---|---|
| *Tropic hormones acting on pituitary* | | |
| Corticotropin-releasing hormone (CRF) | 41 Amino acids | Stimulates adrenocorticotropic hormone release (ACTH) |
| Gonadotropin-releasing hormone (GnRH) | 10 Amino acids | Stimulates follicle-stimulating hormone (FSH) and luteinizing hormone (LH) release |
| Growth hormone releasing hormone (GHRH) | 44 Amino acids | Stimulates growth hormone (GH) release |
| Thyrotropin-releasing hormone (TRH) | 3 Amino acids | Stimulates thyroid stimulating hormone (TSH) and prolactin release |
| Somatostatin | 14 Amino acids | Inhibits GH release |
| *Hormones acting on other organs* | | |
| Antidiuretic hormone (ADH) | 9 Amino acids | Acts on kidney causing water reabsorption |
| Oxytocin | 9 Amino acids | Lactation, parturition, mood |

## Pituitary gland

Pituitary gland is a small gland situated as the base of the skull. The gland is divided into anterior and posterior lobe. Anterior pituitary lobe produces six hormones, and release of such hormones is under control of hypothalamus though various hormones are also produced by the hypothalamus. Hormones produced by pituitary gland are listed in Table 3. It is important to note that although ADH is synthesized by the hypothalamus, it is secreted from the pituitary gland.

Growth hormone (GH: also known as somatotropin) is the most abundant hormone produced by anterior pituitary, and it stimulates growth of cartilage, bone, and many soft tissues. GH stimulates release of IGF-I (somatomedin C), mostly from the liver, and IGF-I is partly responsible for the activity of GH but this hormone has glucose lowering and other anabolic activities. Conditions that cause deficiency of IGF-I include Laron syndrome in children, liver cirrhosis in adults, age-related cardiovascular and neurological diseases, and intrauterine growth restriction [3]. Although IGF-I plays important roles in adults, similar hormone insulin-like growth factor-II (IGF-II, somatomedin A) concentration is high in embryonic and neonatal tissues. Both IGF-I and IGF-II share 45 amino acid positions and approximately 50% amino acid homology with insulin. GH is essential for proper growth of children, and it also plays an important role in adults in maintaining healthy bones, muscles, and metabolism. GH hormone

**Table 3** Characteristics of hormones released by anterior pituitary.

| Hormone | Composition | Action |
|---|---|---|
| Adrenocorticotropic hormone (ACTH) | 39 Amino acids | Stimulates glucocorticoid secretion by adrenal cortex |
| Follicle-stimulating hormone | Contains alpha and beta subunit, but alpha subunit of FSH, LH, TSH, and hCG is same (92 amino acids) The beta subunit confers uniqueness FSH: beta chain: 117 amino acids | Stimulates development of ovarian follicle in female, and in male stimulate spermatogenesis |
| Luteinizing hormone (LH) | LH: beta subunit: 121 amino acids | In female, LH surge stimulates ovulation |
| Thyroid stimulating hormone (TSH) | TSH beta subunit: 112 amino acids | Stimulates thyroid gland to produce T3 and T4 |
| Growth hormone (GH) | 191 Amino acids | Produces growth mediated via IGF-I |
| Prolactin (PRL) | 199 Amino acids | Initiation and maintenance of lactation |

*Insulin-like growth factor I (IGF-I) is produced mainly in the liver.*

deficiency or excess is rarely encountered clinical. Another hormone similar in structure of GH is prolactin, which plays an important role in lactation.

In general, deficiency of pituitary hormones may be selective or multiple or panhypopituitarism (Simmonds' disease also known as Sheehan's syndrome or postpartum hypopituitarism). Deficiency of GH causes dwarfism which can be treated with recombinant human GH replacement therapy. Clinical features are related to the deficiency of the peripheral endocrine gland. Causes of hypopituitarism include:

- Congenital: For example, Kallmann syndrome (isolated GnRH deficiency causing delayed or absent puberty)
- Infections
- Vascular: Sheehan's syndrome (postpartum necrosis), pituitary apoplexy, etc.
- Tumors: For example, pituitary or hypothalamic tumors, craniopharyngioma
- Trauma
- Surgery
- Infiltrative diseases such as sarcoidosis, hemochromatosis
- Radiation
- Empty sella syndrome

Hyperpituitarism is most often due to pituitary tumors affecting GH-secreting cells, prolactin-secreting cells, and adrenocorticotropic hormone (ACTH)-secreting cells. GH-secreting tumors affecting individuals before closure of epiphysis result in gigantism and after closure result in acromegaly. Prolactin-secreting tumors cause hyperprolactinemia. A high level of prolactin (prolactinoma) inhibits action of follicle-stimulating hormone (FSH) and luteinizing hormone (LH) resulting in hypogonadism and infertility. ACTH-secreting tumors cause Cushing's syndrome. Proper endocrine testing for diagnosis of hypopituitarism includes measuring concentrations of various hormones in serum and plasma, including thyroid-stimulating hormone (TSH), prolactin, LH, FSH, T4, and cortisol. For diagnosis of hypopituitarism stimulation test with gonadotropin-releasing hormone, thyrotropin-releasing hormone and insulin induced hypoglycemia (triple stimulation test) are useful. Following stimulation, serum or plasma levels of FSH, LH, TSH, PRL, GH, and cortisol are measured.

Endocrine tests for hyperpituitarism include measurement of hormone levels and suppression test using glucose (oral glucose tolerance). Administration of glucose with rise in blood glucose should suppress anterior pituitary hormones in normal individuals.

## Thyroid gland

The thyroid gland produces two hormones, thyroxine (T4) and triiodothyronine (T3). Four steps are involved in the synthesis of these hormones:

- Inorganic iodide from the circulating blood is trapped (iodide trapping)
- Iodide is oxidized to iodine (oxidation)
- Iodine is added to tyrosine to produce monoiodotyrosine and diiodotyrosine. This is referred to as organification
- One monoiodotyrosine is coupled with one diiodotyrosine to yield T3 and two diiodotyrosine are coupled to yield T4 (coupling)

Both T3 and T4 are bound to thyroglobulin and stored in the colloid. Free (unbound) T4 is the primary secretory hormone from the thyroid gland, and T4 is converted in peripheral tissue (liver, kidney, and muscle) to T3 by 5′-monodeiodination. T3 is the physiologically active hormone. T4 can also be converted to reverse T3 by 3′-monodeiodination. This form of T3 is inactive. Majority (99%) of the T3 and T4 in circulation are found to thyroxine-binding globulin (TBG), albumin, and thyroxine-binding prealbumin. T3 binds to thyroid hormone nuclear receptor on target cells to cause modified gene transcription.

In a normal individual, there is a tightly coordinated feedback mechanism between hypothalamus, pituitary, and thyroid glands. Thyrotropin-releasing hormone (TRH), a tripeptide (smallest hormone molecule known) produced in the hypothalamus, stimulates pituitary to synthesize and secrete thyroid-stimulating hormone (TSH) that finally stimulates thyroid gland to produce thyroid hormones. There is a negative feedback mechanism where a fall in blood thyroid hormone stimulates hypothalamus to secrete TRH. Abnormalities of enzymes involved in synthesis of thyroid hormone may cause hypothyroidism with increased TSH secretion and goiter. This is dyshormonogenetic goiter. Dyshormonogenetic goiter may be associated with nerve deafness, referred to as Pendred syndrome. Thyroid disorders are common with an estimated 3% population suffering from such disorders. Thyroid disorders are also more common in women than in men with an estimated 4.1 women per 1000 women develop hypothyroidism every year while the prevalence among men is 0.6 per 1000 adults. In addition, 0.8 women per 1000 develop hyperthyroidism every year [4]. Moreover, thyroid disorders are more common in older people than in younger people.

## Thyroid function tests

Most commonly ordered thyroid function test is TSH followed by T4 (total or free) and T3 (total or free). More recently, FT4 and FT3 tests are ordered more frequently than total T4 and total T3 tests. TSH is used as a screening test for thyroid status. It is elevated in primary hypothyroidism and suppressed in thyrotoxicosis. Basic interpretation of TSH, FT4 (free T4), and FT3 (free T3) in various thyroid diseases is summarized in Table 4. However, there are several situations when interpretation of thyroid function tests may be confusing:

- There are situations where thyroid hormone-binding proteins may be low or high causing alteration of total T3 and T4 levels. However, free T3 and T4 and TSH levels should be normal. Pregnancy and oral contraceptive pills raise concentrations of thyroid-binding proteins.

**Table 4** Interpretation of basic thyroid function tests (TSH).

| Thyroid disorder | TSH | Free T4 | Free T3 |
|---|---|---|---|
| Primary hypothyroidism | Increased | Low | Low |
| Secondary hypothyroidism (lack of TSH from pituitary) | Low | Low | Low |
| Thyrotoxicosis | Low | High | High |
| T3 toxicosis | Low | Normal | High |

Hypoproteinemia due to cirrhosis of liver, nephrotic syndrome, etc. may cause lower concentration of thyroid-binding proteins.

- Amiodarone can reduce peripheral conversion of T4 to T3. Free T4 levels may be high, but TSH level could be normal. Amiodarone can also cause both hypothyroidism and hyperthyroidism because amiodarone contains iodine molecule.
- Seriously ill patients may have reduced production of TSH with low T4 and reduced conversion of T4 to T3 with increased conversion of T4 to reverse T3. These patients are however euthyroid. This is referred to as sick euthyroid syndrome.
- TSH level may increase with advanced age without any thyroid disease. Slightly elevated TSH level after of age 65–70 years is normal [5].

TSH, T4, and T3 tests can be performed using automated analyzers and immunoassays. Reverse T3 analysis may also be performed under certain circumstances, but due to low volume of this test, most hospital laboratories send this test to a reference laboratory. In addition to these tests, measuring free T4 (FT4) and free T3 (FT3) is useful for diagnosis of thyroid disorder in certain patients. Although FT4 and FT3 can be detected by direct method such as dialysis and ultrafiltration, most clinical laboratories utilize indirect immunoassays based methods, which can be adopted in automated chemistry analyzers. Both two-step immunoassays and one-step immunoassays are commercially available for determination of FT4 and FT3. An indirect way of estimating FT4 is free thyroxine index (FT4I), which is calculated by multiplying total T4 with value of T3 uptake. T3 uptake assay is a measure of number of available free binding sites on thyroxine-binding globulin (TBG) and is expressed as a percentage value. Commercial kits are available for such measurement. Free thyroxine index usually correlates with FT4 concentration.

Patients with suspected thyroid disease may also be tested for thyroxine-binding globulin, which has greatest affinity for T4 as well as antithyroid antibodies. Most common antithyroid antibody is thyroid peroxidase antibody (TPOAb). This antibody is against thyroid peroxidase (originally described as thyroid microsomal antigen), an enzyme found in thyroid gland which plays an important role in the production of thyroid hormones. In addition, there are antibodies against thyroglobulin and TSH receptor. TSH receptor antibodies may inhibit binding of TSH to the receptor or stimulate the receptor. When such antibody stimulates the TSH receptor, it may cause thyrotoxicosis such as Graves' disease. Thyroid antibody testing is essential in establishing a diagnosis of thyroid dysfunction which is autoimmune in nature. Enzyme-linked immunosorbent assay (ELISA) and chemiluminescent-based immunoassays are commercially available for the determination of these antibodies. Other analytical methods are also available.

Various interferences have been reported in thyroid function tests. For TSH, usually third generation assays are used, which are sensitive and can detect levels as low as 0.02 mU/L, and normal range is usually defined as 0.5–5.5 mU/L; however, there are recommendations to lower the upper limit significantly. Ultrasensitive TSH assays (fourth generation) with lower limit of detection compared to third generation assays are also available. Measurement of TSH can suffer interference from heterophilic antibody and rheumatoid factor causing falsely elevated results. In addition, TSH value can be falsely elevated due to the presence of macro-TSH, an autoimmune complex between anti-TSH IgG antibody and TSH (molecular weight approximately 150 kDa; prevalence: 0.6%–1.6%). Loh et al. reported a very high TSH level of 122 mIU/L in a patient who showed a normal FT4 level. The falsely elevated TSH value was due to macro-TSH [6]. Macro-TSH is considered biochemically inactive but falsely elevated TSH due to the presence of macro-TSH, and normal FT4 and FT3 may cause diagnostic confusion because such feature is present in patients with sub-clinical hypothyroidism. An increased recovery of TSH upon dilution may indicate the presence of macro-TSH or interference from heterophilic antibody as well as rheumatoid factor. Polyethylene glycol precipitation or gel filtration chromatography may also be used to identify and eliminate such interference. In addition, thyroid autoantibodies (mostly against T3 and T4) can also interfere with thyroid function tests. In general, automated one-step immunoassays are known to be more susceptible to thyroid autoantibodies interference than two-step immunoassays where an additional washing step is incorporated. The presence of anti-T4 antibody may falsely increase FT4 result using one-step immunoassay.

Biotin can falsely lower TSH value if presence in excess amount but falsely increase FT4 and FT3 thus mimicking Grave's disease if immunoassays incorporate biotinylated antibody in assay design [7]. Please see Chapter 21 for in-depth discussion on biotin interferences in immunoassays. Several strongly protein bound drugs can affect thyroid hormone binding causing decreases T4 and T3 levels. In addition, thyroglobulin autoantibodies (may be present in patients with thyroid cancer) also interfere with thyroglobulin measurement [8].

## Hypothyroidism

Hypothyroidism by definition is the failure of thyroid gland to produce sufficient hormone to meet daily requirement of such hormones to maintain normal metabolic functions. Hypothyroidism can be due to thyroid gland failure (primary gland failure) or insufficient stimulation of thyroid gland due to dysfunctional hypothalamus (producing TRH) or pituitary (producing TSH). Types of hypothyroidism include:

- Primary hypothyroidism (primary disease of thyroid gland): Causes include autoimmune thyroiditis, Hashimoto's thyroiditis, surgery/radiation, dyshormonogenesis, antithyroid drugs, drugs therapy with amiodarone and advanced age
- Secondary hypothyroidism (lack of TSH from pituitary)
- Peripheral resistance to thyroid hormones

Autoimmune thyroid disease especially Hashimoto's thyroiditis is the most common etiology of hypothyroidism in the United States. In Hashimoto's thyroiditis, the principal biochemical characteristics is the presence of thyroid autoantibodies including thyroid peroxidase antibody (TPOAb) and thyroglobulin antibody in sera of patients with this disease. Both antibodies are present in higher concentration in female patients than in male patients and belong to IgG class. These antibodies have high affinity against respective antigen, thus causing destruction of thyroid gland. Lymphocytic infiltration with follicle formation is seen within the thyroid gland. The gland is enlarged and the patient is hypothyroid. During the initial phase, there may be transient hyperthyroidism, referred to as Hashitoxicosis which usually lasts for 1–2 months. However, there is a rare case report of a 21-year-old male who had Hashitoxicosis of 2-year duration before converting to Hashimoto's hypothyroidism [9]. Hashimoto's thyroiditis is considered to be a common autoimmune disease. It is currently accepted that genetic susceptibility, environmental factors, and immune disorders contribute to its development. Moreover, high iodine intake and deficiencies of selenium and iron may also contribute in the pathogenesis of this disease [10].

Iatrogenic forms of hypothyroidism may occur after thyroid surgery, neck irradiation, or drug therapy (amiodarone, lithium, tyrosine kinase inhibitors, etc.) including radioiodine therapy. In addition, transient hypothyroidism may occur due to postpartum thyroiditis. Best laboratory assessment of thyroid function is measuring serum or plasma TSH level. If TSH is elevated, FT4 should be measured. Elevated serum TSH with low FT4 indicates primary hypothyroidism, but elevated TSH with normal FT4 indicates subclinical hypothyroidism. In secondary hypothyroidism, both TSH and FT4 should be low. If FT3 is measured instead of FT4, similar pattern should be observed (Table 4). Thyroid antibody testing is useful if autoimmune hypothyroidism such as Hashimoto's thyroiditis is suspected. Thyroid hormone requirement increases in pregnancy and such requirement lasts through pregnancy. Therefore, dosage must be adjusted carefully if a pregnant woman is receiving levothyroxine. Myxedema coma is a rare medical emergency due to severe manifestation of hypothyroidism, and such condition is usually observed in older women with a history of primary hypothyroidism. Change of mental status, severe hypothermia, and even psychosis may be present in a patient suffering from myxedema coma. If not treated timely, outcome may be fatal [11].

## Hyperthyroidism

Clinical hyperthyroidism also known as thyrotoxicosis is due to excess thyroid hormones in circulation causing various clinical symptoms. The prevalence is 2% for women and 0.2% for men, and as many as 15% of all cases of hyperthyroidism is encountered in patients over 60 years of age. Causes of hyperthyroidism include:

- Graves' disease
- Toxic nodular (single or multiple) goiter
- Thyroiditis (e.g., due to viral infection)
- Drugs induced
- Excess TSH (e.g., due to pituitary tumor)

Graves' disease, the most common cause of hyperthyroidism, is an autoimmune disease of the thyroid where IgG antibodies bind with TSH receptor causing stimulation of thyroid gland and overproduction of thyroid hormones. These antibodies are also referred to as long-acting thyroid stimulator. An infiltrative ophthalmopathy (exophthalmos, lid lag, and lid retraction) may be observed in about 50% patients with Grave's disease. TSH receptor autoantibodies are usually measured in serum or plasma for diagnosis of Grave's disease because over 90% patients have detectable level of these antibodies. In addition, thyroid-stimulating autoantibodies may also be present in patients with Grave's disease. Graves's disease is a complex disease where genetic predisposing is modified by environmental factors. There is an association between polymorphism of human lymphocyte antigen (HLA) genes (HLA-DRB1*3 allele) with young age diagnosis of Grave's disease [12].

Another cause of hyperthyroidism is toxic nodular goiter, which is more common in area of the world where iodine deficiency is common. It is more common in patients older than 40 years. Toxic adenoma is due to autonomously functioning nodules which is usually found in younger people. Thyroiditis can be acute where thyroid hormone leaks out from the inflamed gland most likely due to viral illness. Postpartum thyroiditis is also an acute condition which may occur 3–6 months after delivery but usually resolves. Amiodarone-induced hyperthyroidism is due to high iodine content (37%) of amiodarone. Iodine-induced hyperthyroidism can occur due to excess iodine in diet or exposure to radiographic contrast media. Rare cause of hyperthyroidism includes metastatic thyroid cancer, ovarian tumor that produces thyroid hormone (Struma ovarii), and trophoblastic tumors that produce chorionic gonadotropin and activate TSH receptors. In addition, TSH-secreting pituitary tumor can cause hyperthyroidism. The first screening test for a patient with suspected hyperthyroidism without any evidence of pituitary disease is TSH. If TSH is undetectable or very low, hyperthyroidism should be suspected.

**Table 5** Interpretation of thyroid autoantibody tests.

| Thyroid antibody | Associated with thyroid disease |
| --- | --- |
| Thyroid peroxidase antibody | Hashimoto thyroiditis |
| Thyroglobulin antibody | Hashimoto thyroiditis, thyroid cancer |
| TSH receptor autoantibody | Graves's disease |
| Thyroid-stimulating autoantibody | Grave's disease |

Antithyroid antibodies are elevated in Graves' disease (Table 5). Radionuclide uptake and scan can differentiate high uptake in Grave's disease versus low uptake in thyroiditis [13].

Thyroid storm is a life-threatening condition that may develop if hyperthyroidism is untreated or may be precipitated by trauma or infection in a patient with hyperthyroidism. Thyroid storm is a medical emergency and accompanied by elevated blood pressure and high heart rate. Laboratory findings usually include very low or undetectable level of TSH accompanied by very high thyroid hormone levels such as FT4 and total T3.

## Disorders of parathyroid glands

The parathyroid gland produces parathyroid hormone (PTH) which along with calcitonin (produced by the thyroid gland) and vitamin D regulates calcium metabolism. PTH is an 84 amino acid hormone secreted by the chief cells of the parathyroid and increases calcium levels in blood by following mechanism:

- Increased osteoclastic activity in bone
- Increased synthesis of 1,25-dihydroxycholecalciferol (vitamin D3)
- Increased renal reabsorption of calcium
- Increased intestinal absorption of calcium

Calcitonin which is secreted by the parafollicular C cells of the thyroid essentially has the opposite action to that of PTH. The primary source of vitamin D is photoactivation of 7-dehydrocholesterol in the skin to cholecalciferol. Exposure to UV radiation of sunlight (UVB radiation: 290–320 nm) for 15 min in midday sun is enough to produce enough cholecalciferol, which is then converted to 25-hydroxycholecalciferol in the liver and then to 1,25-dihydroxycholecalciferol in the kidney. Vitamin D promotes absorption of calcium from the gut and helps in bone mineralization. Deficiency or lack of 1,25-dihydroxycholecalciferol in children causes rickets and osteomalacia in adults. In chronic kidney disease, the kidney lacks the ability to covert 25-hydroxycholecalciferol to 1,25-dihydroxycholecalciferol and vitamin deficiency may occur. Both vitamin D and PTH are responsible for calcium metabolism, and causes of hypercalcemia include:

- Primary hyperparathyroidism
- Tertiary hyperparathyroidism
- Malignancy
- Excess vitamin D
- Excessive calcium intake (milk alkali syndrome)
- Drugs therapy such as thiazide diuretics therapy

Hyperparathyroidism is a common cause of hypercalcemia, and hyperparathyroidism can be primary (due to adenomas or hyperplasia of parathyroid glands), secondary (due to compensatory hypertrophy of parathyroid glands due to hypocalcemia as seen in chronic kidney disease), or tertiary where after a long period of secondary hyperparathyroidism the parathyroid glands develop autonomous hyperplasia and hyperparathyroidism persists even when hypocalcemia is corrected.

Various causes of hypocalcemia include:

- Hypoparathyroidism which could be congenital (DiGeorge syndrome) secondary to hypomagnesemia (magnesium is required for PTH secretion) or due to parathyroidectomy
- Chronic kidney disease
- Vitamin D deficiency
- Resistance to PTH (pseudohypoparathyroidism)
- Drugs (e.g., calcitonin, bisphosphonates)
- Acute pancreatitis
- Malabsorption

Hypoparathyroidism is an uncommon disorder characterized by low serum calcium, increased serum phosphorus, and deficient production of PTH. Hypoparathyroidism can be divided into primary hypoparathyroidism, which is due to intrinsic defects within the parathyroid glands primarily due to genetic causes, but secondary or acquired forms of hypoparathyroidism due to impaired function of parathyroid glands are by far the most common etiologies. Anterior neck surgery is the most common cause of acquired hypoparathyroidism and is responsible for about 75% of cases. The next most common acquired cause in adults is thought to be autoimmune disease, affecting either the parathyroid glands alone or multiple other endocrine glands. Although the diagnosis of hypoparathyroidism is usually straightforward once serum calcium, phosphorus, and PTH levels are known, determining the cause of nonsurgical hypoparathyroidism may be challenging. Pseudohypoparathyroidism (PHP), a disorder of PTH resistance, is an even less common disease characterized also by low serum calcium, increased serum phosphorus, but with increased circulating levels of PTH [14]. Pseudohypoparathyroidism (PHP) is a hereditary disorder. These patients in addition to hypocalcemia also have short stature, short metacarpals, and intellectual impairment. Pseudohypoparathyroidism is due to the fact that the peripheral organs are unresponsive to the action of PTH. Pseudohypoparathyroidism is subdivided into different subtypes with different genetic mutations that include deletions, small mutations, or methylation (loss of function) near the GNAS locus on chromosome 20q 13.3. The latter normally mediates the action of G-protein-coupled receptors via the transcription of a signaling protein called Gs alpha [15].

Pseudo-pseudohypoparathyroidism is an inherited condition that causes short stature, round face, and short hand bones and is genetically related to pseudohypoparathyroidism type 1a (PHP-1a); however, people with pseudo-pseudohypoparathyroidism do not show resistance to PTH, while people with PHP-1a show PTH resistance. Obesity is characteristic for PHP-1a and may be severe, while obesity is less prominent and may be absent among people with pseudo-pseudohypoparathyroidism. Approximately 10% patients also have learning disability. Both PHP-1a and pseudo-pseudohypoparathyroidism are caused by mutations that affect the function of the GNAS gene. However, people who inherit the mutation from their mother develop PHP-1a, whereas those who inherit the mutation from their father develop pseudo-pseudohypoparathyroidism [16]. Pseudo-pseudohypoparathyroidism affects bone formation and causes joints and other soft tissues in the body to harden. Therefore, this disease can cause bone, joint, and nerve damage producing painful stimuli.

## Adrenal glands

The adrenal glands consist of a cortex and a medulla. The cortex has three zones: zona glomerulosa, zona fasciculata, and zona reticularis. Zone glomerulosa is responsible for secreting mineralocorticoids (aldosterone), while zona fasciculata is responsible for secreting glucocorticoids. Finally, zona reticularis is responsible for producing sex steroids. Adrenal medulla produces catecholamines. Steroid hormones are synthesized by adrenal glands from cholesterol, while sex steroid hormones are synthesized in the gonad (Scheme 1). Major actions of glucocorticoids include:

- Gluconeogenesis and glycogen deposition
- Fat deposition
- Protein catabolism
- Sodium retention

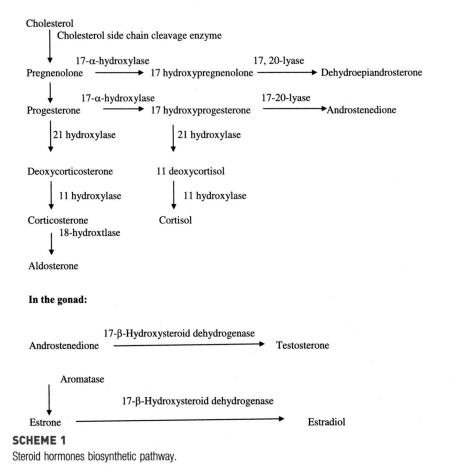

**SCHEME 1**
Steroid hormones biosynthetic pathway.

- Loss of potassium
- Increase in circulating neutrophils and decrease in circulating eosinophils and lymphocytes

Major actions of mineralocorticoids include:

- Sodium and water retention in the distal tubule
- Loss of potassium

Congenital adrenal hyperplasia is most often due to lack of 21-hydroxylase enzyme causing decreased production of deoxycorticosterone and aldosterone as well reduced levels of deoxycortisol and cortisol. Adrenocorticotropic hormone (ACTH) level is also high, and as a result, 17-hydroxypregnenolone and 17-hydroxyprogesterone are produced in higher concentrations, which lead to increased production of dehydroepiandrosterone, androstenedione, and testosterone. Female child will have virilization effect and may also have ambiguous genitalia. Male child will have features of precocious puberty.

# Cushing's syndrome

Cushing's syndrome is a hormonal disorder characterized by high circulating level of cortisol. Major causes of Cushing's syndrome can be subclassified under two broad categories:

- Adrenocorticotropic hormone (ACTH)-dependent disorders which include Cushing's disease (ACTH-secreting pituitary adenoma), ectopic ACTH-producing tumor (such as lung cancer) and secondary to ACTH administration.
- Non-ACTH dependent: Adrenal gland tumor or secondary to glucocorticoid therapy (prednisolone, dexamethasone).

It is important to note that Cushing disease, caused by a pituitary tumor or an excessive growth in the pituitary gland, is the major cause of the Cushing syndrome. Cushing syndrome is characterized by facial (moon face) and torso obesity, high blood pressure, stretch marks on the belly, weakness, osteoporosis, and facial hair growth in females. In general, 66%–70% patients show the presence of ACTH-secreting pituitary adenoma. Other causes include adrenal liaison, usually a unilateral adenoma (20%–27% patients), and ectopic ACTH secretion by neuroendocrine tumors (5%–10% patients) [17]. The most common ectopic ACTH-producing tumors are thoracic (bronchial and thymic) and gastroenteropancreatic neuroendocrine tumors, followed by medullary thyroid carcinoma, and small cell lung cancer. Pheochromocytoma, a rare usually benign tumor that develops in an adrenal gland, may also cause Cushing's syndrome [18]. Very rarely (only 75 cases reported), nonpituitary tumors may gain ability to secrete corticotropin-releasing hormone (CRH) causing hypersecretion of ACTH and subsequent stimulation of adrenal gland to produce excess cortisol. CRH-producing ectopic tumors may also simultaneously produce ACTH. CRH-producing tumors may be bronchial carcinoid tumor, thymic carcinoid tumor, pancreatic tumor, or medullary thyroid carcinoma [19]. Cushing's syndrome is fairly rare. It is more often found in women than in men and most commonly occurs between the ages of 20 and 40. Treatment of Cushing's syndrome varies according to the cause of the condition, but surgical removal of tumor is very effective. If surgical removal is not possible, radiation therapy and/or drug therapy is initiated. Certain drugs that block synthesis of cortisol (ketoconazole, fluconazole, metyrapone, mitotane, etc.) or drugs that reduce secretion of ACTH from pituitary (pasireotide, etc.) are used in treating patients with Cushing syndrome.

Diagnosis of Cushing's syndrome first requires confirmation of hypercortisolism and then the cause of high cortisol. Investigations that are useful for the diagnosis of Cushing's syndrome include:

- Measurement of 24-h urinary free cortisol (values are elevated in Cushing's syndrome).

- Loss of circadian rhythm (measurement of cortisol at 9 a.m. and midnight should show loss of circadian rhythm as evidenced by higher midnight cortisol values compared with 9 a.m. values in patients with Cushing's syndrome).
- Overnight dexamethasone suppression test: patients take 1 mg of dexamethasone at bedtime and serum cortisol is measured following morning. Cushing syndrome patients should still show elevated level of cortisol.
- Low- and high-dose dexamethasone suppression tests: discussed at the end of the chapter.
- More recently, late night or midnight salivary cortisol has been introduced for diagnosis of Cushing's syndrome. It has gained popularity because no venipuncture is needed, and patient can collect specimen at home and store it because salivary cortisol is stable.

It is important to note that renal impairment may falsely lower urinary free cortisol (it is called urinary free cortisol because cortisol level in urine is a measure of free or unbound cortisol in circulation). False-positive test result may occur in dexamethasone suppression test due to treatment with cytochrome P 450 liver enzyme inducing drugs, such as carbamazepine, phenytoin, phenobarbital, rifampicin, meprobamate, aminoglutethimide, methaqualone, and troglitazone. These drugs may significantly increase clearance of dexamethasone causing false-positive result. Moreover, measurement of plasma ACTH levels is useful in differential diagnosis because ACTH levels are low in patients with autonomous adrenal diseases, normal or elevated in patients with Cushing's disease, and elevated in ectopic ACTH syndrome. In rare situation when ACTH value is low normal, corticotropin-releasing hormone (or desmopressin) stimulation test may be useful because desmopressin stimulates ACTH and eventually cortisol release in Cushing's disease [20].

Pseudo-Cushing's syndrome is caused by conditions such as alcoholism, severe obesity, and polycystic ovary syndrome that can activate hypothalamic-pituitary-adrenal axis causing Cushing's like syndrome. Although Cushing's syndrome is rare, pseudo-Cushing's syndrome may be observed more often, but it may be difficult to distinguish between these two conditions because diagnostic tests may provide similar results. However, for alcoholics, pseudo-Cushing's syndrome may resolve spontaneously after cessation of alcohol consumption [21].

## Conn's syndrome

Conn's syndrome is most often due to an adenoma secreting aldosterone from the adrenal cortex. Clinical symptoms include hypertension (due to sodium and water retention) and hypokalemia. Therefore, it is imperative to measure serum electrolytes in a hypertensive patient if there is any suspicion of secondary hypertension. Other tests that may be helpful for diagnosis of Conn's syndrome include:

- Aldosterone to rennin ration (ARR): In Conn's syndrome, the ratio is increased.
- Rennin levels are low in Conn's syndrome.
- Plasma potassium and urinary potassium: In Conn's syndrome, hypokalemia in serum and increased loss of potassium in urine are observed.
- Saline suppression test: Here, aldosterone levels are measured before and after administration of normal saline. Normal individuals should have lower aldosterone levels with the influx of sodium, but in patients with Conn's syndrome, aldosterone level may not change.

## Hypoadrenalism including Addison's disease

Adrenal insufficiency may be primary, secondary, or tertiary. Primary adrenal insufficiency (hypoadrenalism) may be acute or chronic. Primary acute hypoadrenalism may be due to hemorrhagic destruction of adrenal glands (Waterhouse Frederickson syndrome). However, most commonly observed chronic hypoadrenalism is due to Addison's disease or primary adrenocortical insufficiency, which is characterized by inability of the adrenal cortex to produce sufficient amounts of glucocorticoids and mineralocorticoids. The disease often manifests itself between the second and fourth decades of life and affects women more often than men. Secondary hypoadrenalism is due to lack of proper secretion of ACTH from pituitary due to hypothalamus-pituitary dysfunction. Tertiary hypoadrenalism is due to lack of corticotropin-releasing hormone (CRH). Causes of Addison's disease include:

- Congenital adrenal hyperplasia due to enzyme defect
- Autoimmune disease
- Postsurgery
- Tuberculosis
- Sarcoidosis

Congenital adrenal hyperplasia is a group of inherited autosomal recessive disorders due to defect of any of five enzymes responsible for cortisol biosynthesis. Most common cause is deficiency of 21-hydroxylase (90%–95% of all cases) followed by 11-β-hydroxylase. The enzyme 21-hydroxylase converts progesterone into deoxycorticosterone and in addition also converts 17-hydroxyprogesterone into 11-deoxycortisol. If this enzyme is deficient, precursors of cortisol accumulate in blood especially 17-hydroxyprogesterone but to a lesser extent androstenedione. Peripheral conversion of androstenedione into testosterone is responsible for androgenic symptoms associated with congenital adrenal hyperplasia. Autoimmune Addison's disease is a rare disorder with symptoms typically develops over months to years due to appearance of autoantibodies to the key enzyme 21-hydroxylase. Usually, a patient demonstrates a period of compensated or preclinical disease with elevated ACTH and renin before development of symptomatic adrenal failure. The strongest genetic associations to autoimmune Addison's disease are found within certain HLA class II alleles [22]. Addison's disease may be diagnosed by establishing low levels of cortisol as well ACTH stimulation tests. These are discussed at the end of the chapter in endocrine testing section.

## Dysfunctions of gonads

Proper function of hypothalamic-pituitary-gonadal axis is vital for normal function of reproductive system of both men and women. In men GnRH, LH and FSH are secreted in pulsatile pattern with higher levels in early morning hours and lower levels at late evening. LH is also essential for men because it stimulates testosterone production by testes. Circulating testosterone is also a precursor for dihydrotestosterone and estradiol. Level of testosterone in male may start declining after age 50 termed as andropause. A healthy neonate female possesses approximately 400,000 primordial follicle each containing immature ovum and during each menstrual cycle one ovum attains maturity approximately during midcycle (approximately day 14). Therefore, only approximately 400 immature ova attain maturity during reproductive cycle of a woman. In a normal individual, most estrogen is secreted by the ovarian follicle and the corpus luteum, but during pregnancy, placenta produces most estrogen. Progesterone is secreted by corpus luteum, but during pregnancy, placenta is responsible for producing majority of progesterone. Progesterone is important for maintenance of pregnancy.

Hypogonadism may be broadly divided into two categories, hypergonadotropic hypogonadism and hypogonadotropic hypogonadism. Examples of hypergonadotropic hypogonadism include:

- Gonadal agenesis
- Gonadal dysgenesis (e.g., Turner's syndrome, Klinefelter's syndrome)
- Steroidogenesis defect
- Gonadal failure (e.g., mumps, radiation, chemotherapy, autoimmune diseases, granulomatous diseases)
- Chronic diseases (e.g., liver failure, renal failure)

Examples of hypogonadotropic hypogonadism include:

- Hypothalamic lesions (e.g., tumors, infections, Kallmann syndrome)
- Pituitary lesions (e.g., adenomas, Sheehan's syndrome, Sarcoidosis, hemochromatosis)

Polycystic ovary syndrome is most common endocrinological dysfunction in women affecting 6%–10% of women during reproductive age. This syndrome has diverse clinical implications including reproductive dysfunctions (infertility, hirsutism, and hyperandrogenism) and metabolic dysfunctions (type 2 diabetes mellitus or impaired glucose tolerance, insulin resistance, and adverse cardiovascular event), as well as psychological component such as poor quality of life, depression, and anxiety. Polycystic ovary syndrome is a polygenic disease with both genetic and environmental component, and obesity-induced insulin resistance is known to exacerbate all features of polycystic ovary syndrome. Modest weight loss may improve many features of this syndrome [23].

Investigations of hypogonadism may include:

- Measurement of basal levels of testosterone, estrogens, FSH, and LH.
- Measurement of urinary FSH, LH: This can be done in males as FSH, and LH levels in males may be undetectable in blood due to the pulsatile nature of secretion.
- GnRH stimulation test: In individuals with hypogonadotropic hypogonadism, this will help to distinguish hypothalamic causes from pituitary lesions. If administration of GnRH results in increased FSH and LH levels, this means pituitary is functional.
- Clomiphene stimulation test: Clomiphene has antiestrogenic effect. Estrogen inhibits the release of GnRh from the hypothalamus. Administration of clomiphene in normal individuals will result in release of GnRH, which in turn will cause levels of FSH and LH to rise.
- hCG (human chorionic gonadotropin) stimulation test: This is done in males with low testosterone levels. hCG binds to LH receptors and stimulates testosterone production by the Leydig cells.

## Pancreatic endocrine disorders

The most common endocrine pancreatic disorder is diabetes mellitus and is discussed in Chapter 7. Other endocrine disorders of the pancreas include:

- Islet cell tumors
- Nonfunctioning islet cell tumors
- Insulinoma (see Chapter 7)
- Gastrinoma
- VIPoma
- Glucagonoma
- Somatostatinoma

Gastrinoma causes increased secretion of gastric acid which result in multiple recurrent duodenal ulcers, which is referred as Zollinger-Ellison syndrome. VIPoma produces excessive vasoactive intestinal polypeptides (VIP), which cause watery diarrhea (Verner-Morrison syndrome). Glucagonomas are rare tumors from the alpha islet cells. Features include diabetes mellitus, migratory necrolytic dermatitis, and deep vein thrombosis. Somatostatinomas are rare tumors derived from the delta islet cells. Features include diabetes mellitus and gallstones. Nonfunctioning tumors may produce a mass effect and biliary obstruction. Approximately 25% of islet cell tumors form part of multiple endocrine neoplasia (MEN) type 1.

## Multiple endocrine neoplasia

This condition is caused by occurrence of simultaneous or metachronous tumors involving multiple endocrine glands. The subtypes are:

- MEN type 1: parathyroid adenoma or hyperplasia with pituitary adenoma and pancreatic islet cell tumor
- MEN type 2a: adrenal tumor with medullary carcinoma of thyroid and parathyroid hyperplasia
- MEN type 2b: type 2a with marfanoid habitus, intestinal and visceral ganglioneuromas

MEN type 1 is due to defect in the gene menin located on chromosome 11. Menin normally suppresses a transcription factor (JunD), and lack of this suppression results in oncogenesis. Patients with MEN 1 have one defective menin gene and one wild gene. When the wild type undergoes somatic mutation, it results in tumorigenesis. MEN 2a and 2b are due to mutations of RET proto-oncogene located on chromosome 10. Pheochromocytoma is a rare tumor of adrenal gland (usually in adrenal medulla) tissue and may be classified as a multiple endocrine neoplasia category of disease. This tumor is responsible for the release of too much epinephrine and norepinephrine that control heart rate, metabolism, and blood pressure, but rarely pheochromocytoma is cancerous. Surgical correction is the best therapy to cure this disease although drug therapy may also be used depending on the clinical judgment of the clinician.

## Endocrine testing: Suppression and stimulation tests

Endocrine testing consists of measuring blood levels of various hormones as well using suppression and stimulation tests. If a high level of hormone is observed, then suppression test is more appropriate to see if hormone level can be suppressed by using an appropriate agent. If the reason behind high levels is physiological, then hormones levels would be suppressed. If not, the underlying cause is pathological. Similarly, if initial hormone level is

Table 6 Common endocrine tests.

| Endocrine test | Analytes measured | Interpretation |
| --- | --- | --- |
| Glucose tolerance test (GTT) for hyperpituitarism | Measure basal levels of FSH, LH, TSH, ACTH, cortisol, GH and then reanalyze these analytes after administration of oral glucose | With true hyperpituitarism, basal levels will be high and will not reduce |
| Triple bolus (insulin, GnRH, TRH) | Measure basal levels of FSH, LH, TSH, prolactin, cortisol, GH and reanalyze these analytes after administration of insulin, GnRH, and TRH | With true hypopituitarism, basal levels will be low and will not rise |
| Overnight dexamethasone suppression test (1 mg dexamethasone at bed time) For Cushing's syndrome | Measure basal cortisol at 8–9 a.m. and then next morning after receiving dexamethasone | Normal patients should have cortisol below 5 µg/dL, but patients with Cushing's should not show any suppression of morning cortisol level |
| Low-dose dexamethasone test (0.5 mg q6h for 2 days) for Cushing's syndrome | Measure basal cortisol level and reanalyze cortisol level after 48 h | True Cushing' syndrome patient will have high levels and will not reduce |
| High-dose dexamethasone suppression test (2 mg q6h for 2 days). Done after positive low-dose dexamethasone suppression test to differentiate Cushing's disease from other causes | Measure basal cortisol level and after 48 h | Cushing's disease patient will show 50% or more reduction of cortisol level. Other causes of Cushing's syndrome will not |
| Short ACTH (250 µg) stimulation test for hypoadrenalism | Measure cortisol level before and after | True hypoadrenalism patients will have low basal levels and will not rise |

low, then stimulation test is performed. If underlying condition is physiological levels should rise with stimulation challenge. Common suppression and stimulation tests are summarized in Table 6.

Glucose tolerance test is performed most commonly with 75 g of oral glucose dose for diagnosis of gestational diabetes as well as diabetes mellitus if glucose levels are in borderline zone. However, glucose tolerance test is also useful in diagnosis of hyperpituitarism. Most commonly, this test is used for diagnosis of acromegaly. Most common cause of acromegaly is growth hormone (GH)-secreting pituitary adenoma, and surgical removal of tumor is the first choice for therapy. Immediate postoperative GH level is a good indicator for successful surgery, but oral glucose tolerance test performed a week later where GH level is suppressed below 1 µg/L is a good predictor for long-term remission of acromegaly [24]. Although glucose tolerance test is used most commonly for diagnosis of acromegaly, it can also be used in general in diagnosis of pituitarism by measuring base level of any of combination of hormones, including LH, FSH, ACTH, cortisol, and GH. Following oral administration of glucose, values of these hormones should be suppressed, but with true hyperpituitarism, basal levels will be high and would not be suppressed following administration of oral glucose.

For diagnosis of hypopituitarism especially dysfunction of anterior pituitary, bolus test (also known as dynamic pituitary function test) is used where three hormones, including insulin, gonadotropin-releasing hormone (GnRH), and thyrotropin-releasing hormone (TRH), are injected in bolus into a patient's vein to stimulate anterior pituitary gland. Before bolus injection, baseline levels of cortisol, GH, prolactin, TSH, LH, and FSH are measured. After bolus administration, insulin-induced hypoglycemia should increase levels of cortisol and GH, while TRH should increase levels of TSH and prolactin, while levels of LH and FSH should be increased due to administration of GnRH. Serum glucose value is also measured to ensure hypoglycemia is induced by insulin. However, in a patient with hypopituitarism, levels of these hormones should stay low at baseline values despite administration of these hormones by bolus injection.

The Endocrine Society recommends initial use of one test with high diagnostic accuracy (urine cortisol, late night salivary cortisol, 1 mg overnight or 48-h, 2 mg/day dexamethasone suppression test) for screening of Cushing's syndrome. Dexamethasone is a potent glucocorticoid that suppresses the nocturnal rise in ACTH level thus suppresses 8 a.m. cortisol level in a normal individual. Overnight, 1 mg dexamethasone suppression test (low-dose suppression test) is the most commonly used screening test due to ease and convenience. Dexamethasone 1 mg is administered orally between 11 PM and midnight, and serum cortisol levels are drawn the next morning between 8 and 9 a.m. A cortisol value <5 µg/dL (preferably <1.8 µg/dL) is considered as the normal response, but a value over 5 µg/dL may indicate Cushing's syndrome. In fertile women, false-positive test results in dexamethasone suppression test are often due to the use of oral contraceptives. By elevating cortisol-binding globulin, these contraceptives increase the total serum cortisol concentration. In another low-dose dexamethasone suppression test, 0.5 mg dexamethasone is administered every 6 h for 2 days. Blood is drawn around 9 a.m. for baseline cortisol on day 1 before dexamethasone administration. Then, dexamethasone is administered over 48 h in doses of 0.5 mg, beginning at 9 a.m. on day 1, at 6-h intervals for a total of eight doses. Serum cortisol is measured 6 h after the last dose of dexamethasone at 9 a.m. in the morning prior to food ingestion. In patients with Cushing's syndrome, no suppression of cortisol is observed [25].

High-dose dexamethasone suppression test is useful to differentiate Cushing's syndrome caused by adrenal tumors and nonendocrine ACTH-secreting tumor from Cushing's disease. This test is usually performed after 1 mg dexamethasone suppression test. In overnight high-dose dexamethasone suppression test, baseline morning serum cortisol is measured and later dexamethasone 8 mg is administered orally between 11 PM and midnight. Repeat serum cortisol is drawn the next morning (between 8 and 9 a.m.). In 2-day, 8 mg high-dose dexamethasone suppression test, on day 1, a baseline morning serum cortisol or 24-h urinary free cortisol from the day before is obtained. Then, oral dexamethasone 2 mg every 6 h is given for 2 days with simultaneous collection of a urine sample for urine free cortisol. Serum cortisol levels are checked 6 h after the last dose (9 a.m.). In patients with Cushing's syndrome, no suppression of morning cortisol level should be observed, but in patients with Cushing's disease, 50% or more reduction of serum cortisol should be observed.

Dexamethasone suppression test can also be conducted by giving dexamethasone intravenously rather than orally, but such test is less common. Moreover, oral dexamethasone suppression test can be conducted using urine, but due to requirement of collecting all urine specimens over 3 days which is inconvenient, this procedure is also less commonly used.

The combined dexamethasone suppression-corticotropin-releasing hormone test is used to distinguish Cushing's syndrome from pseudo-Cushing's states in patients with mild hypercortisolism based on the rationale that glucocorticoid suppression of the HPA axis can be overcome by corticotropin-releasing hormone stimulation in Cushing disease but not in pseudo-Cushing syndrome. For this test, dexamethasone 0.5 mg is administered orally every 6 h for a total of eight doses over 2 days. Then, 2 h after the last dose of dexamethasone, an intravenous CRH dose of 1 μg/kg is administered and serum cortisol is drawn 15 min later. Serum cortisol levels over 1.4 μg/dL at 15 min suggest Cushing's disease [26].

Although glucose tolerance test is sometimes considered as a gold standard for evaluating hypothalamus-pituitary-adrenal function in diagnosis of adrenal insufficiency, ACTH stimulation test (also known as cosyntropin test) is commonly used to evaluate functional capacity of adrenal glands. ACTH stimulation test measures the ability of the adrenal cortex to respond to exogenously administered ACTH by producing appropriate amount of cortisol. The Endocrine Society guideline considers a short ACTH stimulation test the "gold standard" for the diagnosis of adrenal insufficiency. According to the guidelines, cortisol levels should be measured after an intravenous administration of synthetic ACTH (tetracosactrin: 1–24 amino acid sequence of human ACTH; generic name cosyntropin) at the standard dose of 250 μg (0.25 mg) for adults and children over 2 years of age, a dose of 15 μg/kg for newborns, and a dose of 125 μg for children under 2 years, respectively. In this test, normal individual should show two- to threefold increases in serum cortisol (a gradual increase with time) after administration of exogenous synthetic ACTH. A value over 20 μg/dL 30–60 min following administration of synthetic ACTH is considered as a normal response, but a subnormal cortisol response (<18 μg/dL;

<500 nmol/L) 30 or 60 min after stimulation test is considered as a positive test and indicates an increased possibility of primary or secondary adrenal insufficiency [27]. There are some variations of this protocol. Synthetic ACTH may be given intramuscularly, and serum cortisol may be measured as early as 20 min or serially at specific time interval. Sometimes long-acting ACTH stimulation test using 1 mg of synthetic ACTH is used for differentiation between primary and secondary hypoadrenalism.

However, 250 μg of synthetic ACTH is supra-physiological and is capable of generating false-negative result in patients with milder form of adrenal insufficiency. As a result, it has been suggested that a more suitable adrenal evaluation can be obtained, by using a lower dose of synthetic ACTH (1 μg), and by evaluating cortisol levels at 20–30 min after synthetic ACTH injection [28]. Another alternative approach to test function of hypothalamus-pituitary-adrenal is administration of metyrapone, an inhibitor of 11β-hydroxylase enzyme that converts 11-deoxycortisol to cortisol. Under normal condition, reduced cortisol level in plasma stimulates ACTH release, and concentration of 11-deoxycortisol in serum increases significantly but a lack of response suggests primary adrenal failure.

## Summary/key points

- Endocrine activity can be classified as autocrine, paracrine, and classical endocrine activity. In autocrine activity, chemicals produced by a cell act on the cell itself. In paracrine activity, chemicals produced by a cell act locally. However, in classical endocrine activity, chemicals produced by an endocrine gland act at a distant site after their release in the circulation, and these chemicals are termed as hormones. Most classical hormones are secreted into the systemic circulation. However, hypothalamic hormones are secreted into the pituitary portal system.
- Receptors for hormones may be cell surface or membrane receptors or nuclear receptors.
- Hormone secretion may be continuous or intermittent. Thyroid hormone secretion is continuous. Thus, levels may be measured at any time to assess hormonal status. Secretion of follicle-stimulating hormone (FSH), luteinizing hormone LH and GH are pulsatile. Therefore, a single measurement may not reflect hormonal status. Some hormones exhibit biological rhythms. Cortisol exhibits circadian rhythm where levels are highest in the morning and lowest during late night. The menstrual cycle is an example of a longer biological rhythm where different level of a hormone is observed during a specific part of the cycle.
- Certain hormone levels are elevated during stress. These include adrenocorticotropic hormone (ACTH) as well as cortisol, growth hormone (GH), prolactin, adrenaline, and noradrenaline.

- Certain hormone levels are increased during sleep such as growth hormone (GH) and prolactin.
- The hypothalamus produces thyrotropin-releasing hormone (TRH), corticotrophin hormone (CRH), gonadotropin-releasing hormone (GnRH), growth hormone-releasing hormone (GRH), and somatostatin (growth hormone inhibitory hormone). These hormones act on the anterior pituitary resulting in release of various other hormones, including thyroid-stimulating hormone (TSH), adrenocorticotropin (ACTH), FSH, LH, and GH. Somatostatin inhibits the release of GH. Dopamine (also known as prolactin inhibitory hormone) is also a neurotransmitter which is also produced by the hypothalamus. Dopamine can inhibit GH secretion.
- The supra optic and paraventricular nuclei of the hypothalamus produce antidiuretic hormone (ADH; also known as vasopressin) and oxytocin. These hormones are stored in the posterior pituitary and act on certain body parts rather than acting on pituitary like other tropic hormones. ADH acts on the collecting ducts of the renal tubules and causes absorption of water.
- Growth hormone (GH: also known as somatotropin) is the most abundant hormone produced by anterior pituitary and it stimulates growth of cartilage, bone, and many soft tissues. GH stimulates release of insulin-like growth factor-I (IGF-I; somatomedin C), mostly from the liver.
- Hyperpituitarism is most often due to pituitary tumors affecting GH-secreting cells, prolactin-secreting cells, and adrenocorticotropic hormone (ACTH)-secreting cells. GH-secreting tumors affecting individuals before closure of epiphysis result in gigantism and after closure result in acromegaly. Prolactin-secreting tumors cause hyperprolactinemia. A high level of prolactin (prolactinoma) inhibits action of follicle-stimulating hormone (FSH) and luteinizing hormone (LH), resulting in hypogonadism and infertility. ACTH-secreting tumors cause Cushing's syndrome.
- For diagnosis of hypopituitarism, stimulation test with gonadotropin-releasing hormone, thyrotropin-releasing hormone, and insulin induced hypoglycemia (triple stimulation test) is useful. Following stimulation, serum or plasma levels of FSH, LH, TSH, PRL, GH, and cortisol are measured.
- Endocrine tests for hyperpituitarism include measurement of hormone levels and suppression test using glucose (oral glucose tolerance). Administration of glucose with rise in blood glucose should suppress anterior pituitary hormones in normal individuals.
- Four steps are involved in the synthesis of thyroid hormones: inorganic iodide from the circulating blood is trapped (iodide trapping), oxidation of iodide to iodine, addition of iodine to tyrosine to produce

monoiodotyrosine and diiodotyrosine which is referred to as organification, and finally one monoiodotyrosine is coupled with one diiodotyrosine to yield T3 and two diiodotyrosine are coupled to yield T4 (coupling).

- Free (unbound) T4 is the primary secretory hormone from the thyroid gland, and T4 is converted in peripheral tissue (liver, kidney, and muscle) to T3 by 5′-monodeiodination. T3 is the physiologically active hormone. T4 can also be converted to reverse T3 by 3′-monodeiodination. This form of T3 is inactive. Majority (99%) of the T3 and T4 in circulation are found to thyroxine-binding globulin (TBG), albumin, and thyroxine-binding prealbumin.
- Dyshormonogenetic goiter may be associated with nerve deafness, referred to as Pendred's syndrome.
- There are situations where thyroid hormone-binding proteins may be low or high causing alteration of total T3 and T4 levels. However, free T3 and T4 and TSH levels should be normal. Pregnancy and oral contraceptive pills raise concentrations of thyroid-binding proteins. Hypoproteinemia, which occurs in cirrhosis of liver, nephrotic syndrome, etc., may cause lower concentration of thyroid-binding proteins.
- Amiodarone can reduce peripheral conversion of T4 to T3. Free T4 levels may be high, but TSH level could be normal. Amiodarone can also cause both hypothyroidism and hyperthyroidism because amiodarone contains iodine molecule.
- Seriously ill patients may have reduced production of TSH with low T4 and reduced conversion of T4 to T3 with increased conversion of T4 to reverse T3. Patients are however euthyroid. This is referred to as sick euthyroid syndrome.
- Measurement of TSH can suffer interference from heterophilic antibody and rheumatoid factor causing falsely elevated results. Rarely autoantibodies to TSH develop clinically but such autoantibodies can also falsely increase TSH result. A rare interference in TSH assay is due to macro-TSH, an autoimmune complex between anti-TSH IgG antibody and TSH.
- Primary hypothyroidism (primary disease of thyroid gland): Causes include autoimmune thyroiditis, Hashimoto's thyroiditis, surgery/radiation, dyshormonogenesis, antithyroid drugs, drugs therapy with amiodarone, and advanced age.
- Secondary hypothyroidism may be due to lack of TSH from pituitary or peripheral resistance to thyroid hormones.
- Causes of hyperthyroidism include Graves' disease, toxic nodular (single or multiple) goiter, thyroiditis (e.g., due to viral infection), drugs induced, and excess TSH (e.g., due to pituitary tumor).

- PTH (parathyroid hormone) is an 84 amino acid hormone secreted by the chief cells of the parathyroid and increases calcium levels in blood by increasing osteoclastic activity in bone, increasing synthesis of 1,25-dihydroxycholecalciferol (vitamin D3), increasing renal reabsorption of calcium, and increasing intestinal absorption of calcium.
- Calcitonin which is secreted by the parafollicular C cells of the thyroid essentially has the opposite action to that of PTH.
- Hyperparathyroidism is a common cause of hypercalcemia, and hyperparathyroidism can be primary (due to adenomas or hyperplasia of parathyroid glands), secondary (due to compensatory hypertrophy of parathyroid glands due to hypocalcemia as seen in chronic kidney disease), or tertiary where after a long-period of secondary hyperparathyroidism the parathyroid glands develop autonomous hyperplasia and hyperparathyroidism persists even when hypocalcemia is corrected.
- Hypoparathyroidism refers to low levels of PTH being secreted from the parathyroid glands, whereas pseudohypoparathyroidism refers to inability of PTH to exert its function due to receptor defect. Pseudohypoparathyroidism is a hereditary disorder and patients in addition to hypocalcemia also have short stature, short metacarpals, and intellectual impairment. Pseudo-pseudohypoparathyroidism patients actually have no abnormality of PTH or parathyroid. Only the somatic features seen in pseudohypoparathyroidism are present, but these patients do not have cognitive impairment as seen is patients with pseudohypoparathyroidism.
- The adrenal glands consist of a cortex and a medulla. The cortex has three zones: zona glomerulosa, zona fasciculata, and zona reticularis. Zone glomerulosa is responsible for secreting mineralocorticoids (aldosterone), while zona fasciculata is responsible for secreting glucocorticoids. Finally, zona reticularis is responsible for producing sex steroids. Adrenal medulla produces catecholamines.
- Congenital adrenal hyperplasia is most often due to lack of 21 hydroxylase enzyme causing decreased production of deoxycorticosterone and aldosterone as well reduced levels of deoxycortisol and cortisol. Adrenocorticotropic hormone (ACTH) level is also high, and as a result, 17-hydroxypregnenolone and 17-hydroxyprogesterone are produced in higher concentrations, which lead to increased production of dehydroepiandrosterone, androstenedione, and testosterone. Female child will have virilization effect and may also have ambiguous genitalia. Male child will have features of precocious puberty.
- Major causes of Cushing's syndrome can be subclassified under two broad categories:

- – Adrenocorticotropic hormone (ACTH)-dependent disorders which include Cushing's disease (ACTH-secreting pituitary tumor), ectopic ACTH producing tumor (such as lung cancer), and secondary to ACTH administration
  - – Non-ACTH dependent: Adrenal tumor or secondary to glucocorticoid administration
- ACTH-dependent Cushing's syndrome is more common (70%–80% of all cases) compared with non-ACTH-dependent disorders. Among ACTH-dependent disorders, Cushing's disease is observed more frequently.
- Investigations that are useful for the diagnosis of Cushing's syndrome include measurement of 24-h urinary free cortisol (values are elevated in Cushing's syndrome), loss of circadian rhythm (measurement of cortisol at 9 AM and midnight should show loss of circadian rhythm as evidenced by higher midnight cortisol values compared to 9 a.m. values in patients with Cushing's syndrome), overnight dexamethasone suppression test (patients take 1 mg of dexamethasone at bedtime and serum cortisol is measured following morning. Cushing syndrome patients should still show elevated level of cortisol), as well as low- and high-dose dexamethasone suppression tests.
- Pseudo-Cushing's syndrome is caused by conditions such as alcoholism, severe obesity, and polycystic ovary syndrome that can activate hypothalamic-pituitary-adrenal axis causing Cushing's like syndrome.
- Conn's syndrome is most often due to an adenoma-secreting aldosterone from the adrenal cortex. Clinical symptoms include hypertension (due to sodium and water retention) and hypokalemia. Therefore, it is imperative to measure serum electrolytes in a hypertensive patient if there is any suspicion of secondary hypertension. Other tests that may be helpful for diagnosis of Conn's syndrome include aldosterone to rennin ration (ARR: in Conn's syndrome, the ratio is increased), plasma potassium and urinary potassium (in Conn's syndrome, hypokalemia in serum and increased loss of potassium in urine are observed), and saline suppression test (aldosterone levels are measured before and after administration of normal saline. Normal individuals should have lower aldosterone levels with the influx of sodium, but in patients with Conn's syndrome, aldosterone level may not change). Rennin levels are also low in Conn's syndrome.
- Adrenal insufficiency may be primary, secondary, or tertiary. Primary adrenal insufficiency (hypoadrenalism) may be acute or chronic. Primary acute hypoadrenalism is most commonly due to hemorrhagic destruction of adrenal glands (Waterhouse Frederickson syndrome). Most commonly observed chronic hypoadrenalism is due to Addison's

disease, which is related to progressive dysfunction of adrenal glands due to a local disease process or systematic disorder. Secondary hypoadrenalism is due to lack of ACTH from pituitary due to hypothalamus-pituitary dysfunction. Tertiary hypoadrenalism is due to lack of corticotropin-releasing hormone (CRH).

- Causes of Addison's disease include congenital adrenal hyperplasia due to enzyme defect, autoimmune disease, postsurgery, tuberculosis, and sarcoidosis.
- Hypogonadism may be broadly divided into two categories, hypergonadotropic hypogonadism and hypogonadotrophic hypogonadism. Examples of hypergonadotropic hypogonadism include gonadal agenesis, gonadal dysgenesis (e.g., Turner's syndrome, Klinefelter's syndrome), steroidogenesis defect, gonadal failure (e.g., mumps, radiation, chemotherapy, autoimmune diseases, granulomatous diseases), and chronic diseases (e.g., liver failure, renal failure).
- Examples of hypogonadotrophic hypogonadism include hypothalamic lesions (e.g., tumors, infections, Kallmann's syndrome) and pituitary lesions (e.g., adenomas, Sheehan's syndrome, Sarcoidosis, hemochromatosis).
- Gastrinomas cause increased secretion of gastric acid which result in multiple recurrent duodenal ulcers, which is referred as Zollinger-Ellison syndrome. VIPomas produce excessive vasoactive intestinal polypeptides (VIP), which cause watery diarrhea (Verner Morrison syndrome). Glucagonomas are rare tumors from the alpha islet cells. Features include diabetes mellitus, migratory necrolytic dermatitis, and deep vein thrombosis. Somatostatinomas are rare tumors derived from the delta islet cells. Features include diabetes mellitus and gallstones. This condition is caused by occurrence of simultaneous or metachronous tumors involving multiple endocrine glands. The subtypes are MEN type 1 (parathyroid adenoma or hyperplasia with pituitary adenoma and pancreatic islet cell tumor), MEN type 2a (adrenal tumor with medullary carcinoma of thyroid and parathyroid hyperplasia), and MEN type 2b (type 2a with marfanoid habitus, intestinal and visceral ganglioneuromas).
- Glucose tolerance test is useful in the diagnosis of hyperpituitarism.
- For diagnosis of hypopituitarism, bolus test (also known as dynamic pituitary function test) is used, where three hormones including insulin, gonadotropin-releasing hormone (GnRH), and thyrotropin-releasing hormone (TRH) are injected as a bolus into a patient's vein to stimulate anterior pituitary gland. Before bolus injection baseline levels of cortisol, GH, prolactin, TSH, LH, and FSH are measured. After bolus administration, insulin-induced hypoglycemia should increase levels

of cortisol and GH, while TRH should increase levels of TSH and prolactin, while levels of LH and FSH should be increased due to administration of GnRH. Serum glucose value is also measured to ensure hypoglycemia induced by insulin. In a patient with hypopituitarism, levels of these hormones should stay low at baseline values despite administration of these hormones by bolus injection.

- In patients with Cushing's syndrome, no suppression of cortisol level is observed following administration of dexamethasone.
- High-dose dexamethasone suppression test is useful to differentiate Cushing's syndrome caused by adrenal tumors and nonendocrine ACTH-secreting tumor from Cushing's disease.
- Although glucose tolerance test is sometimes considered as a gold standard for evaluating hypothalamus-pituitary-adrenal function in adrenal insufficiency, ACTH stimulation test (also known as cosyntropin test) is also used to evaluate functional capacity of adrenal glands in a patient with suspected adrenal insufficiency. Sometimes long-acting ACTH stimulation test using 1 mg of synthetic ACTH is used for differentiation between primary and secondary hypoadrenalism.
  In primary hypoadrenalism, there will be no rise in serum cortisol. However, in secondary hypoadrenalism, there will be gradual increase in serum cortisol.

# References

[1] Amaral FGD, Cipolla-Neto J. A brief review about melatonin, a pineal hormone. Arch Endocrinol Metab 2018;62:472–9.

[2] Bosch OJ, Young LJ. Oxytocin and social relationships: from attachment to bond disruption. Curr Top Behav Neurosci 2018;35:97–117.

[3] Pauche JE, Castilla-Cortazar I. Human conditions of insulin-like growth factor-1 (IGF-1) deficiency. J Transl Med 2012;10:224.

[4] Todd CH. Management of thyroid disorders in primary care challenges and controversies. Postgrad Med J 2009;85:655–9.

[5] Livingston EH. Subclinical hypothyroidism: a review. JAMA 2019;322:180.

[6] Loh TP, Kao SL, Halsall DJ, Toh SA, et al. Macro-thyrotropin: a case report and review of literature. J Clin Endocrinol Metab 2012;97:1823–8.

[7] Favresse J, Burlacu MC, Maiter D, Gruson D. Interferences with thyroid function immunoassays: clinical implications and detection algorithm. Endocr Rev 2018;39:830–50.

[8] Dufour DR. Laboratory tests of thyroid function: use and limitations. Endocrinol Metab Clin North Am 2007;36:579–94.

[9] Shahbaz A, Aziz K, Umair M, Sachmechi I. Prolonged duration of Hashitoxicosis in a patient with Hashimoto's thyroiditis: a case report and review of literature. Cureus 2018;10(6), e2804.

[10] Hu S, Rayman MP. Multiple nutritional factors and the risk of Hashimoto's thyroiditis. Thyroid 2017;27:597–610.

[11] Gaitonde D, Rowley K, Sweeney LB. Hypothyroidism: an update. Am Fam Physician 2012;86:244–51.

[12] Jurecka-Lubieniecka B, Ploski R, Kula D, Krol A, et al. Association between age at diagnosis of Grave's disease and variants in genes involved in immune response. PLoS One 2013;8(3): e59349.

[13] Reid JR, Wheeler SF. Hyperthyroidism: diagnosis and treatment. Am Fam Physician 2005;72:623–30.

[14] Clarke BL, Brown EM, Collins MT, Jüppner H. Epidemiology and diagnosis of hypoparathyroidism. J Clin Endocrinol Metab 2016;101:2284–99.

[15] Najim MS, Ali R, Awad M, Omer A. Pseudohypoparathyroidism presenting with seizures: a case report and literature review. Intractable Rare Dis Res 2020;9:166–70.

[16] Bastepe M. The GNAS locus and pseudohypoparathyroidism. Adv Exp Med Biol 2008;626:27–40.

[17] Hirsch D, Shimon I, Manisterski Y, Aviran-Barak N, et al. Cushing's syndrome: comparison between Cushing's disease and adrenal Cushing's. Endocrine 2018;62:712–20.

[18] Okumura T, Takayama S, Nishio SI, Miyakoshi T, et al. ACTH-producing thymic neuroendocrine tumor initially presenting as psychosis: a case report and literature review. Thorac Cancer 2019;10:1648–53.

[19] Nakhjavani M, Amirbaigloo A, Rabizadeh S, Rotondo F, et al. Ectopic Cushing's syndrome due to corticotropin releasing hormone. Pituitary 2019;22:561–8.

[20] Bansal V, El Asmar N, Selman WR, Arafah BM. Pitfalls in the diagnosis and management of Cushing's syndrome. Neurosurg Focus 2015;38(2):E4.

[21] Besemer F, Pireira AM, Smit JW. Alcohol induced Cushing syndrome: hypercortisolism cause by alcohol abuse. Neth J Med 2011;69:318–23.

[22] Hellesen A, Bratland E, Husebye ES. Autoimmune Addison's disease – an update on pathogenesis. Ann Endocrinol (Paris) 2018;79:157–63.

[23] Azziz R. Polycystic ovary syndrome. Obstet Gynecol 2018;132:321–36.

[24] Kim EU, Oh MC, Lee EJ, Kim SH. Predicting long term remission by measuring immediate postoperative growth hormone levels and oral glucose tolerance test in acromegaly. Neurosurgery 2012;70:1106–13.

[25] Nieman LK, Biller BM, Findling JW, Newell-Price J, et al. The diagnosis of Cushing's syndrome: an Endocrine Society clinical practice guideline. J Clin Endocrinol Metab 2008;93:1526–40.

[26] Dogra P, Vijayashankar NP. Dexamethasone suppression test. In: StatPearls [Internet]. Treasure Island, FL: StatPearls Publishing; 2020.

[27] Bornstein SR, Allolio B, Arlt W, Barthel A, et al. Diagnosis and treatment of primary adrenal insufficiency: an Endocrine Society clinical practice guideline. J Clin Endocrinol Metab 2016;101:364–89.

[28] Mongioì LM, Condorelli RA, Barbagallo F, Cannarella R, et al. Accuracy of the low-dose ACTH stimulation test for adrenal insufficiency diagnosis: a re-assessment of the cut-off value. J Clin Med 2019;8(6):806.

# CHAPTER 10

# Liver diseases and liver function tests

## Liver physiology

Liver is the largest internal organ of the body, approximately 1.2–1.5 kg in weight. Liver performs multiple functions essential to sustain life and is the principal site for synthesis of all circulating proteins, except gamma globulins. A functioning normal liver produces 10–12 g of albumin daily and the half-life of albumin is approximately 3 weeks. When liver function is impaired over a prolonged period, albumin synthesis is impaired, which will result in low levels of albumin. Thus, hypoalbuminemia is a common finding in chronic liver disease. However, significant reduction in serum albumin level will not be observed in patients with acute liver failure.

In addition to albumin, all clotting factors with the exception of Factor VIII are produced in the liver. Therefore, as expected when liver function is significantly impaired, in either acute or chronic setting, clotting factor levels will be low. As a result, coagulation tests such as prothrombin time (PT) is prolonged.

Liver is the site of urea production. In severe liver disease, such as fulminant hepatic failure, urea levels may be low. Liver also stores about 80 g of glycogen. Liver releases glucose into the circulation by glycogenolysis and gluconeogenesis. Again in severe liver disease, hypoglycemia may be apparent due to impaired glycogenolysis and gluconeogenesis.

Liver also plays a major role in synthesis of various lipoproteins including very-low-density lipoprotein (VLDL) and high-density lipoprotein (HDL). Hepatic lipase removes triglycerides from intermediate density lipoprotein (IDL) to produce low-density lipoprotein (LDL). Liver is also a site for cholesterol synthesis. Cholesterol is esterified with fatty acids by the action of enzyme lecithin cholesterol acyl transferase (LCAT). In liver disease, LCAT activity may be reduced resulting in increased ratio of cholesterol to cholesteryl ester. This may alter membrane structure with formation of target cells, as seen in liver disease.

**207**

Clinical Chemistry, Immunology and Laboratory Quality Control. https://doi.org/10.1016/B978-0-12-815960-6.00018-2

Bile acids are also synthesized in the liver from cholesterol and are excreted as bile salts. The primary bile acids, cholic acid, and chenodeoxycholic acid are converted into secondary bile acids by bacterial enzymes in the intestine. The secondary bile acids are deoxycholic and lithocholic acid. In liver diseases, decreased production of bile acids may result in fat malabsorption.

Liver is the site for bilirubin metabolism. Heme, derived from the breakdown of hemoglobin, is converted to biliverdin and finally into bilirubin, which is water soluble and called unconjugated bilirubin. Unconjugated bilirubin binds with serum proteins, most commonly albumin. The unconjugated bilirubin is taken up by the liver and with the help of the enzyme UDP (uridine-5′-phospho), glucuronosyltransferase, is converted to conjugated bilirubin (bilirubin conjugated with glucuronide). This conjugation takes place in the smooth endoplasmic reticulum of the hepatocyte. Conjugated bilirubin is water soluble and is excreted in bile. In the clinical laboratory, conjugated bilirubin is measured as direct bilirubin while subtracting total bilirubin from direct bilirubin value provides concentration of unconjugated bilirubin, which is also referred as indirect bilirubin. In the intestine, bacterial enzymes break down conjugated bilirubin and releases free bilirubin, which is reduced to urobilinogen. Urobilinogen bound to albumin is excreted in the urine. Some urobilinogen is converted to stercobilinogen in the intestine and is excreted in stool. Thus, in normal urine, only urobilinogen is present and in normal stool stercobilinogen is present. In obstructive (cholestatic) jaundice, conjugated bilirubin regurgitates into blood because it is water soluble and it is excreted into the urine. This is called choluria or presence of bile in urine. In obstructive jaundice, lesser amount of conjugated bilirubin is taken by the intestine and as a result less amount of stercobilinogen is found in stool (pale stools). Distribution of bilirubin, urobilinogen, and stercobilinogen in various diseases are summarized below:

- In individuals with hemolytic anemia excess breakdown of hemoglobin causes unconjugated hyperbilirubinemia. Urobilinogen in urine and stercobilinogen in stool may also be increased.
- In hepatocellular jaundice uptake and conversion of unconjugated bilirubin into conjugated bilirubin is also reduced resulting in unconjugated hyperbilirubinemia. However, amounts of urobilinogen in urine and stercobilinogen in stool are not increased.
- In cholestatic jaundice, conjugated hyperbilirubinemia is usually observed. Because conjugated bilirubin is water soluble, it is excreted in urine (choluria). However, urobilinogen and stercobilinogen quantities are reduced.

Major functions of the liver are summarized in Table 1.

**Table 1** Physiological function of the liver.

| Liver function | Comments |
| --- | --- |
| Protein synthesis | Low albumin is a feature of chronic liver disease |
| Synthesis of clotting factors | Factor VIII is not produced in the liver. Significant liver dysfunction results in prolonged PT |
| Urea synthesis from ammonia and carbon dioxide | Low urea level in fulminant hepatic failure |
| Liver releases glucose by glycogenolysis and gluconeogenesis | Hypoglycemia in fulminant hepatic failure |
| Lipid metabolism | Liver plays important role in lipoprotein and lipid metabolism |
| Bilirubin synthesis | Hemolytic and hepatocellular jaundice results in unconjugated hyperbilirubinemia. Cholestatic jaundice results in increase in conjugated bilirubin with resultant choluria |
| Bile acid synthesis | Liver dysfunction may result in malabsorption |

## Liver function tests and interpretations

Conventional liver function test (LFT) consists of measuring following analytes:

- Serum bilirubin (total)
- Serum albumin
- Prothrombin time (PT)
- Liver enzymes: alanine aminotransferase (ALT, formerly called SGPT), aspartate aminotransferase (AST, formerly called SGOT) and alkaline phosphatase measured in serum or plasma

If necessary, the physician can order total bilirubin, as well as direct (conjugated) and indirect (unconjugated) bilirubin levels. In addition to the enzymes mentioned earlier, the physician can also order additional enzymes, including gamma glutamyl transpeptidase (GGT) and lactate dehydrogenase (LDH) measured in serum or plasma. Another enzyme 5 nucleotidase may also be used along with other biochemical markers of liver dysfunction. This enzyme is a glycoprotein generally disseminated throughout the tissues of the body localized in cytoplasmic membrane, catalyzing release of inorganic phosphate from nucleoside-5-phosphates. The normal range is 0–15 U/L. Elevated levels are observed in patients with obstructive jaundice, parenchymal liver disease, and hepatic metastases. Elevation of 5 nucleotidase is also found in acute infective hepatitis, as well as chronic hepatitis. Ceruloplasmin is synthesized in the liver and is an acute phase protein. Sometimes ceruloplasmin test is also ordered as a part of liver function test. Serum ceruloplasmin levels are elevated in the chronic active liver disease but lowered in the Wilson's disease because copper bind to ceruloplasmin. If ceruloplasmin synthesis is impaired, then copper tends to accumulate causing Wilson's disease.

Breakdown of hepatocytes results in release of aminotransferases (also referred to as transaminases) such as ALT (alanine aminotransferase) and AST (aspartate aminotransferase) into the blood. ALT is a cytosol enzyme and is more specific for liver disease. AST is primarily a mitochondrial enzyme, which is also found in heart, muscle, kidney, and brain. ALT has a longer half-life than AST. In acute liver injury, AST levels are higher than ALT; however, after 24–48 h, ALT levels should be higher than AST.

Alkaline phosphatase (ALP) is found in liver, bone, intestine, and placenta. ALP is located in the canalicular and sinusoidal membrane of the liver. Production of alkaline phosphatase is increased during cholestasis (intrahepatic or extrahepatic). However, it is important to determine if the source of ALP is liver or other organs. Gamma glutamyl transpeptidase (GGT), a microsomal enzyme, and 5 nucleotidase are solely produced by the biliary epithelium. If either level is raised with ALP, then it can be assumed that the source of the ALP is liver. GGT is a well-established marker for alcohol abuse. In men over age of 40, consumption of even eight standard drinks per week may elevate serum GGT level, whereas those below 40 may show significant change in GGT level only after consumption of 14 standard drinks per week [1].

In acute liver disease, both total protein and serum albumin concentrations are unaltered. In chronic liver disease, total protein may be low or high. If total protein is high, it is most likely due to polyclonal hypergammaglobulinemia. Serum albumin is likely to be low in chronic liver disease and serves as a guide to the severity of the liver disease.

PT reflects integrity of the extrinsic and common pathway activity of the coagulation cascade. Clotting factors with the exception of Factor VIII are produced in the liver. Factor VII is one of the factors involved in the extrinsic pathway and has the shortest half-life among all clotting factors (4–6 h). Therefore, within a short period after significant liver dysfunction PT is prolonged and the magnitude of prolongation is correlated with severity of liver dysfunction. Important points regarding interpretation of liver function tests include:

- In acute liver disease without cholestasis, levels of ALT and AST are significantly elevated and ALP is raised but usually less than three times of upper limit of normal value. Thus, ALT and AST levels exceeding 500 U/L is a common finding in acute liver disease.
- In acute cholestasis, ALT and AST are raised but levels are not very high. ALP is usually more than three times of upper limit of normal value with parallel increase in GGT levels.
- In acute liver disease, total protein and albumin levels are unaltered.

- PT is the best test to assess extent of liver dysfunction.
- In chronic liver disease, albumin levels are low but total protein may be elevated.
- Elevated ALP with normal bilirubin, ALT, AST may be seen in patients with hepatic metastasis or bone metastasis. Patients with hepatic metastasis will also have elevated GGT.
- Children showing only elevated ALP is most likely related to osteoblastic activity in their growing bones.
- Isolated elevated ALP in the elderly is a characteristic feature of Paget's disease.
- Very high isolated ALP levels can be seen in primary biliary cirrhosis.
- Normal liver function test except elevated GGT is a characteristic of excessive alcohol intake. Certain drugs (e.g., warfarin, anticonvulsants) may also produce similar observation.
- An AST to ALT ratio greater than 2:1 is suggestive of alcoholic liver disease.
- AST and ALT if greater than 25 times upper limit of normal: a feature of acute viral hepatitis or toxin-related hepatitis.
- AST and ALT if greater than 50 times upper limit of normal: a feature of ischemic hepatitis.

Blunt abdominal trauma is a common reason for presentation in an emergency department and liver injury is common among patients with blunt abdominal injury. Modest elevation of ALT and AST are observed in blunt traumatic liver injury. Serum AST level >100 U/L and serum ALT level >80 U/L after blunt abdominal trauma may be associated with liver laceration. WBC (white blood cell) count may also be elevated [2]. In another report, the authors observed that in patients with abdominal trauma, abnormal hepatic transaminase and LDH levels are associated with liver injury. Alanine aminotransferase ≤76 U/L, aspartate aminotransferase <130 U/L, and LDH ≤410 U/L are predictive of low-grade liver injury, and patients with serum liver levels below these levels can be managed conservatively [3]. Abnormal patterns of liver function tests are summarized in Table 2.

## Jaundice: An introduction

Jaundice is defined as yellow discoloration of sclera and skin and is associated with hyperbilirubinemia (serum bilirubin >3 mg/dL). Common causes of Jaundice include:

- Congenital hyperbilirubinemia
- Hemolytic (prehepatic) jaundice
- Hepatocellular jaundice
- Cholestatic jaundice (obstructive)

**Table 2** Pattern of abnormal liver function test in various liver diseases.

| Liver disease | Abnormal liver function tests |
| --- | --- |
| Pre hepatic jaundice (hemolytic anemia) | Total and unconjugated bilirubin is high. Liver enzymes (ALT, AST, ALP, and GGT) are normal. PT and proteins are normal |
| Hepatocellular jaundice | Total bilirubin is high. Liver enzymes are elevated. ALT and AST may be very high (in thousands); ALP is less than three time normal. Total protein and albumin may be normal. PT may be prolonged if liver damage is significant |
| Cholestatic jaundice | Total bilirubin and conjugated bilirubin are high. Liver enzymes are elevated (typically mildly). ALP is more than three times normal. PT may be prolonged due to lack of absorption of fat soluble vitamin K |
| Acute liver disease | PT is the best marker for assessment of extent of acute liver disease. Total protein and albumin levels are typically normal |
| Chronic liver disease | Albumin level is decreased. Total protein could be elevated if hypergammaglobulinemia is present, liver enzymes may or may not be elevated |
| Liver metastasis | Only ALP and GGT may be elevated |
| Bone disease or metastasis | Only ALP may be elevated |
| Alcoholic liver disease | Only GGT is elevated |

## Congenital hyperbilirubinemia

There are three common causes of congenital hyperbilirubinemia: Gilbert's syndrome, Crigler-Najjar syndrome, and Dubin-Johnson syndrome.

Gilbert's syndrome is the most common familial hyperbilirubinemia affecting 2%–7% of the population and this disease is transmitted as autosomal dominant. The cause of Gilbert's syndrome is mutation of the *UGT1A1* gene that codes for uridine diphosphate glucuronosyltransferase (UDP-glucuronosyltransferase) enzyme essential for glucuronidation of bilirubin (conjugated bilirubin) for excretion. The most common allele in Gilbert syndrome is *UGT1A1\*28* and activity of UDP-glucuronosyltransferase is reduced by approximately 30% in patients with Gilbert syndrome. In addition, there is evidence for reduced hepatic uptake of unconjugated bilirubin. As a result, serum bilirubin is elevated due to increased concentration of unconjugated bilirubin. However, this condition is benign and often recognized during routine liver function test with the unconjugated hyperbilirubinemia as the only abnormality in serum. Usually patients are asymptomatic. In addition, when a patient with Gilbert syndrome recovers from viral hepatitis, serum bilirubin may be elevated for a prolonged time with values between 2 and 6 mg/dL. Viral illness such as influenza may also cause prolonged hyperbilirubinemia. No specific treatment is necessary except reassuring the patient that this is a benign condition. Interestingly high bilirubin may protect against cardiovascular disease and patients with Gilbert syndrome may be at a lower risk of developing cardiovascular disease [4].

Crigler-Najjar syndrome is rare inborn error of metabolism with an estimated incidence of 0.6–1.0 per million live births [5]. Crigler-Najjar syndrome is a rare autosomal recessive condition caused by complete absence of UDP-glucuronosyltransferase enzyme in Type 1 or severe deficiency of this enzyme in Type II Crigler-Najjar disorder (also known as Arias syndrome). For Crigler-Najjar syndrome Type I, a child must receive a copy of defective gene from both parents to develop this severe disease, which is usually incompatible with life. Fortunately Type I syndrome is affecting only 1 in 1 million new born babies born worldwide. High level of unconjugated bilirubin is apparent at birth that leads to kernicterus. Few therapy options are available, including liver transplant. In type II disorder approximately 20% of UDP-glucuronosyltransferase enzyme activity is retained, and these patients can be treated with enzyme inducers such as phenobarbital and usually present with milder hyperbilirubinemia than Type I disorder. These patients may not have any brain damage and may live a normal life.

It is important to note that Gilbert syndrome, Crigler-Najjar syndrome Type I and Type 2, all are associated with defective gene that leads to lower than normal activity of enzyme UDP-glucuronosyltransferase that conjugates bilirubin. Crigler-Najjar syndrome Type I is the most severe form with absence of any enzymatic activity. Crigler-Najjar syndrome Type II is the intermediate form where UDP-glucuronosyltransferase may be reduced up to 80%, while in Gilbert syndrome only 30% reduction in enzymatic activity is observed.

Dubin-Johnson (autosomal recessive) and Rotor's syndrome (possibly autosomal dominant) are due to impaired excretion of conjugated bilirubin from the hepatocytes. Both conditions result in conjugated hyperbilirubinemia. In Dubin-Johnson syndrome, melanin pigment may be found within the hepatocytes. Various congenital hyperbilirubinemias are summarized in Table 3.

**Table 3** Various congenital hyperbilirubinemias.

| Congenital disease | Mode of transmission | Comments |
|---|---|---|
| Gilbert's syndrome | Autosomal dominant | Reduced levels of UDP glucuronosyltransferase leading to elevated unconjugated bilirubin that increases with fasting |
| Crigler-Najjar type I | Autosomal recessive | Total absence of UDP-glucuronosyltransferase, may be fatal |
| Crigler-Najjar type II | Autosomal dominant | Reduced levels of UDP-glucuronosyltransferase but can be treated with enzyme inducer such as phenobarbital |
| Dubin-Johnson | Autosomal dominant | Impaired excretion of conjugated bilirubin causing conjugated hyperbilirubinemia. Melanin pigment found within hepatocytes |
| Rotor | Autosomal dominant | Impaired excretion of conjugated bilirubin |

## Hemolytic (prehepatic) jaundice

Hemolytic jaundice is due to hemolytic anemias. In hemolytic jaundice, increased unconjugated hyperbilirubinemia and reticulocytosis are accompanied by normal liver enzyme levels. However, lactate dehydrogenase (LDH) level may be high due to increased destruction of red blood cells. For the same reason serum haptoglobins may be low because haptoglobins bind to free hemoglobin released due to hemolysis. Clinical conditions that result in increased in vivo hemolysis are RBC membrane defects (e.g., hereditary spherocytosis), RBC enzyme defects (e.g., G6PD deficiency), hemoglobinopathies, thalassemia, PNH, autoimmune hemolytic anemias, and others. Hemolytic jaundice may be observed due to hematoma reabsorption or after blood transfusion. In addition, in conditions wherever there is ineffective erythropoiesis, the bone marrow destroys red cells rather than releasing them into the circulation. This may also result in unconjugated hyperbilirubinemia.

## Hepatocellular jaundice

Hepatocellular jaundice may be seen in patients with acute hepatitis or chronic liver disease. Common causes of acute hepatitis include viral hepatitis, drugs, and alcohol. Important issues regarding viral hepatitis include:

- Hepatitis B virus is a DNA virus but hepatitis A, C, D, and E are all RNA viruses.
- Hepatitis A and E are transmitted by the fecal-oral route. Hepatitis B, C, and D are transmitted parenterally, vertically, and during sex.
- Hepatitis B, C, and D can cause chronic liver disease.

With acute infection with hepatitis A and E, most patients recover. The mortality rate with hepatitis A is less than 1%, while with hepatitis E the mortality rate is 1%–2%. However, the mortality rate increases to 10%–20% in patients with hepatitis E in pregnant women. Approximately 90% or more of patients with acute infection with hepatitis B eventually clear the virus and achieve immunity. The remainders are at risk of developing chronic liver disease and possibly hepatocellular carcinoma. In the past, 50%–70% of patients with hepatitis C infection fail to clear the virus, and these patients are at risk for chronic liver disease and possibly hepatocellular carcinoma. Although interferon, ribavirin, or combination was used in the past for treating hepatitis C, current guidelines recommend the use of interferon-free combinations of direct-acting antiviral agents (DAAs such as simeprevir, sofosbuvir, and daclatasvir) for the treatment of HCV infection [6]. Other antiviral agents are also available for treating hepatitis C, for example, grazoprevir/elbasvir combination [7]. These treatments are effective in clearing the virus. Hepatitis D infection occurs either as coinfection with hepatitis B or as superinfection in a hepatitis B infected, hepatitis B surface antigen (HBsAg) positive patient. Concurrent infection of hepatitis B along with hepatitis D results in poor outcome of hepatitis B infection in a patient. Testing for hepatitis virus infection in patients is an important function of clinical laboratory.

## Chronic liver disease

Common causes of chronic liver disease include chronic alcohol abuse and chronic infection with hepatitis C and hepatitis B viruses. Other causes include autoimmune liver disease, primary biliary cirrhosis, hemochromatosis, Wilson's disease, and alpha-1 antitrypsin deficiency.

In chronic liver disease, moderate-to-severe hypoalbuminemia is commonly observed, but other liver function tests may be normal or abnormal depending on severity of illness. Serum gamma-globulins may be increased along with increased IgA levels. This feature is manifested in serum protein electrophoresis as polyclonal hypergammaglobulinemia with beta gamma bridging as IgA travels at the junction of the beta and gamma bands.

In autoimmune liver disease, antinuclear antibody (ANA), antismooth muscle actin, antisoluble liver antigen, and antiliver/kidney microsomal antibodies may be present. IgG levels may be raised.

Primary biliary cirrhosis is mostly seen in women. Antimitochondrial antibodies are found in most patients. IgM levels are often raised. Pruritus preceding other clinical features is a characteristic finding, as well as secondary hyperlipidemia.

Hemochromatosis is iron overload with organ damage. On the other hand, hemosiderosis is iron overload without organ damage. Primary or hereditary hemochromatosis is transmitted as autosomal recessive. Secondary hemochromatosis is seen in individuals who require multiple lifelong RBC transfusion and without effective iron chelation. Hemochromatosis is a multisystemic disease with bronze discoloration of skin (due to melanin deposition) and diabetes mellitus. Diabetes occurs due to the damage to the islets of Langerhans as a result of iron deposition. Thus, this disease is also referred to as bronze diabetes. Hypogonadism due to pituitary dysfunction is the most common endocrine feature. Hemochromatosis is also associated with cardiomyopathy with resultant heart failure and cardiac arrhythmias. Pseudo gout due to deposition of calcium pyrophosphate dihydrate (CPPD) crystals is also a common feature. Tests for iron stores such as ferritin, as well as liver iron content, will be increased. Other causes of increased liver iron content include alcohol intake.

Wilson's disease is transmitted as autosomal recessive disorder. Normally, copper is incorporated into apo-ceruloplasmin to form ceruloplasmin, but this process is defective in Wilson's disease. The unbound free copper will be secreted in the urine (thus urinary copper is high) and may also be deposited in certain tissues including the liver, basal ganglia, and cornea. Features of chronic liver disease, as well as extrapyramidal features, are thus easily explained. Quantification of liver copper may be done and should be high. Other causes of increased liver copper include chronic cholestasis. Kayser-Fleischer ring is greenish brown pigment at the sclera corneal junction due to deposition of copper in the Descemet's membrane in the cornea. Wilson's disease is an important cause of acute liver disease in the young people.

Alpha-1 antitrypsin deficiency can cause liver disease, as well as panacinar emphysema. Serum alpha-1 antitrypsin levels are low, which could be evident in serum protein electrophoresis. Genetic variants of alpha-1 antitrypsin are characterized by their electrophoretic mobilities as medium (M), slow (S), and very slow (Z). Normal individuals are MM. Homozygotes are ZZ and heterozygotes are either MZ or SZ. Various chronic liver diseases are summarized in Table 4.

**Table 4** Chronic liver diseases.

| Chronic liver disease | Comments |
|---|---|
| Autoimmune liver disease | Associated with other autoimmune diseases. Extra hepatic features such as polyarthritis, pleurisy, and glomerulonephritis may be present. In addition, ANA, antismooth muscle actin, antisoluble liver antigen, and antiliver/kidney microsomal antibodies may be present. Polyclonal hypergammaglobulinemia may be present due to increased IgG |
| Alcoholic liver disease | Alcoholic fatty liver may be present in many people drinking chronically but alcoholic hepatitis and especially liver cirrhosis are serious medical conditions associated with heavy drinking for a prolonged period of time. Classical cases are micronodular cirrhosis and IgA level may be elevated |
| Primary biliary cirrhosis | Seen primarily in women where pruritus precedes other features and secondary hyperlipidemia may be observed. Antimitochondrial antibody is found in most patients and IgM levels are often high |
| Hemochromatosis | Primary (hereditary) disease is transmitted as autosomal recessive; also called bronze diabetes with endocrine dysfunction, cardiomyopathy and arthropathy. Liver iron content is increased |
| Wilson's disease | Transmitted as autosomal recessive where copper is deposited in the liver. Observed in young age and Kayser-Fleischer ring is seen as greenish brown pigment at the sclera cornea junction. Low serum ceruloplasmin is observed |
| Alpha-1 antitrypsin | This deficiency can cause liver disease and also panacinar emphysema. Serum alpha-1-antitrypsin level is low. Genetic variants are characterized by their electrophoretic mobility as medium (M), slow (S) and very slow (Z). Normal people are MM, homozygotes are ZZ, while heterozygotes may be MZ or SZ |

## Cholestatic jaundice

Cholestatic jaundice can be classified into two broad categories: intrahepatic and extrahepatic cholestasis. Concentration of conjugated bilirubin in serum is elevated in cholestatic jaundice. Intrahepatic cholestasis may be due to:

- Primary biliary cirrhosis
- Primary sclerosing cholangitis
- Viral hepatitis B or C infection (cholestatic phase)
- Dubin-Johnson syndrome, Rotor syndrome
- Cholestatic disease of pregnancy

Extrahepatic cholestasis may be related to gallstone in the common bile duct, malignancy such as pancreatic tumor, pancreatic pseudocyst, pancreatitis, or secondary to surgical procedure [8].

## Alcohol and drug-induced liver disease

Alcohol abuse may produce a spectrum of liver disease which includes fatty changes in the liver, alcoholic hepatitis, and eventually cirrhosis of the liver. Fatty change due to excessive alcohol intake is reversible. Heavy drinking for as little as a few days may produce fatty change in the liver (steatosis), which is reversed after abstinence. However, drinking heavily for a longer period may cause more severe alcohol-related liver injuries such as alcoholic hepatitis and cirrhosis of the liver. In general, women are more susceptible to alcoholic liver diseases than men. The diagnosis of alcoholic hepatitis is a serious medical condition because approximately 70% of such patients may progress to liver cirrhosis, a major cause of death worldwide. In the United States, it is estimated that there are 2 million people who are suffering from alcohol-related liver diseases. In alcoholic hepatitis, hepatocellular necrosis takes place. Cytoplasmic inclusions called Mallory bodies may be seen. Liver cirrhosis is the seventh leading cause of death among young and middle-aged adults and approximately 10,000–24,000 deaths from liver cirrhosis annually may be attributable to alcohol abuse [9]. Alcoholic cirrhosis may be complicated by hepatocellular carcinoma.

The liver is the major site of drug metabolism. Drugs are converted into more water soluble form through drug metabolism so that drug metabolites can be excreted in bile or urine. Drugs that cause liver damage may do so in dose-dependent or dose-independent manner. An example of a drug causing dose-dependent hepatotoxicity is acetaminophen. When ingested a large proportion of acetaminophen undergoes conjugation with glucuronide and sulfate. The remainder is metabolized by microsomal enzymes to produce toxic derivatives. These are detoxified by conjugation with glutathione. Ingestion of large amounts of acetaminophen will result in excess toxic derivatives and saturation of glutathione that result in liver damage. Interestingly, alcoholics may experience acetaminophen toxicity from a therapeutic dose of acetaminophen [10]. However, acetaminophen toxicity can be treated with N-acetylcysteine that can restore liver glutathione supply. Other drugs may also cause liver damage such as azathioprine, methotrexate, chlorpromazine, erythromycin, and even statins in some individuals. Reyes's syndrome is a potentially fatal disease observed mostly in children. If aspirin is given to an infant or a child, it may cause Reye's syndrome due to inhibition of beta oxidation of fatty acids in the mitochondria and uncoupling of oxidative phosphorylation. There is diffuse microvesicular fatty infiltration of liver. Mortality rate is high.

## Liver disease in pregnancy

Hyperemesis gravidarum is a condition in pregnancy usually observed in first trimester, which is associated with nausea, vomiting and dehydration (morning sickness). The cause of this condition is still controversial, but it may be related to hormonal changes during pregnancy mostly likely elevated concentration of human chorionic gonadotropin. This condition may also be accompanied by mild jaundice along with mild elevation of liver enzymes.

Intrahepatic cholestasis of pregnancy (cholestasis of pregnancy) is a liver-specific disorder characterized by maternal pruritus (itching) observed usually in the third trimester. The etiology of this disease is not fully elucidated but may occur due to cholestatic effect of reproductive hormones such as estrogen. The mechanism by which this condition leads to fetal complication is not also understood. Severe pruritus may or may not accompany by jaundice, and this condition may resolve after pregnancy but may reappear during subsequent pregnancy [11].

Acute fatty liver of pregnancy is a rare but potentially life-threatening condition that usually occurs in the third semester with a mean gestational age of 35–36 weeks (range 28–40 weeks) or may also be observed in early postpartum period. Although exact etiology is not known, this disease may be linked to an abnormality in fetal fatty acid metabolism. However, diagnosis is challenging because this condition may appear similar to conditions encountered in preeclampsia, viral hepatitis, or cholestasis of pregnancy. Supportive care and expeditious delivery are required to minimize poor maternal and fetal outcome [12]. Fulminant hepatic failure in late pregnancy is also a very serious condition, which may be potentially fatal.

HELLP syndrome (H; hemolysis, EL: elevated liver enzyme, LP: low platelet count) is a serious complication of pregnancy occurring most commonly in patients with severe preeclampsia or eclampsia. Unconjugated hyperbilirubinemia without encephalopathy may also be observed in HEELP syndrome. HEELP syndrome usually develops around 37th week of gestation or following delivery.

## Liver disease in neonates and children

Physiological jaundice is observed in neonates due to decreased activity of UDP-glucuronosyltransferase enzyme leading to unconjugated hyperbilirubinemia and may affect approximately 65% newborn in the first week of life [13]. Breast milk may have inhibitors to this enzyme causing unconjugated hyperbilirubinemia, which is also referred as breast milk jaundice. Physiological jaundice can be treated with phototherapy but if it persists for more than 2 weeks after birth, possible pathological cause of such jaundice must be investigated. Neonatal hepatitis can occur due to infection with cytomegaly virus (CMV) or rubella or toxoplasma. Biliary atresia may also cause jaundice in infants. Metabolic disorders such as tyrosinemia and galactosemia may also cause jaundice. Progressive familial intrahepatic cholestasis (PFIC) refers to a group of familial cholestatic conditions caused by defective biliary epithelial transporters. The clinical features usually appear first in childhood with progressive cholestasis and hepatic failure. A patient may eventually need a liver transplant. These heterogeneous groups of conditions are inherited in autosomal recessive fashion. Alagille syndrome is a genetic disorder affecting liver, heart, kidney, and other organs, and problems associated with this disorder first appears in infancy or early childhood. The disorder is inherited in autosomal dominant manner with estimated prevalence of 1 in every 100,000 live births. Facial dysmorphism, cardiac abnormalities, and cholestasis are common features of this disorder.

## Nonalcoholic fatty liver disease

Nonalcoholic fatty liver disease (NAFLD) is divided into nonalcoholic fatty liver (NAFL) and nonalcoholic steatohepatitis (NASH). In NAFL, there is hepatic steatosis without inflammation. In NASH, there is hepatic steatosis with hepatic inflammation.

The major risk factors for NAFLD are [14]:

- Obesity
- Diabetes mellitus (type 2)
- Dyslipidemia
- Metabolic syndrome (syndrome X)

Most patients with NAFLD are asymptomatic. Sometimes patients with NASH may complain of fatigue, malaise, and abdominal discomfort.

Individuals with NAFLD may have normal or elevated liver enzymes. Normal enzymes do not exclude a diagnosis. Typically, ALT and AST are two to five times the upper limit of normal and AST to ALT ratio is less than one. The alkaline phosphatase may also be elevated, typically two to three times the upper limit of normal. Serum bilirubin and albumin levels are typically normal. Patients with NAFLD may have elevated serum ferritin levels. It has been shown that individuals with serum ferritin levels 1.5 times greater than upper limit of normal have higher NAFLD disease activity score [15]. The steatosis is demonstrated by imaging study or by biopsy. NAFLD may progress to cirrhosis and is an important cause of cryptogenic cirrhosis [16].

## Macro liver enzymes

On rare occasions, isolated and unexplained elevated levels of liver enzymes such as AST is observed, which is due to binding of AST with serum IgG. This bound enzyme is referred as macro AST. Although macro-AST is benign in nature, macro-AST is not usually recognized by clinicians because AST values are falsely elevated and as a result unnecessary diagnostic tests are ordered. Most commonly used laboratory methods for detection of macro-enzymes are precipitation with polyethylene glycol, ultracentrifugation, and gel filtration chromatography. Binding of ALT with IgG has also been reported. Another enzyme that may bind to IgG is creatinine kinase (CK), giving rise to macro CK. This phenomenon of macroenzymes is seen more frequently in elderly population, but macro-AST may also be seen in children. There is also an association between macroenzyme with autoimmune diseases. Macro-AST may also be detected in patients with chronic hepatitis or malignancy. A 50-year-old female patient with no clinical symptoms showed persistent AST elevation in range of 368–532 U/L and was referred to the author's hospital. Initial laboratory results showed an elevation of AST serum activity (494 U/L) with the rest of liver function tests and enzymes in normal range: ALT 28 U/L, gamma glutamyl transpeptidase 14 U/L, creatine kinase 146 U/L, lactate dehydrogenase 179 U/L, amylase 45 U/L, total protein 7.2 g/dL, albumin 4.1 g/dL, bilirubin 1.02 mg/dL, and alkaline phosphatase 92 U/L. Viral serological markers for hepatitis, cytomegalovirus, and Epstein-Barr virus were negative. Abdominal ultrasonography showed a liver image of normal size and contours. Further investigations for autoimmune hepatitis, Wilson's disease, hemochromatosis, myopathies, celiac, and thyroid diseases were negative. Precipitation of serum with polyethylene glycol reduced AST activity of 469 U/L to 4.4 U/L, confirming the presence of macro-AST in the patient [17].

## Delta bilirubin

When hepatic excretion of bilirubin glucuronides is impaired as in cholestatic jaundice, the glucuronides are found in blood and they bind with albumin. The fraction of delta bilirubin may be as high as 90% of the total bilirubin in certain instances. The half-life of conjugated bilirubin is about 4 h. In contrast the half-life of delta bilirubin is much longer similar to half-life of albumin (21 days). Because in bilirubin is covalently attached to albumin (molecular weight 66.5 kDa), delta bilirubin due to large mass is not found in urine. When there is significant delta bilirubin serial measurement of bilirubin may demonstrate slow decline of bilirubin levels (due to long half-life of albumin), which may not correlate with the clinical picture.

## Laboratory measurement of bilirubin and other tests

For determination of bilirubin, it is important to protect the specimen from light because conjugated and unconjugated bilirubin is photo oxidized. If specimen is stored in the refrigerator, bilirubin is stable up to 3 days and if stored at $-70°C$ in the dark, specimen may be stable up to 3 months. For measuring conjugated bilirubin (direct bilirubin), serum or plasma is acidified with hydrochloric acid and then mixed with diazotized sulfanilic acid to produce azobilirubin. Then the reaction is stopped by using ascorbic acid and the solution is made alkaline where azobilirubin provides a more intense blue color, which is measured by colorimetry. This is called direct bilirubin, which is the concentration of conjugated bilirubin in serum or plasma. For determination of total bilirubin, caffeine is added to serum or plasma in order for less reactive unconjugated bilirubin to react with diazotized sulfanilic acid. The solution after incubation is made alkaline for colorimetric measurement. This is referred as "total bilirubin," and subtracting total bilirubin from direct bilirubin provides the value of unconjugated bilirubin, which is also referred as "indirect bilirubin." This method is referred as Jendrassik and Grog method. It is usually assumed that direct bilirubin measures mostly conjugated bilirubin species, mono and di-conjugated bilirubin, as well as delta bilirubin (bilirubin tightly bound to albumin), while total bilirubin method measured both conjugated and unconjugated bilirubin.

Paraproteins may interfere with serum bilirubin measurement. In addition, paraproteins may also interfere with the estimation of several other biochemical parameters including albumin, creatinine, phosphate, calcium, and iron in human serum.

In neonates, heal puncture is painful and distressing. Therefore, bilirubin can be monitored by using transcutaneous bilirubin analyzer such as BiliChek, which is a handheld fiberoptic device that measures three wavelengths by spectral reflectance to measure bilirubin, melanin, and hemoglobin. This method accounts for differences in skin pigment.

Liver biopsy is often done to establish a diagnosis of cirrhosis. Liver biopsy is a procedure not without risks. There is significant interest in development of tests, which will allow clinicians to avoid performing a liver biopsy. One such test is measuring levels of procollagen type (III) peptide (PIIINP) in blood. Levels are increased in cirrhosis. However, levels can also be increased in inflammation and necrosis.

## Summary/key points

- Hypoalbuminemia is a common finding in chronic liver disease. In addition to albumin, all clotting factors with the exception of Factor VIII are produced in the liver. In severe liver disease, such as fulminant hepatic failure urea levels may also be low. Liver releases glucose into the circulation by glycogenolysis and gluconeogenesis. In severe liver disease, hypoglycemia may be apparent due to depletion of glycogen supply.
- Liver is the site for bilirubin metabolism. Heme, derived from the breakdown of hemoglobin, is converted to biliverdin and finally into to bilirubin, which is water insoluble unconjugated bilirubin. Unconjugated bilirubin can also bind with serum proteins, most commonly albumin. The unconjugated bilirubin is also taken up by the liver and with the help of the enzyme UDP (uridine-5'-phospho), glucuronosyltransferase is converted to conjugated bilirubin (bilirubin conjugated with glucuronide). This conjugation takes place in the smooth endoplasmic reticulum of the hepatocyte. Conjugated bilirubin is water soluble and is excreted in bile and in clinical laboratory measured as direct bilirubin while subtracting total bilirubin from direct bilirubin value provides concentration of unconjugated bilirubin, which is also referred as indirect bilirubin. In the intestine, bacterial enzymes hydrolyses conjugated bilirubin and releases free bilirubin, which is reduced to urobilinogen. Urobilinogen bound to albumin is excreted in the urine. Some urobilinogen is converted to stercobilinogen in the intestine and is excreted in stool. Thus in normal urine, urobilinogen is only present and in normal stool stercobilinogen is present. In obstructive (cholestatic) jaundice, conjugated bilirubin regurgitates into blood and as it is water soluble passes into the urine. This is called choluria or presence of bile in urine. In obstructive jaundice, less conjugated bilirubin is taken by the intestine and as a result less amount of stercobilinogen is found in stool (pale stools). Normal individuals

have mostly unconjugated bilirubin in their blood, urobilinogen in their urine, and stercobilinogen in their stool.

- Breakdown of hepatocytes results in release of aminotransferases (also referred to as transaminases) such as ALT and AST into the blood. ALT is a cytosol enzyme and more specific for the liver disease. AST is primarily a mitochondrial enzyme which is also found in heart, muscle, kidney, and brain.
- Alkaline phosphatase (ALP) is found in liver, bone, intestine, and placenta. ALP is located in the canalicular and sinusoidal membrane of the liver. Production of alkaline phosphatase is increased during cholestasis (intrahepatic or extrahepatic) resulting in elevated activity of ALP in serum; however, it is important to determine if the source of ALP is liver or other organs. If alkaline phosphatase is raised and the question is whether the source of this enzyme is from the liver or not, measurement of GGT or 5′ nucleotidase levels can be used to determine if the source of ALP is liver or not because both GGT and 5′ nucleotidase are solely produced by the biliary epithelium. Gamma glutamyl transferase (or gamma-glutamyl transpeptidase, GGT) is a microsomal enzyme.
- In acute liver disease without cholestasis, levels of ALT and AST are significantly elevated and ALP is raised but usually less than three times normal. Therefore, ALT and AST levels exceeding 500 U/L is a common finding in acute liver disease.
- In acute cholestasis, ALT and AST are raised but levels are not very high. ALP is usually more than three times normal with parallel increase in GGT levels.
- In acute liver disease, total protein and albumin levels are unaltered.
- PT is the best test to assess extent of liver dysfunction.
- In chronic liver disease, albumin levels are low but total protein may be elevated.
- Elevated ALP with normal bilirubin, ALT, AST may be seen in patients with hepatic metastasis or bone metastasis. Patients with hepatic metastasis may also have elevated GGT.
- Children showing only elevated ALP is most likely related to osteoblastic activity in their growing bones.
- Isolated elevated ALP in the elderly is a characteristic feature of Paget's disease.
- Very high isolated ALP levels can be seen in primary biliary cirrhosis.
- Normal liver function test except elevated GGT is a characteristic of excessive alcohol intake. Certain drugs (e.g., warfarin, anticonvulsant) may also produce similar observation.
- Gilbert's syndrome is the most common familial hyperbilirubinemia affecting 2%–7% of the population, and this disease is transmitted as

autosomal dominant. The cause of Gilbert's syndrome is mutation of the UGT1A1 gene that codes uridine diphosphate glucuronosyltransferase (UDP-glucuronosyltransferase) enzyme essential for glucuronidation of bilirubin (conjugated bilirubin) for excretion. In addition, there is evidence for reduced hepatic uptake of unconjugated bilirubin. As a result, serum bilirubin is elevated due to increased concentration of unconjugated bilirubin.

- Crigler-Najjar syndrome is a rare autosomal recessive condition caused by complete absence of UDP-glucuronosyltransferase enzyme in Type 1 or severe deficiency of this enzyme in Type II Crigler-Najjar disorder.
- Dubin-Johnson (autosomal recessive) and Rotor's syndrome (possibly autosomal dominant) are due to impaired excretion of conjugated bilirubin from the hepatocytes. Both conditions result in conjugated hyperbilirubinemia.
- Hepatitis B virus is a DNA virus but hepatitis A, C, D, and E are all RNA viruses.
- Hepatitis A and E are transmitted by the fecal-oral route. Hepatitis B, C, and D are transmitted parenterally, vertically, and during sex.
- Hepatitis B, C, and D can cause chronic liver disease. However, with acute infection with hepatitis A and E, most patients recover, and mortality rate with hepatitis A is less than 1% while with hepatitis E, the mortality rate is 1%–2%. However, the mortality rate increases to 10%–20% in patients with hepatitis E in pregnant women.
- Approximately 90% or more of patients with acute infection with hepatitis B eventually clear the virus and achieve immunity. The remainders are at risk of developing chronic liver disease and possibly hepatocellular carcinoma. In contrast, 50%–70% of patients with hepatitis C infection fail to clear the virus, and these patients are at risk for chronic liver disease and possibly hepatocellular carcinoma. Hepatitis D infection occurs either as coinfection with hepatitis B or as superinfection in a hepatitis B infected, hepatitis B surface antigen (HBsAg) positive patient. However, concurrent infection of hepatitis B along with hepatitis D results in poor outcome of hepatitis B infection in a patient.
- In chronic liver disease moderate to severe hypoalbuminemia is commonly observed but other liver function tests may be normal or abnormal depending of severity of illness. However, serum gamma-globulins may be increased along with increased IgA levels. This feature is manifested in serum protein electrophoresis as polyclonal hypergammaglobulinemia with beta gamma bridging as IgA travels at the junction of the beta and gamma bands.

- In autoimmune liver disease anti-nuclear antibody (ANA), antismooth muscle actin, antisoluble liver antigen, and antiliver/kidney microsomal antibodies may be present. IgG levels may be raised.
- Primary biliary cirrhosis is mostly seen in women. Antimitochondrial antibodies are found in most patients. IgM levels are often raised. Pruritus preceding other features is a characteristic finding, as well as secondary hyperlipidemia.
- Primary or hereditary hemochromatosis is transmitted as autosomal recessive.
- Hemochromatosis is a multisystemic disease with bronze discoloration of skin (due to melanin deposition) and diabetes mellitus. This disease is also referred to as bronze diabetes.
- Hypogonadism due to pituitary dysfunction is the most common endocrine feature, which may cause cardiomyopathy with heart failure and cardiac arrhythmias. Pseudo gout due to deposition of calcium pyrophosphate dihydrate (CPPD) crystals is a common feature. Iron stores, as well as liver iron content, may also be increased. Other causes of increased liver iron content include alcohol intake and iron overload due to chronic transfusion.
- Wilson's disease is transmitted as autosomal recessive disorder. Normally, copper is incorporated into apo-ceruloplasmin to form ceruloplasmin, but this process is defective in Wilson's disease. The unbound free copper may be secreted in the urine (urinary copper is high) and may also be deposited in certain tissues, including the liver, basal ganglia, and cornea. Quantification of liver copper may be done and should be high. Other causes of increased liver copper include chronic cholestasis. Kayser-Fleischer ring is greenish brown pigment at the sclera corneal junction due to deposition of copper in the Descemet's membrane in the cornea. Wilson's disease is an important cause of acute liver disease in the young people.
- Alpha-1 antitrypsin deficiency can cause liver disease, as well as panacinar emphysema. Serum alpha-1 antitrypsin levels are low, which could be evident in serum protein electrophoresis. Genetic variants of alpha-1 antitrypsin are characterized by their electrophoretic mobilities as medium (M), slow (S), and very slow (Z). Normal individuals are MM. Homozygotes are ZZ and heterozygotes are either MZ or SZ.
- The liver is the major site of drug metabolism. Drugs are converted into more water soluble form through drugs metabolism so that drug metabolites can be excreted in bile or urine. Drugs that cause liver damage may do so in dose-dependent or dose-independent manner. An example of a drug causing dose-dependent hepatotoxicity is acetaminophen. Interestingly, alcoholics may experience acetaminophen toxicity from a

therapeutic dose of acetaminophen. However, acetaminophen toxicity can be treated with *N*-acetylcysteine that can restore liver glutathione supply.

- HELLP syndrome is accompanied by hemolysis, elevated liver enzymes, and low platelet and is a serious complication of pregnancy occurring most commonly in patients with severe preeclampsia or eclampsia. Unconjugated hyperbilirubinemia without encephalopathy may also be observed in HEELP syndrome. HEELP syndrome usually develops around 37 weeks of gestation or following delivery.

- On rare occasions isolated and unexplained elevated levels of liver enzymes such as AST are observed, which is due to binding of AST with serum IgG. This bound enzyme is referred as macro AST. Binding of ALT with IgG has also been reported. Another enzyme that may bind to IgG is creatinine kinase (CK), giving rise to macro CK. This phenomenon of macroenzymes is seen more frequently in elderly population, but macro-AST may also be seen in children. There is an association between macroenzyme with autoimmune diseases. Macro-AST may also be detected in patients with chronic hepatitis or malignancy.

# References

[1] Tynjala J, Kangastupa P, Laatikainen T, Aalto M, et al. Effect of age and gender on the relationship between alcohol consumption and serum GGT: time to recalibrate goals for normal ranges. Alcohol 2012;47:558–62.

[2] Lee WC, Kuo LC, Cheng YC, Chen CW, et al. Combination of white blood cell count with liver enzymes in the diagnosis of blunt liver laceration. Am J Emerg Med 2010;28:1024–9.

[3] Bilgic I, Gelecek S, Akgun AE, Ozmen MM. Predictive value of liver transaminases levels in abdominal trauma. Am J Emerg Med 2014;32:705–8.

[4] Schwertner HA, Vitek L. Gilbert syndrome UGT1A1*28 allele and cardiovascular disease risk: possible protective effects and therapeutic applications of bilirubin. Atherosclerosis 2008;198:1–11.

[5] Ebrahimi A, Rahim F. Crigler-Najjar syndrome: current perspectives and the application of clinical genetics. Endocr Metab Immune Disord Drug Targets 2018;18:201–11.

[6] Lawitz E, Poordad F, Gutierrez JA, Beumont M, et al. Simeprevir, daclatasvir, and sofosbuvir for hepatitis C virus-infected patients: long-term follow-up results from the open-label, phase II IMPACT study. Health Sci Rep 2020;3:e145.

[7] Boyd A, Miailhes P, Chas J, Valantin MA, et al. Grazoprevir/elbasvir for the immediate treatment of recently acquired HCV genotype 1 or 4 infection in MSM. J Antimicrob Chemother. 2020;75:1961–1968.

[8] Winger J, Mchelfelder A. Diagnostic approach to the patients with jaundice. Prim Care Clin Office Pract 2011;38:469–82.

[9] DeBarkey SF, Stinson FS, Grant BF, Dufour MC. Surveillance report #41. Liver cirrhosis mortality in the United States 1970–1993. National Institute of Alcohol Abuse and Alcoholism; 1996.

[10] Prescott LF. Paracetamol, alcohol and the liver. Br J Clin Pharmacol 2000;49:291–301.

[11] Greenes V, Willamson C. Intrahepatic cholestasis of pregnancy. World J Gastroenterol 2009;15:2049–66.

[12] Ko HH, Yoshida E. Acute fatty liver of pregnancy. Can J Gastroenterol 2006;20:25–30.

[13] Jangaard KA, Curtis H, Goldbloom RB. Estimation of bilirubin using BiliChek, a transcutaneous bilirubin measurement device: effect of gestational age and use of phototherapy. Paediatr Child Health 2006;11:79–83.

[14] Manne V, Handa P, Kowdley KV. Pathophysiology of nonalcoholic fatty liver disease/nonalcoholic steatohepatitis. Clin Liver Dis 2018;22:23–37.

[15] Kowdley KV, Belt P, Wilson LA, et al. Serum ferritin is an independent predictor of histologic severity and advanced fibrosis in patients with nonalcoholic fatty liver disease. Hepatology 2012;55:77.

[16] Caldwell SH, Crespo DM. The spectrum expanded: cryptogenic cirrhosis and the natural history of non-alcoholic fatty liver disease. J Hepatol 2004;40:578.

[17] González Raya A, Coca Zúñiga R, Martín Salido E. Isolated elevation of aspartate aminotransferase (AST) in an asymptomatic patient due to macro-AST. J Clin Lab Anal 2019;33, e22690.

# CHAPTER 11

# Renal function tests

## Basic functions of kidneys

Kidneys are paired organ system located in the retroperitoneal space and weighing approximately 150 g each. Renal blood supply represents roughly 25% of the cardiac output. The functional unit of the kidney is the nephron. The components of each nephron include the glomerulus, proximal tubule, loop of Henle, distal tubule, and the collecting duct. Kidneys have three very important physiological roles:

- Excretory function: Removing undesirable end products of metabolism, excess inorganic ions ingested in diet, and removing drugs and toxins from the body through urine formation.
- Regulatory function: Maintaining proper acid-base balance and homeostasis.
- Endocrine function: Kidney can be regarded as an endocrine organ producing certain hormones and is also responsible for activation of several hormones.

Kidneys are responsible for urine formation and secretion of undesired end products of metabolism from the body, including urea formed from protein catabolism and uric acid produced from nucleic acid metabolism. The glomerulus is the site of filtration. The basement membrane of the capillaries serves as a barrier to passage of large proteins into the glomerular filtrate.

Glomerular filtration, a relatively nonselective process, is a physiological function of kidney nephrons. The ultrafiltrate, which appears in the lumen of the proximal convoluted tubule, is composed of water and solutes that can pass through the filtering membrane of the capillaries. Under normal physiological conditions, the large molecules such as proteins and blood cells do not pass through the capillary wall. In general, the glomerular filtration rate (GFR) is relatively stable. Homeostasis is maintained by varying the amount of water and solute reabsorbed by the different segments of the nephron beyond the glomerulus. Overall, the balance of pressures in the glomerular capillaries causes

**229**

filtration of 20% of the plasma entering the kidney (filtration fraction 20%). Normal renal plasma flow is average, 625 mL/min. Therefore, normal average GFR is approximately 125 mL/min (20% of 625 mL/min). About 99% of the glomerular filtrate is reabsorbed in the tubules, so normal urine production is approximately 1 mL/min. Typically, the volume of the glomerular filtrate in 1 day ranges from 150 to 200 L. This volume is reduced to 1–2 L of urine per day.

Small molecules such as electrolytes and small organic molecules are freely filtered by the glomerular capillary membrane. The filterability factor significantly decreases as the molecular weight increases to a level comparable to, or exceeding, the molecular weight of albumin (molecular weight 66.5 kDa). This is the reason why under normal conditions, albumin does not appear in urine. Electric charge of the molecule also plays a role during glomerular filtration. For example, the diameter of the albumin (6 nm) is somewhat smaller than the pores of the glomerular membrane (8 nm). Nonetheless, negative charge of albumin produces an electrostatic repulsion with the basement membrane causing very poor filtration of albumin in healthy individuals. However, if loss of these negative charges occurs during disease, such as poorly controlled diabetes, some of the larger molecular weight proteins, like albumin, are filtered and appear in the urine even in the absence of noticeable changes in kidney histology. Glomerular capillary hydrostatic pressure, which is approximately 60 mmHg, has a direct relation with GFR, where an increase in the glomerular capillary hydrostatic pressure also increases GFR [1].

Approximately two-third of the filtrate volume is reabsorbed in the proximal tubule. Moreover, 90% of hydrogen ion secreted by the kidney takes place at the proximal tubule. Further reabsorption of water and solutes takes place in the more distal parts of the nephron. The loop of Henle is the site of where urine is concentrated. At the distal tubule sodium and chloride are reabsorbed, while potassium and hydrogen ions are excreted. The proper function of distal tubule is essential in maintaining plasma acid-base and electrolyte homeostasis. The collecting duct is the site of further water reabsorption, which occurs under the influence of antidiuretic hormone (ADH).

The kidneys also produce two hormones; erythropoietin and renin. Erythropoietin is produced in response to renal hypoxia and acts on the bone marrow to stimulate erythropoiesis. Renin is produced by the juxtaglomerular apparatus. Renin converts angiotensinogen released by the liver into angiotensin I, which is then converted into angiotensin II in the lungs by angiotensin-converting enzyme (ACE). Angiotensin II is a vasoconstrictor and also stimulates release of aldosterone from adrenal cortex. This is defined as "renin-angiotensin-aldosterone" system. Aldosterone, a mineralocorticoid acts on the distal tubules and collecting ducts of the nephron and causes retention of water and sodium, as well as excretion of potassium and hydrogen ions.

The kidneys are responsible for producing the active form of vitamin D, a fat soluble vitamin essential for absorption of calcium. Vitamin D deficiency can cause osteomalacia in adults and rickets in children. Human skin is capable of synthesizing inactive form of vitamin D (cholecalciferol; vitamin D3) from 7-dehydrocholesterol in the presence of sunlight (solar ultraviolet radiation: 290–315 nm; reaches earth between 10 a.m. and 3 p.m.). This is why vitamin D is also called "sunshine vitamin." Very few foods naturally contain vitamin D and as a result many foods are fortified with inactive form vitamin D. Then inactive form of vitamin D obtained either from skin exposure to the sun or food is first converted into 25-hydroxy vitamin D (25-hydroxycholecalciferol) in the liver by the action of enzyme vitamin D-25 hydroxylase. Finally, the kidneys (proximal tubular epithelial cells) convert this form of vitamin D into the active form, which is called 1,25-dihydroxyvitamin D (1,25-dihydroxycholecalciferol) by the action of the enzyme 25-hydroxylcholecalciferol-1α-hydroxylase. The biologically active form of vitamin D is 1,25-dihydroxycholecalciferol, which plays an important role in absorption of calcium from the gastrointestinal tract. The enzyme 25-hydroxylcholecalciferol-1α-hydroxylase is stimulated by parathyroid hormone (PTH) and inhibited by high blood levels of calcium and phosphate. Although 1,25-dihydroxyvitamin D is the bioactive form of vitamin D, the best laboratory parameter to monitor vitamin D status of a patient is to measure 25-hydroxy vitamin D. A serum 25-hydroxy vitamin D level of 30 ng/mL or greater is considered as an adequate level [2]. Prostaglandins are synthesized by the action of cyclooxygenase enzyme acting on arachidonic acid. This enzyme is present in many organs including kidneys. Kidney is also a site of degradation of hormones such as insulin and aldosterone.

## Glomerular filtration rate

The GFR is the sum of the filtration rates in all of the functioning nephrons. Therefore, the GFR gives an estimate of the number of functioning nephrons. The normal value for GFR depends upon age, sex, and body size, with an average value of 125 mL/min. In general, for men GFR is estimated to be approximately 130 and for females approximately 120 mL/min/1.73 m$^2$, with considerable variation even among normal individuals [3].

In patients with kidney disease, a reduction in GFR implies either progression of the underlying disease or the development of a superimposed and often reversible problem, such as decreased renal perfusion due to volume depletion. The level of GFR also has prognostic implications in patients with chronic kidney disease (CKD), where these patients are staged, in part, according to GFR. However, there is no exact correlation between the loss of kidney mass (i.e., nephron loss) and the loss of GFR. The kidney adapts to the loss of some nephrons by compensatory hyperfiltration and/or increasing solute and water reabsorption in the remaining normal nephrons. Thus, an individual who has lost one-half of total kidney mass will not necessarily have one-half the normal amount of GFR. These concepts have important consequences:

- A stable GFR does not necessarily imply stable kidney function.
- An increase in GFR may indicate improvement in the kidney disease or may imply a counterproductive increase in filtration (hyperfiltration) due to hemodynamic factors.
- Some patients who have true underlying renal disease may go unrecognized because they have a normal GFR.

Glomerular filtration is one of the major functions of the kidney. Neutral molecules show much higher glomerular permeability than highly negatively charged molecules. GFR is a measure of the functional capacity of the kidney and is an important parameter to asses kidney function. GFR can be estimated by using the formula:

$$GFR = (Ua \times V)/Pa$$

where $Ua =$ the concentration of a solute in urine, $V =$ volume of urine in mL// min, and $Pa =$ concentration of the same solute in plasma. However, this formula is often corrected to consider the body surface area as follows:

$$GFR = (Ua \times V)/Pa \times 1.73/A$$

where $A$ is the body surface area in meter square. Standard body surface is $1.73 \, m^2$.

Body surface area of most adults is between 1.6 and 1.9. The formula for calculating GFR using any analyte that is freely filtered through glomeruli is only valid if the solute is in stable concentration in plasma and is inert (neutral charge), freely filtered at the glomerulus. This compound must not be secreted, reabsorbed, synthesized, or metabolized by the kidneys. Estimation of GFR by inulin clearance is considered the gold standard. However, in routine clinical practice creatinine clearance is one, which is more practical. In addition, cystatin C has been introduced as an alternative to creatinine clearance. For example, if serum creatinine is 1.0 mg/dL (0.01 mg/mL), urine creatinine is 1 mg/mL, volume of urine is 60 mL in 1 h, and body surface area is 1.70, then GFR for these patients using creatinine clearance should be:

$$GFR \, (\text{creatinine clearance}) = 1.0 \, mg/mL \times 1 \, mL/min/0.01 \, mg/mL \times 1.73/1.70$$
$$= 101.7 \, mL/min/1.73 \, m^2$$

In order to calculate GFR based on creatinine clearance, a 24-h urine collection is recommended, which should be from one morning void to next day morning void. This is difficult in real practice and usually GFR is estimated from using a formula.

## Creatinine clearances and estimated GFR

Creatinine is produced in the body at a constant rate and is freely filtered and not reabsorbed although a small amount of creatinine is secreted by the tubules. Thus, it is a convenient marker for estimation of GFR. Because collection of 24 h urine is difficult, estimation of GFR can be done using values of plasma creatinine concentration and relevant formulas. However, it is also important to consider the age, sex, and race of the patient when performing such calculation. Cockroft-Gault formula is widely used for calculating GFR.

Cockroft-Gault formula:

$$\text{Creatinine clearance} = \frac{(140 - \text{Age in years}) \times \text{Weight in kg}}{\text{Serum creatinine}(\mu\text{mol}/\text{L})} \times 1.23 \text{ if male or } 1.04 \text{ if female}$$

Another form of formula, which is also commonly used which produces same result:

$$\text{Creatinine clearance} = \frac{(140 - \text{Age in years}) \times \text{Weight in kg}}{0.814 \times \text{Serum creatinine}(\mu\text{mol}/\text{L})} \times 0.85 \text{ if female}$$

If serum creatinine is given in mg/dL as often the case with US laboratories, then this equation can be further modified so that GFR can be calculated directly using creatinine concentration as expressed in mg/dL:

$$\text{Creatinine clearance} = \frac{(140 - \text{Age in years}) \times \text{Weight in kg}}{72 \times \text{Serum creatinine}(\text{mg}/\text{dL})} \times 0.85 \text{ if female}$$

Conversion factor for converting serum creatinine given in mg/dL into $\mu$mol/L is 88.4. Therefore, $88.4 \times 0.814$ is 71.9 which can be considered as 72 as used in the modified equation. Alternatively, if serum creatinine is expressed in $\mu$mol/L, it can be multiplied by 0.011 to get the creatinine concentration in mg/dL.

Cockroft-Gault formula was modified to MDRD formula by Modification of Diet in Renal Disease Study Group as follows:

$$\text{Estimated GFR } (\text{mL}/\text{min}/1.73\,\text{m}^2) = 186 \times (\text{plasma creatinine in mg}/\text{dL})^{-1.154} \times \text{Age}^{-0.203} \times \text{F}$$

F: 0.742 if female: 1.21 if African American.

In children usually, Schwartz formula is used to estimate GFR:

$$\text{Estimated GFR} = \frac{k \times \text{Height(cm)}}{\text{Serum creatinine (mg/dL)}}$$

For a preterm baby, value of $k$ is 0.33 in the first years of life, but for full-term infants, it is 0.45. For infants and children up to age 12, value of $k$ is assumed as 0.55.

## Biochemical assessment of renal function

Measurement of plasma creatinine, urea, and cystatin C are all available to assess adequacy of renal function.

### Creatinine

Creatine is synthesized in the kidneys, liver, and the pancreas and then transported in blood to other organs, especially brain and muscle where it is phosphorylated to phosphocreatine. Phosphocreatine, a high energy compound important for muscle function, can break down into creatinine. Creatinine is the waste product derived from creatine and phosphocreatine and creatinine production is related to the muscle mass of an individual. Women usually excrete 1.2 g of creatinine per day, while men excrete 1.5 g/day. Dietary intake of meat also affects amount of creatinine produced daily. Therefore, serum creatinine levels are affected by:

- Gender
- Age
- Weight
- Lean body mass
- Dietary protein intake

Plasma creatinine is inversely related to the GFR. However, GFR can decrease by 50% before plasma creatinine concentration rises above the normal range. Therefore, a normal plasma creatinine does not necessarily mean normal renal function.

Changes in creatinine levels can also occur without change in renal function, such as change in muscle mass. In healthy pregnancy, GFR is increased as early as the first trimester, as compared with nonpregnant values, and the kidneys continue to function at a higher rate throughout gestation. In contrast, kidney function is decreased in hypertensive pregnancy [4]. Therefore in normal pregnancy, levels of creatinine typically fall.

## Urea (blood urea nitrogen) and uric acid

Although serum creatinine and cystatin C are more commonly used to evaluate renal function, urea (often called blood urea nitrogen), and uric acid measurements also have some clinical values. Urea is the resultant of catabolism of proteins and amino acids. This takes place in the liver where first ammonia is formed, which is eventually converted into urea. The kidneys are the primary route of excretion of urea accounting for over 90% excretion of urea. Minor loss of urea takes place through the gastrointestinal tract and skin. Urea is freely filtered at the glomerulus and is subsequently not reabsorbed or secreted at the tubules. However, measurement of urea levels is inferior in assessing renal function, compared to creatinine levels because serum or plasma concentration of urea may be increased in the following situations:

- Dehydration
- Hypoperfusion of the kidneys
- High protein diet
- Protein catabolism
- Steroid administration

Under similar situation, serum creatinine is not elevated (normal range: 0.5–1.2 mg/dL). However, measuring urea level along with creatinine is of clinically relevance. Urea level in blood is usually measured as BUN (blood urea nitrogen) with a normal level between 6 and 20 mg/dL. Following criteria are usually used to interpret BUN/creatinine ratio:

- BUN/creatinine ratio for normal individuals usually is from 12:1 to 20:1. For example, if BUN is 15 mg/dL and creatinine is 1.1 mg/dL, then BUN/creatinine ratio is 13.6:1
- BUN/creatinine ratio below 10:1 may indicate intrinsic renal disease. BUN/creatinine ratio above 20:1 may be hypoperfusion of kidney, including prerenal failure.

## Cystatin C

Cystatin C is a low-molecular-weight protein (13.3 kDa), which can be used for calculating GFR. In contrast to creatinine plasma concentrations of Cystatin C are unaffected by sex, diet, or muscle mass. It is considered to be superior marker for estimation of GFR than creatinine. In addition, in pediatric population, it has been documented that plasma cystatin C is a better marker of kidney function than creatinine. The reference value of plasma cystatin C is considered as 0.5–1.0 mg/L.

However, whether cystatin C is a superior marker in diabetic patients is unclear. In addition, steroid use may affect cystatin C levels, therefore limiting its use in transplant recipients. Although cystatin C appears to be more accurate for the assessment of GFR than serum creatinine in certain populations, whether measurement of cystatin C levels will improve patient care is at present unknown. However, it is reasonable to use cystatin C-based estimated GFR in patients with clear reductions in muscle mass. Many authors have compared calculated GFR using creatinine (Cockroft-Gault formula), MDRD formula, and with calculated GFR using serum cystatin concentration. Several formulas have been proposed to estimate GFR based on serum cystatin C levels. Three common formulas are provided in this chapter but for in-depth discussion, review published by Rosenthal is useful [5].

Le Bricon formula: $GFR = (78/cystatin\ C,\ mg/L) + 4$
Hoek formula: $GFR = (80.35/cystatin\ C,\ mg/L) - 4.32$
Larsson formula: $GFR = 77.24/cystatin\ C^{-1.263}$ (cystatin C in mg/L)

Hojs et al. commented that estimated GFR based on serum cystatin C is concentration is superior to calculated GFR based on serum creatinine concentration in renal compromised patients with estimated GFR around 60 mL/min/1.73 m$^2$ [6].

## Assessment of glomerular integrity

If the glomerulus is damaged, then the following situations may occur:

- Proteinuria
- Hematuria (the RBCs that are present in such cases are dysmorphic)
- Red cells casts

## Tests for renal tubular function

Evidence of the following implies tubular damage:

- Glucose in the urine with normal blood glucose
- Amino acid in the urine
- Beta-2 or alpha-2 microglobulin

Specific tests to check for evidence of distal tubular function are fluid deprivation test (to assess renal concentrating ability) and tests for urinary acidification (to diagnose renal tubular acidosis (RTA)).

## Acute renal failure

Acute renal failure (ARF) means abrupt deterioration of renal function. It is usually reversible and is considered as a medical emergency. This is because of life-threatening biochemical disturbances. ARF may also arise in a background of CKD. There are three broad categories of ARF: prerenal, renal, and postrenal.

In prerenal type, renal perfusion is impaired. This subtype if untreated may convert to acute tubular necrosis (renal type of ARF). There are many examples of renal type of ARF and noteworthy ones include

- Acute tubular necrosis (ATN)
- Acute interstitial nephritis (most often due to drugs)
- Rapidly progressive glomerulonephritis

Postrenal ARF is typically due to obstruction (e.g., stones). An important clinical issue with an individual with ARF and history of hypovolemia (e.g., due to persistent vomiting, diarrhea, or blood loss) is whether the patient has prerenal ARF or has ATN developed. In such situation, BUN/creatinine ratio and fractional excretion of sodium in the urine may be helpful.

## BUN/creatinine ratio

In critically ill patients with renal hypoperfusion but intact tubular function (prerenal azotemia), BUN concentration may increase out of proportion to serum creatinine concentration and BUN/creatinine ratio may exceed 20:1. However, critically ill patients are also prone to accelerated protein catabolism and that may also increase BUN/creatinine ratio without prerenal azotemia [7].

## Fractional excretion of sodium

Fractional excretion of sodium, a measure of percent of filtered sodium that is excreted in urine, is also useful in evaluating renal function. It is useful to differentiate prerenal type of ARF from renal type.

The formula is:

$$\text{Fractional excretion of sodium} = \frac{\text{Urine sodium} \times \text{Serum creatinine}}{\text{Serum sodium} \times \text{Urine creatinine}} \times 100$$

A value less than 1% is indicative of prerenal disease (prerenal azotemia) where reabsorption of all or most of all filtered sodium is an appropriate response to renal hypoperfusion. However, a value over 3% (some authors suggest a value over 2%) is indicative of acute tubular narcosis or urinary tract obstruction. Values between 1% and 2% may be observed in either disease.

## Chronic kidney disease

Kidney function depends on age. The GFR is low at birth, reaching adult levels at approximately 2 years of age. Renal function declines after the age of 40 years and declines even further after the age of 65 years. Approximately, 19 million Americans older than 20 years have CKD. In addition, an estimated 435,000 individuals have end-stage renal disease. Early diagnosis of CKD may delay or even prevent end-stage renal disease.

CKD is defined by the presence of kidney damage or decreased kidney function for 3 or more months. Kidney damage refers to pathologic abnormalities, whether established via renal biopsy or imaging studies, or concluded from markers such as urinary sediment abnormalities or increased rates of urinary albumin excretion.

Decreased kidney function refers to a decreased GFR, which is usually estimated eGFR using serum creatinine and any one of the available equations. GFR is generally considered to be the best index of overall kidney function, and declining GFR is the hallmark of progressive kidney disease. Kidney damage is identified in most cases by the presence of one of the following clinical markers:

- Albuminuria: albuminuria is the most frequently assessed marker of kidney damage. Albuminuria reflects increased glomerular permeability to macromolecules. Albuminuria reflects primary kidney disease or kidney involvement in systemic disease and may represent widespread endothelial dysfunction, due to hypertension, diabetes, hypercholesterolemia, smoking, obesity, and other disorders.
- Although a number of different measurement methods have been used to assess and define albuminuria, the albumin-to-creatinine ratio (ACR) in a random "spot" urine has many advantages [8]. The generally accepted threshold for an abnormally elevated ACR is 30 mg/g (3.4 mg/mmol) or greater and such values should be considered part of the definition of CKD. Individuals with a urine ACR > 30 mg/g (or equivalent) have a significantly increased risk for all-cause and cardiovascular mortality, ESRD, AKI, and CKD progression compared with those who have a lower ACR, even when eGFR is normal [9].
- Urinary sediment abnormalities—Urinary sediment abnormalities such as red or white blood cell casts may indicate the presence of glomerular injury or tubular inflammation.
- Imaging abnormalities—Kidney damage may be detected by the presence of imaging abnormalities such as polycystic kidneys, hydronephrosis, and small and echogenic kidneys.
- Pathologic abnormalities—A kidney biopsy may reveal evidence for glomerular, vascular, or tubulointerstitial disease.

**Table 1** Various stages of renal disease based on National Kidney Foundation Guidelines.

| Glomerular filtration rate | Stage |
|---|---|
| >90 mL/min/1.73 m$^2$ | 0: If no proteinuria/hematuria present/no risk factor |
| >90 mL/min/1.73 m$^2$ | 1: With proteinuria or hematuria |
| 60–89 mL/min/1.73 m$^2$ | 2: Mild disease with decreased GFR |
| 30–59 mL/min/1.73 m$^2$ | 3: Moderate chronic kidney disease |
| 15–29 mL/min/1.73 m$^2$ | 4: Severe chronic kidney disease |
| <15 mL/min/1.73 m$^2$ | 5: End-stage renal disease or end state renal failure |

- Kidney transplantation—Patients with a history of kidney transplantation are assumed to have kidney damage whether or not they have documented abnormalities on kidney biopsy or markers of kidney damage.

The Kidney Disease Outcomes Quality Initiative from National kidney foundation has developed guidelines for detection and evaluation of CKD [10]. Following criteria are adopted to define CKD:

- If creatinine clearance is above 90 mL/min/1.73 m$^2$ or preferable above 100, in the absence of any abnormal finding (imaging study or laboratory based tests), a normal kidney function can be assumed.
- If a kidney damage for 3 months is present (structural damage of kidney based on imaging study or functional damage based on laboratory tests), then it is usually assumed that CKD is present. However, if GFR is below 60 mL/min for 3 or more months, it can also be assumed that the patient has CKD.
- End-stage renal disease or end-stage renal failure is defined as GFR below 15 mL/min or if the patient is receiving dialysis.

However, creatinine clearance may not always reflect true nature of CKD [3]. Various stages of renal disease based on National Kidney Foundation Guidelines are listed in Table 1 [11].

## Proteinuria

Molecules less than 15 kDa pass freely into urine through the glomerular filtration, whereas selected few proteins with molecular weight between 16 and 69 kDa also can be filtered by the kidney. The molecular weight of albumin, the major protein found in serum is 66.5 kDa, and as expected a very small amount of albumin is also found in urine of normal individuals. Glomerular filtration of protein depends on several factors including molecular weight of the protein, concentration in serum, charge of the protein, as well as hydrostatic pressure. Although 90% of these proteins are reabsorbed (smaller proteins are effectively absorbed by renal tubule), following proteins may pass through the glomerular filtration process:

- Albumin
- Alpha-1 acid glycoprotein
- Alpha-1 microglobulin
- Beta-2 microglobulin
- Gamma trace protein
- Retinol-binding protein

Total urinary protein excretion in the normal adult is less than 150 mg/day and consists of mostly albumin and Tamm Horsfall protein (secreted from ascending limb of Loop of Henle). In normal individuals, low-molecular-weight proteins and small amounts of albumin are filtered by the glomerulus. Most of the filtered albumin enters the proximal tubule where it is almost completely reabsorbed. The net result is the normal daily protein excretion of less than 150 mg, of which approximately 4–7 mg is albumin. Proteinuria with minor injury (typically only albumin is lost in urine) can be due to vigorous physical exercise, congestive heart failure, pregnancy, therapy with certain drugs, high fever, and alcohol abuse. However, early renal disease may be reflected by lesser degrees of proteinuria, particularly increased amounts of albuminuria.

The normal rate of albumin excretion is less than 20 mg/day. Persistent albumin excretion between 30 and 300 mg/day (20–200 μg/min) is called moderately increased albuminuria (formerly called "microalbuminuria"). In patients with diabetes, this is usually indicative of incipient diabetic nephropathy. In nondiabetics, the presence of moderately increased albuminuria is associated with an increased risk for cardiovascular disease.

Albumin excretion above 300 mg/day (200 μg/min) is considered overt proteinuria or severely increased albuminuria (formerly called "macroalbuminuria"), the level at which the standard dipstick becomes positive. At this level, much of the protein in the urine consists of albumin.

Isolated proteinuria is defined as proteinuria without abnormalities in the urinary sediment, including hematuria, or a reduction in GFR, as well as the absence of hypertension or diabetes. In most cases of isolated proteinuria, the patient is asymptomatic, and the presence of proteinuria is discovered incidentally by use of a dipstick during routine urinalysis. Most patients with benign causes of isolated proteinuria excrete less than 1–2 g/day.

The extent of proteinuria can be assessed by quantifying the amount of protein-uria, as well as expressing as protein to creatinine ratio. The normal total protein/creatinine ratio is <0.2 for adults. However, for children 6 months to 2 years old, the ratio is <0.5 and for children over 2 years of age is <0.25. Total protein/creatinine ratio is also useful in grading proteinuria.

- Protein/creatinine ratio: 0.2–1.0, low-grade proteinuria
- Protein/creatinine ratio: 1.0–5.0, moderate proteinuria
- Protein/creatinine ratio: >5.0, nonselective proteinuria

## Types of proteinuria

The types of proteinuria are

- Glomerular proteinuria
- Tubular proteinuria
- Mixed glomerular and tubular proteinuria
- Overflow proteinuria
- Postrenal proteinuria

### Glomerular proteinuria

Glomerular proteinuria is due to increased filtration of macromolecules (such as albumin) across the glomerular capillary wall. This is a sensitive marker for the presence of glomerular disease. Glomerular proteinuria can be subclassified as: selective (albumin and transferrin in urine) and nonselective (all proteins are present). In glomerular proteinuria, the major protein present is always albumin. In mild glomerular proteinuria, total protein concentration is usually within 1500 mg/24 h urine but with moderate glomerular proteinuria, total protein level may be 1500–3000 mg/24 h. In the case of nonselective protein-uria, total protein in urine often exceeds 3000 mg/24 h.

### Tubular proteinuria

Low-molecular-weight proteins that are generally under 25,000 (e.g., beta-2 microglobulin) Daltons in comparison to the 665,000 Da molecular weight of albumin are filtered across the glomerulus and are then almost completely reabsorbed in the proximal tubule. Interference with proximal tubular reab-sorption, due to a variety of tubulointerstitial diseases or even some primary glomerular diseases, can lead to increased excretion of these smaller proteins. Tubular proteinuria is often not diagnosed clinically, since the dipstick for pro-tein is not highly sensitive for the detection of proteins other than albumin and because the quantity of nonalbumin proteins excreted is relatively low.

### Mixed glomerular and tubular proteinuria

In the mixed type proteinuria, both albumin and small molecular weight proteins such as alpha-1 microglobulin and beta-2 microglobulin are present [12].

### Overflow proteinuria

Increased excretion of low-molecular-weight proteins can occur with marked overproduction of a particular protein, leading to increased glomerular filtration and excretion. This is almost always due to immunoglobulin light chains in multiple myeloma but may also be due to lysozyme (in acute myelomonocytic leukemia), myoglobin (in rhabdomyolysis), or free hemoglobin (in intravascular hemolysis) that is not bound to haptoglobin. In these settings, the filtered load is increased to a level that exceeds the normal proximal reabsorption capacity. Patients with myeloma kidney also may develop a component of tubular proteinuria since the excreted light chains may be toxic to the tubules, leading to diminished reabsorption.

### Postrenal proteinuria

Inflammation in the urinary tract, which can occur with urinary tract infection, can give rise to increase in urinary protein excretion, although the mechanism is unclear. The excreted proteins are often nonalbumin (often IgA or IgG), and only small amounts are excreted. Leukocyturia is frequently present in such patients.

## Methods to detect proteinuria

Two semiquantitative methods are available to screen patients for proteinuria. These are the standard urine dipstick and the precipitation of urine proteins with sulfosalicylic acid (SSA). Neither method is quantitative, and if abnormal proteinuria is suggested by either technique, proteinuria should be quantified using a timed urine collection.

The standard urine dipstick primarily detects albumin but is relatively insensitive to nonalbumin proteins. Thus, a positive dipstick usually reflects glomerular proteinuria. SSA can be used for detection of tubular or overflow proteinuria. The dipstick is very specific but not sensitive to low levels of albumin excretion. The lower limit of detection is a urine albumin concentration of approximately 10–20 mg/dL. Thus, patients with moderately increased albuminuria (formerly called "microalbuminuria") will not usually be identified by this method unless the urine is highly concentrated.

In contrast to the urine dipstick, which primarily detects albumin, SSA detects all proteins in the urine at a sensitivity of 5–10 mg/dL. Use of SSA is primarily indicated in patients who present with acute kidney injury, a benign urinalysis, and a negative or trace dipstick, a setting in which myeloma kidney should be excluded. A significantly positive SSA test in conjunction with a negative dipstick usually indicates the presence of nonalbumin proteins in the urine, most often immunoglobulin light chains. Patients with persistent proteinuria should undergo a quantitative measurement of total protein excretion. The gold standard for measurement of protein excretion is a 24-h urine collection, with the normal value being less than 150 mg/day.

However, major limitations of measuring protein excretion in a 24-h urine collection include the following:

- It is cumbersome for patients.
- It is often collected incorrectly. (Over- and undercollections are common.)

Due to the limitations of a 24-h urine collection, a number of shortcuts have been proposed. These shortcuts generally involve measuring the ratio of protein to creatinine (or the urine albumin-to-creatinine ratio (UACR)) in urine specimens of less than 24 h duration. The accuracy of the spot UACR is important among patients with proteinuric kidney disease. Treatment decisions, such as the initiation or discontinuation of immunosuppressive agents and choice of antihypertensive therapy, are often based upon the degree of proteinuria in a given patient.

On a population level, the 24-h urine protein excretion and the spot UACR are reasonably well correlated. However, in individual patients with kidney disease, the spot UPCR often does not accurately predict the result of a 24-h urine, particularly in patients with lower degrees of protein excretion.

## Nephrotic syndrome

Nephrotic syndrome consists of the following:

- Massive proteinuria (in adults, proteinuria >3.5 g/day)
- Hypoalbuminemia (this is due to proteinuria coupled with increased protein catabolism in the kidneys)
- Generalized edema
- Hyperlipidemia

Causes of nephrotic syndrome include primary glomerular disease and systematic disease. Primary glomerular disease may be characterized by:

- Minimal change glomerulonephritis (commonest cause of nephrotic syndrome in children)
- Membranous glomerulonephritis (commonest cause of nephrotic syndrome in adults)
- Focal segmental glomerulosclerosis
- Membranoproliferative glomerulonephritis
- IgA nephropathy

Systemic disease may be characterized by:

- Diabetes mellitus
- Amyloidosis
- Drugs: penicillamine, street heroin
- Infections: malaria, syphilis, HIV, hepatitis B
- Malignancy

## Nephritic syndrome

Nephritic syndrome consists of hematuria, oliguria, and azotemia, as well as hypertension. Causes of nephritic syndrome include primary glomerular disease such as acute postinfection glomerulonephritis and IgA nephropathy. Systemic disease such as systemic lupus erythematosus may also cause nephritic syndrome.

## Hematuria

Hematuria may be visible to the naked eye (called gross hematuria) or detectable only on examination of the urine sediment by microscopy (called microscopic hematuria). Gross hematuria is suspected due to the presence of red or brown urine. Gross hematuria with passage of clots usually indicates a lower urinary tract source but can be seen with some forms of intrarenal bleeding (e.g., kidney cancer).

Hematuria is responsible if the red to brown color is seen only in the urine sediment, with the supernatant being clear. If, on the other hand, it is the supernatant that is red to brown, then the supernatant should be tested for heme (hemoglobin or myoglobin) with a urine dipstick:

- A red to brown supernatant that is negative for heme (hemoglobin or myoglobin) is a rare finding that can be seen in several conditions, including porphyria.
- A red to brown supernatant that is positive for heme is due to myoglobinuria or hemoglobinuria.

Microscopic hematuria refers to blood detectable only on examination of the urine sediment by microscopy. The urine sediment (or direct counting of RBC per mL of uncentrifuged urine) is the gold standard for the detection of microscopic hematuria (which is defined as three RBCs or more per high-power field). Dipsticks for heme detect one to two RBCs per high-power field and are therefore at least as, or more, sensitive as urine sediment examination, but they result in more false-positive tests due to the following:

- Semen
- An alkaline urine with a pH greater than 9 or contamination with oxidizing agents used to clean the perineum

- The presence of myoglobinuria or hemoglobinuria

Thus, a positive dipstick test must always be confirmed with microscopic examination of the urine. False-negative dipstick tests have been reported in patients ingesting large amounts of vitamin C; the clinical relevance of this observation is unknown.

## Glomerular versus nonglomerular bleeding

The identification of the glomeruli as the source of bleeding can optimize the subsequent evaluation. Signs of glomerular bleeding include RBC casts, a dysmorphic appearance of some red blood cells (RBCs), and, in patients with gross hematuria, a brown cola-colored urine. The presence of RBC casts is virtually diagnostic of glomerulonephritis or vasculitis, although such casts are infrequently seen in acute interstitial nephritis. The absence of these casts, however, does not exclude glomerular hematuria. Various renal diseases are summarized in Table 2.

**Table 2** Various renal disorders.

| Disease | Comments |
|---|---|
| Acute renal failure (ARF) | ARF is sudden deterioration of renal function, which can be broadly divided into prerenal, renal, and postrenal subtypes. In prerenal subtype, there is associated with hypoperfusion of the kidneys. Renal causes include glomerulonephritis and interstitial nephritis. Postrenal causes are related to obstructive uropathy |
| Acute interstitial nephritis | Tubulointerstitial is damaged due to various agents such as drugs, infections, and immunological injuries |
| Acute tubular necrosis (ATN) | Necrosis of renal tubules may be related to hypoperfusion and hypoxia. Patient undergoes three phases: oliguric phase, polyuric phase, and finally phase of recovery |
| Chronic renal failure (CRF)/chronic kidney disease (CKD) | Defined as chronic and progressive loss of renal function<br>Based on the GFR, it is divided into five stages (stage 5 with the lowest GFR, see Table 1) |
| Nephrotic syndrome | Defined as proteinuria (>3 g/day), hypoalbuminemia, hypercholesterolemia, and edema. Most common cause of nephritic syndrome in adults is membranous glomerulonephritis and common cause of nephrotic syndrome in children is minimal lesion |
| Nephritic syndrome | Defined as oliguria, hematuria with hypertension and edema. Acute diffuse glomerulonephritis is the leading cause of nephritic syndrome |
| Renal tubular acidosis | A group of disorders characterized by normal anion gap metabolic acidosis with inappropriately high urine pH (>5.5 in early morning urine). Type I (distal) is associated with decreased hydrogen ion secretion at the distal tubule. Type II is associated with increased loss of bicarbonate from the proximal tubule. In type IV (type III is discontinued), there is hyporeninemic hypoaldosteronism |

## Fanconi syndrome

This is a generalized disorder of tubular function characterized by glycosuria, aminoaciduria, phosphaturia, and acidosis. Causes of Fanconi syndrome include:

- Cystinosis, galactosemia, hereditary fructose intolerance, glycogen storage disease
- Tyrosinemia
- Paraproteinemia
- Nephrotoxicity (nephrotoxic drugs and heavy metal)

## Renal tubular acidosis

This is where patient develops metabolic acidosis due to defects in the renal tubules. There are various types:

- Type 1 distal; classical: Defect in hydrogen ion secretion, may cause hypokalemia.
- Type 2 proximal: Impairment of bicarbonate reabsorption; part of Fanconi syndrome.
- Type 3: It is rarely used as a classification because it is now thought to be a combination of type 1 and type 2.
- Type 4: Also known as hyperkalemic RTA: This type is due to low aldosterone causing hyperkalemia. Diuretics used to treat congestive heart failure such as spironolactone or eplerenone, blood pressure lowering drug such as angiotensin-converting enzyme (ACE) inhibitors and angiotensin receptor blockers (ARBs), antibiotics such as trimethoprim, pentamidine, heparin, nonsteroidal antiinflammatory drugs (NSAIDs), and some immunosuppressants may cause this disorder.

Generally, in metabolic acidosis, there is concurrent hyperkalemia, with the exception of RTA types 1 and 2.

## Renal calculi

There are multiple factors that predispose an individual to renal stones.

These include:

- Dehydration
- Urinary tract infection
- Alkaline urine
- Hypercalciuria

- Hyperuricosuria
- Hyperoxaluria
- Urinary obstruction

Hyperoxaluria can be primary (inherited) or secondary, which is due to increased ingestion or increased intestinal absorption of oxalate.

## Laboratory measurements of creatinine and related tests

Plasma creatinine may be measured using chemical methods or enzymatic methods. Most chemical methods utilize the Jaffe reaction. In this method, creatinine reacts with picrate ion in an alkaline medium to produce an orange-red complex. The Jaffe reaction is not entirely specific for creatinine. Substances such as ascorbic acid, high glucose, cephalosporins, and ketone bodies can interfere with this method. High bilirubin (both conjugated and unconjugated) may falsely lower the creatinine value (negative interference) measured by using Jaffe reaction. Enzymatic methods are also available for serum creatinine determination. Enzymes commonly used for creatinine determination are creatininase (also called creatinine deaminase) and creatinine hydrolase (also called creatinine aminohydrolase). Although enzymatic creatinine methods are subjected to less interference than Jaffe method, nevertheless interferences in enzymatic methods have also been reported. The reference method for creatinine measurement is isotope dilution mass spectrometry. Liu et al. reported that although enzymatic methods were less affected than creatinine determination using Jaffe reaction in patients undergoing hemodialysis, the gold standard for creatinine determination is isotope dilution mass spectrometry, which is free from interferences [13]. Nevertheless, both Jaffe method and enzymatic methods are commonly used in clinical laboratories. In one study, the authors concluded that both Jaffe and enzymatic methods were found to meet the analytical performance requirements in routine use. However, enzymatic method was found to have better performance in low creatinine levels [14].

Blood urea can be measured by both chemical and enzymatic methods. Most chemical methods are based on "Fearon Reaction," where urea reacts with diacetyl forming diazine, which absorbs at 540 nm. Enzymatic methods are based on hydrolysis of urea by the enzyme urease and this reaction generates ammonia. Then ammonia can be measured using Berthelot method or by another enzymatic method such as using glutamate dehydrogenase. Ammonia can also be measured by conductometry.

Measurement of uric acid can be done by either chemical or enzymatic method. A commonly used colorimetric method uses phosphotungstic acid, which is reduced by uric acid in alkaline medium producing a blue color (tungsten blue) that can be measured spectrophotometrically. However, this method is subjected to interferences, including interference from endogenous compounds such as high glucose and ascorbic acid (vitamin C). The enzymatic method based on uricase is more specific.

## Urine dipstick analysis

Urinalysis is a good screening tool for diagnosis of urological conditions such as urinary tract infection, as well as subclinical kidney disease. Urine dipstick analysis is usually the first tests performed during urinalysis followed by microscopic examination. Urine dipsticks are inexpensive paper or plastic devices with various segments (reaction pads) capable of color change if a particular substance of interest is present and such change in color can be compared to a color chart provided by the manufacturer for interpretation of result. Usually test strips can detect the presence of glucose, bilirubin, ketones, blood, protein, urobilinogen, nitrite, and leukocytes in the urine. Specific gravity of urine and pH can also be roughly estimated using a dipstick. Normal specific gravity of urine is between 1.002 and 1.035 and pH between 4.5 and 8.0. On a typical Western diet, urine pH is around 6.0 [15]. The urine dipstick is very sensitive to the presence of red blood cell or free hemoglobin. Negative or trace protein in urine is normal but a value of 1+ should be investigated further. Typically, glucose does not appear in urine unless plasma glucose is over 180–200 mg/dL. Positive nitrite test is indicative of bacteria in urine and urine culture is recommended. In addition, positive test for leukocyte esterase indicates presence of neutrophils (neutrophils produce leukocyte esterase) due to infection or inflammation. However, both false-positive and false-negative tests results may be encountered with urine dipstick analysis. Major interferences include:

- Protein reaction pad of urine dipstick detects albumin in urine but cannot detect Bence Jones proteins. If urine is alkaline false-positive protein test result may occur.
- Hemoglobin test pad can show false-positive result if myoglobin is present.
- Ketone reaction pad based on sodium nitroprusside can detect only acetoacetic acid and is weakly sensitive to acetone but cannot detect beta-hydroxybutyric acid.
- Presence of ascorbic acid (vitamin C) in urine can cause false-negative dipstick test with glucose and hemoglobin. Such interference may occur after taking vitamin C supplement or even fruit juice enriched with vitamin C [16]. Most glucose test strips use glucose-oxidase-based method where ascorbic acid may cause falsely lower values (negative interference). However, in glucometer that uses glucose dehydrogenase, ascorbic acid can cause false-positive result (see Chapter 7).

In one study, urinary vitamin C was detected in 1110 (18.1%) of 5006 samples. However, a high vitamin C concentration (3+) was observed more frequently in the medical check-up group than in the others. The authors concluded that presence of ascorbic acid (vitamin C) is problematic as it may cause false negative with glucose, hemoglobin, and leukocyte esterase dipstick tests [17].

## Summary/key points

- Kidney has three important functions: excretory, regulatory, and endocrine function.
- Kidney produces two important hormones: erythropoietin and renin. Erythropoietin is produced in response to renal hypoxia and acts on the bone marrow to stimulate erythropoiesis. Renin is produced by the juxtaglomerular apparatus. Renin converts angiotensinogen released by the liver into angiotensin I, which is then converted into angiotensin II in the lungs by angiotensin-converting enzyme (ACE). Angiotensin II is a vasoconstrictor and also stimulates release of aldosterone from adrenal cortex.
- Kidney also produces active form of vitamin D (1,25-dihydroxyvitamin D or 1, 25 dihydroxycholecalciferol). Serum vitamin D level over 30 ng/mL is considered adequate. The best laboratory parameter to monitor vitamin D status of a patient is to measure 25-hydroxy vitamin D.
- The basement membrane of the capillaries serves as a barrier to passage of large proteins into the glomerular filtrate. Molecules with a molecular weight of more than 15 kDa are not found in the glomerular filtrate. The loop of Henle is the site of where urine is concentrated.
- GFR can be estimated by using the formula: GFR $= (Ua \times V)/Pa$, where $Ua =$ the concentration of a solute in urine, $V =$ volume of urine in mL// min, and $Pa =$ concentration of the same solute in plasma.

A stable GFR does not necessarily imply stable disease.

- Serum creatinine levels are affected by gender, age, weight, lean body mass, and dietary protein intake mol/L). Cystatin C is a low-molecular-weight protein (13.3 kDa), which can be used for calculating GFR. In contrast to creatinine plasma, concentrations of cystatin C are unaffected by sex, diet, or muscle mass.
- Both creatinine clearance and cystatin C clearance may be used to evaluate glomerular filtration rate but cystatin C may be slightly superior to creatinine.
- Cockroft-Gault formula is widely used for calculating GFR.

Cockroft-Gault formula:

$$\text{Creatinine clearance} = \frac{(140 - \text{Age in years}) \times \text{Weight in kg}}{\text{Serum creatinine}(\mu\text{mol/L})} \times 1.23 \text{ if male or } 1.04 \text{ if female}$$

However, in United States, creatinine concentration is expressed in mg/dL, and this formula can be modified into:

$$\text{Creatinine clearance} = \frac{(140 - \text{Age in years}) \times \text{Weight in kg}}{72 \times \text{Serum creatinine}(\text{mg/dL})} \times 0.85 \text{ if female}$$

Cockroft-Gault formula was modified to MDDR formula by Modification of Diet in Renal Disease Study Group as follows:

- Estimated GFR $(\text{mL/min/1.73 m}^2) = 186 \times (\text{plasma creatinine in mg/dL})^{-1.154} \times \text{Age}^{-0.203} \times \text{F}$.
- In chronic renal disease, creatinine clearance is usually less than $60\,\text{mL/min/1.73 m}^2$ but a value below $15\,\text{mL/min}$ is indicative of end-stage renal disease.
- Fractional excretion of sodium over 3% may indicate acute tubular necrosis but less than 1% may indicate hypoperfusion of kidney.
- BUN/creatinine ratio for normal individuals usually is from 12:1 to 20:1. BUN/creatinine ratio below 10:1 may indicate intrinsic renal disease. BUN/creatinine ratio above 20:1 may be hypoperfusion of kidney including prerenal failure.
- Normally total urinary protein is $<150\,\text{mg/24 h}$ and consists of mostly albumin and Tamm Horsfall protein (secreted from ascending limb of Loop of Henle). Damage to glomerular integrity results in proteinuria, hematuria (the RBCs that are present in such cases are dysmorphic), and red cell casts.
- Persistent albumin excretion between 30 and 300 mg/day (20–200 µg/min) is called moderately increased albuminuria (formerly called "microalbuminuria"). In patients with diabetes, this is usually indicative of incipient diabetic nephropathy.
- Proteinuria can be classified into glomerular proteinuria, tubular proteinuria, mixed proteinuria, overflow, and postrenal. Glomerular proteinuria can be subclassified as selective (albumin and transferrin in urine) and nonselective (all proteins are present). In glomerular proteinuria, the major protein present is always albumin.
- In tubular proteinuria, albumin is a minor component, but proteins with smaller molecular weight such as alpha-1 microglobulin and beta-2 microglobulin are the major proteins found in the urine. In the mixed type proteinuria, both albumin and small molecular weight proteins such as alpha-1 microglobulin and beta-2 microglobulin are present.

- Nephrotic syndrome consists of massive proteinuria (in adults, proteinuria > 3.5 g/day), hypoalbuminemia (this is due to proteinuria coupled with increased protein catabolism in the kidneys), generalized edema, and hyperlipidemia.
- Nephritic syndrome consists of hematuria, oliguria, and azotemia and hypertension.
- Plasma creatinine may be measured using chemical methods or enzymatic methods. Most chemical methods utilize the Jaffe reaction. In this method, creatinine reacts with picrate ion in an alkaline medium to produce an orange-red complex.
- The Jaffe reaction is not entirely specific for creatinine. Substances such as ascorbic acid, high glucose, cephalosporins, and ketone bodies can interfere with this method. High bilirubin (both conjugated and unconjugated) may falsely lower the creatinine value (negative interference) measured by using Jaffe reaction.
- Usually test strips can detect the presence of glucose, bilirubin, ketones, blood, protein, urobilinogen, nitrite, and leukocytes in the urine. Specific gravity of urine and pH can also be roughly estimated using a dipstick.
- Typically, glucose does not appear in urine unless plasma glucose is over 180–200 mg/dL. Positive nitrite test is indicative of bacteria in urine and urine culture is recommended. In addition, positive test for leukocyte esterase indicates presence of neutrophils (neutrophils produce leukocyte esterase) due to infection or inflammation.
- Protein reaction pad of urine dipstick detects albumin in urine but cannot detect Bence Jones proteins. If urine is alkaline, false-positive protein test result may occur.
- Hemoglobin test pad can show false-positive result if myoglobin is present.
- Ketone reaction pad based on sodium nitroprusside can detect only acetoacetic acid and is weakly sensitive to acetone but cannot detect beta-hydroxybutyric acid.
- Presence of ascorbic acid (vitamin C) in urine can cause false-negative dipstick test with glucose and hemoglobin. Such interference may occur after taking vitamin C supplement or even fruit juice enriched with vitamin C. Most glucose test strips use glucose-oxidase-based method where ascorbic acid may cause falsely lower values (negative interference).

# References

[1] Carroll R.G., Abdel-Rahman A. In: Enna S.J., Bylund D.B., editors. Glomerular filtration. xPharm: the comprehensive pharmacology reference; 2007. p. 1–3. https://www.sciencedirect.com/referencework/9780080552323.

[2] Khan KA, Akram J, Fazal M. Hormonal cations of vitamin D and its role beyond just a vitamin: a review article. Int J Med Mol Med 2011;3:65–72.

[3]   Stevens LA, Coresh J, Greene T, Levey AS. Assessing kidney function—measured and estimated glomerular filtration rate. N Engl J Med 2006;354:2473.

[4]   Lopes van Balen VA, van Gansewinkel TAG, de Haas S, Spaan JJ, et al. Maternal kidney function during pregnancy: systematic review and meta-analysis. Ultrasound Obstet Gynecol 2019;54:297–307.

[5]   Rosenthal SH, Bokenkamp A, Hoffmann W. How to estimate GFR serum creatinine, serum cystatin C or equation? Clin Biochem 2007;40:153–61.

[6]   Hojs R, Bevc S, Ekhart R, Gorenjak M, et al. Serum cystatin C based equation compared to serum creatinine based equations for estimation of glomerular filtration rate in patients with chronic kidney disease. Clin Nephrol 2008;70:10–7.

[7]   Rachoin JS, Dahar R, Moussallem C, Milcarek B, et al. The fallacy of the BUN: creatinine ratio in critically ill patients. Nephrol Dial Transplant 2012;27:2248–54.

[8]   Eknoyan G, Hostetter T, Bakris GL, et al. Proteinuria and other markers of chronic kidney disease: a position statement of the national kidney foundation (NKF) and the national institute of diabetes and digestive and kidney diseases (NIDDK). Am J Kidney Dis 2003;42:617.

[9]   Chronic Kidney Disease Prognosis Consortium, Matsushita K, van der Velde M, et al. Association of estimated glomerular filtration rate and albuminuria with all-cause and cardiovascular mortality in general population cohorts: a collaborative meta-analysis. Lancet 2010;375:2073.

[10]  Snyder S, Pendergraph B. Detection and evaluation of chronic kidney disease. Am Fam Physician 2005;72:1723–32.

[11]  National Kidney Foundation. K/DQQI clinical practice guidelines for chronic kidney disease: evaluation, classification and stratification. Am J Kidney Dis 2002;39(Suppl. 2):S1–S266.

[12]  Lillehoj EP, Poulik MD. Normal and abnormal aspects of proteinuria: part I: mechanisms, characteristics and analyses of urinary protein. Part II: clinical considerations. Exp Pathol 1986;29:1–28.

[13]  Liu WS, Chung YT, Yang CY, Lin CC, et al. Serum creatinine determined by Jaffe, enzymatic methods and isotope dilution liquid chromatography-mass spectrometry in patients under hemodialysis. J Clin Lab Anal 2012;26:206–14.

[14]  Küme T, Sağlam B, Ergon C, Sisman AR. Evaluation and comparison of Abbott Jaffe and enzymatic creatinine methods: could the old method meet the new requirements? Clin Lab Anal 2018;32. https://doi.org/10.1002/jcla.22168.

[15]  Patel H. The abnormal urinalysis. Pediatr Clin North Am 2006;53:325–7.

[16]  Brigden ML, Edgell D, McPherson M, Leadbeater A, et al. High incidence of significant urinary ascorbic acid concentrations in west coast population-implications for routine urinalysis. Clin Chem 1992;38:426–31.

[17]  Lee W, Kim Y, Chang S, Lee AJ, Jeon CH. The influence of vitamin C on the urine dipstick tests in the clinical specimens: a multicenter study. J Clin Lab Anal 2017;31. https://doi.org/10.1002/jcla.22080.

# CHAPTER 12

# Inborn errors of metabolism

## Overview of inborn errors of metabolism

Congenital metabolic disorders are a class of genetic diseases which result from lack of or abnormality of an enzyme or its cofactor responsible for a clinically significant block in a metabolic pathway. As a result, abnormal accumulation of a substrate or deficit of the product is observed. In the majority of cases, this is due to single gene defect that encodes a particular enzyme important in the metabolic pathway. All inborn errors of metabolism are genetically transmitted typically in an autosomal recessive or X-linked recessive fashion. Although individual inborn errors of metabolism are rare genetic disorder, over 500 human diseases related to inborn errors of metabolism have been reported. Therefore, collectively inborn errors of metabolism affect more than one baby out of 1000 live births [1].

Children with inherited metabolic disorders most likely appear normal at birth because metabolic intermediates responsible for the disorder are usually small molecules that can be transported by the placenta and then eliminated by mother's metabolism. However, symptoms usually appear due to accumulation of metabolites days, weeks, or months after birth but very rarely few years after birth. Although clinical presentation may vary, infants with metabolic disorders typically present with lethargy, decreased feeding, vomiting, tachypnea (related to acidosis), decreased perfusion, and seizure. With progression of disease, infants may be presented to the hospital with stupor or coma.

Metabolic screening must be initiated in any infant suspected of inborn errors of metabolism, and elevated plasma ammonia level, hypoglycemia, and metabolic acidosis are indications of inborn errors of metabolism. Therefore, presenting clinical features of inborn errors of metabolism although variable may include:

- Failure to thrive, weight loss, delayed puberty, and precocious puberty
- Recurrent vomiting, diarrhea, and abdominal pain
- Neurologic features such as seizures and stroke

Clinical Chemistry, Immunology and Laboratory Quality Control. https://doi.org/10.1016/B978-0-12-815960-6.00027-3

- Organomegaly such as lymphadenopathy and hepatosplenomegaly
- Dysmorphic features
- Cytopenias
- Heart failure
- Immunodeficiency

Currently, newborn screenings are commonly performed in many states to potentially identify any of 40 most commonly encountered inborn errors of metabolism preferably using new technology of tandem mass spectrometry. Common inborn errors of metabolisms are listed in Table 1.

**Table 1** Common inborn errors of metabolism.

| Disorder | Enzyme defect |
| --- | --- |
| **Amino acid metabolism disorders** | |
| Phenylketonuria | Phenylalanine hydroxylase |
| Maple syrup disease | Branched chain alpha keto acid dehydrogenase complex |
| Tyrosinemia type I | Fumarylacetoacetate hydrolase |
| Tyrosinemia type II | Tyrosine aminotransferase |
| Homocystinuria | Cystathionine beta-synthase |
| **Carbohydrate metabolism disorders** | |
| Galactosemia | Galactose-1-phosphate uridyl transferase (most common cause; other enzyme defect may also cause galactosemia) |
| GSD type I (Von Gierke's disease) | Glucose 6 phosphatase |
| GSD type II (Pompe's disease) | Acid alpha glucosidase |
| GSD type V (McArdle disease) | Muscle glycogen phosphorylase |
| Hereditary fructose intolerance | Aldolase B |
| Fructose intolerance (benign) | Fructose kinase |
| Lactose intolerance | Lactase |
| **Urea cycle defect** | |
| Most common cause | Ornithine transcarbamylase or carbamoyl synthase |
| **Organic aciduria** | |
| Methylmalonic acidemia | Methylmalonyl CoA mutase |
| Propionic aciduria | Propionyl-CoA carboxylase |
| Isovaleric aciduria | Isovaleryl-CoA dehydrogenase |
| Glutaric aciduria type I | Glutaryl-CoA dehydrogenase |

**Table 1** Common inborn errors of metabolism—cont'd

| Disorder | Enzyme defect |
|---|---|
| **Fatty acid oxidation disorders** | |
| MCAD deficiency (most common) | Medium-chain acyl coenzyme A dehydrogenase (MCAD) |
| SCAD deficiency | Short-chain acetyl-CoA dehydrogenase deficiency (SCAD) |
| LCAD deficiency | Long-chain acetyl-CoA dehydrogenase deficiency (LCAD) |
| VLCAD deficiency | Very long-chain acetyl-CoA dehydrogenase deficiency (VLCAD) |
| CPT-I deficiency | Carnitine palmitoyl transferase type I (CPT-I) |
| CPT-II deficiency | Carnitine palmitoyl transferase type I (CPT-II) |
| CACT deficiency | Carnitine acylcarnitine translocase (CACT) |
| **Mitochondrial disorders** | |
| Kearns-Sayre syndrome (KSS) | Mitochondrial DNA abnormality |
| **Peroxisomal disorders** | |
| Zellweger syndrome | Peroxisome membrane protein |
| **Lysosomal storage disorders** | |
| Hunter syndrome | Iduronate sulfatase |
| Hurler syndrome | Alpha-L-iduronidase |
| Gaucher disease | Beta-glucocerebrosidase |
| Tay-Sachs disease | Hexosaminidase A |
| Fabry's disease | Alpha-galactosidase A |
| Niemann-Pick disease type A and B | Sphingomyelinase |
| **Purine or pyrimidine metabolic disorders** | |
| Lesch-Nyhan syndrome | Hypoxanthine guanine phosphoribosyltransferase |

## Amino acid disorders

Amino acids are an integral part of proteins and also may act as a substrate for gluconeogenesis. Of 20 amino acids, 9 amino acids are essential as they cannot be synthesized by humans. In a patient with amino acid disorders, accumulation of amino acids in the blood is a common feature, and as expected, increased excretion of amino acids is observed in urine. Common amino acid disorders are phenylketonuria and maple syrup urine disease.

## Phenylketonuria

Phenylketonuria is the most prevalent disorder caused by an inborn error in amino acid metabolism. Nevertheless, phenylketonuria is a rare (on average 1 in 10,000–12,000 live births in Western Europe) inborn error of metabolism characterized by reduced activity of the hepatic enzyme phenylalanine hydroxylase due to mutations in the encoding gene. Phenylalanine hydroxylase is responsible for the conversion of phenylalanine to tyrosine. Reduced enzymatic activity results in elevated phenylalanine and reduced tyrosine level [2]. Accumulated phenylalanine is then converted to phenylpyruvate, which is eventually excreted into urine. Sustained phenylalanine concentration greater than 20 mg/dL (1211 µmol/L) correlates with classical symptoms of phenylketonuria, such as mental retardation, impaired head circumference growth, poor cognitive function, and lighter skin pigmentation. The disease is mild if phenylalanine concentration is in the range of 9.9–19.9 mg/dL (600–1200 µmol/L). The phenylalanine to tyrosine ratio (cutoff: 1.5) is also used for diagnosis of phenylketonuria, and this ratio is helpful in reducing false-positive rates. Treatment is phenylalanine-restricted diet.

## Maple syrup urine disease (MSUD)

Maple syrup urine disease (MSUD) is a rare disease that requires a protein-restricted diet for successful management. This disease is a metabolic disorder caused by a deficiency of the branched chain alpha keto acid dehydrogenase complex resulting in accumulation of branched chain amino acids, including leucine, isoleucine, and valine. The urine of such patient has odor like maple syrup, and this disease is called maple syrup urine disease. Elevated leucine is responsible for brain injury and neurological symptoms in these patients. This disease is inherited in an autosomal recessive manner and usually caused by mutations in any one of the genes: *BCKDHA*, *BCKDHB*, or *DBT*, which represent E1α, E1β, and E2 subunits of the branched-chain α-keto acid dehydrogenase (BCKDH) complex [3]. Newborn screening involves plasma amino acid analysis for diagnosis which can be conducted using dried blood spot.

## Other amino acid disorders

Tyrosinemia type I is caused by a deficiency of fumarylacetoacetate hydrolase, and affected patients may present in childhood to their physicians with acute hepatic failure, coagulopathy, renal dysfunction, growth retardation, and possibly peripheral nerve involvement. Tyrosinemia type II is caused by a deficiency of tyrosine aminotransferase and is an oculocutaneous form of the disease that causes corneal lesions and skin involvement. Treatment involved low tyrosine/phenylalanine diet and may also be treated with nitisinone. Newborn screening involving only tyrosine has certain limitations in diagnosis of tyrosinemia. Analysis of succinylacetone using tandem mass spectrometry may be helpful for diagnosis [4].

Homocystinuria is due to cystathionine beta-synthase deficiency, and patients have developmental delay as well as may present with ocular, skeletal, vascular, and central nervous system abnormalities. Newborn diagnosis is based on high methionine and high homocysteine levels.

# Carbohydrate metabolism disorders

Carbohydrate metabolism disorders may include deficiencies of enzymes involved in the metabolism of glycogen, galactose, and fructose. These diseases can be broadly subclassified as diseases causing liver dysfunction, disease affecting muscle and liver, and diseases affecting only muscle. Galactosemia, glycogen storage diseases, hereditary fructose intolerance, and fructose 1,6-diphospahte deficiency are common examples of carbohydrate metabolism disorders.

## Galactosemia

Three enzymes are involved in metabolism of galactose, and deficiency of any of these three enzymes can cause galactosemia. Most common form of galactosemia is caused by lack of the enzyme galactose-1-phosphate uridyl transferase affecting an estimated 1 in every 55,000 newborns. In these patients, galactose-1-phosphate accumulates, which is degraded to galactonate and galactitol causing early cataract formation in eyes. If untreated, these children may develop intellectual disability, speech problem, and other dysfunctions. Diagnosis can be established by measuring galactose-1-phosphate uridyl transferase activity in dried blood spot.

## Glycogen storage disease (GSD)

There are multiple types of glycogen storage disease depending on the exact deficient enzyme. They are numbered according to their discovery, and numbers are not useful in separating these disorders according to clinical symptoms. It is noteworthy to mention Type I (Von Gierke's disease) which is due to the absence of glucose 6 phosphatase, Type II (Pompe's disease) due to the absence of acid alpha glucosidase, and Type V (McArdle disease) due to the absence of muscle glycogen phosphorylase. In general, patients with glycogen storage disease I, III, VI, and IX present with hepatomegaly and hypoglycemia, while patients with glycogen storage disease IV often experience liver failure prior to symptom of hypoglycemia. Patients with glycogen storage disease II, V, and VII primarily have muscle dysfunction. The onset of glycogen storage II disease may be early childhood, but type V and VII patients often present to clinics during their adolescence with complains of exercise intolerance accompanied usually by myoglobinuria.

## Fructose intolerance

Hereditary fructose intolerance is a rare (1 in 20,000 birth) recessive inherited disorder of carbohydrate metabolism due to catalytic deficiency of aldolase B (fructose biphosphate or liver aldolase), and these patients show impaired fructose metabolism when exposed to fructose or sucrose during infancy through diet. Persistence ingestion of fructose and sucrose can cause severe liver and kidney damage that may be associated with seizure, coma, and even death. Early diagnosis is essential for good prognosis because these individuals can live a normal life by avoiding fruits and sweets containing fructose. The diagnosis can be confirmed by measuring particular enzyme activity on a liver biopsy. DNA analysis is also available for diagnosis [5]. The most common mutation in fructose metabolism is due to lack of fructokinase, which is the first step in the metabolism of dietary fructose. This condition, however, is asymptomatic and excess fructose is excreted in urine (fructosuria).

## Lactose intolerance

Lactose intolerance also called lactase deficiency is due to insufficient level of lactase which hydrolyses lactose into glucose and galactose. Therefore, ingestion of milk and dairy products results in bloating, abdominal cramps, diarrhea, and related symptoms. This is a common condition that may develop later in life or may be manifested in early childhood. These individuals can live a normal life by either avoiding milk or dairy products or consuming lactase free milk or take lactase enzyme as supplement. Lactose intolerance is transmitted in either autosomal recessive (Caucasian population) or autosomal dominant (Asian population) fashion.

## Urea cycle disorders

Degradation of amino acids results in the formation of ammonia as a waste product which then enters urea cycle in the liver and in the first step ammonia combines with carbon dioxide to form carbamoyl phosphate. Finally, in urea cycle, ammonia is converted into urea for excretion by kidneys. Patients with urea cycle disorders present with hyperammonemia which may be encountered either the neonatal period or later. This urea cycle utilizes six enzymes: carbamoylphosphate synthetase 1, ornithine transcarbamylase, argininosuccinate synthetase, argininosuccinate lyase, arginase 1, and N-acetylglutamate synthase. In addition, at least two transporter proteins are essential to urea cycle function. Congenital defects of any one of the enzymes or transporters of the urea cycle may cause the disease. Severity and age of onset depend on residual enzyme or transporter function and are related to the respective gene mutations. Ornithine transcarbamylase deficiency, an X-linked disorder (occurring most commonly in males), is the most common urea cycle defect. The strategy for therapy is to prevent the irreversible toxicity of high-ammonia exposure to the brain [6].

Newborns with urea cycle disorder develop high levels of ammonia after a protein feed. Only patients with arginase deficiency, a defect in the last step of the urea cycle, do not present with hyperammonemia but instead present with neurological dysfunctions. It is often difficult to diagnose urea cycle defect although hyperammonemia in a sick neonate is an indication. Determination of orotic acid concentration can help differentiating ornithine transcarbamylase deficiency (elevated orotic acid) from carbamyl phosphatase synthetase deficiency (normal or low orotic acid level). Urea cycle disorders are treatable cause of hyperammonemia in infants and pediatric age group. Presentation in adolescence or adult life of urea cycle disorder is rare.

## Organic acid disorders (organic aciduria)

Organic acid disorders are a group of inborn errors of metabolism due to enzyme deficiency in the amino acid degradation pathways including defects in metabolism of branched chain amino acids (leucine, isoleucine, and valine) as well as other amino acids including homocysteine, tyrosine, methionine, threonine, lysine, and tryptophan. As a result, toxic organic acids accumulate in circulation and are eventually excreted in urine causing organic aciduria. More than 25 disorders are known. However, division of organic acid disorders and amino acid metabolism disorders are somehow arbitrary because phenylketonuria and maple syrup urine disease also cause organic aciduria. However, amino acid disorders are traditionally diagnosed by amino acid analysis in blood, while organic acidurias are traditionally diagnosed by urine organic acid analysis by gas chromatography/mass spectrometry or tandem mass spectrometry. Alternatively, dried blood spot in a newborn can be analyzed to establish diagnosis of organic aciduria.

Methylmalonic acidemia is due to defect in methylmalonyl-CoA mutase enzyme, which is involved in metabolism of branched chain amino acids. Vitamin $B_{12}$ is also required for this conversion. Mutations leading to defects in vitamin $B_{12}$ metabolism or in its transport frequently result in the development of methylmalonic acidemia.

In healthy individuals, the enzyme propionyl CoA carboxylase converts propionyl CoA to methylmalonyl CoA. In individuals with propionic acidemia, this pathway is blocked causing conversion of excessive propionyl CoA to propionic acid leading to propionic acidemia. Isovaleric acidemia is caused by a deficiency of isovaleryl-CoA dehydrogenase, which is involved in metabolism of leucine. Glutaric aciduria type I is a rare organic aciduria due to deficiency of glutaryl-CoA dehydrogenase, which is involved in catabolism of lysine, hydroxylysine, and tryptophan [7].

## Fatty acid oxidation disorders

Mitochondrial fatty acid oxidation is a major pathway for energy production during fasting and strenuous that may cause hypoglycemic condition. However, fatty acid must be transported to mitochondria prior to oxidation, and either transport defect of fatty acid or defect in any enzyme involved in fatty acid oxidation pathway may cause fatty acid oxidation disorders which are inherited in autosomal recessive pattern. Patients with fatty acid oxidation disorders usually have features of hypoglycemia without ketosis during episodes of decreased carbohydrate intake. Most common fatty acid oxidation disorder is due to deficiency of medium-chain acyl coenzyme A dehydrogenase (MCAD), but long-chain acetyl-CoA dehydrogenase deficiency (LCAD), very long-chain acetyl-CoA dehydrogenase deficiency (VLCAD), and short-chain acetyl CoA dehydrogenase deficiency (SCAD) have also been reported.

Carnitine is essential for transporting long-chain fatty acid because long-chain fatty acids cannot pass through mitochondrial membrane although short- and medium-chain fatty acids can pass through mitochondrial membrane. Carnitine palmitoyl transferase type I (CPT-I) is responsible for attaching carnitine to long-chain fatty acid molecule, and then carnitine acylcarnitine translocase (CACT) transports the resulting molecule into the mitochondria and finally palmitoyl transferase type II (CPT-II) removes the carnitine and releases fatty acid for beta oxidation inside the mitochondria which produces energy. The free carnitine is transported back into circulation for binding with more fatty acid. Defect in any of these carnitine transporting enzymes can also cause fatty acid oxidation defect in an individual. Newborn screening can identify different types of fatty acid oxidation disorders, and treatment of fatty acid oxidation is primarily aimed at maintaining blood glucose by feeding at regular intervals and diet high in carbohydrate but low in fat. Some patients may also need carnitine supplement.

## Mitochondrial disorders

Organic acids, fatty acids, and amino acids are metabolized to acetyl CoA within the mitochondria. Acetyl CoA combines with oxaloacetate to form citric acid, which is oxidized in the Krebs cycle (also known as citric acid cycle). Mitochondrial diseases are a clinically and genetically heterogeneous group of disorders that result from dysfunction of the mitochondrial electron transport chain and oxidative phosphorylation due to pathogenic variants in mitochondrial DNA (mitochondria has their own DNA) or nuclear DNA encoding mitochondrial proteins. In addition to a wide range of cellular perturbations, dysfunctional mitochondria are unable to generate sufficient energy to meet the needs of various organs, particularly these with high energy demand, such as the nervous system, skeletal and cardiac muscles, kidneys, liver, and endocrine system. Energy deficiency in various organs leads to the variable manifestations observed in mitochondrial diseases, including cognitive impairment, epilepsy, cardiac and skeletal myopathies, nephropathies, hepatopathies, and endocrinopathies [8].

Patients with mitochondrial disorders may present with hypoglycemia with ketosis. Mitochondrial disorders may affect muscle alone. Examples of mitochondrial disorders include cytochrome $c$ oxidase deficiency and Kearns Sayre syndrome (KSS). KSS is a syndrome that is characterized by isolated involvement of the muscles controlling eyelid movement (levator palpebrae, orbicularis oculi) and those controlling eye movement (extra-ocular muscles). This results in ptosis and ophthalmoplegia, respectively. KSS involves a triad of the eye changes with bilateral pigmentary retinopathy and cardiac conduction abnormalities.

## Peroxisomal disorders

Peroxisomes are cellular organelles which play important role in beta oxidation of very long-chain fatty acids, degradation of phytanic acid by alpha-oxidation, degradation of hydrogen peroxide as well as synthesis of bile acids and plasmalogen, an important component of cell membranes and myelin. Examples of peroxisomal disorders are Zellweger syndrome and adrenoleukodystrophy. Zellweger syndrome is due to biogenesis defect, and as a result, all peroxisomal enzymes are deficient making it a very severe disorder. However, neonatal adrenoleukodystrophy is milder than Zellweger syndrome.

## Lysosomal storage disorders

Lysosomes are cellular organelles that contain more than 30 acid hydrolases which can degrade unwanted complex molecules such as mucopolysaccharides, sphingolipids, and glycoproteins into molecules which could be used by the body again. Therefore, lysosomes can be regarded as recycling center of the body. Lysosomal storage diseases are a group of over 70 diseases that are characterized by lysosomal dysfunction, most of which are inherited as autosomal recessive manner except Hunter syndrome, Danon disease, and Fabry disease which are inherited by X-linked manner. These disorders are individually rare but collectively affect 1 in 5000 live births. Lysosomal storage diseases are due to defect in lysosomal enzymes, enzyme receptors, membrane proteins, activator proteins, or transporters. In lysosomal storage disorders, accumulation of few complex lipids occurs which in normal individuals should be degraded. Such accumulation results in impaired lysosomal function such as delivery of nutrients to cells. Therefore, cellular starvation due to lack of proper nutrients finally causes organ dysfunction.

Lysosomal storage disorders typically present in infancy and childhood, although adult-onset forms also occur. Most such diseases have a progressive neurodegenerative clinical course, although symptoms in other organ systems are frequent. Several lysosomal storage diseases can be treated with approved, disease-specific therapies that are mostly based on enzyme replacement [9]. Common examples of lysosomal storage disorders include:

- Mucopolysaccharidoses: Examples of such disorders are Hunter syndrome, Hurler syndrome, Sanfilippo syndrome, and Scheie syndrome
- Sphingolipidoses: For example, Gaucher disease, Tay-Sachs, Fabry disease, Niemann-Pick disease
- Glycoproteinoses: For example, mannosidosis
- Mucolipidosis

Mucopolysaccharidoses are a group of metabolic disorders due to deficiency of lysosomal enzymes responsible for the breakdown of polysaccharide chain (glycosaminoglycan). Gaucher disease is the most common form of lysosomal storage disease which is due to deficiency of the enzyme glucocerebrosidase, leading to accumulation of glucocerebroside. Gaucher's disease has three common clinical subtypes:

- Type I (or nonneuropathic type) is the most common form of the disease. It is seen most often in Ashkenazi Jews. Features are apparent early in life or in adulthood and include hepatosplenomegaly. Neurological features are not seen. Depending on disease onset and severity, type 1 patients may live well into adulthood. Many individuals have a mild form of the disease or may not show any symptoms.
- Type II (or acute infantile neuropathic Gaucher's disease) typically begins within 6 months of birth. Neurological features are prominent and most children will die at a very early age.
- Type III (the chronic neuropathic form) can begin at any time in childhood or even in adulthood. It is characterized by slowly progressive but milder neurologic symptoms compared to the acute or type 2 version.

Tay Sachs disease has a higher frequency in Ashkenazi Jews and caused by deficiency of hexosaminidase. Niemann-Pick type A disease is a fatal disorder of infancy (life expectancy: 2–3 years) due to accumulation of sphingomyelin as a result of mutation in sphingomyelin phosphodiesterase 1 gene encoding enzyme acid sphingomyelinase. In type A disease, activity of this enzyme is almost completely absent. In type B disease, some activity of this enzyme is preserved. However, in type C disease, accumulation of nonesterified cholesterol takes place.

## Purine or pyrimidine metabolic disorders

Purine and pyrimidine nucleotides are part of the DNA, RNA as well ATP, and nicotinamide adenine dinucleotide (NAD). Examples of purine and pyrimidine disorders include Lesch-Nyhan disease or syndrome and adenosine deaminase deficiency.

Lesch-Nyhan disease is a rare monogenic disorder and is transmitted as X-linked recessive fashion. These patients have high risk of developing gout due to overproduction of uric acid as a result of deficiency of the enzyme hypoxanthine guanine phosphoribosyltransferase. The patients with classical phenotype present with overproduction of uric acid, sever motor dysfunction resembling patients with dystonic cerebral palsy, intellectual deficiency, and self-injurious behavior. The mildest form of this disease includes only overproduction of uric acid. In between classical (extreme) and mild form, there is an intermediate form of this disease where patients experience some motor and cognitive dysfunction but no self-injurious behavior [10].

## Disorders of porphyrin metabolism

The various porphyrias are due to abnormalities in the enzymes involved in the synthesis of heme resulting in accumulation of intermediate compounds. Porphyrins consist of four pyrrole rings, and the precursors in the formation of the pyrrole rings are glycine and succinyl CoA which combine in the presence of delta-aminolevulinate synthase to form delta aminolevulinic acid inside the mitochondria which is then transported into cytosol for further transformation. Then, two molecules of delta aminolevulinic condense to form one pyrrole ring (mono-pyrrole porphobilinogen). Then, porphobilinogen molecules are cyclized to form hydroxymethylbilane which is eventually converted into coproporphyrinogen III which enters into mitochondria for further transformation into protoporphyrinogen IX by the action of coproporphyrinogen oxidase enzyme. Finally, protoporphyrinogen IX is converted into protoporphyrin and ferrous iron is incorporated in the molecule by the action of ferrochelatase to form heme molecule inside the mitochondria. In porphyria, the heme intermediates accumulate due to partial deficiency in certain enzymes involved in heme biosynthesis. Porphyrias can be subclassified under two broad categories:

- Porphyrias involving skin lesions and photosensitivity
- Porphyrias with neurovisceral dysfunction such as neuropathy, convulsions, psychiatric disorders, acute abdomen, hypertension, and tachycardia

Porphyrias are caused by well-characterized enzyme defects in the complex heme biosynthetic pathway and are divided into categories of acute vs nonacute

**Table 2** Various types of porphyrias.

| Enzyme | Function of enzyme | Type of porphyria for missing enzyme |
| --- | --- | --- |
| ALA (delta-aminolevulinate) synthase | Forms ALA from glycine and succinyl CoA | Not known |
| ALA dehydratase | Converts ALA to porphobilinogen (PBG) | ALA dehydrase deficiency porphyria |
| PBG deaminase | Converts PBG to hydroxymethylbilane | Acute intermittent porphyria (AIP) |
| Uroporphyrinogen synthase | Converts hydroxymethylbilane to uroporphyrinogen III | Congenital erythropoietic porphyria |
| Uroporphyrinogen decarboxylase | Converts uroporphyrinogen III to coproporphyrinogen III | Porphyria cutanea tarda and hepatoerythropoietic porphyria |
| Coproporphyrinogen oxidase | Converts coproporphyrinogen III to protoporphyrinogen | Hereditary coproporphyria (HCP) |
| Protoporphyrinogen oxidase | Converts protoporphyrinogen to protoporphyrin | Variegate porphyria (VP) |
| Ferrochelatase | Adds iron to protoporphyrin to form heme | Erythropoietic protoporphyria (EPP) |

or hepatic vs erythropoietic porphyrias. Acute hepatic porphyrias (acute intermittent porphyria, variegate porphyria, hereditary coproporphyria, and aminolevulinic acid dehydratase deficient porphyria) are characterized by overproduction of porphyrin precursors, producing often serious abdominal, psychiatric, neurologic, or cardiovascular symptoms. Patients with variegate porphyria and hereditary coproporphyria can present with skin photosensitivity. Acute porphyrias are inherited in autosomal dominant fashion. The non-acute porphyrias are porphyria cutanea tarda, erythropoietic protoporphyria, X-linked protoporphyria, and the rare congenital erythropoietic porphyria. They lead to the accumulation of porphyrins that cause skin photosensitivity and occasionally severe liver damage. Diagnosis relies on measurement of increased urinary 5-aminolevulinic acid (in patients with aminolevulinic acid dehydratase deficient porphyria) or increased 5-aminolevulinic acid and porphobilinogen (in patients with other acute porphyrias) [11]. Various types of porphyrias are listed in Table 2.

## Newborn screening and evaluation

Newborn screening tests are routinely performed to identify approximately 40 disorders. The tests and methods may vary from state to state and country to country. False-positive and false-negative screening tests can occur. Clinical evaluation includes a detailed history, including family history, physical examination, and laboratory evaluation. Laboratory evaluation may include initial tests and specialized tests. Initial tests may include complete blood count, serum levels of glucose, ammonia, creatinine, urea, uric acid, electrolytes, muscle enzymes such as creatine kinase, aldolase as well as liver function tests.

Urinalysis is also helpful. However, specialized tests are needed for diagnosis of inborn errors of metabolism and these tests include:

- Quantitative plasma amino acid profile (for urea cycle disorders and disorders of amino acid metabolism).
- Urine organic acids (for diagnosis of various acidurias).
- Serum pyruvate and lactate (lactic acidosis is seen in mitochondrial disorders as well disorders of carbohydrate metabolism and glycogen storage diseases).
- Acylcarnitine profile (used for fatty acid oxidation disorders).

More recently, new platform technology such as tandem mass spectrometry (MS/MS) is used widely in developed countries for newborn screening. New developments in tandem mass spectrometry coupled with electrospray detection (ESI) allow rapid and high-throughput screening for a large number of inborn errors of metabolism from a single dried blood spot specimen after extraction [12]. Next-generation DNA sequencing (NGS) also has the potential to improve the diagnostic and prognostic utility of newborn screening programs [13].

## Summary/key points

- Phenylketonuria is due to deficiency of phenylalanine hydroxylase which converts phenylalanine into tyrosine, and as a result, phenylalanine cumulates into the circulation which is converted to phenylpyruvate, which is eventually excreted into urine, hence the name phenylketonuria. Phenylketonuria is an autosomal recessive disorder.
- Maple serum urine disease is a metabolic disorder caused by a deficiency of the branched chain alpha keto acid dehydrogenase complex resulting in accumulation of branched chain amino acids, including leucine, isoleucine, and valine. The urine of such patient has the same odor as maple syrup.
- Most common form of galactosemia is caused by lack of the enzyme galactose-1-phosphate uridyl transferase affecting an estimated 1 in every 55,000 newborns. In these patients, galactose-1-phosphate accumulates.
- The most common mutation in fructose metabolism is due to lack of fructokinase, which is the first step in the metabolism of dietary fructose. This condition, however, is asymptomatic, and excess fructose is excreted in urine (fructosuria).
- Lactose intolerance also called lactase deficiency is due to insufficient level of lactase which hydrolyses lactose into glucose and galactose. Therefore, ingestion of milk and dairy products results in bloating, abdominal cramps, diarrhea, and related symptoms.

- Ornithine transcarbamylase deficiency, an X-linked disorder (occurring most commonly in males), is the most common urea cycle defect. Newborns with urea cycle disorder develop high levels of ammonia after a protein feed.
- Organic acid disorders are a group of inborn errors of metabolism due to enzyme deficiency in the amino acid degradation pathways including defects in metabolism of branched chain amino acids (leucine, isoleucine, and valine) as well as other amino acids including homocysteine, tyrosine, methionine, threonine, lysine, and tryptophan. As a result, toxic organic acids accumulate in circulation and are eventually excreted in urine causing organic aciduria.
- Most common fatty acid oxidation disorder is due to deficiency of medium-chain acyl coenzyme A dehydrogenase (MCAD).
- Organic acids, fatty acids, and amino acids are metabolized to acetyl CoA within the mitochondria. Acetyl CoA combines with oxaloacetate to form citric acid, which is oxidized in the Krebs cycle (also known as citric acid cycle). If there is a defect in energy producing pathway especially during oxidative phosphorylation, that causes disease termed as mitochondrial disorder or mitochondrial disease. Patients with mitochondrial disorders may present with hypoglycemia with ketosis. Although mitochondrial disorders may affect muscle alone, multiple organ involvement such as brain, heart, kidney, and liver may also be seen. Mitochondrial diseases are due to mutation of mitochondrial DNA (mitochondria has their own DNA), and all mitochondrial DNAs are derived from ovum so that these diseases are maternally inherited.
- Peroxisomes are cellular organelles which play important role in beta oxidation of very long-chain fatty acids, degradation of phytanic acid by alpha-oxidation, degradation of hydrogen peroxide, as well as synthesis of bile acids and plasmalogen, an important component of cell membranes and myelin. Examples of peroxisomal disorders are Zellweger syndrome and adrenoleukodystrophy.
- Lysosomal storage diseases are a heterogenous group of more than 70 disorders due to defect in lysosomal enzymes, enzyme receptors, membrane proteins, activator proteins, or transporters. Common examples of lysosomal storage disorders include:
  - Mucopolysaccharidoses: Examples of such disorders are Hunter syndrome, Hurler syndrome, Sanfilippo syndrome, and Scheie syndrome
  - Sphingolipidoses: For example, Gaucher disease, Tay-Sachs, Fabry disease, Niemann-Pick disease
  - Glycoproteinoses: For example, mannosidosis
  - Mucolipidosis

- Gaucher disease is the most common form of lysosomal storage disease which is due to deficiency of the enzyme glucocerebrosidase, leading to accumulation of glucocerebroside. Gaucher's disease has three common clinical subtypes:
  - Type I (or nonneuropathic type) is the most common form of the disease. It is seen most often in Ashkenazi Jews. Features are apparent early in life or in adulthood and include hepatosplenomegaly. Neurological features are not seen. Depending on disease onset and severity, type 1 patients may live well into adulthood. Many individuals have a mild form of the disease or may not show any symptoms.
  - Type II (or acute infantile neuropathic Gaucher's disease) typically begins within 6 months of birth. Neurological features are prominent and most children will die at a very early age.
  - Type III (the chronic neuropathic form) can begin at any time in childhood or even in adulthood. It is characterized by slowly progressive but milder neurologic symptoms compared to the acute or type 2 version.
- Tay Sachs disease has a higher frequency in Ashkenazi Jews and caused by deficiency of hexosaminidase. Niemann-Pick type A disease is a fatal disorder of infancy (life expectancy: 2–3 years) due to accumulation of sphingomyelin as a result of mutation in sphingomyelin phosphodiesterase 1 gene encoding enzyme acid sphingomyelinase.
- Examples of purine and pyrimidine disorders include Lesch-Nyhan disease or syndrome and adenosine deaminase deficiency. Lesch-Nyhan disease is a rare monogenic disorder and is transmitted as X-linked recessive fashion. These patients have high risk of developing gout due to overproduction of uric acid as a result of deficiency of the enzyme hypoxanthine guanine phosphoribosyltransferase.
- Acute porphyrias such as acute intermittent porphyria variegate porphyria and hereditary coproporphyria are inherited in autosomal dominant fashion. Acute life-threatening neurovisceral attacks seen in these three porphyrias are similar in nature. Nonacute porphyrias include congenital erythropoietic porphyria, porphyria cutanea tarda, and erythropoietic protoporphyria.
- Newborn screening tests are routinely performed to identify approximately 40 disorders. The tests and methods may vary from state to state and country to country. False-positive and false-negative screening tests can occur. Clinical evaluation includes a detailed history, including family history, physical examination, and laboratory evaluation. Laboratory evaluation may include initial tests and specialized tests. Initial tests may include complete blood count, serum levels of glucose, ammonia, creatinine, urea, uric acid, electrolytes,

muscle enzymes such as creatine kinase, aldolase, as well as liver function tests. Urinalysis is also helpful. However, specialized tests are needed for diagnosis of inborn errors of metabolism, and these tests include quantitative plasma amino acid profile (for urea cycle disorders and disorders of amino acid metabolism), urine organic acids (for diagnosis of various acidurias), serum pyruvate and lactate (lactic acidosis is seen in mitochondrial disorders as well disorders of carbohydrate metabolism and glycogen storage diseases), and acylcarnitine profile (used for fatty acid oxidation disorders).

## References

[1] Alfadhel M, Al-Thihli K, Moubayed H, Eyaid W, et al. Drug treatment of inborn errors of metabolism: a systematic review. Arch Dis Child 2013;98:454–61.

[2] Blau N, Van Spronsen FJ, Levy HL. Phenylketonuria. Lancet 2010;376:1417–27.

[3] Ali EZ, Ngu LH. Fourteen new mutations of BCKDHA, BCKDHB and DBT genes associated with maple syrup urine disease (MSUD) in Malaysian population. Mol Genet Metab Rep 2018;17:22–30.

[4] Allard P, Greiner A, Korson MS, et al. Newborn screening for hepatorenal tyrosinemia by tandem mass spectrometry: analysis of succinylacetone extracted from dried blood spot. Clin Biochem 2004;37:1010–5.

[5] Ferri L, Caciotti A, Cavicchi C, Rigoldi M, et al. Integration of PCR sequencing analysis with multiplex ligation dependent probe amplification for diagnosis of hereditary fructose intolerance. JIMD Rep 2012;6:31–7.

[6] Matsumoto S, Häberle J, Kido J, Mitsubuchi H, et al. Urea cycle disorders-update. J Hum Genet 2019;64:833–47.

[7] Kolker S, Christensen E, Leonard J, Greenberg C, et al. Diagnosis and management of glutaric aciduria type I-revised recommendations. J Inherit Metab Dis 2011;34:677–94.

[8] El-Hattab AW, Zarante AM, Almannai M, Scaglia F. Therapies for mitochondrial diseases and current clinical trials. Mol Genet Metab 2017;122:1–9.

[9] Platt FM, d'Azzo A, Davidson BL, Neufeld EF, Tifft CJ. Lysosomal storage diseases. Nat Rev Dis Primers 2018;4(1):27.

[10] Torres RJ, Puig JG, Jinnah HA. Update on the phenotypic spectrum of Lesch-Nyhan disease and its attenuated variants. Curr Rheumatol Rep 2012;14:189–94.

[11] Stölzel U, Doss MO, Schuppan D. Clinical guide and update on porphyrias. Gastroenterology 2019;157:365–81.

[12] Ozben T. Expanded newborn screening and confirmatory follow-up testing for inborn errors of metabolism detected by tandem mass spectrometry. Clin Chem Lab Med 2013;51:157–76.

[13] van Campen JC, Sollars ESA, Thomas RC, Bartlett CM, et al. Next generation sequencing in newborn screening in the United Kingdom National Health Service. Int J Neonatal Screen 2019;5(4):40.

# CHAPTER 13

# Tumor markers

## Introduction to tumor markers

Screening for early diagnosis of malignancies has resulted in lower mortalities for certain diseases, such as breast cancer and cervical cancer. Tumor markers are mostly proteins associated with a malignancy. These may be detected in a solid tumor, in lymph nodes, in bone marrow, peripheral blood or body fluids. A tumor may be specific when it results from a fusion protein in which an oncogene is translocated and fused to an active promoter of another gene. Unspecific markers include oncofetal proteins expressed by more than one cancers.

Other than assays for measuring these proteins, immunohistochemistry, fluorescent in situ hybridization (FISH) and PCR are the most common methods utilized to detect these proteins. In this chapter, we are going to discuss the tumor markers that are commonly measured in clinical laboratories using blood specimens.

Most tumor markers are produced by normal cells as well as by cancer cells, but in the process of developing cancer, concentrations of these markers are elevated many folds compared to very low concentrations of these markers observed in blood under noncancerous condition. So far, more than 25 different tumor markers have been characterized and are used clinically for diagnosis and monitoring of treatment. Some tumor markers are elevated with only one type of cancer, whereas others are associated with two or more cancer types. Although theoretically any type of biological molecule can act as tumor marker, in practice, most markers are either proteins or glycoproteins. However, low molecular weight substances, for example, vanillylmandelic acid and homovanillic acid are used as markers for the diagnosis of neuroblastoma. In addition, nucleic acids (both DNA and RNA) are currently being evaluated as possible tumor markers. More recently, patterns of gene expression and changes to DNA are also under intense investigation be used as tumor markers. These types of markers are measured specifically in tumor tissues. In contrast, tumor-specific, circulating cell-free

Clinical Chemistry, Immunology and Laboratory Quality Control. https://doi.org/10.1016/B978-0-12-815960-6.00026-1

DNA can be measured in blood and is a promising source of biomarkers for minimally invasive serial monitoring of treatment responses in cancer management [1]. Most of the traditionally used markers are probably not involved in tumorigenesis but are likely to be byproducts of malignant transformation.

## Clinical uses of tumor markers and common tumor markers

Tumor markers can be used for one of five purposes:

- Screening a healthy population or a high-risk population for the probable presence of cancer
- Diagnosis of cancer or of a specific type of cancer
- Evaluating prognosis in a patient
- Predicting potential response of a patient to therapy
- Monitoring recovery of a patient during receiving surgery, radiation, or chemotherapy

Tumor markers were first developed to test for cancer in people without symptoms, but very few markers are effective in achieving this goal. Today, the most widely used tumor marker in the clinical setting is prostate-specific antigen (PSA). In addition, only a few markers that are now available have clinically useful predictive values for cancer at an early stage only when patients at high risk are tested. Tumor markers are not the gold standard for diagnosis of a cancer. In most cases, a suspected cancer can only be diagnosed by a biopsy. Alpha-fetoprotein (AFP) is an example of a tumor marker that can be used to aid in diagnosis of cancer, especially hepatocellular carcinoma (HCC). However, the level of AFP can also be increased in some liver diseases, although when it reaches a certain threshold, it is usually indicative of hepatocellular carcinoma.

Some types of cancer grow and spread faster than others while some cancers also respond well to various therapies. Sometimes the level of a tumor marker can be useful in predicting the behavior and outcome for certain cancers. For example, in testicular cancer, very high levels of a tumor marker, such as human chorionic gonadotropin (hCG) or AFP may indicate an aggressive cancer with poor survival outcome. Patients with these high levels may require very aggressive therapy even at the initiation of cancer therapy. Certain markers found in cancer cells can be used to predict if a certain treatment is likely to produce a favorable outcome or not. For example, in breast and stomach cancer, if the cells have too much of a protein termed as human epidermal growth factor receptor 2 (HER2), drugs such as trastuzumab (Herceptin) can be helpful if used during chemotherapy. However, with normal expression of HER2, these drugs may not produce expected therapeutic benefits. Tumor markers are also used to identify recurrence of certain tumors after successful therapy. Certain tumor markers may be useful for further evaluation of a patient after completion of the treatment when there is no obvious sign of cancer in the body. Commonly measured tumor markers in clinical laboratories include:

- Prostate specific antigen (PSA) for prostate cancer
- CA-125 (carbohydrate antigen or cancer antigen 125) for ovarian cancer
- CA 19-9 (carbohydrate antigen 19-9) for pancreatic and gastro-intestinal cancers
- Carcinoembryonic antigen (CEA) for colon and rectal cancers
- Alpha fetoprotein (AFP) for certain germ cell tumors and HCC
- $\beta_2$-Microglobulin for multiple myeloma
- Human chorionic gonadotropin (hCG) for gestational trophoblastic tumors and some germ cell tumors
- CA-15-3 (cancer antigen 15-3), marker for breast cancer
- CA 72-4 (cancer antigen 72-4), marker for colorectal cancer
- Calcitonin for thyroid (medullary) cancer
- Thyroglobulin for thyroid cancer
- HER2 for breast cancer

Less commonly monitored tumor markers include:

- CYFRA 21-1 (cytokeratin fragment), marker for lung cancer
- HE4 for ovarian cancer
- Squamous cell carcinoma antigen, marker of squamous cell lung cancer
- Neuron specific enolase, marker for lung cancer
- Chromogranin A, marker for neuroendocrine tumor
- Thymidine kinase for multiple myeloma, chronic lymphocytic leukemia

Clinical utility of common tumor markers are listed in Table 1. Common causes of elevated levels of tumor markers in the absence of cancer are listed in Table 2.

## Prostate-specific antigen (PSA)

Prostate-specific antigen (PSA) is a serine protease belonging to the kallikrein family. PSA is a single-chain glycoprotein containing 237 amino acid and four carbohydrate side chains (molecular weight: 28,430). PSA is expressed by both normal and neoplastic prostate tissue. Under normal conditions, PSA is produced as a proantigen (proPSA) by the secretory cells that line the prostate glands and secreted into the lumen, where the propeptide moiety is removed to generate active PSA. The active PSA can then undergo proteolysis to generate inactive PSA, of which, a small portion then enters the bloodstream and circulates in an unbound state (free PSA). Alternatively, active PSA can diffuse

**Table 1** Tumor markers commonly tested in clinical laboratories.

| Marker | Application | Normal range in serum (adults) |
|---|---|---|
| Prostatic-specific antigen (PSA) | Prostate carcinoma | 0–4 ng/mL |
| Cancer antigen 125 (CA 125) | Ovarian and fallopian carcinoma | 0–35 U/mL |
| Cancer antigen 15-3 (CA 15-3) | Breast cancer | 0–30 U/mL |
| Cancer antigen 19-9 (CA 19-9) | Pancreatic and ovarian cancer | 0–35 U/mL |
| CA-72-4 | Colorectal cancer | <6 U/mL |
| Alpha-fetoprotein | Hepatoblastoma, hepatocellular carcinoma, and germ cell tumors | <15 ng/mL |
| Carcinoembryonic antigen (CEA) | Colorectal, gastric, pancreatic, lung, and breast carcinomas | <2.5 ng/mL, nonsmokers <br> <5 ng/mL, smokers |
| Beta-2-microglobulin (B2M) | Multiple myeloma and lymphoma | Up to 3 µg/mL |
| Human chorionic gonadotropin (hCG) | Choriocarcinoma and testicular carcinoma | Men: <5 mIU/mL <br> Nonpregnant women: <5 mIU/mL <br> Postmenopausal woman: <10 mIU/L |
| Thyroglobulin | Thyroid cancer | Adult euthyroid: 3–42 ng/mL <br> Level should be very low or undetectable after DTC |

**Table 2** Common causes of elevation of level of various tumor markers in the absence of neoplasia.

| Tumor marker | Common causes of elevated levels |
|---|---|
| Prostate-specific antigen | Prostatitis/benign prostatic hyperplasia, breast cancer/ cyst, heterophilic antibody |
| Alpha-fetoprotein | Hepatobiliary disease, pneumonia, pregnancy, autoimmune disease, heterophilic antibody |
| CA-125 | Hepatobiliary disease, pulmonary disease, renal failure, hypothyroidism, endometriosis, pregnancy, autoimmune disease, skin disease, cardiovascular disease, heterophilic antibody |
| Carcinoembryonic antigen (CEA) | Hepatobiliary disease, renal failure, hypothyroidism, gastrointestinal disease, pancreatitis, endometriosis, autoimmune disease, heterophilic antibody |
| CA-19-9 | Hepatobiliary disease, renal failure, pulmonary disease, pancreatitis, gastrointestinal disease, endometriosis, heterophilic antibody |
| CA-15-3 | Vitamin B12 deficiency, effusion, renal failure, |
| Beta-2-microglobulin | Renal failure, autoimmune disease, cerebral lesion |
| hCG or beta-hCG | Renal failure, pregnancy, autoimmune disease, heterophilic antibody |
| CA-72-4 | Hepatobiliary disease, renal failure, effusion, pancreatitis, gastrointestinal disorder |
| CYFRA 21-1 | Hepatobiliary disease, renal failure, effusion, pulmonary disease |
| Chromogranin A | Cardiovascular disease, viral infection, prostatitis/benign prostatic hyperplasia, gastrointestinal disease, heterophilic antibody |

directly into the circulation where it is rapidly bound to alpha-1-antichymotrypsin (ACT) and alpha-2-macroglobulin [2]. In men with a normal prostate, the majority of free PSA in the serum reflects the mature protein that has been inactivated by internal proteolytic cleavage. In contrast, free PSA is relatively decreased in patients with prostate cancer. Thus the percentage of free or unbound PSA is lower in the serum of men with prostate cancer. Therefore the ratio of free to total PSA or complexed PSA (cPSA) is a means of distinguishing between prostate cancer and BPH in a patient with elevated PSA. Causes of elevated PSA include:

- Benign prostatic hyperplasia (BPH)
- Prostate cancer
- Prostatic inflammation/infection
- Perineal trauma

Studies in the 1980s confirmed that serum total PSA could be used as a screening tool to identify men with prostate cancer because elevated serum PSA is clearly a more sensitive marker that digital rectal examination. However, between 20% and 50% of men with newly diagnosed prostate cancers may have serum PSA values below 4.0 ng/mL (upper end of normal is usually considered as 4.0 ng/mL) indicating that PSA lacks specificity as a tumor marker. In general, patients with PSA below 4.0 ng/mL are more likely have prostate cancer which is confined to the organ. These patients have a better prognosis than patients with prostate cancer who showed levels above 4.0 ng/mL [3].

Prostatitis with or without active infection is an important cause of an elevated PSA, and levels as high as 75 ng/mL have been reported in the literature. Thus, many physicians often initially treat a man with an isolated elevated serum PSA with antibiotics for a presumed diagnosis of prostatitis and then obtain a repeat serum PSA for further clinical evaluation. The percent-free PSA may be less affected by the presence of inflammation particularly when the total serum PSA is less than 10 ng/mL. However, free to total ratio of serum PSA may be unable to distinguish chronic inflammation from prostate cancer, as both conditions may lower the percentage of free PSA. This would be expected because inflammation leads to elevated serum PSA in a similar fashion as prostate cancer through disruption of the basal membrane and increased leakage of "immature" PSA into the blood stream.

Any perineal trauma can also increase the serum PSA. Prostate massage and digital rectal examination may cause minor transient elevations which may be clinically insignificant. Mechanical manipulation of the prostate by cystoscopy, prostate biopsy, or TURP (transurethral resection of the prostate) may more significantly elevate the serum PSA. Vigorous bicycle riding has been reported to cause substantial elevations in the serum PSA, but this is not a consistent finding. Sexual activity can minimally elevate the PSA (usually in the 0.4 to 0.5 ng/mL range) for approximately 48 to 72 h after ejaculation.

Emerging concepts regarding PSA testing that may help refine the interpretation of an elevated concentration include PSA density, and PSA velocity, free versus complexed or bound PSA. These modifications would presumably be most useful for prostate cancer screening when the total PSA is between 2.5 and 10.0 ng/mL the range in which decisions regarding further diagnostic testing are most difficult. To more directly compensate for BPH and prostate size, transrectal ultrasound (TRUS) has been used to measure prostate volume. Serum PSA is then divided by prostate volume to obtain PSA density, with higher PSA density values (greater than 0.15) being more suggestive of prostate cancer while lower values are more suggestive of BPH. Another approach has been to assess the rate of PSA change over time (the PSA velocity). An elevated serum PSA that continues to rise over time is more likely to reflect prostate cancer than the one that is consistently stable. For practical purposes, the clinical usefulness of PSA velocity is in part limited by intrapatient variability in the serum PSA; at least three consecutive measurements should be performed. A longer time over which values are continuously measured can be useful in reducing the general variation in the PSA measurements.

Prostate cancer is associated with a lower percentage of free PSA in the serum compared to PSA values observed in benign conditions. The percentage of free PSA has been used to improve the sensitivity of cancer detection when total PSA is in the normal range (<4 ng/mL) and also to increase the specificity of cancer detection when total PSA in the "gray zone" (4.1–10 ng/mL). In this latter group (PSA between 4.1 and 10 ng/mL), the lower the value of free PSA, the greater the likelihood that an elevated PSA representing cancer rather than BPH. As with PSA, there is no absolute free/total cut-off that can completely differentiate prostate cancer from BPH. The optimal cut-off value for free PSA is unclear and depends upon whether optimal sensitivity or specificity is sought. The higher the cut-off value, the greater is the sensitivity but the lower is the specificity. Free PSA could be useful for risk stratification in men with prostate cancer. A lower percentage of free to total PSA may be associated with a more aggressive form of prostate cancer.

Assays for alpha-1-antichymotrypsin (ACT)-complexed PSA (cPSA) have been implemented that could theoretically provide a similar enhanced degree of specificity compared to free to total PSA ratio. Most but not all reports suggest that cPSA outperforms both total PSA and the ratio of free to total PSA, with similar sensitivity but a higher specificity. According to one study, for men with total PSA in the diagnostic gray zone (4.0 and 10.0 ng/mL), the use of cPSA alone would have missed only one of the 36 men with cancer who would be diagnosed with prostate cancer using both total PSA and biopsies. Interestingly, free to total PSA alone would also have also missed one cancer, but eliminated biopsy in only 20 men compared to 34 men where biopsy could be eliminated using cPSA assay alone. The utility of cPSA in men with a lower total PSA (2 to 4 ng/mL) is under investigation as there are conflicting data as to whether cPSA improves specificity compared with free to total PSA ratio [4]. Complexed PSA has been approved for the monitoring of men with prostatic carcinoma. The utility of complexed PSA for screening is uncertain and is not routinely used at this time for clinical practice on a regular basis.

PSA is initially produced as proPSA and this form can preferentially leak into the blood stream in men with prostate cancer. One specific isoform of proPSA is [-2]proPSA, which is unbound and potentially higher in concentration in men with prostate cancer. Based upon this observation, there has been growing interest in using the ratio of [-2]proPSA to free PSA (expressed as percent [-2]proPSA or %[-2]proPSA) for screening of prostate cancer [5].

## False-positive and unexpected PSA results

False-positive PSA test results may be encountered causing confusion regarding the diagnosis of prostate cancer. Kilpelainen et al., based on screening of 61,604 men in Europe, observed 17.8% false-positive PSA results. However, men who tested false-positive with one PSA screening test were more prone to be diagnosed with prostate cancer in the future [6]. Nevertheless, the major cause of false elevation of PSA is the presence of heterophilic antibody in the serum. In fact, the presence of heterophilic antibody in the specimen not only may cause false elevation of PSA but also false elevation of other tumor markers. Falsely elevated PSA due to interference of heterophilic antibody may result in inappropriate and unnecessary treatment for prostate cancer. Morgan and Tarter commented that human antimouse IgG heterophilic antibody if present in patient's serum can interfere with serum PSA assay, and if PSA is detectable after radical prostatectomy and the likelihood of incomplete resection or systematic disease is low, unexpected PSA result due to the presence of heterophilic antibody must be considered [7].

Although most reports show falsely elevated PSA most commonly due to the presence of heterophilic antibody, there are few reports of falsely lower PSA due to presence of inhibitory factors in the serum. A 63-year-old man was treated with prostatectomy for high-risk prostate cancer and was found to have a rising PSA after approximately 3 years following surgery. He subsequently transferred his care to a different health system and was found to have an undetectable PSA. He was eventually found to have an elevated PSA once again after the particular assay at this institution was changed. The reason for negative interference could be due to the presence of an inhibitory factor in serum such as anti-PSA antibody. Other serum factors that have been shown to interfere with immunoassays include complement, lysozyme, and paraprotein. Because significant negative interference was also observed with Roche PSA assay for this patient which is affected by biotin, such interference cannot be ruled out because the patient was taking vitamin supplements [8]. Biotin is known to cause positive interference with competitive immunoassays but negative interference with sandwich immunoassays (where biotinylated antibodies are used). Because PSA immunoassays utilize sandwich format, negative interference is possible due to high serum biotin level. Please see Chapter 22 for more detail.

## Newer biomarkers of prostate cancer

More recently, newer biomarkers are emerging for the diagnosis of prostate cancer. Prostate-specific antigen 3 has been regarded as highly specific to prostate. A urine assay is approved by the FDA. Gene fusions of the transmembrane protease serine 2 (TMPRSS2) and the E26 transformation-specific ERG oncogene have been identified as a common event in human prostate cancer. Therefore, quantitative determination of the TMPRSS2-ERG fusion gene product in urine has promise as a biomarker of prostate cancer. In addition, DNA-based biomarkers such as polymorphisms in the *GSTP1* gene may improve its predictive value. More recently, aberrant microRNA expression in prostate cancer has been reported. Preliminary results indicate the utility of circulating and urinary microRNAs in the detection and prognosis of prostate cancer [9].

## Cancer antigen 125 (carbohydrate antigen 125: CA-125)

CA-125 also known as mucin 16 or MUC16 is a glycoprotein. CA-125 in humans is encoded by the *MUC16* gene. CA-125 is used as a tumor marker because CA-125 concentrations may be elevated in the blood of some patients with ovarian cancer but also in some benign condition. CA-125 levels in serum are elevated in approximately 50% of women with early-stage disease and in over 80% of women with advanced ovarian cancer. Monitoring CA-125 serum levels is also useful for determining the response of a patient to ovarian cancer therapy as well as for predicting a patient's prognosis after treatment. In general, persistence of high levels of CA-125 during therapy is associated with poor survival rates in patients. Also, an increase in CA-125 levels in a patient during remission is a strong predictor of recurrence of ovarian cancer. Monitoring of serum CA-125 levels has value in patients with fallopian tube cancer, a relatively rare disease. The pretreatment serum CA-125 concentration appears to have prognostic value in patients with fallopian tube cancer, and following initial treatment, it is a sensitive marker for monitoring of recurrence and response to chemotherapy. However, the specificity of CA-125 is limited, because CA-125 levels are elevated in approximately 1% of healthy women and fluctuate during the menstrual cycle. CA-125 is also increased in a variety of benign and malignant conditions, including:

- Endometriosis
- Uterine leiomyoma
- Cirrhosis with or without ascites
- Pelvic inflammatory disease
- Cancers of the endometrium, breast, lung, and pancreas
- Pleural or peritoneal fluid inflammation due to any cause

## False-positive CA-125

Meigs' syndrome (association of ovarian fibroma, pleural effusion, and ascites) may also cause marked elevation of CA-125. Abnormally high values of both CA-125 and CA-19-9 have been reported in women with benign tumors. Nagata et al. reported that three patients with endometriosis showed elevated levels of CA-125 and two patients with dermoid cyst showed elevated levels of CA-19-9. Therefore, the authors recommended that tumor marker values should be considered along with bimanual examination, ultrasound, and CT scan for diagnosis of ovarian tumors [10]. Sometimes, F(ab')2 fragments of the murine monoclonal antibody OC-125 are administered to patients with ovarian cancer because OC-125 is directed against the CA-125 antigen present on the surface of human ovarian cancers. Exposure to such antibody may lead to development of an immune response causing the presence of HAMA (human antimouse monoclonal antibody; also, broadly termed as heterophilic antibody), which may interfere in an unpredictable manner with the determination of CA-125 using serum specimens in such patients.

Measurable CA-125 concentrations can also be observed in patients without any cancer. CA-125 concentrations are known to rise in patients with severe congestive heart failure and the elevations correlate with the severity of disease and elevations of a specific marker of heart failure, for example, B-type natriuretic peptide (BNP). In the menstrual phase of the cycle in women, CA-125 values may be elevated causing false positive test results. CA-125 may also increase after abdominal surgery, chronic obstructive pulmonary disease, active tuberculosis, and lupus erythematous. During pregnancy, CA-125 concentrations increase 10 weeks after gestation and remain high throughout the pregnancy. During the terminal phase of pregnancy, the CA-125 concentration may be as high as twice the upper limit of the reference range.

## Emerging biomarkers of ovarian cancer

Serum epididymis protein-4 (HE4; molecular weight 23–27 kDa) level is a useful biomarker for the management of ovarian and endometrial cancer patients.

HE4 has been shown to be expressed and secreted by ovarian carcinoma cells. This marker can be measured in both serum and urine. The ROMA (the risk of ovarian malignancy algorithm) score that utilizes serum HE-4 and CA-125 level in calculating the score can also be used for improved sensitivity and specificity for diagnosis of ovarian cancer [11].

## Cancer antigen-19-9 (carbohydrate antigen: 19-9: CA 19-9)

CA 19-9, also called cancer antigen 19-9 or sialylated Lewis (a) antigen, is a tumor marker used primarily in the management of pancreatic cancer. Guidelines from the American Society of Clinical Oncology discourage the use of CA-19-9 as a screening test for cancer particularly pancreatic cancer because the test may be falsely negative in many cases, or abnormally elevated in people with no cancer at all (false-positive). However, in individuals with pancreatic masses, CA-19-9 can be useful in distinguishing between cancer and other pathology of the gland. The reported sensitivity and specificity of CA 19-9 for pancreatic cancer are 80% to 90%, respectively and these values are closely related to tumor size. The accuracy of CA 19-9 to identify patients with small surgically resectable cancers is limited. The specificity of CA 19-9 is limited because CA 19-9 is frequently elevated in patients with cancers other than pancreatic cancer and various benign pancreaticobiliary disorders. As a result of all of these issues, CA 19-9 is not recommended as a screening test for pancreatic cancer.

The degree of elevation of CA 19-9 (both at initial presentation and in the postoperative setting) is associated with long-term prognosis. Furthermore, in patients who appear to have potentially resectable disease, the magnitude of the CA 19-9 level can also be useful in predicting the presence of radiographically occult metastatic disease. The rates of unresectable disease among all patients with a CA 19-9 level $\geq$130 units/mL versus <130 units/mL were 26% and 11%, respectively. Among patients with tumors in the body/tail of the pancreas, more than one-third of those who had a CA 19-9 level $\geq$130 units/mL had unresectable disease.

Serial monitoring of CA 19-9 levels (once every 1 to 3 months) is useful for further monitoring of patients after potentially curative surgery and for those who are receiving chemotherapy for advanced disease. Elevated CA 19-9 levels usually precede the radiographic appearance of recurrent disease, but confirmation of disease progression should be pursued with imaging studies and/or biopsy. CA-19-9 can be elevated in many types of gastrointestinal cancer, such as colorectal cancer, esophageal cancer, and hepatocellular carcinoma. Apart from cancer, elevated levels may also occur in pancreatitis, cirrhosis, and diseases of the bile ducts. It can be elevated in people with obstruction of the bile duct.

However, in patients who lack the Lewis antigen (a blood type protein on red blood cells), which is about 10% of the Caucasian population; CA-19-9 is not expressed even in those with large tumors. This is due to deficiency of the fucosyltransferase enzyme which is needed to produce CA-19-9 as well as the Lewis antigen. The use of a combined index of serum CA 19-9 and CEA (CA 19-9 + [CEA x 40]) has also been proposed for screening of cholangiocarcinoma.

Interference of heterophilic antibodies causing false-positive CA-19-9 result has been documented and usually treating the specimen with heterophilic antibody blocking agents can eliminate such interference. Heavy tea consumption may also falsely elevate CA19-9 levels. A 52-year-old woman who was a heavy consumer of tea showed CA 19-9 level of 1432 U/mL (normal <37 U/mL) but no cancer was detected. She stopped drinking tea and her level returned to 42 U/ML. A rechallenge test was then attempted. The patient restarted tea consumption as previously. Four weeks later, CA19-9 increased to 745 UI/mL followed by a fall to 25 UI/mL 1 month after withdrawal. Follow up one year later revealed no clinical abnormalities [12].

## Alpha-fetal protein (AFP, α-fetoprotein)

AFP, sometimes called alpha-1-fetoprotein or alpha-fetoglobulin, is a protein encoded in humans by the *AFP* gene which is located on the q arm of chromosome 4 (4q25). AFP is a major plasma protein produced by the yolk sac and the liver during fetal development and is considered as the fetal form of albumin. The half-life of AFP is approximately 5 to 7 days. Following effective cancer therapy, normalization of the serum AFP concentration over 25–30 days is indicative of an appropriate decline. However, it is essentially undetectable in the serum in normal men. The upper limit of normal serum AFP concentration is less than 10–15 µg/L. Many tissues regain the ability to produce this oncofetal protein while undergoing malignant degeneration, but serum AFP concentrations above 10,000 µg/L are most commonly observed in patients with nonseminomatous germ cell tumors (NSGCTs) or hepatocellular carcinoma. In men with NSGCTs, AFP is produced by yolk sac (endodermal sinus) tumors, and less often due to embryonal carcinomas. As with β-hCG (beta-human chorionic gonadotropin), the frequency of an elevated serum AFP increases with advancing clinical stage of the tumor, from 10% to 20% in men with stage I tumors to 40% to 60% of those with disseminated NSGCTs. By definition, pure seminomas do not cause an elevated serum AFP. However, molecular studies have demonstrated AFP mRNA in minute quantities in pure seminoma and several case reports have documented pure seminoma with borderline elevations in serum AFP (10.4 to 16 ng/mL). Higher serum AFP concentrations are considered diagnostic of a nonseminomatous component of the tumor (especially yolk sac elements) or hepatic metastases. If the presence of an elevated serum AFP is confirmed, patients should be treated as if they had an NSGCT.

Serum AFP is the most commonly used marker for diagnosis of hepatocellular carcinoma (HCC). Serum levels of AFP do not correlate well with other clinical features of HCC, such as size, stage, or prognosis. Elevated serum AFP may also be seen in patients without HCC such as acute or chronic viral hepatitis. AFP may be slightly elevated in patients with liver cirrhosis due to chronic hepatitis C infection. A significant rise in serum AFP in a patient with cirrhosis should raise concern that HCC may be developed. It is generally accepted that serum levels greater than 500 μg/L (upper limit of normal in most laboratories is between 10 and 20 μg/L) in a high-risk patient is diagnostic of HCC. However, HCC is often diagnosed at a lower AFP level in patients undergoing screening. Not all tumors secrete AFP and serum concentrations could be normal in up to 40% of patients with small HCCs. In a study of 357 patients with hepatitis C and without HCC, 23% had an AFP >10.0 μg/L. Elevated levels were associated with the presence of stage III or IV fibrosis, an elevated international normalized ratio, and an elevated serum aspartate aminotransferase level [13]. AFP levels are however normal in the majority of patients with fibrolamellar carcinoma, a variant of HCC. Despite the issues inherent in using AFP for the diagnosis of HCC, it has emerged as an important prognostic marker, especially in patients being considered for liver transplantation. Patients with AFP levels >1000 μg/L have an extremely high risk of recurrent disease following the transplant, irrespective of the tumor size seen on imaging.

## False-positive AFP

Elevations of serum AFP in the absence of liver cancer can occur from tumors of the gastrointestinal tract, particularly gastric adenocarcinoma, or from liver damage (e.g., cirrhosis, hepatitis, or drug or alcohol abuse). Lysis of tumor cells during the initiation of chemotherapy may result in a transient increase in serum AFP. Elevated serum AFP occurs in pregnancy with tumors of gonadal origin (both germ cell and nongerm cell) and in a variety of other malignancies, of which gastric cancer is the most common. Values up to 140 times the upper reference range have been observed in cases of hereditary tyrosinemia type 1, with 71% patients showing levels twice the upper references range, indicating liver damage and the possibility of hepatomas and risk of HCC development [14]. As expected, heterophilic antibodies, if present in the specimen, can also cause falsely elevated AFP concentration. AFP level may be falsely elevated due to the presence of rheumatoid factor in the specimen. Interestingly, Wang et al. reported negative interference of rheumatoid factor in the chemiluminescent microparticle immunoassay of AFP using ARCHITECT analyzer (Abbott Laboratories, Abbott Park, IL) [15]. However, such negative interference is rare.

Concentration of AFP may also be increased in pregnant women with systematic lupus erythematous.

## Carcinoembryonic antigen (CEA)

CEA is a glycoprotein involved in cell adhesion which is normally produced during fetal development, but the production of CEA stops before birth. Therefore, it is not usually present in the blood of healthy adults, although levels are raised in heavy smokers. CEA is a glycosyl phosphatidyl inositol (GPI)-cell surface anchored glycoprotein whose specialized sialofucosylated glycoforms serve as functional colon carcinoma L-selectin and E-selectin ligands, which may be critical to the metastatic dissemination of colon carcinoma cells. It is found in the sera of patients with colorectal carcinoma (CRC), gastric carcinoma, pancreatic carcinoma, lung carcinoma, and breast carcinoma. Patients with medullary thyroid carcinoma also have higher levels of CEA compared to healthy individuals (above 2.5 ng/mL). However, CEA blood test is not reliable for diagnosing cancer or as a screening test for early detection of cancer. Most types of cancer do not produce a high level of CEA. Elevated CEA levels should return to normal after successful surgical resection or within 6 weeks of starting treatment, if cancer treatment is successful. However, due to lack of both sensitivity and specificity, serum CEA is not a useful screening tool for CRC but in patients with established disease, the absolute level of the serum CEA correlates with disease burden and is of prognostic value. Furthermore, elevated preoperative levels of CEA should return to baseline after complete resection; residual disease should be suspected if they do not. Serum levels of the tumor marker carcinoembryonic antigen (CEA) should be routinely measured preoperatively in patients undergoing potentially curative resections for CRC for two reasons:

- Elevated preoperative CEA levels that do not normalize following surgical resection imply the presence of persistent disease and the need for further evaluation.
- Preoperative CEA values are of prognostic significance. CEA levels ≥5.0 ng/mL are associated with an adverse impact on survival that is independent of tumor stage.

As a single analyte, serum levels of carcinoembryonic antigen (CEA) are neither sufficiently sensitive nor specific to diagnose cholangiocarcinoma. Many conditions other than cholangiocarcinoma can increase serum levels of CEA. Noncancer-related causes of an elevated CEA include gastritis, peptic ulcer disease, diverticulitis, liver disease, chronic obstructive pulmonary disease, diabetes, and any acute or chronic inflammatory state.

## False-positive CEA

As expected, false-positive CEA test results can occur due to the presence of heterophilic antibodies in the specimen. However, CEA concentrations can also be elevated in nonneoplastic conditions. Renal failure and fulminant hepatitis can falsely increase CEA values. CEA concentrations may be also elevated in patients receiving hemodialysis. Patients with hypothyroidism may also have elevated levels of CEA, correlated with the duration of hypothyroidism. CEA levels may also be raised in some nonneoplastic conditions like ulcerative colitis, pancreatitis, cirrhosis, chronic obstructive pulmonary disease (COPD), Crohn's disease, as well as in smokers.

## $\beta_2$-Microglobulin

Beta-2-microglobulin ($\beta$2-microglobulin) is a component of MHC (major histocompatibility complex) class I molecules, present on all nucleated cells (excludes red blood cells). In humans, the $\beta_2$-microglobulin protein is encoded by the *B2M* gene. For the diagnosis of multiple myeloma, the serum $\beta_2$-microglobulin level is one of the prognostic factors incorporated into the International Staging System. The serum $\beta_2$-microglobulin level is elevated (i.e., >2.7 mg/L) in 75% of patients at the time of diagnosis. Patients with high values have inferior survival. The prognostic value of serum $\beta_2$-microglobulin levels in myeloma is probably due to two factors:

- High levels are associated with greater tumor burden.
- High levels are also associated with renal failure, which carries an unfavorable prognosis.

In lymphoma, $\beta_2$-microglobulin levels usually correlate with disease stage and tumor burden in patients with CLL (chronic lymphocytic leukemia), with increasing levels associated with a poorer prognosis. Beta-2-microglobulin may be regulated, at least in part, by exogenous cytokines. The source of these elevated cytokines in CLL is unclear, although IL-6, which inhibits apoptosis in CLL cells, may be released from vascular endothelium. However, $\beta_2$-microglobulin levels also rise with worsening renal dysfunction leading some investigators to suggest a measure of $\beta_2$-microglobulin adjusted for the glomerular filtration rate (GFR). This GFR-adjusted $\beta_2$-microglobulin requires validation in prospective confirmatory studies. The plasma $\beta_2$-microglobulin concentration is increased in dialyzed patients, with a level ranging from 30 to 50 mg/L, much higher than the normal value of 0.8 to 3.0 mg/L. Infection with AIDS virus, hepatitis, and active tuberculosis may also elevate level of $\beta_2$-microglobulin.

# Human chorionic gonadotropin (hCG)

Human chorionic gonadotropin (hCG) is a hormone composed of alpha- and beta-subunits and the beta-subunit is specific for hCG (beta-hCG) and provides functional specificity. Beta-hCG is synthesized in large amounts by placental trophoblastic tissue and in much smaller amounts by the hypophysis and other organs, such as testicles, liver, and colon. Therefore elevated levels of beta-hCG are observed during pregnancy, produced by the developing placenta after conception, and later by the placental component syncytiotrophoblast. The major application of hCG testing is detection of pregnancy using urine specimens, and often point of care devices are used to qualitatively measure hCG (positive or negative). In pregnancy, hCG immunoreactivity in urine is due to the presence of intact hCG, as well as hyperglycosylated hemoglobin, nicked hCG, beta-subunit of hCG, the core fragment of beta-hCG, and others. As expected, these forms are also present in serum. However, many over-the-counter pregnancy tests do not measure hyperglycosylated hCG, which accounts for most of the total hCG at the time of missed menses. Clinical tests for pregnancy may only detect total hCG levels $\geq 20$ mIU/mL. Therefore, when following hCG levels to negative ($<1$ mIU/mL) in women with gestational trophoblastic disease, it is important to use a sensitive hCG test that detects both regular and other forms of hCG. False-negative results using qualitative tests for hCG in urine due to increased concentration of core fragment of beta-hCG have also been reported [16].

At levels of hCG above 500,000 mIU/mL, a "hook effect" can occur resulting in an artifactually low value for hCG (i.e., 1 to 100 mIU/mL). This is because the sensitivity of most hCG tests is set to the pregnancy hCG range (i.e., 27,300 to 233,000 mIU/mL at 8 to 11 weeks of gestation); therefore when an extremely high hCG concentration is present, both the capture and tracer antibodies used in assays become saturated, preventing the binding of the two to create a sandwich. For this reason, a suspected diagnosis of gestational trophoblastic disease must be communicated to the laboratory so that the hCG assay could also be performed at 1:1000 dilution.

Molar pregnancy (hydatidiform mole) is a nonmalignant tumor that arises from the trophoblast in early pregnancy after an embryo fails to develop and molar pregnancy is known to produce high amounts of beta-hCG. However, in urine pregnancy test, false-negative beta-hCG result may be observed due to hook effect because very large amount of beta-hCG may be present. Dilution of the specimen is essential to further investigate such false-negative result. Another approach is to perform a serum beta-hCG test.

# Causes and evaluation of persistent low level of hCG

Determining the clinical value of a low level of hCG can be challenging. It is important to determine if the hCG represents an actual early pregnancy (intrauterine or ectopic), active gestational trophoblastic disease (complete or partial mole, invasive mole, choriocarcinoma), quiescent gestational trophoblastic disease, a laboratory false-positive (also called phantom hCG), or a physiologic artifact (pituitary hCG). For example, a false-positive hCG test result or pituitary hCG is commonly found in women who also have a history of gestational trophoblastic disease. Unless tumor is evident, it is essential to exclude these possibilities before initiating chemotherapy for assumed persistence of disease. Persistent low-level positive hCG results can be defined as hCG levels varying by no more than two-fold over at least a 3-month period in the absence of tumor on imaging studies.

## False-positive hCG

The capture and tracer antibodies used for hCG testing may be goat, sheep, or rabbit polyclonal antibodies or mouse, goat, or sheep monoclonal antibodies. Humans extensively exposed to animals or certain animal by-products can develop human antibodies against animal antibodies which are collectively called heterophilic antibodies, and human antimouse antibody (HAMA) is a common example of heterophilic antibody. Individuals with recent exposure to mononucleosis are prone to develop heterophilic antibodies. False-positive hCG test due to presence of heterophilic antibody in the serum specimen has been well documented in the literature. Such false-positive results in the absence of pregnancy have led to many men and women misdiagnosed with cancer, confusion and misunderstanding, and needless surgery and chemotherapy. Since heterophilic antibodies are found mainly in serum, plasma or whole blood, but not in urine, such interference is absent in the analysis of urine specimen for the same analyte. This gives an excellent way to detect the interference for analytes which may be present in both matrices. Although among tumor markers, serum hCG assay is mostly affected by the presence of heterophilic antibody, false-positive test results with other tumor markers including PSA, CA-125, CA-19-9, CEA, alpha-fetoprotein, and even beta-2-microglobulin. In fact, interference of heterophilic antibody is the major problem in the assay of various cancer markers. Even IgM lambda antibody to *Escherichia coli* can produce false-positive test results with determination of various tumor markers as well as troponin I. The prevalence of heterophilic antibody in the general population is difficult to estimate as published literature reports indicated the prevalence of heterophilic antibody from 1% to 11.7% [17]. There are two main methods for identifying false-positive hCG:

- The most readily available approach is to show the absence of hCG in the patient's urine, since large molecules like heterophilic antibodies fail to cross the glomerular basement membrane. A true hCG elevation should be present in both serum and urine.
- A second useful way of identifying a false-positive serum hCG result is to send the serum to two laboratories using different commercial assays. If

the assay results vary greatly or are negative in one or both alternative tests, then a false-positive hCG can be presumed.

## CA-15-3 and HER2

Breast cancer is by far the most common cancer in women worldwide where more than 1.2 million cases are detected every year affecting 10%–12% of women and responsible for approximately 500,000 deaths per year. Several tumor markers are used in diagnosis and management of patients with breast cancer but the most commonly used tumor marker is CA 15-3 (cancer antigen 15-3) which is also known as MUC-1. This tumor marker, a high molecular mass glycoprotein, is a transmembrane mucin expressed by most glandular epithelial cells and was first discovered in human breast milk where this protein is shed from lactating mammary epithelial cells. CA 15-3 was identified on the surface of many types of cancer including breast, and ovarian cancer, as well as lung, pancreatic, and prostate cancer, but the major application is in diagnosis and management of breast cancer. The molecules detected by CA 15-3 assay are the shed or soluble forms of the MUC-1 protein and there are many commercially available immunoassays for measuring CA 15-3 in serum or plasma. Although a major application of CA 15-3 is in monitoring patients with metastatic disease, this marker should not be used alone but should be measured in conjunction with diagnostic imaging, clinical history, and physical examination. Moreover, CA 15-3 levels may be elevated in other types of advanced cancer including ovarian, pancreatic, gastric, and lung cancer.

Human epidermal growth factor receptor 2 (HER 2) (also known as cluster of differentiation 340 or protooncogene Neu, encoded by the *ERBB2* gene) is a 185 kDa glycoprotein consisting of three domains; a 105 kDa extracellular domain (ECD), a transmembrane lipophilic segment, and an intracellular domain with tyrosine kinase activity. The ECD portion can be released by cleavage from HER2 receptor and shed into serum. Overexpression of HER2 receptor is observed in 20%–30% breast cancer patients and such overexpression is associated with an aggressive tumor subtype, reduced survival, and possible treatment with monoclonal antibody trastuzumab or other types of therapy targeted against the HER 2 receptor protein. HER 2 can be measured by various methods but most commonly employed methods are immunohistochemistry and fluorescence in situ hybridization (FISH) using breast biopsy specimen.

## CA 72-4

CA 72-4 (cancer antigen 72-4) is a marker for colorectal cancer. In general, about 5% women show CA 72-4 values over the reference range. In addition,

3% to 7% patients with pancreatitis show CA72-4 value over the normal range. Interestingly, approximately 50% of patients suffering from Familial Mediterranean Fever demonstrate CA72-4 values above the reference range [18].

## Markers for thyroid cancer

Thyroid nodules are frequently found. Although they are often palpable, many are discovered incidentally during unrelated radiographic studies. Ten to fifteen percentages of thyroid nodules represent thyroid malignancy [19]. Thyroglobulin, a glycoprotein (molecular weight: 660 kDa), is produced exclusively in the thyroid gland where it serves as a precursor of thyroid hormones. Small amounts of thyroglobulin is detected in sera of normal individuals but elevated thyroglobulin concentrations are encountered in various thyroid disorders such as goiter, Graves' disease, thyroiditis as well as in differentiated thyroid cancer (DTC). However, measurement of thyroglobulin is not recommended during initial evaluation of suspicious thyroid nodules due to overlap of thyroglobulin levels in patients with benign thyroid disorders and DTC. However, thyroglobulin measurement is useful postoperatively in follow-up of DTC patients to monitor residual or recurrent disease. After successful total thyroidectomy, levels of thyroglobulin should decrease and may even be undetectable. Therefore, postoperative detection of thyroglobulin may indicate recurrence of thyroid cancer.

Usually, thyroglobulin is measured 6 to 12 months after surgery but for high-risk patients, more frequent monitoring is recommended. There is also a growing trend to measure thyroglobulin in needle washout fluid following fine needle aspiration biopsy of suspicious lymph node as an adjunct to cytological examination on suspected cases of metastatic DTC. Thyroglobulin in serum can be measured using immunoassays or liquid chromatography combined with tandem mass spectrometry (LC-MS/MS). Immunoassays need functional sensitivity of <1 ng/mL. Although first generation assays have functional sensitivity of 0.5 to 1.0 ng/mL, 2nd generation assays have functional sensitivity of 0.1 ng/mL or less which is adequate for monitoring of patients with DTC after surgery. However, major problem of immunoassays is interference from circulating antibody against thyroglobulin (antithyroglobulin autoantibody) and heterophilic antibody. The presence of antithyroglobulin autoantibody (may be present in 10% of general population but up 30% patients with DTC) might cause falsely low/undetectable thyroglobulin level thus masking disease or false-positive values due to presence of heterophilic antibody which may be mistaken for residual or recurrent disease. However, LC-MS/MS assays are free from such interferences. Assays are also available for measuring antithyroglobulin antibody. Therefore if an immunoassay is used for measuring thyroglobulin, it is recommended to test for the presence of antithyroglobulin antibody [20].

Medullary thyroid carcinoma is an uncommon malignancy arising from parafollicular C-cells and accounts for 2% of thyroid cancers. However, this type of cancer may be more aggressive than well-differentiated thyroid cancer because it has a greater propensity to metastasize to regional lymph nodes and distant sites. An increased level of serum calcitonin is produced in medullary thyroid carcinoma. Therefore, preoperative serum calcitonin levels serve to confirm the diagnosis to confirm medullary thyroid carcinoma and detectable postoperative serum calcitonin levels confirm the presence of persistent or recurrent tumor, often before evident clinical disease. Disease progression and dissemination correlates with serum calcitonin and lesser doubling times of serum calcitonin correlate with increased mortality [21].

## Less commonly monitored cancer biomarkers

Cytokeratins are epithelial markers whose expression is not lost during malignant transformation. A serum marker, called cytokeratin fraction 21-1 (CYFRA 21-1), has shown a promising diagnostic value for diagnosis of nonsmall-cell lung cancer and possibly squamous carcinoma of head and neck. CYFRA 21-1 is a cytoplasmic protein fragment of cytokeratin 19 (CK-19) which is found in various epithelial malignancies. This soluble debris can be released into the blood after tumor cell death, thus exhibiting a close relationship with tumor cell necrosis. Its level could be a proper indicator of necrosis degree [22]. However, patients with pulmonary disease without cancer may show elevated levels of CYFRA 21-1. Patients receiving dialysis, patients with renal failure and liver disease may also show elevated levels of this biomarker.

Chromogranin A is a marker for various neuroendocrine diseases such as gastroenteropancreatic tumor, endocrine tumors, and bronchial carcinoid. It may also be useful in the diagnosis of pheochromocytoma along with urine/plasma metanephrine determination. However, the assay for chromogranin A may be affected by the presence of heterophilic antibodies in the serum.

Squamous cell carcinoma antigens (SCCA) are members of the serpin family of endogenous serine protease inhibitors. The first such marker, SCCA1 was originally identified in squamous cell carcinoma of the uterine cervix. Later a similar molecule with 92% amino acid homology with SCCA1 and SCCA2 was identified. High levels of SCCA are often observed with poorly differentiated and advanced metastatic squamous cell carcinoma. Elevated levels of SCCA have been reported in cancer of epithelial (cervix, lung, head, and neck) and endodermal (liver) and possibly also in advanced grade as well as high-grade human breast cancer. In the clinical laboratory, circulating SCCA (both SCCA1 and SCCA2) can be detected by immunoassays or by using immunohistochemistry. Tissue biopsy may also be used for measuring this tumor marker.

More recently, circulating cell-free DNA assays are getting attention as viable tumor markers that can be analyzed in blood (liquid biopsy). Most cell-free DNA is released from the bone marrow and white blood cells, whereas cell-free DNA in cancer patient is derived from necrotic and apoptotic cancer cells. In general, cell-free DNA contains heterogenic defects, such as single-nucleotide mutations and methylation changes, identical to the primary tumor and/or metastases. The half-life varies from 16 min to 2.5 h, thus providing a dynamic monitoring of disease burden. In general, the overall cell-free DNA level is significantly higher in cancer patients than healthy volunteers. Nevertheless, the biggest obstacle is the detection limit of the assay as tumor-related cell-free DNA may represent <0.1% of total cell-free DNA. Moreover, several factors, such as the type of matrix, storage conditions, or particular handling of blood samples may affect cell-free DNA concentrations and fragmentation values. The epidermal growth factor receptor (EGFR) is a transmembrane tyrosine kinase receptor that plays a central role in regulating cell division and death. Mutations and copy number changes in the gene encoding EGFR that lead to overexpression of the protein have been associated with a number of different cancers. The FDA recently approved the first such liquid biopsy test for EGFR mutation in patients with nonsmall-cell lung cancer. This noninvasive method allows the identification of specific mutants such as EGFR T790M in patients with nonsmall-cell lung cancer being treated with gefitinib [23].

## Summary/key points

- Tumor markers can be used for one of five purposes including screening a healthy population or a high-risk population for the probable presence of cancer, diagnosis of cancer, or of a specific type of cancer, evaluating prognosis in a patient, predicting potential response of a patient to therapy and monitoring the recovery of a patient during receiving surgery, radiation, or chemotherapy.
- Commonly measured tumor markers in clinical laboratories include prostate-specific antigen (PSA) for prostate cancer, human chorionic gonadotropin (hCG) for gestational trophoblastic tumors and some germ cell tumors, alpha-fetoprotein (AFP) for certain germ cell tumors, and HCC, CA-125 (carbohydrate antigen or cancer antigen 125) for ovarian cancer, CA 19-9 (carbohydrate antigen 19-9) for pancreatic and gastrointestinal cancers, and carcinoembryonic antigen (CEA) for colon and rectal cancers.
- PSA is expressed by both normal and neoplastic prostate tissue. Under normal conditions, PSA is produced as a proantigen (proPSA) by the secretory cells that line the prostate glands and secreted into the lumen,

where the propeptide moiety is removed to generate active PSA. The active PSA can then undergo proteolysis to generate inactive PSA, of which a small portion then enters the bloodstream and circulates in an unbound state (free PSA). Alternatively, active PSA can diffuse directly into the circulation where it is rapidly bound to protease inhibitors, including alpha-1-antichymotrypsin (ACT) and alpha-2-macroglobulin.

- In men with a normal prostate, the majority of free PSA in the serum reflects the mature protein that has been inactivated by internal proteolytic cleavage. In contrast, free PSA is relatively decreased in patients with prostate cancer. Thus, the percentage of free or unbound PSA is lower in the serum of men with prostate cancer. This finding has been used in the use of the ratio of free to total PSA and complexed PSA (cPSA) as a means of distinguishing between prostate cancer and BPH as a cause of an elevated PSA. Causes of elevated PSA include benign prostatic hyperplasia (BPH), prostate cancer, prostatic inflammation/infection, and perineal trauma.
- Emerging concepts regarding PSA testing that may help refine the interpretation of an elevated concentration include PSA density, PSA velocity, and free versus complexed or bound PSA.
- These modifications would presumably be most useful for prostate cancer screening when the total PSA is between 2.5 and 10.0 ng/mL, the range in which decisions regarding further diagnostic testing are most difficult.
- The major cause of false elevation of PSA is the presence of heterophilic antibody in the serum. False-negative result due to the presence of inhibitory factors in serum include anti-PSA antibody is uncommon.
- More recently, newer biomarkers, such as prostate-specific antigen 3 measured in urine, quantitative determination of the TMPRSS2-ERG fusion gene product in urine as well as DNA-based biomarkers are emerging. Preliminary results indicate the utility of circulating and urinary microRNAs in the detection and prognosis of prostate cancer.
- CA-125 concentrations may be elevated in the blood of some patients with specific types of cancers such as ovarian cancer. CA-125 is also increased in a variety of benign and malignant conditions, including endometriosis, uterine leiomyoma, cirrhosis with or without ascites, pelvic inflammatory disease, cancers of the endometrium, breast, lung, and pancreas and pleural or peritoneal fluid inflammation due to any cause.
- Serum epididymis protein-4 (HE4; molecular weight 23–27 kDa) level is a useful biomarker for the management of ovarian and endometrial cancer patients. HE4 has been shown to be expressed and secreted by ovarian carcinoma cells. This marker can be measured in both serum and urine. The ROMA (the risk of ovarian malignancy algorithm) score that utilizes serum HE-4 and CA-125 level in calculating the score can also be used for improved sensitivity and specificity for the diagnosis of ovarian cancer.

- Serum AFP is the most commonly used marker for the diagnosis of hepatocellular carcinoma (HCC). Serum levels of AFP do not correlate well with other clinical features of HCC, such as size, stage, or prognosis. Elevated serum AFP may also be seen in patients without HCC such as acute or chronic viral hepatitis. AFP may be slightly elevated in patients with liver cirrhosis due to chronic hepatitis C infection. A significant rise in serum AFP in a patient with cirrhosis should raise concern that HCC may be developed. It is generally accepted that serum levels greater than 500 µg/L (upper limit of normal in most laboratories is between 10 and 20 µg/L) in a high-risk patient is diagnostic of HCC. However, HCC is often diagnosed at a lower AFP level in patients undergoing screening.
- CEA (carcinoembryonic antigen) is found in the sera of patients with colorectal carcinoma (CRC), gastric carcinoma, pancreatic carcinoma, lung carcinoma, and breast carcinoma. Patients with medullary thyroid carcinoma also have higher levels of CEA compared to healthy individuals (above 2.5 ng/mL). However, CEA blood test is not reliable for diagnosing cancer or as a screening test for early detection of cancer. Serum levels of CEA should be routinely measured preoperatively in patients undergoing potentially curative resections for CRC for two reasons:
- Elevated preoperative CEA levels that do not normalize following surgical resection imply the presence of persistent disease and the need for further evaluation.
- Preoperative CEA values are of prognostic significance. CEA levels ≥5.0 ng/mL are associated with an adverse impact on survival that is independent of tumor stage.
- CA 19-9, also called cancer antigen 19-9 or sialylated Lewis (a) antigen, is used primarily in the management of pancreatic cancer. Guidelines from the American Society of Clinical Oncology discourage the use of CA-19-9 as a screening test for cancer, particularly pancreatic cancer. However, in individuals with pancreatic masses, CA-19-9 can be useful in distinguishing between cancer and other pathology of the gland.
- Laboratory tests for hCG are essentially very sensitive and specific for the diagnosis of trophoblast-related conditions including pregnancy and the gestational trophoblastic diseases. However, serum hCG may be falsely elevated due to the presence of heterophilic antibody. Because heterophilic antibodies due to large molecular mass are absent in urine, negative urine beta-hCG result in conjunction with elevated serum hCG value confirms the interference of heterophilic antibody in serum hCG measurement.
- At levels of hCG above 500,000 mIU/mL, a "hook effect" can occur resulting in an artifactually low value for hCG (i.e., 1 to 100 mIU/mL).

This is because the sensitivity of most hCG tests is set to the pregnancy hCG range (i.e., 27,300 to 233,000 mIU/mL at 8 to 11 weeks of gestation); therefore when an extremely high hCG concentration is present, both the capture and tracer antibodies used in assays become saturated, preventing the binding of the two to create a sandwich. For this reason, a suspected diagnosis of gestational trophoblastic disease must be communicated to the laboratory so that the hCG assay could also be performed at 1:1000 dilution to get true hCG value.

- Breast cancer is by far the most common cancer in women worldwide. Several tumor markers are used in diagnosis and management of patients with breast cancer but the most commonly used tumor marker is CA 15-3 (cancer antigen 15–3) which is also known as MUC-1.
- Human epidermal growth factor receptor 2 (HER 2) (also known as cluster of differentiation 340 or protooncogene Neu, encoded by the *ERBB2* gene) is a 185 kDa glycoprotein. Overexpression of HER2 receptor is observed in 20%–30% breast cancer patients and such overexpression is associated with an aggressive tumor subtype, reduced survival, and possible treatment with monoclonal antibody trastuzumab or other types of therapy targeted against the HER 2 receptor protein. HER 2 can be measured by various methods but most commonly employed methods are immunohistochemistry and FISH (Fluorescence in situ hybridization) using breast biopsy specimen.
- Thyroglobulin is produced by thyroid gland and is a precursor of thyroid hormones. Thyroglobulin measurement is not recommended during initial evaluation of suspicious thyroid nodules due to overlap of thyroglobulin levels in patients with benign thyroid disorders and DTC. However, thyroglobulin measurement is useful postoperatively in follow-up of DTC patients to monitor residual or recurrent disease. After successful total thyroidectomy, levels of thyroglobulin should decrease and may even be undetectable.
- Calcitonin is a marker for medullary thyroid carcinoma.
- Cytokeratins are epithelial markers whose expression is not lost during malignant transformation. CYFRA 21-1 is a cytoplasmic protein fragment of cytokeratin 19 (CK-19) which is a marker of nonsmall-cell lung cancer.
- Chromogranin A is a marker for various neuroendocrine diseases such as gastroenteropancreatic tumor, endocrine tumors, and bronchial carcinoid. It may also be useful in diagnosis of pheochromocytoma along with urine/plasma metanephrine determination.
- Squamous cell carcinoma antigens (SCCA) are members of the serpin family of endogenous serine protease inhibitors. High levels of SCCA are often observed with poorly differentiated and advanced metastatic squamous cell carcinoma. In the clinical laboratory, circulating SCCA

(both SCCA1 and SCCA2) can be detected by immunoassays or by using immunohistochemistry. Tissue biopsy may also be used for measuring this tumor marker.

- More recently, circulating cell-free DNA assays are getting attention as viable tumor markers that can be analyzed in blood (liquid biopsy).

## References

[1] Kustanovich A, Schwartz R, Peretz T, Grinshpun A. Life and death of circulating cell-free DNA. Cancer Biol Ther 2019;20:1057–67.

[2] Lilja H, Christensson A, Dahlén U, Matikainen MT, et al. T. Prostate-specific antigen in serum occurs predominantly in complex with alpha 1-antichymotrypsin. Clin Chem 1991;37:1618–25.

[3] Hudson MA, Bahnson RR, Catalona WJ. Clinical use of prostate specific antigen in patients with prostate cancer. J Urol 1989;142:1011–7.

[4] Tanguay S, Bégin LR, Elhilali MM, Behlouli H, et al. Comparative evaluation of total PSA, free/total PSA, and complexed PSA in prostate cancer detection. Urology 2002;59:261–5.

[5] Sokoll LJ, Sanda MG, Feng Z, Kagan J, et al. A prospective, multicenter, National Cancer Institute early detection research network study of [-2]proPSA: improving prostate cancer detection and correlating with cancer aggressiveness. Cancer Epidemiol Biomarkers Prev 2010;19:1193–200.

[6] Kilpelainen TP, Tammela TL, Roobol M, Hugosson J, et al. False positive screening results in the European randomized study of screening for prostate cancer. Eur J Cancer 2011;47:2698–705.

[7] Morgan BR, Tarter TH. Serum heterophilic antibodies interfere with prostate specific antigen test and result in over treatment in a patient with prostate cancer. J Urol 2001;166:2311–2.

[8] Loudas NB, Killeen AA, Palamalai V, Weight CJ, et al. Falsely undetectable prostate-specific antigen (PSA) due to presence of an inhibitory serum factor: a case report and review of pertinent literature. Am J Case Rep 2019;20:1248–52.

[9] Tosoian JJ, Ross AE, Sokoll LJ, Partin AW, et al. Urinary biomarkers for prostate cancer. Urol Clin N Am 2016;43:17–38.

[10] Nagata H, Takahashi K, Yamane Y, Yoshino K, et al. Abnormally high values of CA-125 and CA 19-9 in women with benign tumors. Gynecol Obstet Invest 1989;28:156–68.

[11] Dochez V, Caillon H, Vaucel E, Dimet J, et al. Biomarkers and algorithms for diagnosis of ovarian cancer: CA125, HE4, RMI and ROMA, a review. J Ovarian Res 2019;12(1):28.

[12] Howaizi M, Abboura M, Krespine C, Sbai-Idrissi MS, et al. A new cause for CA19.9 elevation: heavy tea consumption. Gut 2003;52:913–4.

[13] Hu KQ, Kyulo NL, Lim N, Elhazin B, et al. Clinical significance of elevated alpha-fetoprotein (AFP) in patients with chronic hepatitis C, but not hepatocellular carcinoma. Am J Gastroenterol 2004;99(5):860–4.

[14] Phaneuf D, Lambert M, Laframboise R, Mitchell G, et al. Type 1 hereditary tyrosinemia: evidence for molecular heterogeneity and identification of a casual mutation in a French Canadian patient. J Clin Invest 1992;90:1185–92.

[15] Wang H, Bi X, Xu L, Li Y. Negative interference by rheumatoid factor in alpha-fetoprotein chemiluminescent microparticle immunoassay. Ann Clin Biochem 2017;54:55–9.

[16] Hronowski AM, Cervinski M, Stenman UH, Woodworth A, et al. False negative results in point-of-care qualitative human chorionic gonadotropin (hCG) devices due to excess hCG beta core fragment. Clin Chem 2009;55:1389–94.

[17] Koshida S, Asanuma K, Kuribayashi K, Goto M, et al. Prevalence of human anti-mouse antibodies (HAMAs) in routine examination. Clin Chim Acta 2010;411:391–4.

[18] Trape J, Filella X, Alsina-Donadeu M, Juan-Pereira L, et al. Increased plasma concentrations of tumor markers in the absence of neoplasia. Clin Chem Lab Med 2011;49:1605–20.

[19] Nixon AM, Provatopoulou X, Kalogera E, Zografos GN, Gounaris A. Circulating thyroid cancer biomarkers: current limitations and future prospects. Clin Endocrinol (Oxf) 2017;87:117–26.

[20] Algeciras-Schimnich A. Thyroglobulin measurement in the management of patients with differentiated thyroid cancer. Crit Rev Clin Lab Sci 2018;55:205–18.

[21] Yip DT, Hassan M, Pazaitou-Panayiotou K, Ruan DT, et al. Preoperative basal calcitonin and tumor stage correlate with postoperative calcitonin normalization in patients undergoing initial surgical management of medullary thyroid carcinoma. Surgery 2011;150:1168–77.

[22] Liu L, Xie W, Xue P, Wei Z, et al. Diagnostic accuracy and prognostic applications of CYFRA 21-1 in head and neck cancer: a systematic review and meta-analysis. PLoS One 2019;14(5), e0216561.

[23] Oellerich M, Schutz E, Beck J, Kanzow P, et al. Using circulating cell free DNA to monitor personalized cancer therapy. Crit Rev Clin Lab Sci 2017;54:205–18.

# CHAPTER 14

# Therapeutic drug monitoring

## What is therapeutic drug monitoring?

There are over 6000 prescription and nonprescription (over-the-counter) drugs available for clinical use in the United States. Most drugs have a wide therapeutic index (the difference in the therapeutic and toxic drug levels) and do not require therapeutic drug monitoring. For example, acetaminophen has a therapeutic range between 5 and 20 µg/mL, and toxicity is encountered at a concentration of 150 µg/mL and above. Therefore, acetaminophen therapy does not require therapeutic drug monitoring unless in a case of suspected overdose. In contrast, digoxin has a therapeutic range of 0.8–1.5 ng/mL, but toxicity can be observed at level 2.0 ng/mL or even above 1.5 ng/mL in some patients and less commonly within the therapeutic range. Therefore, digoxin therapy requires routine monitoring. The highlights of the current state of therapeutic drug-monitoring practices include:

- Approximately 20–26 prescription drugs are frequently monitored in a majority of hospital-based laboratories because these drugs have a narrow therapeutic index.
- In addition, 25–30 drugs are subjected to therapeutic drug monitoring less frequently. Usually therapeutic drug monitoring of these drugs are offered in large academic medical centers and national reference laboratories.
- The goal of therapeutic drug monitoring is to optimize pharmacological responses of a drug while avoiding adverse effects.
- For most drugs, serum or plasma is used for therapeutic drug monitoring and measuring the trough level of the drug (15–30 min prior to the next dose), except immunosuppressants (cyclosporine, tacrolimus, sirolimus, and everolimus), where trough drug levels are measured in whole blood. Interestingly, another immunosuppressant drug mycophenolic acid is monitored in serum or plasma.
- For aminoglycosides and vancomycin, both peak and trough concentrations in serum or plasma may be measured.

Clinical Chemistry, Immunology and Laboratory Quality Control. https://doi.org/10.1016/B978-0-12-815960-6.00002-9

## Drugs that require therapeutic drug monitoring

As mentioned earlier, only a small fraction of prescription drugs require therapeutic drug monitoring, because for most prescription drugs there is a wider difference between therapeutic and toxic concentrations. The characteristics of a drug where therapeutic drug monitoring is beneficial include:

- Difficulty in interpreting the therapeutic range or a low toxicity of a drug based on clinical evidence alone.
- Narrow therapeutic range.
- Toxicity of a drug may lead to hospitalization, irreversible organ damage, and even death, but an adverse drug reaction may be avoided by therapeutic drug monitoring.
- There is a correlation between serum or whole blood concentration of the drug and its therapeutic response or toxicity.

## Free versus total drug monitoring

A drug may be bound to a serum protein, and only the unbound (free) drug is pharmacologically active. Protein binding of drugs may vary from 0% (not bound to a protein at all) to over 99%. Usually when a drug concentration is measured in serum/plasma, it is the total drug concentration (free drug + protein bound drug). However, if a drug is <80% protein bound, the total concentration can adequately predict free drug concentration and direct measurement of the free drug may not be necessary. However, for a strongly protein-bound drug (>80%), direct measurement of the free drug may be clinically beneficial in patients with hypoalbuminemia, uremia, liver disease, and critically ill patients. Important issues regarding free drug monitoring include:

- Usually free phenytoin is the most commonly measured free drug in clinical laboratories although monitoring free valproic acid and carbamazepine may also be beneficial.
- Measurement of other strongly protein-bound anticonvulsants such as valproic acid and carbamazepine may be useful.
- Free mycophenolic acid concentration (an immunosuppressant) may be useful in uremic patients.

## Therapeutic drug monitoring benefits

In general, many drugs are used as a prophylactic to prevent clinical symptoms. Therefore, noncompliance has serious clinical consequences. Therapeutic drug monitoring is very helpful in identifying such noncompliant patients. Mattson et al. commented that undetected, subtherapeutic level or variable drug levels are indicators of noncompliance to a medication where therapeutic drug monitoring is available [1]. Chandra et al. commented that poor patient compliance is one of the major causes of nonresponsiveness to antiepileptic drug therapy [2]. Therapeutic drug monitoring can greatly reduce chances of treatment failure by personalization of drug dosage based on drug level in the serum/plasma or whole blood. Benefits of therapeutic drug monitoring include:

- Identifying noncompliance
- Personalizing the drug dosage for maximum therapeutic benefit and avoiding drug toxicity
- Identifying clinically significant drug-drug, drug-food and drug-herb interactions
- Identifying a nonresponder to a drug

If a patient does not respond to a drug, despite the drug level in the therapeutic range, it may be an indication that the patient is a nonresponder to that drug. A different drug choice must be made to treat the patient.

## Basic pharmacokinetics

When a drug is administered orally, it undergoes several steps in the body that determine the concentration of that drug in the serum/plasma or whole blood. These steps include:

- Liberation: The release of a drug from the dosage form (tablet, capsule, extended-release formulation).
- Absorption: Movement of a drug from the site of administration (for drugs taken orally) to blood circulation. Many factors affect this stage, including gastric pH, presence of food particles, as well as the efflux mechanism if present in the gut. First-pass metabolism plays an important role in determining bioavailability of a drug administered orally.
- Distribution and protein binding: Movement of a drug from the blood circulation to tissues/target organs. Drugs may also be bound to serum proteins ranging from zero protein binding to 99% protein binding.
- Metabolism: Chemical transformation of a drug to the active and inactive metabolites. Liver is responsible for metabolism of many drugs, although drugs may also be metabolized by a nonhepatic path or are subjected to minimal metabolism.
- Excretion: Elimination of the drug from the body via a renal, biliary, or pulmonary mechanism.

Liberation of a drug after oral administration depends on the formulation of the drug. Immediate-release formulation releases the drugs at once from the dosage form when administered, while the same drug may also be available in a sustained-release formulation. Absorption of a drug depends on the route of administration. Generally, oral administration is the route of choice but if a drug has poor bioavailability or undergoes extensive first-pass metabolism, intramuscular or intravenous administration may be needed. Rectal administration is a less common route of administration. Although the rate of drug absorption is usually lower after rectal administration compared to oral administration, for certain drugs, rectal absorption is higher than oral absorption due to avoidance of the hepatic first-pass metabolism. These drugs include lidocaine, morphine, metoclopramide, ergotamine, and propranolol. Local irritation is a possible complication of rectal drug delivery [3]. When a drug is administered by direct injection, it enters the blood circulation immediately.

When a drug enters the blood circulation, it is distributed throughout the body into various tissues, and the pharmacokinetic parameter is called the volume of distribution ($V_d$). This is the hypothetical volume to account for all drugs in the body and is also termed as the apparent volume of distribution.

$V_d$ = Dose/plasma concentration of drug

The amount of a drug that interacts with the receptor or target site is usually a small fraction of the total drug administered. Muscle and fat tissues may serve as a reservoir for lipophilic drugs. For neurotherapeutics, penetration of the blood-brain barrier is essential. Drugs usually undergo chemical transformation (metabolism) before elimination. Drug metabolism may occur in any tissues including the blood. For example, plasma butylcholinesterase metabolizes drugs such as succinylcholine. The role of metabolism is to convert lipophilic nonpolar molecules to water-soluble polar compounds for excretion in urine. Many drugs are metabolized in the liver in two phases by various enzymes. Major steps in drug metabolism involve:

- Phase I step, which involves manipulation of a functional group of a drug molecule in order to make the molecule more polar by oxidation or reduction of a functional group in the drug molecule or hydrolysis. Cytochrome P 450 mixed function family of enzymes (CYP) play major role in Phase I reaction although other enzymes may also be involved.
- Phase II step may involve acetylation (adding an acetate group to the drug molecule), sulfation (adding an inorganic sulfate), methylation (adding a methyl group), amino acid conjugation or glucuronidation (adding sugar such as glucuronic acid) in order to increase the polarity of the drug metabolite.

Cytochrome P 450 mixed function oxidase family of enzymes (CYP) play a major role in the Phase I metabolism of many drugs. These enzymes are found in abundance in the liver, but also may be found in other organs such as the gut. NADPH (nicotinamide adenine dinucleotide phosphate) is a required cofactor for CYP-mediated biotransformation, and oxygen serves as a substrate. At present, 57 human genes are known to encode CYP isoforms; of these, at least 15 are associated with xenobiotic metabolism. CYP isoenzymes are named according to sequence homology: amino acid sequence similarity >40% assigns the numeric family (e.g., CYP1, CYP2); >55% similarity determines the subfamily letter (e.g., CYP2C, CYP2D); isoforms with >97% similarities are given an additional number (e.g., CYP2C9, CYP2C19) to distinguish them. The major CYP isoforms responsible of metabolism of drugs include CYP1A2, CYP2B6, CYP2C9, CYP2C19, CYP2D6, CYP2E1, and CYP3A4/CYP3A5. However, CYP3A4 is the predominant isoform of the CYP family (almost 30%) usually responsible for metabolism of 30–50% of all marketed drugs. In addition to CYP, other enzymes are also involved in the Phase I metabolism.

One of the enzymes that play a vital role in Phase II metabolism is uridine-5 phosphate glucuronyl transferase (UDP-glucuronyl transferase). This enzyme is responsible for conjugation of glucuronic acid with the drug molecule in the phase II metabolism, thus inactivating the drug. This enzyme is mostly found in the liver but may also be present in other organs. Major enzymes involved in drug metabolism are summarized in Table 1. Genetic predisposition may determine the activities of such drug metabolizing enzymes, which ultimately affect the drug metabolism. This is an important topic known as pharmacogenomics. Please see Chapter 20 for an introduction to this

**Table 1** Major enzymes involved in drug metabolism.

| Reaction type | Phase | Name of enzyme |
| --- | --- | --- |
| Oxidation | Phase I | Cytochrome P450<br>Alcohol dehydrogenase<br>Aldehyde dehydrogenase<br>Monoamine oxidase |
| Reduction | Phase I | Various reductases |
| Hydrolysis | Phase I | Butylcholinesterase, epoxide hydrolase<br>Amidases |
| Glucuronidation | Phase II | Glucuronosyltransferase |
| Acetylation | Phase II | N-Acetyltransferase |
| Methylation | Phase II | Methyltransferase |
| Amino acid conjugation | Phase II | Glutathione transferase |
| Sulfation | Phase II | Sulfotransferase |

important subject. The half-life of a drug is the time required for the serum concentration to be reduced by 50%. Key points regarding steady-state concentration of a drug include:

- With repeatedly administered doses, the steady state is reached after approximately five half-lives.
- Therapeutic drug monitoring is recommended when a drug reaches a steady state.

The half-life of a drug can be calculated from the elimination rate constant ($K$) of a drug.

Half-life $= 0.693/K$

The elimination rate constant can be easily calculated from the serum concentrations of a drug at two different time points using the formula where $Ct_1$ is the concentration of the drug at a time point $t_1$, and $Ct_2$ is the concentration of the same drug at a later time point $t_2$:

$$K = \frac{\ln Ct_1 - \ln Ct_2}{t_2 - t_1}$$

A drug may also undergo extensive metabolism before fully entering the blood circulation. This process is called first-pass metabolism. If a drug undergoes significant first-pass metabolism, then the drug should not be delivered orally, such as lidocaine. Renal excretion is a major pathway for the elimination of drugs and their metabolites. Drugs may also be excreted via other routes, such as biliary excretion. A drug excreted in bile may also be reabsorbed from the gastrointestinal tract, or a drug conjugate may be hydrolyzed by the bacteria of the gut, liberating the original drug, which can return into the blood circulation. Enterohepatic circulation may prolong the effects of a drug, for example, mycophenolic acid. Cholestatic disease, where normal bile flow is reduced, may reduce bile clearance of the drug, causing drug toxicity.

## Effect of gender and pregnancy on drug metabolism and disposition

Men and women may show differences in response to certain drugs. In addition, pregnancy may also significantly alter the metabolism and disposition of certain drugs. Gender difference affects bioavailability, distribution, metabolism, and elimination of drugs due to variations between men and women in body weight, blood volume, gastric emptying time, drug protein binding, activities of drug-metabolizing enzymes, drug transporter function, and excretion activity. Other important gender differences in drug metabolisms include:

- Hepatic metabolism of drugs by Phase I (via CYP1A2 and CYP2E1) and Phase II (by glucuronyl transferase, methyltransferases, and dehydrogenases) reactions appear to be faster in males than females although metabolisms of drugs by other enzymes appear to be the same.
- Women may have higher activity of CYP3A4.

In general, women are also more susceptible to adverse effects of drugs than men. Women are at increased risk of QT prolongation with many antiarrhythmic drugs, which may even lead to a critical condition, such as torsade de pointes, compared to men even at same levels of serum drug concentrations.

Drug therapy in pregnant women usually focuses on potential tetragonic effects of the drug, and therapeutic drug monitoring during pregnancy aims to improve individual dosage improvement, taking into account pregnancy-related changes in drug disposition. Gastrointestinal absorption and bioavailability of many drugs vary in pregnancy due to changes in gastric secretion and small intestine motility. Elevated concentrations of various hormones in pregnancy, such as estrogen, progesterone, placental growth hormone, and prolactin, could be related to altered drug metabolism observed in pregnant women. The renal excretion of unchanged drugs is increased in pregnancy. In general, dosage adjustments are required for anticonvulsants, lithium, digoxin, certain beta blockers, ampicillin, cefuroxime, and certain antidepressants in pregnant women [4].

## Effect of age on drug metabolism and disposition

In the fetus, CYP3A7 is the major hepatic enzyme, but CYP3A5 may also be present in significant levels in half of all children. However, in adults CYP3A4 is the major functional cytochrome P 450 enzyme responsible for metabolism of many drugs. CYP1A1 is also present during organogenesis while CYP2E1 may be present in some second trimester fetuses. After birth, hepatic CYP2D6, CYP2C8/9, and CYP2C18/19 are activated. CYP1A2 becomes active during the fourth to fifth month after birth [5]. Neonates and infants have increased total body water to body fat ratio compared to adults whereas the reverse is observed in the elderly people. These factors may affect the volume of distribution of drugs depending on their lipophilic character. Moreover, altered plasma binding of drugs may be observed in both neonates and some elderly people due to low albumin, thus increasing the fraction of pharmacologically active free drug. General features of drug metabolizing capacity as a function of age include:

- Neonates and infants (0–4 months) may metabolize drug slower than adults because of the lack of mature (fully functioning) drug-metabolizing enzymes.
- Renal function at the time of birth is reduced by more than 50% of the adult value but then increases rapidly in the first 2–3 years of life. However, activities of drug-metabolizing enzymes are higher in children than adults. Therefore, children may need a higher per kilogram dosage of a drug than an adult.
- Activities of drug-metabolizing enzymes decrease with advanced age. In addition, renal clearance of drugs may also start to decline with age. Therefore, careful adjustment of dosage is needed to treat elderly patients.
- Elderly patients (>70 years of age) may also have lower albumin. Therefore, protein binding of strongly protein-bound drugs may be impaired.

## Drug metabolism and disposition in uremia

Renal disease causes impairment in the clearance of many drugs by the kidney. Correlations have been established between creatinine clearance and the clearance of digoxin, lithium, procainamide, aminoglycoside, and several other drugs, but creatinine clearance does not always predict renal excretion of all drugs. Moreover, elderly patients may have unrecognized renal impairment, and caution should be exercised when medications are prescribed to elderly patients. Serum creatinine remains normal until GFR (glomerular filtration rate) has fallen by at least 50%. Nearly half of the older patients have normal serum creatinine but reduced renal function. Dose adjustments based on renal function is recommended for many medications in elderly patients, even for medications that exhibit large therapeutic windows [6]. Other characteristics of drug metabolism and disposition in elderly patients include:

- Patients with chronic renal failure show reduced activities of cytochrome P 450 enzymes. Therefore, metabolism of many drugs mediated by these enzymes is significantly reduced.
- Chronic renal failure can also significantly reduce nonrenal clearance of many drugs, including drugs that are metabolized by Phase II reactions and drug transporter proteins, such as P-glycoprotein and organic anion-transporting polypeptides.
- Renal disease also causes impairment of drug protein binding, because uremic toxins compete with drugs for binding to albumin. In addition, uremic patients may also have hypoalbuminemia. Such interaction leads

to increases in the pharmacologically active free drug concentration, especially for classical anticonvulsants such as phenytoin, carbamazepine, and valproic acid. Therefore, monitoring free phenytoin, free valproic acid, and to some extent free carbamazepine is recommended in uremic patients in order to avoid drug toxicity.

## Drug metabolism and disposition in liver disease

Liver dysfunction not only reduces the clearance of a drug metabolized through hepatic enzymes or biliary mechanisms, but also affects plasma protein binding due to reduced synthesis of albumin and other drug-binding proteins. Even mild to moderate hepatic disease may cause an unpredictable effect on drug metabolism. Portal-systemic shunting present in patients with advanced liver cirrhosis may cause a significant reduction in first-pass metabolism of high extraction drugs, thus increasing bioavailability as well as risk of drug overdose and toxicity. Important points to remember regarding drug metabolism and disposition in patients with liver disease include:

- Although the Phase I reaction involving cytochrome P-450 enzymes may be significantly impaired in liver disease, the Phase II reaction (glucuronidation) seems to be unaffected.
- Patients with liver disease often suffer from hypoalbuminemia. Therefore, protein bindings of strongly protein-bound drugs are impaired causing elevation of pharmacologically active free fraction of the drug.
- Nonalcoholic fatty liver disease is the most common chronic liver disease. This type of liver disease also affects the activity of drug-metabolizing enzymes in the liver with the potential to produce adverse drug reactions from the standard dosage.
- Mild-to-moderate hepatitis infection may also decrease drug clearance.

## Effect of cardiovascular disease on drug metabolism and disposition

Cardiac failure is often associated with disturbances in cardiac output, influencing the extent and pattern of tissue perfusion, sodium and water metabolism, and gastrointestinal motility that eventually may affect absorption and disposition of many drugs. Hepatic elimination of drugs via oxidative Phase I metabolism is impaired in patients with congestive heart failure due to decreased blood supply in the liver. Theophylline metabolism is reduced in patients with severe cardiac failure, and dose reduction is strongly recommended. Digoxin clearance is also decreased. Quinidine plasma levels may also be high in these patients due to lower volume of distribution [7]. Therefore, therapeutic drug monitoring is crucial in avoiding drug toxicity in these patients.

## Thyroid dysfunction and drug metabolism

Patients with thyroid disease may have an altered drug disposition, because thyroxine is a potent activator of the cytochrome P 450 enzyme system. The key points on the effect of thyroid dysfunction on drugs metabolisms are listed below:

- Drug metabolism is increased in patients with hyperthyroidism due to excessive levels of thyroxine.
- Drug metabolism is decreased in patients with hypothyroidism due to lower levels of thyroxine.
- Amiodarone is an antiarrhythmic drug associated with thyroid dysfunction, because due to high iodine content, amiodarone inhibits 5-deiodinase activity. Screening of thyroid disease before amiodarone therapy and periodic monitoring of thyroid function are recommended for patients treated with amiodarone.

## Effect of food, alcohol consumption, and smoking on drug disposition

Drug-food interactions may be pharmacokinetic or pharmacodynamic in nature. Certain foods can affect absorption of certain drugs and can also alter the activity of enzymes that metabolize drugs, especially CYP3A4. It has been documented that the intake of charcoal-broiled food or cruciferous vegetables induces the metabolism of multiple drugs. The most important interaction between fruit juice and a drug involves consumption of grapefruit juice. It was reported in 1991 that a single glass of grapefruit juice caused a two- to threefold increase in the plasma concentration of felodipine, a calcium channel blocker after oral intake of a 5 mg tablet, but a similar amount of orange juice showed no effect [8]. Subsequent investigations demonstrated that the pharmacokinetics of approximately 40 other drugs are also affected by intake of grapefruit juice [9]. The main mechanism for enhanced bioavailability of drugs after intake of grapefruit juice is as follows:

- Furanocoumarins found in grapefruit juice inhibits CYP3A4 in the small intestine, thus inhibiting the metabolism of drugs in the small intestine and increasing the concentration of available drugs. Grapefruit juice does not inhibit liver CYP3A4. Therefore, if the drug is injected, no change in pharmacokinetics is observed. Because calcium channel blockers are commonly administered orally, patients should avoid consuming grape fruit juice.
- Grapefruit juice also inhibits P-glycoprotein, thus inhibiting its drug efflux metabolism, which indirectly increases the bioavailability of a drug that is a substrate for P-glycoprotein.

- Common drugs that interact with grapefruit juice include alprazolam, carbamazepine, cyclosporine, erythromycin, methadone, quinidine, simvastatin, and tacrolimus.

There are two types of interactions between alcohol and a drug: pharmacokinetic and pharmacodynamic. Pharmacokinetic interactions occur when alcohol interferes with the hepatic metabolism of a drug. Pharmacodynamic interactions occur when alcohol enhances the effect of a drug, particularly in the central nervous system. The package insert of many antibiotics and other drugs states that the medication should not be taken with alcohol due to drug-alcohol interactions. Fatal toxicity may occur from alcohol and drug overdoses due to pharmacodynamic interactions. In a Finnish study, it was found that median amitriptyline and propoxyphene concentrations were lower in alcohol-related fatal cases compared to cases where no alcohol was involved. The authors concluded that when alcohol is present, a relatively small overdose of a drug may cause fatality [10].

Approximately 4800 compounds are found in tobacco smoke, including nicotine and carcinogenic compounds such as polycyclic aromatic hydrocarbons (PAHs) and N-nitroso amines. PAHs induce CYP1A1, CYP1A2, and possibly CYP2E1 and may also induce Phase II metabolism. Key points regarding the effect of smoking on drug metabolism include:

- Cigarette smoke, not nicotine, is responsible for the alteration of drug metabolism.
- Increased theophylline metabolism in smokers due to induction of CYP1A2 is well documented.

In one study, the half-life of theophylline was reduced by almost twofold in smokers compared to nonsmokers [11]. Significant reductions in drug concentrations with smoking have been reported for caffeine, chlorpromazine, clozapine, flecainide, fluvoxamine, haloperidol, mexiletine, olanzapine, proprandol, and tacrine due to increased metabolism of these drugs. Smokers may therefore require higher doses than nonsmokers in order to achieve pharmacological responses [12]. Warfarin disposition in smokers is also different than in nonsmokers. One case report described an increase in INR (international normalization ratio) to 3.7 from a baseline of 2.7–2.8 in an 80-year-old man when he stopped smoking. Subsequently, his warfarin dose was reduced by 14% [13].

## Therapeutic drug monitoring of various drug classes: General considerations

For a meaningful interpretation of a serum drug concentration, time of specimen collection should be noted along with the time and date of the last dose and route of administration of the drug. This is particularly important for aminoglycosides, because without knowing the time of specimen collection, the serum drug concentration cannot be interpreted. The information needed for proper interpretation of drug levels for the purpose of therapeutic drug monitoring is listed in Table 2. Reference ranges of various therapeutic drugs are provided with the result. Therapeutic ranges of common drugs are given in Table 3. However, therapeutic ranges may vary slightly between different laboratories due to variations in the patient population.

**Table 2** Information required for interpretation of results of therapeutic drug monitoring.

| Patient-related information required on the request |
| --- |
| Name of the patient<br>Hospital identification number<br>Age<br>Gender (if female pregnant?)<br>Race |
| **Other essential information** |
| Time of last dosage<br>Type of and number of specimen (serum, whole blood urine, saliva, other body fluid)<br>Identification of peak versus trough specimen (for aminoglycosides and vancomycin only)<br>Special request (such as free phenytoin) |
| **Essential information needed for interpretation of results** |
| Dosage regimen<br>Other drugs the patient is receiving<br>Concentration of the drug<br>Pharmacokinetic parameters of the drug<br>Is the patient critically ill or suffer from hepatic, cardiovascular, or renal disease?<br>Albumin level, creatinine clearance |

**Table 3** Therapeutic level of commonly monitored drugs.

| Drug class/drug | Recommended therapeutic range (trough) |
| --- | --- |
| **Anticonvulsants** | |
| Phenytoin | 10–20 µg/mL |
| Carbamazepine | 4–12 µg/mL |
| Phenobarbital | 15–40 µg/mL |
| Primidone | 5–12 µg/mL |
| Valproic acid | 50–100 µg/mL |
| Clonazepam | 10–75 ng/mL |
| Lamotrigine | 3–14 µg/mL |

**Table 3** Therapeutic level of commonly monitored drugs—cont'd

| Drug class/drug | Recommended therapeutic range (trough) |
| --- | --- |
| **Cardioactive drugs** | |
| Digoxin | 0.5–1.5 ng/mL |
| Procainamide | 4–10 µg/mL |
| N-Acetyl procainamide | 4–8 µg/mL |
| Quinidine | 2–5 µg/mL |
| Lidocaine | 1.5–5 µg/mL |
| **Antiasthmatic** | |
| Theophylline | 10–20 µg/mL |
| Caffeine | 5–15 µg/mL |
| **Antidepressants** | |
| Amitriptyline + Nortriptyline | 120–250 ng/mL |
| Nortriptyline | 50–150 ng/mL |
| Doxepin + Nordoxepin | 150–250 ng/mL |
| Imipramine + Desipramine | 150–250 ng/mL |
| Lithium | 0.8–1.2 mEq/L |
| **Immunosuppressants** | |
| Cyclosporine[a] | 100–400 ng/mL |
| Tacrolimus[a] | 5–15 ng/mL |
| Sirolimus[a] | 4–20 ng/mL |
| Everolimus[a] | 3–8 ng/mL |
| Mycophenolic acid | 1–3.5 µg/mL |
| **Antineoplastic** | |
| Methotrexate | Varies with therapy type |
| **Antibiotics** | |
| Amikacin | 20–35 µg/mL, Peak<br>4–8 µg/mL, Trough |
| Gentamicin | 5–10 µg/mL, Peak<br><2 µg/mL, Trough |
| Tobramycin | 5–10 µg/mL, Peak<br><2 µg/mL, Trough |
| Vancomycin | 20–40 µg/mL, Peak<br>5–15 µg/mL, Trough |

Therapeutic ranges based on published literature, books, and ranges adopted by reputed national reference laboratories such as Mayo Medical Laboratories and ARUP laboratories. However, therapeutic ranges vary widely among different patient populations and each institute should establish their own guidelines. These values are for the purpose of providing examples only.

[a]Monitored in whole blood instead of serum or plasma.

Usually therapeutic drug monitoring should be ordered after a drug reaches its steady state. It usually takes at least five half-lives after initiation of a drug therapy to reach the steady state. For example, the half-life of digoxin is 1.6 days, and the steady state of digoxin is reached after 7 days of therapy. However, for a drug with a shorter half-life, for example, valproic acid (half-life 11–17 h), it takes only 3 days to reach the steady state. Preanalytical errors can contribute significantly to get an erroneous result for therapeutic drug monitoring.

## Preanalytical issues in therapeutic drug monitoring

The preferred specimen for most therapeutic drug monitoring is trough specimens (exceptions are vancomycin and aminoglycosides, which may require both peak and trough monitoring). Strictly speaking, trough samples should be collected immediately before the administration of the next dose. Many laboratories allow a window of 30–60 min for collection, but even 30 min may introduce inaccuracies of 10%–15% for drugs with short half-lives such as aminoglycosides and caffeine. When a drug is effective over a very narrow concentration range or toxicity is a concern, peak concentrations may be monitored. This type of collection is challenging to time and requires consideration of absorption and distribution phases. The time of collection is influenced by drug formulation, dosing route, and other factors. For the aminoglycosides, peak samples are used to judge efficacy when using conventional dosing regimens, and trough samples are used to judge the potential for toxicity for either conventional or once daily dosing (ODD) regimens. In addition, when using an ODD protocol, samples used to adjust the next dose are collected at defined points within the dosing cycle. Some laboratories call these "random" samples since they are neither peak nor trough and the time of collection may vary, but this terminology is not technically correct since the collection time must be controlled and known—i.e., not truly random. The therapeutic ranges used in the conventional dosing protocols must never be applied to samples collected from a patient receiving these drugs on ODD protocols. For these dosing protocols, pharmacy services use nomograms to compare the drug concentration and the time since initiation of the dose to assess efficacy and modify subsequent dosing intervals.

Lithium and digoxin present challenges due to their long absorption and distribution phases. Samples for lithium should be collected no earlier than 10–12 h after dosing. Samples for digoxin measurement should be collected no earlier than 6–8 h after dosing, preferably after 12 h.

Specimens for therapeutic drug monitoring must be collected using proper blood collection tubes. Plastic gel-barrier tubes (plasma or serum separators) are frequently used for clinical chemistry analyses, but these devices may cause falsely lower concentrations for certain drugs due to absorption by the gel.

For early generation gel-barrier tubes, drug adsorption was an insidious and unpredictable problem, with significant lot-to-lot variation. As a result, many clinical laboratories moved to gel-free tubes. The introduction of a new generation of plastic gel-barrier tubes has laboratories once again considering these as an option. The report from Bush and colleagues [14] was encouraging in that they found <10% loss for only two of the drugs tested, carbamazepine and phenytoin, and only after contact for 7 days.

More recently, Schrapp et al. explored the impact of Becton Dickinson (BD) PST II (plasma separator tube) and Barricor separator tubes on the stability of 167 therapeutic compounds and common drugs of abuse in plasma samples using liquid chromatography combined with tandem mass spectrometry (LC-MS/MS). In 2016, BD released a new heparin collection tube with a novel technology of separation, a mechanical separator called BD Vacutainer Barricor. Instead of a gel, it uses a high-density plastic connected on the elastomer top that stretches during centrifugation, finally creating a seal. Blood cells are allowed to flow around this separator, reducing the time of centrifugation and providing high-quality plasma. The study showed the Barricor nongel tubes cause less drug interference than conventional serum separator gel tubes. Therefore, Barricor nongel tubes are recommended for collecting blood for therapeutic drug monitoring [15].

## Therapeutic drug monitoring of anticonvulsants

Phenytoin, phenobarbital, primidone, ethosuximide, valproic acid, and carbamazepine are considered as conventional anticonvulsant drugs. All of these antiepileptic drugs have a narrow therapeutic range requiring therapeutic drug monitoring. Phenytoin, carbamazepine, and valproic acid are also strongly bound to serum proteins. Therefore, for a selected patient population, monitoring free phenytoin, free valproic acid, and to some lesser extent free carbamazepine is clinically useful. However, free phenobarbital monitoring is not required, because this drug is only moderately bound to serum protein. Free phenytoin is the most commonly ordered free drug monitoring request in our hospital. Monitoring free phenytoin (and also free valproic acid) is recommended in the following patients:

- Uremic patients
- Patients with liver disease
- Pediatric population (small children often show impaired protein binding or altered disposition)
- Pregnant women
- Critically ill patients, elderly patients, and patients with hypoalbuminemia

In addition, monitoring free phenytoin concentration is useful in patients where a drug-drug interaction is suspected. Several strongly protein-bound drugs such as valproic acid, nonsteroidal antiinflammatory drugs (aspirin, ibuprofen, naproxen, tolmetin, etc.), and certain antibiotics (ceftriaxone, nafcillin, oxacillin, etc.) can displace phenytoin from the protein-binding site causing an elevated free phenytoin level.

Carbamazepine is metabolized to carbamazepine 10, 11-epoxide, which is an active metabolite. Although in the normal population, epoxide concentrations may be 10%–14% of total carbamazepine concentration, patients with renal failure may show an over 40% epoxide concentration relative to the carbamazepine concentrations. Monitoring active metabolite concentration using chromatographic method may be useful in these patients as there is no immunoassay available for monitoring epoxide levels. Combined valproic acid and carbamazepine therapy tends to increase the epoxide concentration. Monitoring the epoxide level may be helpful if a patient experiences drug toxicity from an elevated epoxide level.

Primidone is an anticonvulsant which is metabolized to phenobarbital, another anticonvulsant. Although pharmacological activities of primidone are partly due to phenobarbital, primidone itself has anticonvulsant activity.

For routine therapeutic drug monitoring of classical anticonvulsants, immunoassays are commercially available and can be easily adopted on various automated analyzers. Since 1993, new antiepileptic drugs have been approved: eslicarbazepine acetate, felbamate, gabapentin, lacosamide, lamotrigine, levetiracetam, oxcarbazepine, pregabalin, rufinamide, stiripentol, tiagabine, topiramate, vigabatrin, and zonisamide. In general, these antiepileptic drugs have better pharmacokinetic profiles, improved tolerability in patients, and are less involved in drug interactions compared to traditional anticonvulsants. However, felbamate is a very toxic drug with a risk of fatal aplastic anemia, and the use of this drug is reserved for few patients where benefit may override the risk. Therapeutic drug monitoring of some of these new anticonvulsants is not needed, although a few drugs may benefit from therapeutic drug monitoring. Therapeutic drug monitoring of lamotrigine in all patients and levetiracetam and pregabalin in patients with renal impairment has clinical utility. Monitoring active metabolite of oxcarbazepine (10-hydroxycarbazepine) has some justification. In addition, therapy with lamotrigine, zonisamide, and topiramate may also benefit from therapeutic drug monitoring. Usually chromatographic techniques are employed for therapeutic drug monitoring of these newer anticonvulsants. These methods are usually free from interferences. However, there are commercially available immunoassays for lamotrigine, zonisamide, and topiramate.

# Therapeutic drug monitoring of cardioactive drugs

Therapeutic drug monitoring of several cardioactive drugs, including digoxin, procainamide, lidocaine, and quinidine, is routinely performed in clinical laboratories due to the established correlation between serum drug concentrations and the pharmacological response of these drugs. Moreover, drug toxicity can be mostly avoided by therapeutic drug monitoring. Digoxin monitoring is challenging for the following reasons:

- Digoxin has a very narrow therapeutic window, and there are overlaps between the therapeutic and toxic ranges. The classical therapeutic window of 0.5–1.5 ng/mL is problematic. Although digoxin toxicity is common with digoxin level >2 ng/mL, some patients may experience digoxin toxicity at a level of 1.5 ng/mL or with digoxin concentration within the therapeutic range.
- Digoxin immunoassays are affected by both endogenous and exogenous factors (see Chapter 15).
- Digoxin overdose can be treated with Digibind or DigiFab. For these patients, progress of therapy must be monitored by measuring the free digoxin level, because total digoxin level may be misleading due to interference of Digibind/DigiFab with digoxin immunoassays.

Procainamide is metabolized to an active metabolite N-acetyl procainamide (NAPA). During therapeutic drug monitoring of procainamide, NAPA should also be monitored, because NAPA contributes to toxicity of procainamide. In patients with renal insufficiency, NAPA concentration increases in blood due to impaired renal clearance. Immunoassays are available for monitoring both procainamide and NAPA.

Important points to remember regarding therapeutic drug monitoring of lidocaine and quinidine include:

- Lidocaine cannot be given orally due to high first-pass metabolism. However, tocainide, an analog of lidocaine, can be administered orally.
- Lidocaine is strongly bound to alpha-1 acid glycoprotein, but monitoring free lidocaine is not usually conducted.
- Lidocaine is metabolized into monoethylglycinexylidine (MEGX), and this conversion can be used as a liver function test (MEGX test). For this, test blood specimens are collected 15 and 30 min after injection of an intravenous bolus of a small lidocaine test dose (1 mg/kg) administered over 2 min. Patients with a MEGX test value of <10 ng/mL have poor liver function. The MEGX test is widely used in assessing liver function of critically ill patients and in transplantation.
- Quinidine is infrequently used today. Although this drug is strongly bound to alpha-1 acid glycoprotein, free quinidine is usually not monitored.

Less frequently monitored cardioactive drugs include tocainide, flecainide, mexiletine, verapamil, propranolol, and amiodarone. Tocainide was developed as an oral analog of lidocaine, because lidocaine cannot be administered orally due to high first-pass metabolism, but tocainide and lidocaine have similar electrophysiological properties.

## Therapeutic drug monitoring of antiasthmatic drugs

Theophylline and caffeine are two antiasthmatic drugs that require therapeutic drug monitoring. Theophylline is a bronchodilator and a respiratory stimulant effective in the treatment of acute and chronic asthma. The drug is readily absorbed after oral absorption, but peak concentration may be observed much later with sustained-release tablets. Theophylline is metabolized by hepatic cytochrome P 450, and altered pharmacokinetics of theophylline in disease states has been reported. In infants, theophylline is partly metabolized to caffeine, but in adults this metabolite is not formed. Theophylline is metabolized to 3-methylxanthine and other metabolites in adults.

Apnea with or without bradycardia is a common medical problem in premature infants. Caffeine is effective in treating apnea in neonates. Because effectiveness of caffeine therapy can be readily observed clinically, therapeutic drug monitoring of caffeine is only indicated when caffeine toxicity is apparent from clinical symptoms, including tachycardia, gastrointestinal intolerance, and jitteriness. In addition, therapeutic drug monitoring of caffeine is also indicated if a neonate is unresponsive to caffeine therapy despite a high dose.

## Therapeutic drug monitoring of antidepressants

Tricyclic antidepressants (TCAs), including amitriptyline, doxepin, nortriptyline, imipramine, desipramine, protriptyline, trimipramine, and clomipramine, were introduced in the 1950s and 1960s. These drugs have a narrow therapeutic window, and therapeutic drug monitoring is essential for efficacy of these drugs and to avoid drug toxicity. The efficacy of lithium in acute mania and for prophylaxis against recurrent episode of mania has been well established. Therapeutic drug monitoring of lithium is essential for efficacy and to avoid lithium toxicity. Important points to remember in therapeutic drug monitoring of antidepressants:

- Although immunoassays are available for determination of tricyclic antidepressants, such assays should be only used for diagnosis of tricyclic overdose (usually a total tricyclic antidepressant concentration >500 ng/mL is considered critical).
- For routine therapeutic drug monitoring of tricyclic antidepressants, chromatographic techniques must be used (high-performance liquid

chromatography or gas chromatography) because only such methods can differentiate between different tricyclics, for example, amitriptyline from its metabolite nortriptyline. Immunoassays can provide a total concentration only, because both amitriptyline and nortriptyline have almost 100% cross-reactivity. This is true for other tricyclic antidepressants.

- A common mistake for therapeutic drug monitoring of lithium is to collect the specimen in a lithium heparin tube which will falsely elevate the true lithium concentration. This must be avoided and either a sodium heparin tube or serum specimen (with no anticoagulant such as a red top tube) must be collected for lithium analysis.

More recently introduced antidepressants are selective serotonin reuptake inhibitors (SSRIs), for example, citalopram, fluoxetine, fluvoxamine, paroxetine, and sertraline. This class of drugs has a wide therapeutic index. Usually most of these drugs do not require routine therapeutic drug monitoring, but some drugs may benefit from infrequent monitoring especially in certain patient populations like children, the elderly, pregnant women, and individuals with intelligence disabilities.

## Therapeutic drug monitoring of immunosuppressants

Therapeutic drug monitoring of all immunosuppressants is required. While cyclosporine, tacrolimus, sirolimus, and everolimus are monitored using whole blood due to high distribution of these drugs in blood cells, another immunosuppressant mycophenolic acid can be monitored using serum or plasma. Important points in monitoring of immunosuppressants are listed below.

- Immunosuppressant drugs cyclosporine and tacrolimus are calcineurin inhibitors, but sirolimus and everolimus (the most recently approved drug, which is a 2-hydroxyethyl derivative of sirolimus) are m-TOR (mammalian target of rapamycin) inhibitors. All these drugs are monitored in whole blood.
- Everolimus was developed to improve pharmacokinetic parameters of sirolimus. The half-life of sirolimus is 60 h, but the half-life of everolimus is 18–35 h.
- Mycophenolic acid is a potent noncompetitive inhibitor of inosine monophosphate dehydrogenase enzymatic activity and thus selectively inhibits lymphocyte proliferation. This is the only immunosuppressant which is monitored in serum or plasma.
- In patients with uremia and hypoalbuminemia, monitoring free mycophenolic acid may be clinically useful.
- Although immunoassays are available for monitoring immunosuppressants, metabolite interferences in the immunoassay are a

significant problem for all immunosuppressant drugs using immunoassays. Chromatographic methods, especially liquid chromatography combined with tandem mass spectrometry (LC-MS/MS), are a gold standard for therapeutic drug monitoring of immunosuppressants.

- Although cyclosporine, sirolimus, everolimus, and mycophenolic acid can be determined by high-performance liquid chromatography combined with ultraviolet detector (HPLC-UV), tacrolimus cannot be monitored by HPLC-UV due to lack of absorption peak in the ultraviolet region.

## Therapeutic drug monitoring of selected antibiotics

Most commonly monitored antibiotics in clinical laboratories are aminoglycosides and vancomycin. The aminoglycoside antibiotics consist of two or more aminosugars joined by a glycosidic linkage to a hexose or aminocyclitol. These drugs are used in the treatment of serious and often life-threatening systemic infections. However, aminoglycosides can produce serious nephrotoxicity and ototoxicity. Aminoglycosides are poorly absorbed from the gastrointestinal tract, and these drugs are administered intravenously or intramuscularly. Children have a higher clearance of aminoglycosides. Patients with cystic fibrosis usually exhibit an altered pharmacokinetics of the antibiotics. After a conventional dose of an aminoglycoside, a patient with cystic fibrosis shows a lower serum concentration compared to a patient not suffering from cystic fibrosis. Although there are several aminoglycosides used in the United States, most commonly monitored aminoglycosides are tobramycin, amikacin, and gentamicin. Aminoglycosides are administered either in traditional dosing (—two to three times a day) or once daily dosing. Other less common types of dosing such as synergy dosing are also practiced. Important points in therapeutic drug monitoring of aminoglycosides are:

- If aminoglycoside therapy is needed for 3 days or less, therapeutic drug monitoring may not be needed.
- During traditional dosing, a peak concentration blood level should be drawn 30–60 min after each dose, and a trough concentration specimen must be drawn 30 min prior to the next dose.
- In once-daily dosing, a larger dose of aminoglycoside is administered compared to traditional dosing. There is no firm established guideline for therapeutic drug monitoring after once-daily dosing. Peak and/or trough concentrations can be monitored. Alternatively the specimen can be drawn 6–14 h after the first dose to calculate dosing intervals using various nomograms.
- During aminoglycoside therapy, serum creatinine must be monitored at least twice a week to ensure there is no significant renal insufficiency.

Therapeutic drug monitoring is also frequently employed during vancomycin therapy (vancomycin is not an aminoglycoside). The drug is excreted in the urine with no metabolism. Vancomycin therapy warrants therapeutic drug monitoring if the patient receives vancomycin for 5 or more days, receives higher dosage of vancomycin, or receives both vancomycin and aminoglycosides. Both peak and trough concentrations should be monitored. Ranges for peak concentrations of 20–40 µg/mL have been widely quoted. A trough concentration of 5–15 µg/mL is recommended because nephrotoxicity and other complications are observed at vancomycin concentrations higher than this level.

Rapid infusion of vancomycin may be associated with pruritus, a rash involving the upper torso, head, and neck, and occasionally hypotension. Known as "red man" or "red neck" syndrome, this phenomenon is caused by nonimmunologically mediated release of histamine and can be avoided by slower administration of vancomycin over at least 60 min. Heparin and vancomycin are incompatible if mixed in intravenous solution or infused one after another through a common intravenous line. Aminophylline, amobarbital, aztreonam, chloramphenicol, dexamethasone, and sodium bicarbonate are also incompatible with vancomycin if mixed in the same container. However, there are other antibiotics which are monitored infrequently. Examples of less frequently monitored antibiotics include ciprofloxacin, chloramphenicol, isoniazid, rifampin, and rifabutin.

## Therapeutic drug monitoring of antineoplastic drugs

Methotrexate is a competitive inhibitor of dihydrofolate reductase, a key enzyme for biosynthesis of nucleic acid. The cytotoxic activity of this drug was discovered in 1955. The use of leucovorin to rescue normal host cells has permitted the higher doses of methotrexate therapy in clinical practice. Methotrexate is used in the treatment of acute lymphoblastic leukemia (ALL), brain tumors, carcinomas of the lung, and other cancers. Most of the toxicities of this drug are related to serum concentrations and pharmacokinetic parameters. Methotrexate is also approved for the treatment of refractory rheumatoid arthritis. Usually low doses of methotrexate are used for treating rheumatoid arthritis (5–25 mg once weekly). Although therapeutic drug monitoring of methotrexate is not indicated in patients receiving low-dose methotrexate for rheumatoid arthritis, therapeutic drug monitoring is essential during high-dose treatment of methotrexate because of frequent adverse reactions leading to leucopenia and thrombocytopenia with high-dose methotrexate therapy.

Pharmacokinetic studies showed that clinical response and toxicity of 5-fluorouracil are related to AUC (area under the curve). Monitoring 5-flurouracil may have clinical benefits. In addition, therapeutic drug monitoring of busulfan may also be valuable. Monitoring busulfan, and tamoxifen especially its active metabolite (4-hydroxy tamoxifen) has clinical utilities.

## Therapeutic drug monitoring of antiretrovirals

Human immunodeficiency virus (HIV) is the virus that causes AIDS (acquired immunodeficiency syndrome). Six classes of drugs are used today to treat people with AIDS, including nucleoside reverse transcriptase inhibitors (NRTIs), such as zidovudine, nonnucleoside reverse transcriptase inhibitors (NNRTIs), which include nevirapine, delavirdine, and efavirenz; and protease inhibitors (PIs), saquinavir, ritonavir, indinavir, nelfinavir, amprenavir, lopinavir, and atazanavir, entry inhibitors such as maraviroc, fusion inhibitors such as enfuvirtide, and integrase inhibitors such as raltegravir. Although some antiretroviral agents do not require therapeutic drug monitoring, patients receiving protease inhibitors may benefit from therapeutic drug monitoring. Currently there is no commercially available immunoassay for any antiretroviral agent. Therefore, therapeutic drug monitoring for these drugs are only available in major academic medical centers and reference laboratories, and most hospital laboratories do not provide service for therapeutic drug monitoring of antiretroviral drugs. Usually LC-MS/MS based methods are used for monitoring.

## Therapeutic drug monitoring using oral fluids

Although majority of therapeutic drug monitoring is conducted using serum, plasma, or whole blood, more recently oral fluid (saliva) is gaining acceptance as an alternative noninvasive matrix for conducting therapeutic drug monitoring. Oral fluid is an ultrafiltrate of blood. Therefore, drug concentrations in oral fluid reflect free (unbound) drug levels in serum for certain drugs. The pH of oral fluid varies from 6.2 to 7.6 (average: 6.7). However, not all drugs can be monitored using oral fluid. Unionized lipophilic drugs are excreted passively in saliva while basic and neutral drugs are found in a higher concentration in saliva than acidic drugs. The parent drug is present in a higher concentration in saliva than a metabolite. Moreover, when a drug's effective permeability is high in intestinal mucosa, it will undergo salivary excretion in the salivary mucosa.

Typically, oral fluid is collected for a period of 1–2 min either directly or following stimulation into the collection tube. Stimulation can be achieved by placing cotton swab impregnated with citric acid for example in Salivette. Oral fluid collection is noninvasive and advantageous in neonates, infants [16], or those

Table 4 Common drugs that can or cannot be monitored in oral fluid.

| Drugs suitable for monitoring using oral fluid | Drugs not suitable for monitoring in oral fluid |
|---|---|
| Phenytoin | Valproic acid |
| Carbamazepine | Quinidine |
| Ethosuximide | Tocainide |
| Primidone | Tricyclic antidepressants |
| Phenobarbital (a mathematical equation is needed | Lithium |
| to convert oral fluid value to serum value) | Methotrexate |
| Lamotrigine | |
| Levetiracetam | |
| Caffeine | |
| Theophylline | |
| Digoxin | |
| Procainamide and NAPA | |
| Disopyramide | |
| Fluconazole | |
| Mycophenolic acid | |

with skin or vascular conditions that impede blood collection by finger stick or venipuncture. However, one disadvantage is limited specimen volume (usually 1 mL or less) of oral fluid compared to blood collection. Common drugs monitored in oral fluid are classical anticonvulsants (phenytoin, carbamazepine, ethosuximide, and possibly phenobarbital. However, valproic acid cannot be monitored using oral fluid. Commonly monitored drugs which are suitable for therapeutic drug monitoring and drugs not suitable for therapeutic drug monitoring are listed in Table 4.

## Therapeutic drug monitoring using dried blood spot

Dried blood spot (DBS) specimens have been used for decades for newborn metabolic testing using heel sticks. However, for children and adults, capillary blood is typically obtained by a finger prick using an automatic lancet. The first blood drop is typically discarded to minimize contamination with tissue fluids. The next blood drops are used to fill premarked circles on filter paper, with one drop in each designated circle. The blood spots automatically dry on the filter paper at ambient temperature and then the dried blood spot is packaged for transport to the clinical laboratory. Upon specimen receipt, the laboratory personnel evaluate the homogeneity of the blood spots and punches out disks of a defined diameter. The analyte(s) of interest are extracted from the disk and then measured. Perhaps the major advantages of DBS sampling relate to specimen storage and transport. Many drugs and drug metabolites are as stable or more stable in DBS compared to specimens collected by venipuncture and then frozen after processing. DBS specimens can be transported with very low infectious risk by routine postal systems without specialized biohazard shipping containers. These characteristics make DBS sampling attractive for specimens collected in remote and/or resource-limited settings.

There are some disadvantages to DBS sampling for TDM. The small amount of sample in DBS necessitates the use of sensitive analytical methods (especially liquid chromatography combined with mass spectrometry or tandem mass spectrometric techniques) and generally precludes use of automated immunoassays commonly used for serum and plasma. In addition, there is limited ability to perform repeated analysis or more than one type of analysis on a blood spot. There are also preanalytical issues with DBS sampling. Improper application of blood to the filter paper may lead to spot inhomogeneity. Analytically, use of DBS sampling for TDM requires extensive validation that should include evaluation of factors such as hemolysis, sample homogeneity, extraction recovery, analyte stability, and impact of varying hematocrits. Hematocrit variation represents perhaps the biggest challenge to drug analysis using DBS [17, 18]. Reference ranges for specimens obtained by DBS sampling need to be determined by detailed validation and correlated with those for plasma, serum, or whole blood. Table 5 summarizes advantages and disadvantages of serum/plasma/whole blood, oral fluid, and dried blood spot as specimens for therapeutic drug monitoring.

**Table 5** Advantages and disadvantages of various specimens used for therapeutic drug monitoring.

| | Serum/plasma or whole blood | Oral fluid | Dried blood spot |
|---|---|---|---|
| Advantage | • Well-studied matrix<br>• High drug level<br>• Can be collected via a catheter<br>• Multiple tubes collected<br>• Results unaffected by hematocrit<br>• Both total and free drug can be measured<br>• Immunoassays are available for most drugs<br>• Established therapeutic range | • Noninvasive<br>• Represent active (free) drug level<br>• Preferred method for infants/children<br>• Does not require trained personnel: lower cost<br>• High compliance | • Less invasive than venipuncture<br>• Specimen can be collected at home and transported to the lab by regular mail<br>• Low overall cost<br>• Most drugs stable at room temperature |

Table 5 Advantages and disadvantages of various specimens used for therapeutic drug monitoring—cont'd

| | Serum/plasma or whole blood | Oral fluid | Dried blood spot |
|---|---|---|---|
| Disadvantage | • Invasive and painful<br>• Risk of hematoma or other complications<br>• Requires trained personnel; higher cost of collection<br>• Poor compliance | • Does not correlate with blood levels for all drugs<br>• Low sample volume (1 mL or less)<br>• Immunoassays are not available. Therefore, liquid chromatography combined with tandem mass spectrometric method is needed<br>• Therapeutic ranges not well established | • Invasive<br>• Very small specimen volume<br>• Unstable drug may degrade<br>• Hematocrit can affect result<br>• Immunoassays are not available. Therefore, liquid chromatography combined with tandem mass spectrometric method is needed<br>• Therapeutic ranges not well established |

## Summary/key points

- Therapeutic drug monitoring is only required for drugs with a narrow therapeutic window where there is a better correlation between serum or plasma (or whole blood) drug level and therapeutic efficacy as well as toxicity.
- Therapeutic drug monitoring should be ordered after a drug reaches its steady state. It usually takes at least five half-lives after initiation of a drug therapy to reach the steady state.
- Usually trough specimen (15–30 min before next dose) is used for therapeutic drug monitoring except for vancomycin and aminoglycosides where both peak and trough levels are monitored.
- Serum or plasma is the preferred specimen for therapeutic drug monitoring except certain immunosuppressants (cyclosporine, tacrolimus, sirolimus, and everolimus are monitored only using whole blood). However, another immunosuppressant drug, mycophenolic acid, is usually monitored using serum or plasma specimen.
- Usually in therapeutic drug monitoring, total drug concentration (free drug + drug bound to protein) is measured. Only free drug is pharmacologically active. If a drug is <80% bound to serum protein, direct monitoring of unbound drug (free drug) is not necessary. However,

for a strongly protein-bound drug, free drug monitoring may be useful for certain patient population where protein binding may be impaired.

- Free phenytoin is the most commonly monitored free drug. It is important to do free phenytoin monitoring in patients with uremia, liver disease, and any condition that may cause hypoalbuminemia such as elderly, critically ill patients, pregnant woman, etc.
- The major CYP isoforms responsible of metabolism of drugs include CYP1A2, CYP2B6, CYP2C9, CYP2C19, CYP2D6, CYP2E1, and CYP3A4/CYP3A5. However, CYP3A4 is the predominant isoform of the CYP family (almost 30%) usually responsible for metabolism of approximately 30–50% of all marketed drugs.
- Neonates and infants (0–4 months) may metabolize drug slower than adults because of the lack of mature (fully functioning) drug-metabolizing enzymes but activities of drug-metabolizing enzymes are higher in children than adults. Therefore, children may need a higher per kilogram dosage of a drug than an adult. However, drug metabolism rate is reduced in elderly and appropriate dosage adjustment is needed. In addition, elderly patients (>70 years of age) may also have lower albumin and protein binding of a strongly protein-bound drug such as phenytoin may be impaired causing elevated concentration of pharmacologically active free drug level.
- Uremia may impair clearance of a drug/metabolite by the kidney while liver disease may impair metabolism of certain drugs. Dosage adjustment may be needed for these patients.
- Drug metabolism is increased in patients with hyperthyroidism due to excessive levels of thyroxine but decreased in patients with hypothyroidism due to lower levels of thyroxine. Amiodarone has high iodine content and may cause thyroid dysfunction. Screening of thyroid disease before amiodarone therapy and periodic monitoring of thyroid function are recommended for patients treated with amiodarone.
- The specimen collected in serum separator gel tubes for therapeutic drug monitoring may falsely reduce concentrations of some drugs such as phenytoin if plasma is in contact with the gel for several hours without analysis. However, if drug levels are measured within 2 h, drug levels are not affected. The new BD Vacutainer Barricor uses a high-density plastic connected on elastomer top that stretches during centrifugation, finally creating a seal. Blood cells are allowed to flow around this separator, reducing the time of centrifugation and providing high-quality plasma. This tube is superior to a gel separator tube for specimen collection intended for therapeutic drug monitoring.
- Grapefruit juice inhibits intestinal cytochrome P-450 and P-glycoprotein, thus increasing bioavailability of many drugs. However, if a drug is given intramuscularly or intravenously, no such interaction is observed because

grapefruit juice does not inhibit liver cytochrome P-450 enzymes. Common drugs that interact with grapefruit juice if taken orally include calcium channel blockers, alprazolam, carbamazepine, cyclosporine, erythromycin, methadone, quinidine, simvastatin, and tacrolimus.

- Increased theophylline metabolism in smokers due to induction of CYP1A2 is well documented.
- Carbamazepine is metabolized to carbamazepine 10, 11-epoxide, which is an active metabolite. Although in the normal population, epoxide concentrations may be 10%–14% of the total carbamazepine concentration, patients with renal failure may show an over 40% epoxide concentration relative to the carbamazepine concentrations. Monitoring active metabolite concentration using chromatographic method may be useful in these patients as there is no immunoassay available for monitoring epoxide levels.
- Procainamide is metabolized to *N*-acetylprocainamide (NAPA) and monitoring both procainamide and NAPA is essential. NAPA accumulates in patients with renal failure causing toxicity where the procainamide level may be within the therapeutic range. Immunoassays are available for therapeutic drug monitoring of both procainamide and NAPA.
- Lidocaine cannot be given orally due to high first-pass metabolism. However, tocainide, an analog of lidocaine, can be administered orally.
- Immunoassays for tricyclic antidepressant (TCA) should be only used in a case of suspected overdose but such assays should not be used for therapeutic drug monitoring because almost all tricyclic antidepressants show significant cross reactivity. In a suspected overdose, total TCA concentration (drug + metabolite) matters but in drug monitoring, individual concentration is important.
- Although immunoassays are available for monitoring immunosuppressants, metabolite interferences in the immunoassay are a significant problem, especially with monitoring of sirolimus and mycophenolic acid. Chromatographic methods, especially liquid chromatography combined with tandem mass spectrometry, are a gold standard for therapeutic drug monitoring of immunosuppressants.
- If aminoglycoside therapy is needed for 3 days or less, therapeutic drug monitoring may not be needed but needed if therapy is continued for over 3 days. During traditional dosing, a peak concentration blood level should be drawn 30–60 min after each dose, and a trough concentration specimen must be drawn 30 min prior to the next dose.
- Therapeutic drug monitoring of antineoplastic drug methotrexate is useful. However, when methotrexate is used as a disease-modifying agent in low dose to treat a rheumatoid arthritis patient, therapeutic drug monitoring is not needed.

- Therapeutic drug monitoring of other antineoplastic drugs such as busulfan, 5-fluoro uracil, tamoxifen, and its active metabolite 4-hydroxy tamoxifen has clinical utility.
- Oral fluid (saliva) is an alternative specimen for therapeutic drug monitoring especially for the pediatric population. Oral fluid drug levels reflect free (unbound) drug levels in serum. Although phenytoin, carbamazepine, ethosuximide, primidone, lamotrigine, levetiracetam, caffeine, theophylline, procainamide, NAPA, disopyramide, mycophenolic acid, and fluconazole can be monitored using oral fluid, other drugs such as valproic acid, quinidine, methotrexate, tricyclic antidepressants, lithium, and tocainide cannot be monitored using oral fluid due to poor correction between oral fluid drug levels and serum/plasma drug levels.
- Therapeutic drug monitoring can be also conducted using a dried blood spot which can be easily collected at home and shipped to the testing center. However, due to limited specimen volume, liquid chromatography combined with mass spectrometry is needed for determining drug levels using a dried blood spot. Moreover, hematocrit may affect results for certain drugs.

## References

[1] Mattson RH, Cramer JA, Collins JF. Aspects of compliance: taking drugs and keeping clinic appointments. Epilepsy Res Suppl 1988;1:111–7.

[2] Chandra RS, Dalvi SS, Karnad PD, Kshirsagar NA, et al. Compliance monitoring in epileptic patients. J Assoc Physicians India 1993;41:431–2.

[3] Babalik A, Babalik A, Mannix S, Francis D, et al. Therapeutic drug monitoring in the treatment of active tuberculosis. Can Respir J 2011;18:225–9.

[4] Jeong H. Altered drug metabolism during pregnancy: hormonal regulation of drug metabolizing enzymes. Expert Opin Drug Metab Toxicol 2010;6:689–99.

[5] Anderson GD. Pregnancy induced changes in pharmacokinetics: a mechanistic based approach. Clin Pharmacokinet 2005;44:989–1008.

[6] Terrell KM, Heard K, Miller DK. Prescribing to older ED patients. Am J Emerg Med 2006;24:468–78.

[7] Lainscak M, Vitale C, Seferovic P, Spoletini I, et al. Pharmacokinetics and pharmacodynamics of cardiovascular drugs in chronic heart failure. Int J Cardiol 2016;224:191–8.

[8] Bailey DG, Spence JD, Munoz C, Arnold JM. Interaction of citrus juices with felodipine and nifedipine. Lancet 1991;337:268–9.

[9] Saito M, Hirata-Koizumi M, Matsumoto M, Urano T, Hasegawa R. Undesirable effects of citrus juice on the pharmacokinetics of drugs: focus on recent studies. Drug Saf 2005;28:677–94.

[10] Koski A, Vuori E, Ojanpera I. Relation of postmortem blood alcohol and drug concentrations in fetal poisonings involving amitriptyline, propoxyphene and promazine. Hum Exp Toxicol 2005;24:389–96.

[11] Zevin S, Benowitz NL. Drug interactions with tobacco smoking: an update. Clin Pharmacokinet 1999;36:425–38.

[12] Kroon LA. Drug interactions and smoking: raising awareness for acute and critical care provider. Crit Care Nurs Clin N Am 2006;18:53–62.

[13] Colucci VJ, Knapp JE. Increase in international normalization ratio associated with smoking cessation. Ann Pharmacother 2001;35:385–6.

[14] Bush V, Blennerhasset J, Wells A, Dasgupta A. Stability of therapeutic drugs in serum collected in vacutainer serum separator tubes containing a new gel (SST II). Ther Drug Monit 2001;23:259–62. Erratum in: Ther Drug Monit 2001;23:738.

[15] Schrapp A, Mory C, Duflot T, Pereira T, et al. The right blood collection tube for therapeutic drug monitoring and toxicology screening procedures: standard tubes, gel or mechanical separator? Clin Chim Acta 2019;488:196–201.

[16] Hutchinson L, Sinclair M, Reid B, Burnett K, Callan B. A descriptive systematic review of salivary therapeutic drug monitoring in neonates and infants. Br J Clin Pharmacol 2018;84:1089–108.

[17] Antunes MV, Charao MF, Linden R. Dried blood spots analysis with mass spectrometry: potentials and pitfalls in therapeutic drug monitoring. Clin Biochem 2016;49:1035–46.

[18] Sharma A, Jaiswal S, Shukla M, Lal J. Dried blood spots: concepts, present status, and future perspectives in bioanalysis. Drug Test Anal 2014;6:399–414.

# CHAPTER 15

# Interferences in therapeutic drug monitoring

## Methodologies used in therapeutic drug monitoring and issues of interferences

Various methods are used for therapeutic drug monitoring in clinical laboratories but immunoassays are most commonly used because such assays require no or minimal (extraction of drugs from whole blood) specimen preparation prior to analysis using automated chemistry analyzers. Moreover, specimens can be analyzed in batch and results can be obtained within 20–30 min. However, immunoassays are not available for less commonly monitored drugs. Therefore, chromatography-based methods are needed for therapeutic drug monitoring of such drugs. These methods include:

- Gas chromatography (GC) with flame ionization or nitrogen detector: Can be used for selective drugs which are relatively volatile such as pentobarbital.
- Gas chromatography combined with mass spectrometry (GC/MS): Most commonly used for analysis of abused drugs but also used for analysis of volatile drugs. This method is very specific and relatively free from interferences.
- High-performance liquid chromatography (HPLC) combined with ultraviolet or fluorescence detector: Widely used for monitoring of drugs where immunoassays are not commercially available. Both polar (less or nonvolatile) and nonpolar (relatively volatile) drugs can be analyzed by this method.
- High-performance liquid chromatography combined with mass spectrometry (LC-MS) or tandem mass spectrometry (LC-MS/MS): Very specific method and considered as the gold standard for therapeutic drug monitoring because this method is virtually free from interferences.

Although immunoassays are widely used in clinical laboratories for therapeutic drug monitoring, interference in immunoassays is a major limitation of such method. The antibody used in an immunoassay may cross-react with another

**325**

Clinical Chemistry, Immunology and Laboratory Quality Control. https://doi.org/10.1016/B978-0-12-815960-6.00011-X

molecule with a similar structure to the analyte molecule, most commonly the drug metabolite in analysis of the parent drug. In addition, other structurally related drugs and even endogenous compounds such as high bilirubin, hemolysis, and elevated lipids may also interfere with immunoassays. Although chromatographic techniques are more labor-intensive, requiring highly experienced medical technologists, and instruments are expensive, chromatography-based methods are relatively free from interferences. Therefore, LC-MS or LC-MS/MS methods are used mostly for therapeutic monitoring of drugs where immunoassays are not commercially available, such as in therapeutic drug monitoring of antiretroviral agents. In addition, chromatographic methods, especially LC-MS/MS, are also used where immunoassays may suffer from significant interferences due to cross-reactivities with metabolites, for example, various immunosuppressants.

## Effect of endogenous factors on therapeutic drug monitoring

Endogenous factors such as bilirubin, hemolysis, and high lipid, if present in a specimen, may interfere with therapeutic drug monitoring of various drugs. Bilirubin is derived from the hemoglobin of aged or damaged red blood cells. Some part of serum bilirubin is conjugated as glucuronides ("direct" bilirubin); the unconjugated bilirubin is also referred as indirect bilirubin. In normal adults, bilirubin concentrations in serum are from 0.3 to1.2 mg/dL (total) and <0.2 mg/dL (conjugated). Usually total bilirubin concentration up to 20 mg/dL is not a problem for most assays used for therapeutic drug monitoring, but a bilirubin level over 20 mg/dL, which is uncommon, may cause some interference in certain immunoassays. Currently, the interference of elevated bilirubin in the colorimetric assay for acetaminophen and salicylate (Trinder salicylate assay) are of major concern. However, immunoassays for acetaminophen and salicylate are free from such interferences [1, 2].

Hemolysis can occur in vivo, during venipuncture and blood collection, or during processing of the specimen. Hemoglobin interference depends on its concentration in the specimen. Serum appears hemolyzed when the hemoglobin concentration exceeds 20 mg/dL. Hemoglobin interference is caused not only by the spectrophotometric properties of hemoglobin, but also by its participation in chemical reaction with sample or reactant components as well. The absorbance maxima of the heme moiety in hemoglobin are at 540–580 nm. However, hemoglobin begins to absorb around 340 nm with absorbance increasing at 400–430 nm as well. Methods which use the absorbance properties of NADH or NADH (340 nm) may thus be affected by hemolysis. Lipids in serum or plasma exist as complexed with proteins called lipoproteins.

Lipoproteins, consisting of various proportions of lipids, range from 10 to 1000 nm in size (the higher the percentage of the lipid, the lower the density and larger the particle size of the resulting lipoprotein). The lipoprotein particles with high lipid contents (such as chylomicrons and VLDL) are micellar and are the main source of assay interference, especially in turbidimetric assays.

## Digoxin immunoassays: So much interference

Digoxin immunoassays are subjected to most interference compared to other immunoassays for therapeutic drugs [3]. Sources of interferences in digoxin immunoassays include:

- Endogenous factors such as digoxin like immunoreactive substances (DLIS).
- Endogenous heterophilic antibody (very rare interference).
- Digibind and DigiFab interferences.
- Digoxin metabolites.
- Interferences from spironolactone, potassium canrenoate, and their common metabolite canrenone.
- Chinese medicine such as Chan Su, Lu-Shen-wan, oleander containing herbs and convallotoxin, the active ingredient of lily of the valley extract interfere with digoxin immunoassays.

Interestingly, some of these interferences can be eliminated by monitoring free digoxin concentrations in the protein free ultrafilter (Table 1).

The presence of endogenous digoxin-like immunoreactive substances (DLIS) was first described in a volume-expanded dog in 1980 [4]. After publication of that initial report, many investigators confirmed the presence of endogenous DLIS in serum and other biological fluids in volume-expanded patients, not limited to patients with uremia, liver disease, essential hypertension, transplant recipients, eclampsia, pregnant women, and preterm babies. Usually high amounts of DLIS are encountered in premature babies. DLIS like digoxin can inhibit Na, K-ATPase. Although DLIS interference with digoxin immunoassays was a significant problem in the past (due to use of polyclonal antibody in assay design), more recently introduced, digoxin immunoassays utilizing more specific monoclonal antibodies against digoxin are relatively free from DLIS interference. Moreover, taking advantage of strong protein binding of DLIS and only 25% protein binding of digoxin, interference of DLIS in digoxin immunoassay can be eliminated by monitoring free digoxin concentration in protein-free ultrafiltrate [5].

The presence of human antianimal antibody, especially those directed against mouse, in serum may cause interference with certain immunoassays.

**Table 1** Various interferences in digoxin immunoassays and possible elimination of interference by monitoring free digoxin.

| Interfering substances | Magnitude of interference in older assays | Magnitude of interferences in newer monoclonal antibody-based assays | Interference eliminated by free digoxin monitoring? |
|---|---|---|---|
| DLIS | + Modest interference | No interference | Yes |
| Heterophilic antibody | +++ (very rare interference) | No report | Yes |
| Digoxin metabolites | + Modest interference | No interference | No |
| Spironolactone | + Modest interference | No interference | Yes |
| Potassium canrenoate | ++ Significant interference | No interference | Yes |
| Digibind/Digifab | Variable and assay dependent | Variable and assay dependent | Yes, free digoxin must be monitored in patients overdosed with digoxin and being treated with Digibind/Digifab |
| Chinese medicines: Chan Su, Lu-Shen-Wan (Bufalin, active ingredient) | +++ Very significant interference | ++ Significant interference | Yes |
| Herbal medicines containing oleander (oleandrin) or lily of the valley extract (convallotoxin) | +++ Very significant interference | ++ Significant interference | Yes |

The clinical use of mouse monoclonal antibody for radio imaging and treatment for certain cancers may cause accumulation of human antimouse antibody (HAMA). Antianimal antibodies are also found among veterinarians, farm workers, or pet owners due to exposure to animals, and these antibodies are broadly classified as heterophilic antibodies. Usually the presence of heterophilic antibodies in serum may interfere with sandwich assays designed for measuring relatively large molecules such as beta-hCG (human chorionic gonadotropin). Nevertheless, Liendo et al. described a case report of a patient with cirrhotic liver disease and atrial fibrillation, treated with spironolactone and digoxin, and showed an elevated digoxin concentration of 4.2 ng/mL. Despite a toxic digoxin level, the patient was asymptomatic and after discontinuation of both drugs, digoxin values over 3.0 ng/mL were observed for approximately 5 weeks in the patient's serum. Because such interference was eliminated by measuring digoxin in protein-free ultrafiltrate (heterophilic antibodies due to large molecular weights are absent in protein-free ultrafiltrate), the authors concluded that the falsely observed digoxin level was due to interference from heterophilic antibodies [6].

The major metabolites of digoxin are digoxigenin, digoxigenin monodigitoxoside, digoxigenin bisdigitoxoside, and dihydrodigoxin. These metabolites exhibit significantly different cross-reactivities against various antidigoxin antibodies. However, due to relatively low levels of digoxin metabolites in serum in comparison to digoxin, effects of metabolite cross-reactivities are minimal on serum digoxin measurements by immunoassays in patients with normal renal function. In contrast, for patients with renal disease, liver disease, or diabetes, the digoxin immunoassays may significantly overestimate digoxin values compared to chromatographic methods [7].

Digibind and DigiFab are Fab fragments of antidigoxin antibody used in treating life-threatening acute digoxin overdose. Digibind was marketed in 1986 while DigiFab was approved for use in 2001. Both products are Fab fragment of antidigoxin antibody. The molecular weight of DigiFab (46,000 Da) is similar to the molecular weight of Digibind (46,200 Da), and both compounds can be excreted in urine. Both Digibind and DigiFab interfere with serum digoxin measurement using immunoassays where values may be falsely elevated or reduced. Therefore, in patients overdosed with digoxin and being treated with Digibind or DigiFab, only the unbound digoxin (free digoxin) measurement is clinically useful because unbound (free) digoxin is responsible for digoxin toxicity. In addition, neither Digibind nor DigiFab is present in the protein-free ultrafiltrate (molecular weight of filter in ultrafiltration devise is usually 30,000 Da) due to their high molecular weight. Therefore, free digoxin measurements by immunoassays are not affected by Digibind or DigiFab.

Although various diuretics interact with digoxin and may increase serum digoxin level, potassium-sparing diuretics such as spironolactone, potassium canrenoate, and eplerenone, not only pharmacokinetically interact with digoxin but also may interfere with serum digoxin measurements using digoxin immunoassays. After oral administration, spironolactone is rapidly and extensively metabolized to several metabolites, including canrenone which is an active metabolite. Potassium canrenoate is also metabolized to canrenone. This drug is not approved for clinical use in the United Sates, but still used clinically in Europe and other countries throughout the world. Because of structural similarities between spironolactone and related compounds with digoxin, these substances interfere with serum digoxin assays, especially assays utilizing polyclonal antibody against digoxin. Although spironolactone and related compounds falsely elevate serum digoxin level, negative interference was observed with the microparticle enzyme immunoassays (MEIA) digoxin assay [8]. Monitoring free digoxin may eliminate some interference. However, relatively new digoxin assays utilizing specific monoclonal antibodies are free from such interferences.

Chinese medicines Chan Su and Lu-Shen-wan contain bufalin which has structural similarity with digoxin. Therefore, if a patient takes such Chinese medicines, a false digoxin level may be observed in serum using immunoassay, although the patient is not taking digoxin. Oleander containing herbal supplements also interferes with serum digoxin measurement because oleandrin, the active component of oleander, has structural similarity with digoxin.

Convallotoxin is the active ingredient of lily of the valley plant. Despite toxicity, extract of lily of the valley plant is used in alternative medicines. Convallotoxin, structurally similar to digoxin, is known to interfere with serum digoxin measurements using immunoassays [9].

Bufalin, the active ingredient of Chan Su and Lu-Shen-wan, oleandrin, the active ingredient of oleander extract and convallotoxin, the active ingredient of lily of the valley extract, are all strongly bound to serum protein. As a result, these compounds are virtually absent in protein-free ultrafiltrate. Therefore, monitoring free digoxin mostly eliminates interferences of these compounds in serum digoxin measurement. Digoxin immunoassays can be used for free digoxin measurement because free digoxin represnts approximately 75% of total digoxin concentration.

## Interferences in analysis of antiepileptics

Monitoring classical anticonvulsants, such as phenytoin, carbamazepine, phenobarbital, and valproic acid, is essential for proper patient management, and immunoassays are commercially available for determination of serum or plasma concentrations of these drugs. Usually phenobarbital and valproic acid immunoassays are robust, and interference is observed only rarely. Overdose with amobarbital or secobarbital may cause a falsely elevated phenobarbital level, but it is observed rarely. However, carbamazepine and phenytoin immunoassays are subjected to interferences (Table 2). Cross-reactivity of carbamazepine 10, 11-epoxide with carbamazepine immunoassays may vary from 0% to

**Table 2** Common interferences in immunoassays for measuring various therapeutic drugs.

| Drug | Interfering compound | Comments |
|---|---|---|
| Carbamazepine | Carbamazepine 10, 11-epoxide (active metabolite) | Cross-reactivity varies: PETINIA: 96% EMIT: 0.4% CEDIA: 10.5% Beckmann SYNCHRON: 7.6% |
| Carbamazepine | Hydroxyzine, cetirizine | Falsely elevated values with PETINIA (Siemen's assay) assay only in severely overdosed patients |
| Phenytoin | Fosphenytoin | Fosphenytoin metabolite in uremic patients may falsely elevate phenytoin level |
| Phenytoin | Paraprotein (IgM) | Falsely lower value with PETINIA |
| Gentamicin, vancomycin | IgM | Negative interference or unable to measure due to precipitation of paraprotein during analysis |
| Cyclosporine, tacrolimus, sirolimus and everolimus | Metabolites | Significant positive bias in immunoassay due to metabolite cross-reactivities. LC-MS/MS is the gold standard |

96%. Therefore, true carbamazepine concentration may be overestimated if a carbamazepine immunoassay with high cross-reactivity toward its active epoxide metabolite is used for therapeutic drug monitoring of carbamazepine, for example, particle enhanced turbidimetric inhibition immunoassay (PETINIA assay: Siemens Diagnostics; 96% cross-reactivity with epoxide). Hydroxyzine and cetirizine are antihistamine drugs which interfere only with the PETINIA carbamazepine immunoassay.

Fosphenytoin, a pro-drug of phenytoin, is rapidly converted into phenytoin after administration. Fosphenytoin, unlike phenytoin, is readily water soluble and can be administered through intravenous (IV) or intramuscular (IM) routes. Unlike phenytoin, fosphenytoin does not crystallize at the injection site and no discomfort is experienced by the patient. Fosphenytoin is not typically monitored clinically because of its short half-life and lack of pharmacological activity. However, phenytoin is monitored in a patient after administration of fosphenytoin, but in this case, monitoring of phenytoin must be initiated after complete conversion of fosphenytoin into phenytoin. In uremic patients after fosphenytoin administration, phenytoin concentrations measured by various immunoassays may be significantly higher than true phenytoin level as determined by high-performance liquid chromatography. Annesley et al. identified a unique oxymethylglucuronide metabolite derived from fosphenytoin in sera of uremic patients and demonstrated that this unusual metabolite was responsible for the cross-reactivity [10].

## Interferences in analysis of tricyclic antidepressants

The major interference in immunoassays for determining tricyclic antidepressant concentrations in serum or plasma is the interference from their metabolites. Other drugs that may interfere with immunoassays for tricyclic antidepressants are listed in Table 3. In general, a tertiary amine tricyclic antidepressant (TCA) is metabolized to secondary amine, and this metabolite usually has almost 100% cross-reactivity with an antibody used for immunoassay

Table 3 Common interferences in immunoassays for tricyclic antidepressants (TCA).

| Interfering substance |
| --- |
| Metabolites of TCAs |
| Phenothiazines and metabolites |
| Carbamazepine and its active metabolite carbamazepine 10, 11-epoxide |
| Quetiapine |
| Cyproheptadine if present in toxic concentrations |
| Diphenhydramine if present in toxic concentration |

for TCA. Therefore, for monitoring tertiary amines, immunoassays in general indicate total concentration of the parent drug along with the active metabolite and TCA immunoassays should be used only for diagnosis of overdose.

Phenothiazines and its metabolites may interfere with TCA immunoassays, and even a therapeutic concentration of such drug may cause falsely elevated serum TCA level. Carbamazepine is metabolized to carbamazepine, 10, 11-epoxide, an active metabolite. Both the parent drug and the epoxide metabolite interfere with immunoassays for TCA due to structural similarities. Another structurally related drug oxcarbazepine also interferes with immunoassays for TCA. Quetiapine may also interfere with TCA immunoassays. However, interference of diphenhydramine (Benadryl) and cyproheptadine with TCA immunoassays occurs only in overdosed patients.

## Interferences in analysis of immunosuppressants

Immunosuppressant drugs cyclosporine, tacrolimus, sirolimus, and everolimus must be monitored in whole blood while mycophenolic acid is the only immunosuppressant which is monitored in serum or plasma. Although immunoassays are commercially available for therapeutic drug monitoring of all immunosuppressants, immunoassays suffer from metabolite cross-reactivities. Therefore, positive bias is commonly observed in therapeutic drug monitoring of immunosuppressants using various immunoassays. The positive bias may vary 10%–40% depending on the immunosuppressant and type of immunoassay compared to the corresponding values obtained by the more specific chromatographic method. Important points regarding monitoring of immunosuppressants include the following:

- Chromatographic methods especially liquid chromatography combined with mass spectrometry (LC/MS) or tandem mass spectrometry (LC/MS/MS) is the gold standard for therapeutic drug monitoring of immunosuppressants.
- Most immunoassays for cyclosporine, tacrolimus, sirolimus and everolimus require specimen pretreatment to extract drugs from whole blood except antibody-conjugated magnetic immunoassay (ACMIA, Siemens Diagnostics) for cyclosporine and tacrolimus which utilize on online mixing and ultrasonic lysis of whole blood. However, ACMIA cyclosporine and tacrolimus immunoassays may show falsely elevated levels due to interference of heterophilic antibody.
- Immunoassays for mycophenolic acid do not require serum pretreatment, because the assay can be run using serum or plasma. Usually acyl glucuronide (minor active metabolite) is responsible for interferences in some immunoassays.

- Low hematocrit may cause interference (false positive results) with the microparticle enzyme immunoassay (MEIA) for tacrolimus (Abbott Laboratories), but newer tacrolimus assay from the same manufacturer on the Abbott Architect analyzer is free from such interference.

## Interferences in analysis of antibiotics

Usually immunoassays used for monitoring various aminoglycosides and vancomycin are robust and relatively free from interferences. There are only few case reports of paraprotein interference in vancomycin immunoassay causing falsely lower values.

Gentamicin is not a single molecule, but a complex of three major ($C_1$, $C_{1a}$, and $C_2$) and several minor components. In addition, the $C_2$ component is a mixture of stereoisomers. Most immunoassay methods can measure a total gentamicin concentration in serum or plasma but are incapable of measuring individual components.

## Interferences of paraprotein in therapeutic drug monitoring

Paraproteins are either IgG or IgM or fragments which are observed most commonly as a result of monoclonal gammopathy of unknown significance (MGUS) or with monoclonal gammopathy. Paraproteins are known to cause both negative and positive interference in various laboratory tests such as blood counts, serum sodium, calcium, phosphorous, high-density lipoprotein (HDL) cholesterol as well as few therapeutic drugs using specific immunoassays. There is a poor correlation between the concentration or type of paraproteins and the likelihood of interference. Paraprotein usually interferes with turbidimetric and nephelometric assays due to precipitation under assay condition causing mostly negative interferences. Negative interference of IgM in serum phenytoin measurement by the PETINIA immunoassays has been reported. The presence of high IgM paraprotein, 18.9 g/L (reference range 0.4–2.3 g/L) was the cause of precipitation during assay, giving high blank readings using Beckman assays and hence gentamicin level cannot be measured in a 93-year-old woman receiving gentamicin. However, Roche assay in a specimen collected 50 h after infusion of gentamicin showed a value of 3.3 µg/mL, indicating that IgM interference was assay-specific [11]. Negative interference of IgM paraprotein in PETINIA vancomycin (both Siemens and Beckman) assay has also been reported [12].

## Summary/key points

- Immunoassays are commonly used for therapeutic drug monitoring. Chromatographic methods used for therapeutic drug monitoring include gas chromatography (GC) with flame ionization or nitrogen detector, gas chromatography combined with mass spectrometry (GC/MS: commonly used for confirming abused drugs), high-performance liquid chromatography (HPLC) combined with ultraviolet or fluorescence detector and high-performance liquid chromatography combined with mass spectrometry (LC-MS) or tandem mass spectrometry (LC-MS/MS). LC-MS/MS is a very specific method and considered the gold standard for therapeutic drug monitoring because this method is virtually free from interferences.
- Interferences in digoxin immunoassays may be observed due to the presence of endogenous factors such as digoxin-like immunoreactive substances (DLIS), heterophilic antibody (very rare interference) in patients receiving spironolactone. However, more recently marketed digoxin immunoassays using very specific monoclonal antibodies are free from interference from DLIS, spironolactone, and related compounds.
- For a patient overdosed with digoxin and being treated with Digibind or DigiFab, only free digoxin should be monitored because pharmacologically active free digoxin correlates better with success of therapy compared to total digoxin level. In addition, Digibind or DigiFab is absent in protein-free ultrafiltrate (molecular weight around 46,000 but filter used for making ultrafiltrate has a molecular weight cut-off of 30,000 Da), and monitoring free digoxin in the protein free ultrafiltrate can be conducted using digoxin immunoassays.
- Chinese medicine such as Chan Su, Lu-Shen-wan, and oleander containing herbs interfere with digoxin immunoassays even with more recently introduced monoclonal antibody-based assays because bufalin found in Chan Su and Lu-Shen-wan and oleandrin, the active component of oleander extract has structural similarity with digoxin.
- Interference of convallotoxin, the active ingredient of extract of lily of the valley plant, in serum digoxin assay has been reported.
- Cross-reactivity of carbamazepine 10, 11-epoxide with carbamazepine immunoassays may vary from 0% to 96%. Therefore, true carbamazepine concentration may be overestimated if a carbamazepine immunoassay with high cross-reactivity toward its active epoxide metabolite is used for therapeutic drug monitoring of carbamazepine, for example, PETINIA assay (PETINIA: particle enhanced turbidimetric inhibition immunoassay Siemens Diagnostics; 96% cross-reactivity with epoxide).

Hydroxyzine and cetirizine are antihistamine drugs which interfere only with the PETINIA carbamazepine immunoassay.

- Fosphenytoin is a prodrug of phenytoin. In uremic patients after fosphenytoin administration, phenytoin concentrations measured by various immunoassays may be significantly higher than true phenytoin level as determined by high-performance liquid chromatography. This is due to the presence of a unique oxymethylglucuronide metabolite derived from fosphenytoin in sera of uremic patients that cross-react with phenytoin immunoassays.

- Phenothiazines and its metabolites, carbamazepine, oxcarbazepine and quetiapine, interfere with immunoassays for tricyclic antidepressants. However, interferences of diphenhydramine (Benadryl) and cyproheptadine with TCA immunoassays occur only in overdosed patients.

- Chromatographic methods, especially liquid chromatography combined with mass spectrometry (LC-MS) or tandem mass spectrometry (LC-MS/MS), are the gold standard for therapeutic drug monitoring of immunosuppressants because all immunoassays for various immunosuppressant suffer significantly from metabolite cross-reactivity and show significant positive bias when compared to values obtained by a LC-MS or LC-MS/MS method.

- There are only few case reports of paraprotein interference in vancomycin immunoassay causing falsely lower values. Negative interference of IgM in phenytoin measurement by the PETINIA assay and negative interference in Beckman gentamicin assay has also been reported.

# References

[1] Polson J, Wians FH, Orsulak P, Fuller D, et al. False positive acetaminophen concentrations in patients with liver injury. Clin Chim Acta 2008;391:24–30.

[2] Dasgupta A, Zaldi S, Johnson M, Chow L, et al. Use of fluorescence polarization immunoassay for salicylate to avoid positive/negative interference by bilirubin in the Trinder salicylate assay. Ann Clin Biochem 2003;40:684–8.

[3] Dasgupta A. Therapeutic drug monitoring of digoxin: impact of endogenous and exogenous digoxin-like immunoreactive substances. Toxicol Rev 2006;25:273–81.

[4] Gruber KA, Whitaker JM, Buckalew VM. Endogenous digitalis-like substances in plasma of volume expanded dogs. Nature 1980;287:743–5.

[5] Dasgupta A, Trejo O. Suppression of total digoxin concentration by digoxin-like immunoreactive substances in the MEIA digoxin assay: elimination of interference by monitoring free digoxin concentrations. Am J Clin Pathol 1999;111:406–10.

[6] Liendo C, Ghali JK, Graves SW. A new interference in some digoxin assays: anti-murine heterophilic antibodies. Clin Pharmacol Ther 1996;60:593–8.

[7] Tzou MC, Reuning RH, Sams RA. Quantitation of interference in digoxin immunoassay in renal, hepatic and diabetic disease. Clin Pharm Ther 1997;61:429–41.

[8] Steimer W, Muller C, Eber B. Digoxin assays: frequent, substantial and potentially dangerous interference by spironolactone, canrenone and other steroids. Clin Chem 2002;48:507–16.

[9] Fink SL, Robey TE, Tarabar AF, Hodsdon ME. Rapid detection of convallatoxin using five digoxin immunoassays. Clin Toxicol (Phila) 2014;52:659–63.

[10] Annesley T, Kurzyniec S, Nordblom G, et al. Glucuronidation of prodrug reactive site: isolation and characterization of oxymethylglucuronide metabolite of fosphenytoin. Clin Chem 2001;46:910–8.

[11] Dimeski G, Bassett K, Brown N. Paraprotein interference with turbidimetric gentamicin assay. Biochem Med (Zagreb) 2015;25:117–24.

[12] LeGatt DF, Blakney GB, Higgins TN, Schnabl KL, et al. The effect of paraproteins and rheumatoid factor on four commercial immunoassays for vancomycin: implications for laboratorians and other health care professionals. Ther Drug Monit 2012;34:306–11.

# CHAPTER 16

# Drugs of abuse testing

## Commonly abused drugs in the United States

National Survey on Drug Use and Health conducted by the SAMHSA (substance abuse and mental health services administration) in 2018 reported that an estimated 19.3 million people (7.8% of American population) suffer from substance use disorder out of which an estimated 10.3 million people (3.7% of population) are opioid abusers. Over the past decade, opioid prescription drug abuse has become a national public health crisis especially among adolescents and young adults. On October 26, 2017, Acting Health and Human Services (HHS) secretary Eric D. Hargan declares opioid crisis as a public health emergency. In response to this epidemic, The Drug Enforcement Administration (DEA) mandated reductions in opioid production by 25% in 2017 and 20% in 2018. The number of prescriptions for opioids declined significantly from 252 million in 2013 to 196 million in 2017 [1]. The Center for Disease Control and Prevention (CDC) has estimated that approximately 68% of the more than 70,200 drug overdose deaths in 2017 involved an opioid (including prescription opioids, illegal opioids such as heroin, and illicitly manufactured fentanyl). On average, 130 Americans die every day from an opioid overdose.

Other than opioid abuse, Americans also abuse marijuana (most commonly abused), cocaine (including crack cocaine), amphetamines, heroin, opioids, benzodiazepines, and many designer drugs including synthetic cathinone (bath salts) and synthetic cannabinoids (Spice, K2, Blonde, etc.). Abuse of prescription drugs (including prescription drugs obtained illegally) is also becoming an epidemic in the United States. The most commonly abuse prescription drugs are benzodiazepines, opioids, and some psychoactive drugs. In general drugs that are commonly encountered can be classified under different categories (Table 1). Key points regarding drug abuse include:

- Marijuana is most popular abused drugs in the United States according to the SAMHSA survey.

**337**

**Table 1** Abused drugs in the United States.

| | |
|---|---|
| Stimulants | Amphetamine, methamphetamine, MDMA (3,4-methylenedioxymethamphetamine), MDA (3,4-methylenedioxyamphetamine), synthetic cathinone (bath salts) |
| Narcotic analgesics | Heroin, codeine, morphine, oxycodone, hydrocodone, hydromorphone, oxymorphone, buprenorphine, meperidine, methadone, fentanyl, and its analogs |
| Sedative hypnotics | Benzodiazepines (e.g., diazepam, alprazolam, clonazepam, lorazepam, temazepam, zolpidem, flunitrazepam), others (e.g., gamma-hydroxybutyrate), barbiturates (less commonly abused) |
| Hallucinogens | Lysergic acid diethylamide (LSD), marijuana, synthetic cannabinoids (K2, Blonde, etc.) |
| Anesthetics | Ketamine, phencyclidine |
| Date rape drugs | Gamma-hydroxybutyric acid (GHB), Rohypnol (flunitrazepam) |

- Abuse of barbiturates is declining, but abuse of various benzodiazepines, especially alprazolam, lorazepam, and temazepam is on rise.
- Opioid abuse epidemic is a public health emergency.

Reports indicate an increase in the rates of hospitalization due to drug abuse in the United States. Excessive consumption of alcohol combined with drug abuse is prevalent in young adults age 18–24. The authors commented that strong efforts are needed to educate public regarding risk of drug overdose, especially if consumed with alcohol [2]. Currently, drug and alcohol testing are conducted in almost all hospital laboratories. Drugs of abuse testing are usually conducted using a urine specimen and commercially available immunoassays. Immunoassays are the first step in drugs of abuse analysis, primarily because they can be easily automated using automated analyzers, especially chemistry analyzers. Alternatively, immunoassay principles can also be adopted in a point of care device, and many such devices are commercially available for rapid drug of abuse testing in physician offices and outpatient clinics. Although immunoassay screening tests performed in clinical laboratories are high complexity tests, point of care drugs of abuse testing devices have exempt status (waived test). Major limitations of immunoassays are false-positive test results due to interferences. For example, labetalol, a beta blocker used for treating hypertension, may cause false-positive amphetamine test result [3].

## Medical versus workplace drug testing

Drugs of abuse testing is usually conducted in two different steps. The first step is screening of urine specimen for the presence of 8–10 common drugs using FDA-approved immunoassays, which can be easily adopted on automated chemistry analyzer. A positive result is considered as presumptive positive and must be confirmed by another more specific analytic technique, such as gas chromatography combined with mass spectrometry (GC/MS), liquid chromatography combined with mass spectrometry (LC-MS), or liquid chromatography combined with tandem mass spectrometry (LC-MS/MS). Tandem mass spectrometry is also called a triple quadrupole mass spectrometry because the system consists of two quadrupole mass spectrometers with a nonmass-resolving radio frequency only quadrupole between them to act as a cell for collision-induced dissociation.

Drug testing may be either for medical or for legal purpose. Results obtained during medical drug testing cannot be used in any legal context because such results are protected by patient's privacy act such as Health Insurance Portability and Accountability Act (HIPPA) regulations. As a result, a physician may choose not to order a confirmatory test for immunoassay positive test result. However, it is important to note that immunoassay positive results are considered as presumptive positive and require further validation by an alternative technique such as GC/MS (commonly used) or LC-MS/MS.

Legal drug testing was initiated by President Reagan who issued executive order number 12564 on September 15, 1986. This executive order directed drug testing for all federal employees who are involved in law enforcement, national security, protection of life and property, public health, and other services requiring high degree of public trust. Following this executive order, the National Institute of Drug of Abuse (NIDA) was given the responsibility of developing guidelines for federal drug testing. Currently SAMHSA, affiliated with Department of Health and Human Services of the Federal Government, is responsible for providing mandatory guidelines for federal work place drug testing. Bush et al. summarized guidelines for legal drug testing [4]. Preemployment or workplace drug testing conducted by private employers usually follow SAMHSA mandated guidelines for federal work place drug testing. However, private employers may also test for additional drugs that are not mandated by SAMHSA guideline, for example, benzodiazepines, barbiturates, and methadone.

In medical drug testing, informed consent may not be taken from a patient. An overdosed patient admitted to the emergency department may not be able to grant an informed consent anyway. In contrast, in workplace or any other legal drug testing program, obtaining an informed consent prior to testing is mandatory. Moreover, a chain of custody must be maintained in legal drug testing. However, a chain of custody is not required for medical drug testing. A chain of custody is a record of the personnel who had the possession of the specimen from the time of collection until time of analysis and reporting of the result. Moreover, after collection, the specimen must be sealed in front of the donor to ensure specimen integrity. A medical review officer (MRO) who is not affiliated with the testing laboratory must also certify a positive drug testing result as true positive in legal drug testing setting. This adds another level of safety for the donor, because the medical review officer ensures that there is no alternative explanation for the positive drug testing result. The three most important points regarding legal drug testing include:

- Initial immunoassay test results must be confirmed by a second method, preferably GC/MS or LC-MS/MS.
- The MRO must review and certify the result, indicating that there is no alternative explanation for a positive drug test.
- Special certification, such as College of American Pathology (CAP) Forensic Drug Testing certification or SAMHSA certification, is needed for operation of a laboratory performing legal drug testing including workplace drug testing.

## SAMHSA versus non-SAMHSA drugs

Commonly tested drugs include both SAMHSA mandated and non-SAMHSA drugs. For drug testing involving federal employee, only SAMHSA mandated drugs are tested. Originally, SAMHSA mandated drug testing of five commonly abused drugs including amphetamine, cocaine (tested as benzoylecgonine, the inactive metabolite), opiates, phencyclidine (PCP), and marijuana tested as 11-nor-9-carboxy $\Delta^9$-tetrahydrocannabinol (THC-COOH), the inactive metabolite of marijuana. In 2015 revision to proposed guidelines, SAMHSA recommended additional testing for oxycodone, oxymorphone, hydrocodone, and hydromorphone [2]. Moreover, heroin metabolite 6-monoacetylmorphine is also tested under SAMHSA protocol at a screening and confirmation cutoff of 10 ng/mL. Therefore, currently SAMHSA mandated drug testing includes:

- Amphetamine, including 3,4-methylenedioxymethamphetamine (MDMA, Ecstasy) and 3,4-methylenedioxyamphetamine (MDA)
- Cocaine, tested as benzoylecgonine
- Opiates (morphine, codeine), heroin metabolite 6-monoacetyl morphine (separate immunoassay with 10 ng/mL cutoff), and drugs such as oxycodone/oxymorphone (single immunoassay can detect both) and hydrocodone/hydromorphone (single immunoassay can detect both)
- Marijuana, tested as marijuana metabolite
- Phencyclidine

Some private employers may test for additional drugs in their workplace drug testing protocols, and such comprehensive drug panel may include barbiturates, benzodiazepines, and methadone. Testing of SAMHSA mandated drugs is summarized in Table 2. Testing of non-SAMHSA mandated drugs is summarized in Table 3.

**Table 2** Screening, confirmation cutoff concentrations for SAMHSA drugs, as well as target analyte for immunoassay.

| Drug class/drug | Target analyte in urine for immunoassays | Screening cutoff | Confirmation cutoff |
|---|---|---|---|
| Amphetamine/methamphetamine[a] | Amphetamine or methamphetamine | 500 ng/mL | 250 ng/mL |
| MDMA/MDA/MDEA | MDMA | 500 ng/mL | 250 ng/mL for all drugs |
| Cocaine | Benzoylecgonine | 150 ng/mL | 100 ng/mL |
| Opiates | Morphine | 2000 ng/mL[b] | 2000 ng/mL for either morphine or codeine |
| Heroin | 6-Monoacetylmorphine | 10 ng/mL | 10 ng/mL |
| Hydrocodone/hydromorphone[c] | Hydrocodone | 300 ng/mL | 100 ng/mL for either hydrocodone or hydromorphone |
| Oxycodone/oxymorphone[c] | Oxycodone | 100 ng/mL | 50 ng/mL for either oxycodone or oxymorphone |
| Marijuana | THC-COOH[d] | 50 ng/mL | 15 ng/mL |
| Phencyclidine | Phencyclidine | 25 ng/mL | 25 ng/mL |

MDMA, 3, 4-methylenedioxymethamphetamine; MDA, 3, 4-methylenedioxyamphetamine; MDEA, 3, 4-methylenedioxyethylamphetamine.

[a]If methamphetamine is confirmed, then its metabolite amphetamine must be present at a concentration of 100 ng/mL or higher.

[b]Private employer may use 300 ng/mL cutoff for screening and confirmation, although SAMHSA guideline recommends 2000 ng/mL cutoff to avoid false-positive results due to consuming poppy seed containing foods.

[c]From new SAMHSA proposed guidelines (2015).

[d]THC-COOH, 11-nor-$\Delta^9$-tetrahydrocannabinol-9-carboxylic acid.

**Table 3** Screening and confirmation cutoff concentrations of non-SAMHSA drugs as well as target analyte for immunoassay.

| Drug class/drug | Target analyte in urine for immunoassays | Screening cutoff | Confirmation cutoff |
|---|---|---|---|
| Barbiturates | Secobarbital | 200/300 ng/mL | 200 ng/mL |
| Benzodiazepines | Oxazepam or nor-diazepam | 200 ng/mL | 200 ng/mL |
| Methadone | Methadone or EDDP (metabolite) | 300 ng/mL | 100 ng/mL |

Although drug testing can be conducted also in blood, oral fluid, sweat, hair, nail, and meconium (in newborn babies), urine is the most commonly used specimen for drug testing. For drug testing in urine, depending on the particular drug, either the parent drug or its metabolite is targeted. For SAMHSA mandated drugs, recommended cutoff concentrations for both immunoassay and GC/MS of various drugs are available (Table 2). In general, such guidelines are also followed in medical drug testing. Usually a drug or its metabolites can only be detected in urine for a limited time after last abuse. However, detection time varies also on the dosage administered as well as characteristics of screening and confirmation assay. For example, marijuana ($\Delta^9$-tetrahydrocannabinol, THC) is tested as marijuana metabolite (THC-COOH) and is detectable in urine for 2–4 days (using 50 ng/mL cutoff), but for more frequent use, it may be detected up to 1 month.

## Detection window of various drugs in urine

Although drug testing can be conducted by using hair, saliva (oral fluid), blood, and sweat, the urine drug test is the most common, consisting of over 90% of drugs of abuse testing. The advantage of the urine specimen is that it can be collected noninvasively (although hair, oral fluid, and sweat specimens can also be collected noninvasively), and drug metabolites can be detected for a longer time period (window of detection) than blood. However, an abused drug may be detected up to 6 months in a hair specimen. Relative windows of detection for SAMHSA and non-SAMHSA drugs using urine specimens are summarized in Table 4. Although hair testing offers a long window of detection for several months after abuse, it takes several days for a drug to be incorporated

**Table 4** Window of detection of abused drugs.

| Drug or drug class | Detection window in urine |
| --- | --- |
| *SAMHSA mandated drugs* | |
| Amphetamines | 2 days |
| MDMA/MDA | 2 days |
| Cannabinoids | 2–3 days single use/30 days chronic abuse |
| Cocaine metabolites | 2 days single use/4 days repeated use |
| Morphine | 2–3 days |
| Heroin | 12 h as 6-monoacetylmorphine (6-acetylmorphine) |
| Hydrocodone/hydromorphone | 3 days |
| Oxycodone/oxymorphone | 3 days |
| Phencyclidine | 14 days |
| *Non-SAMHSA drugs* | |
| Barbiturates | Short acting (secobarbital, etc.) 1 day<br>Long acting (phenobarbital) 21 days |
| Benzodiazepines | Short acting (alprazolam, etc.) 3 days<br>Long acting (diazepam, etc.) 30 days |
| Methadone | 3 days |
| Methaqualone | 3 days |

in hair follicle. Therefore, recent abuse cannot be detected using hair specimen. In contrast, oral fluid testing can detect recent abuse of drug but has much shorter window of detection compared to urine.

Although a parent drug is present in blood, oral fluid, or hair, many drugs are detected through their respective metabolites in urine specimens. For example, cocaine abuse is detected in urine by confirming the presence of benzoylecgonine, the major inactive metabolite of cocaine. However, for hair, cocaine is usually present along with much lower concentration of the metabolite. Immunoassays for detecting various drugs in urine often target a major metabolite of the parent drug, although some drugs are also detected as parent drug in the urine specimen.

## Metabolism of abused drugs/target of immunoassay antibody

Immunoassays are widely used as the first step of drug screening in both medical and workplace drug testing programs. Immunoassays can be easily automated and several drugs can be analyzed using one urine specimen, and results can be directly downloaded in laboratory information system (LIS). In general, competitive immunoassays are used for drugs of abuse testing. Various commercially available immunoassays are available for drug testing including fluorescence polarization immunoassays (FPIA), enzyme multiplied immunoassay technique (EMIT), cloned enzyme donor immunoassay (CEDIA), turbidimetric immunoassay (TIA), and kinetic interaction of microparticle in solution (KIMS). The ONLINE Drugs of Abuse Testing immunoassays marketed by Roche Diagnostics (Indianapolis, IN) are based on KIMS format. In addition, enzyme-linked immunosorbent assay (ELISA) is also available for certain drugs. Immunoassays used for screening of drugs in urine are usually qualitative assay where a specimen can be tested as "positive" if present at a concentration above the cutoff concentration. However, some of the assays come in both qualitative and semiquantitative formats, and in most cases, such formats are defined by assay protocol and calibration. In qualitative formats, the calibration can be simplified to two calibrators, centering on the cutoff point thus providing the most accuracy around that point. The algorithm compares the signal observed with a sample with that of the cutoff calibrator and reports the result as positive or negative. Semiquantitative results can be reported using a calibration curve containing at least three calibrators (often the combination of the zero-calibrator, together with two or more calibrators). However, semiquantitative results may not compare well with values obtained by more sophisticated analytic methods such as GC/MS or LC-MS/MS.

Various abused drugs are mostly metabolized by the liver, and these metabolites are excreted in the urine. Most drugs are metabolized in the liver by cytochrome P 450 mixed function oxidase (CYP 450), while CYP3A4 and CYP2D6, the two isoforms of cytochrome P 450, are responsible for the metabolism of the majority of drugs in the liver. However, some drugs are not extensively metabolized by the liver; for example, amphetamine (although methamphetamine is metabolized to amphetamine). Therefore, antibodies in immunoassays for amphetamine/methamphetamine may target either amphetamine or methamphetamine (Table 2). MDMA (3,4-methylenedioxy-methamphetamine) cross-reacts with antibodies used in amphetamine/methamphetamine immunoassay. However, specific immunoassay is also available for detecting MDMA in urine if present. Chemical structure of amphetamine, methamphetamine, and MDMA are given in Table 1.

Some abused drugs are rapidly metabolized, for example, cocaine is spontaneously hydrolyzed in plasma into benzoylecgonine, an inactive metabolite of cocaine. Another metabolite of cocaine is ecgonine methyl ester. A small amount of unchanged cocaine can also be recovered in urine, and cocaine can also be metabolized by liver enzymes into nor-cocaine. Chemical structure of cocaine is given in Fig. 1. The elimination half-life of cocaine is approximately 45 min. Abusing cocaine and alcohol simultaneously is dangerous, because benzoylecgonine undergoes transesterification in the presence of alcohol (ethyl alcohol) to form cocaethylene, which is facilitated

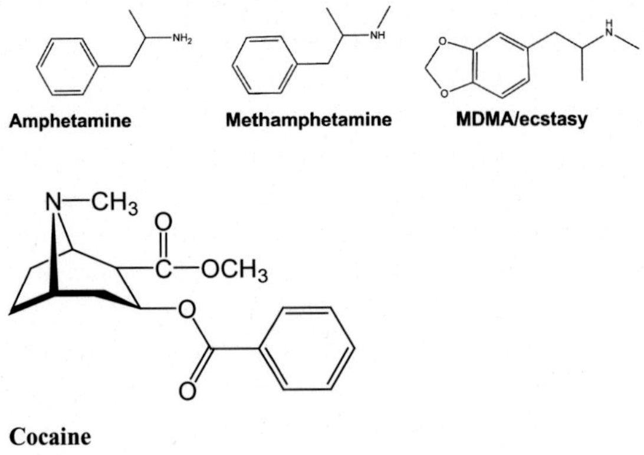

**FIG. 1**

Chemical structure of amphetamine, methamphetamine, MDMA, and cocaine. Note the structural difference between amphetamine and methamphetamine where a methyl group is attached to NH group in the side chain.

by the liver enzyme carboxylesterase. Two important points regarding cocaine metabolism are:

- The antibody used in immunoassays for cocaine usually targets benzoylecgonine.
- Cocaethylene is an active metabolite responsible for life-threatening toxicity if cocaine and alcohol are both abused simultaneously.

Heroin (diacetyl morphine) is first metabolized to 6-acetylmorphine (also called 6-monoacetyl morphine), and then to morphine by hydrolysis of the ester linkage by pseudocholinesterase in serum and by human carboxylesterase-1 and carboxylesterase-2 in the liver.

The 6-acetylmorphine, which can be confirmed by GC/MS or LC-MS/MS, is considered as the biomarker compound for heroin abuse. The majority of morphine is excreted in urine as morphine-3-glucuronide. This metabolite is formed by conjugation in the liver by the action of the liver enzyme uridine diphosphate glucuronosyltransferase. Codeine is metabolized to morphine in liver mostly by CYP2D6. Although morphine is a stronger narcotic analgesic than codeine, morphine is poorly absorbed from the stomach while codeine is well absorbed after oral administration. However, the conversion of codeine to morphine is essential for pain relief. Almost 10% Caucasian population has genetic polymorphism of CYP2D6 making the enzyme less active, and these patients may not get adequate pain relief during codeine therapy. Hydromorphone is also excreted in urine, mostly in conjugated form, but a small part of free hydromorphone can also be recovered in urine. Oxycodone is metabolized to oxymorphone, which is then conjugated in liver. However, the antibody used in opiate immunoassays targets only morphine. Therefore, usually the presence of oxycodone in urine cannot be detected by opiate immunoassays due to low cross-reactivity with the antibody. Specific immunoassay is available for detection of oxycodone in urine, which can also detect oxymorphone due to high cross-reactivity with the assay antibody. Similarly, immunoassays are also available for hydromorphone which can also detect hydrocodone. Chemical structures of morphine, codeine, heroin, and oxycodone are given in Fig. 2.

$\Delta^9$-Tetrahydrocannabinol (marijuana, THC) is the most active component (out of over 60 related compounds identified) of marijuana, which is found in various parts of the cannabis plant (*Cannabis sativa*) including flowers, stem, and leaves. THC is rapidly oxidized by cytochrome P 450 enzymes to active 11-hydroxy $\Delta^9$-tetrahydrocannabinol (11-OH-THC) metabolite and inactive 11-nor-9-carboxy $\Delta^9$-tetrahydrocannabinol (THC-COOH) metabolite. In

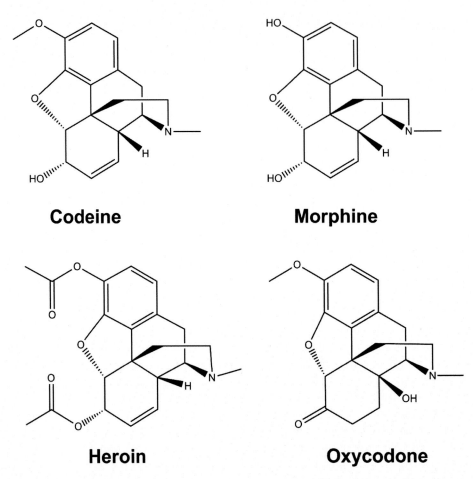

**FIG. 2**

Chemical structure of codeine, morphine, heroin, and oxycodone, and note that heroin is diacetyl morphine where both hydroxyl groups of morphine molecules are acetylated by chemical reaction. Morphine molecule has two hydroxyl group (–OH) where codeine has only one hydroxyl group.

immunoassays for THC, the antibody recognized the inactive metabolite THC-COOH. Chemical structure of THC and THC-COOH is given in Fig. 3.

PCP undergoes extensive metabolism by liver cytochrome P 450 enzymes (especially CYP3A4) into several hydroxy metabolites including *cis*-1-(1-phenyl-4-hydroxycyclohexyl) piperidine, *trans*-1-(1-phenyl-4-hydroxycyclohexyl) piperidine, 1-(1-phenylcyclohexyl)-4-hydroxypiperidine, and 5-(1-phenylcyclohexylamino) pentanoic acid. The elimination half-life of PCP varies significantly in humans (7–57h; average 17h). Despite extensive metabolism of PCP, significant amount of parent drugs is also present in urine PCP immunoassays target PCP. Chemical structure of PCP is given in Fig. 3.

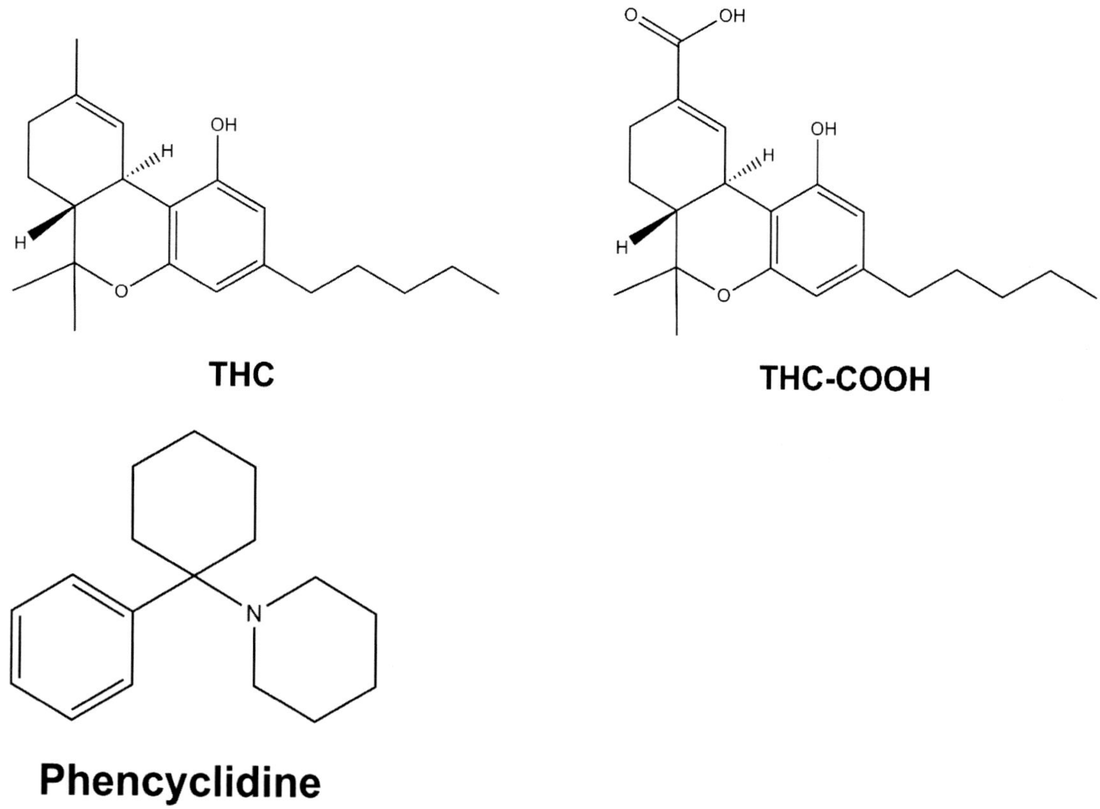

**FIG. 3**
Chemical structures of THC, THC-COOH (THC metabolite), and phencyclidine.

Barbiturates can be ultra-short acting, short acting, or long acting (phenobarbital). Barbiturates are extensively metabolized. However, many barbiturate immunoassays use antibody that recognizes secobarbital. Although there are more than 50 benzodiazepines, approximately 15 various benzodiazepines are used in the United States. Benzodiazepines are also extensively metabolized. The antibody in many benzodiazepine immunoassays targets oxazepam because it is a common metabolite of diazepam and temazepam. Chemical structures of common barbiturates and benzodiazepines are given in Fig. 4.

## Immunoassays versus confirmation

Immunoassays for various drugs have different cutoff concentrations as mandated by SAMHSA guidelines. If the concentration of the drug is below the cutoff concentration, the immunoassay test result is considered as negative. In workplace drug testing, a positive immunoassay test result must be confirmed

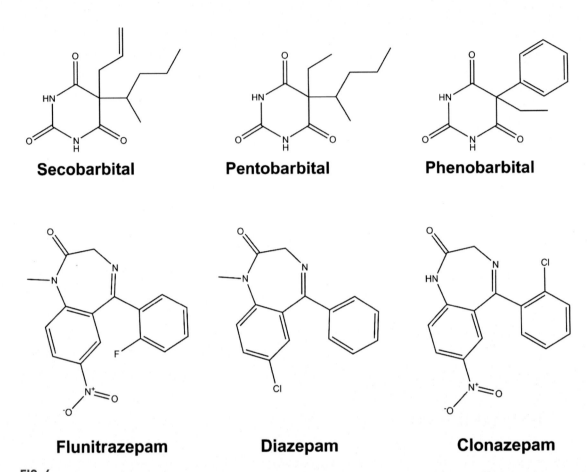

**FIG. 4**
Chemical structure of common barbiturates and benzodiazepines.

by using a different analytical technique preferably GC/MS, but confirmation concentration could be lower than the immunoassay screening cutoff concentration. More recently, LC-MS/MS-based methods are increasingly used for drug confirmation. Important points regarding confirmation step for amphetamine as recommended by SAMHSA include:

- If methamphetamine is confirmed by GC/MS, its metabolite amphetamine must also be confirmed by GC/MS at least at 100 ng/mL concentration. This protocol ensures that no pseudoephedrine or ephedrine present in the urine is converted into methamphetamine during GC/MS procedure (if injector temperature in GC is high, ephedrine/pseudoephedrine may be thermally converted into methamphetamine) as an artifact.

# False-positive immunoassay test results with various abused drugs

Immunoassays are subjected to false-positive test results. Amphetamine immunoassays suffer from more interferences than any other immunoassays used for screening of abused drugs in urine. Many over-the-counter cold and cough medications containing ephedrine or pseudoephedrine may produce false-positive test result with amphetamine/methamphetamine immunoassays. Many sympathomimetic amines found in over-the-counter medications may also cause false-positive test result. In addition, other drugs such as bupropion (mostly metabolites), trazodone, labetalol, and doxepin can also cause false-positive test result (Table 5). Amphetamine and methamphetamine have optical isomers designated as $d$ (or +) for dextrorotatory and $l$ (or −) for levorotatory. The $d$ isomer which is abused is intended targets of immunoassays. Ingestion of medications containing the $l$ isomer can cause false-positive results. For example, Vicks inhaler contains the active ingredient $l$-methamphetamine. In general, recommended use of this inhaler should not cause false-positive test result with amphetamine immunoassay because $l$-methamphetamine has poor cross-reactivity with assay antibody that targets $d$-amphetamine or $d$-methamphetamine. However, extensive use of this product may cause false-positive results for immunoassay screening [5]. A 77-year-old man tested positive for amphetamine using immunoassay. When chiral analysis was performed, the methamphetamine was identified as $l$-isomer, not the $d$-isomer. Further investigation revealed that the decedent frequently

**Table 5** Most commonly encountered drugs in false-positive test results using immunoassays for various abused drugs.

| Immunoassay | Interfering drugs[a] |
| --- | --- |
| Amphetamine | Ephedrine, pseudoephedrine, phentermine, tyramine, methylphenidate, doxepin, labetalol |
| Benzodiazepines | Oxaprozin, sertraline |
| Opiates | Diphenhydramine, rifampin, dextromethorphan, verapamil, fluoroquinolones |
| Phencyclidine | Dextromethorphan, diphenhydramine, ibuprofen, imipramine, ketamine, meperidine, thioridazine, tramadol |
| Tetrahydrocannabinol | Nonsteroidal antiinflammatory drugs, pantoprazole, efavirenz |
| Methadone | Diphenhydramine |

[a]These drugs are most commonly encountered clinically, but this list does not provide names of all drugs that are known to interfere with various immunoassays as based on published literature. Please see any toxicology reference book for in-depth information on this topic.

used Vicks inhaler for his bronchial asthma which lead to positive amphetamine screen as well as GC/MS confirmation [6].

Certain fluoroquinolone antibiotics may cause false-positive test results with opiate immunoassay screening. PCP is rarely abused today, but dextromethorphan present in many over-the-counter cough mixtures may cause false-positive test result with PCP immunoassays. However, GC/MS confirmation is negative indicating the absence of PCP in the urine specimen. Studies indicate that nonsteroidal antiinflammatory drugs may produce false-positive results in immunoassays screening [7]. Use of niflumic acid may cause false-positive test result with testing of marijuana as THC-COOH using immunoassay [8].

Various drugs that may cause false-positive test results with immunoassays used for detecting various drugs of abuse in urine are summarized in Table 5. Important regarding false-positive immunoassay test results include:

- Amphetamine/methamphetamine immunoassay screening tests are subjected to interferences most commonly from over-the-counter cold medications containing ephedrine/pseudoephedrine.
- Although Vick's inhaler contains *l*-methamphetamine which has less cross-reactivity with amphetamine/methamphetamine immunoassay which targets *d*-methamphetamine, excessive use of Vick's inhaler may cause false-positive test result with both immunoassay and GC/MS confirmation. Chiral derivatization is essential to resolve this issue (chiral derivatization distinguishes *d* from *l*-isomer).
- Positive opiate test results may be observed due to therapy with fluoroquinolone antibiotics but GC/MS confirmation should be negative.
- A common cause of positive PCP immunoassay screening result is the presence of dextromethorphan in the specimen. GC/MS confirmation should be negative.

## False-negative test results

Although false-positive test results with immunoassays are more common, false-negative test result (where drug was present in the specimen but was not detected) can also be encountered during immunoassay screening step. Obviously if the concentration of a particular drug is below the cutoff concentration of the immunoassay, test result should be negative. Major reasons of false negative test results during immunoassay screening include:

- Drug concentration may be below the cutoff concentration. This is a problem for detecting various benzodiazepines.

- Cross-reactivity of a drug is poor with the antibody of the assay. Best example is poor cross-reactivity of oxycodone with various opiate immunoassays.
- The drug may not have been metabolized and appeared in urine. Common example is negative urine test for benzoylecgonine (cocaine metabolite) in a patient with acute cocaine overdose.

Various benzodiazepines are common prescription drugs in the United States, and these drugs are also widely abused. Physicians sometimes order urine drug test in their patients to ensure compliance. However, using 200 ng/mL cutoff concentrations for benzodiazepine immunoassays, certain drugs may not be detected after taking recommended dosage due to low concentration of the drug in urine in particular alprazolam, clonazepam, and lorazepam. Clonazepam is metabolized to 7-aminoclonazepam. West et al. reported that when urine specimens collected from subjects taking clonazepam were tested using DRI benzodiazepine immunoassay at 200 ng/mL cutoff, only 38 specimens out of 180 tested positive by the immunoassay (21% positive). However, using liquid chromatography combined with tandem mass spectrometry (LC-MS/MS), 126 specimens out of 180 specimens tested positive (70% positive) when detection limit of LC-MS/MS assay was set at 200 ng/mL, the same cutoff used by the DRI benzodiazepine assay indicating poor detection capability of benzodiazepine immunoassay for clonazepam and its metabolite in urine. The authors concluded that 200 ng/mL cutoff may not be adequate in monitoring patient's compliance with clonazepam therapy [9].

Opiate immunoassays usually utilize a morphine-specific antibody and certain opioids cannot be detected by opiate immunoassays due to low cross-reactivities. These opioids include:

- Oxycodone and oxymorphone (keto opioids)
- Methadone
- Propoxyphene
- Fentanyl and its analogs

A 40-year-old man suffering from severe chronic migraine attack was treated with oxycodone 20 mg dosage twice a day, but the patient was dismissed from the clinic for testing negative for opiate immunoassay screen. On his behalf, a family member contacted a toxicologist who informed that oxycodone may cause false-negative test result in urine opiate drug screen due to poor cross-reactivity of oxycodone with the assay antibody that targets morphine. An aliquot of the original urine specimen was retested using GC/MS, and the presence of oxycodone at 1124 ng/mL was confirmed. This value confirmed that the patient was compliant with oxycodone. The clinician in the pain clinic ordered opiate screening which cannot detect oxycodone and the patient was wrongly discharged from the clinic [10]. Therefore, for monitoring patient taking oxycodone, it is important to order oxycodone screen using specific immunoassay that can detect oxycodone or opioid confirmation using mass spectrometry.

Cocaine overdose may cause fatality, and sometimes the victim die from cocaine overdose, but urine toxicology screen for cocaine metabolite benzoylecgonine may be negative because death may occur quickly from cocaine overdose and sufficient time has not been passed for the metabolite to be excreted in the urine in enough concentration to trigger positive response.

A 44-year-old white man was found dead in his apartment and police transported the body to the coroner's office. The urine toxicology screen using immunoassays (EMIT assays) was negative for benzoylecgonine (cocaine metabolite) using 300 ng/mL cutoff concentration. However, GC/MS analysis of most-mortem heart blood specimen showed a very high cocaine concentration of 18,330 ng/mL, and it was concluded that the cause of death was cocaine overdose [11].

## Derivatization in GC/MS vs direct analysis using LC-MS/MS

Although urine specimens can be directly used for analysis using immunoassays for various abused drugs, GC/MS confirmation test is manual requiring extraction of drugs from the urine specimen using organic solvents or solid-phase extraction using BondElute or another solid-phase extraction column. In general, hexane, acetone, chloroform, methanol, ethyl acetate, 1-chlorobutane, or dichloromethane are usually used for extraction. Some drugs such as amphetamine and methamphetamine can be extracted directly from the urine specimen using an organic solvent. However, some drug such as morphine is present in urine as glucuronide conjugate, which is very water soluble. Therefore, prior to extraction, breaking this conjugate using acid or alkali hydrolysis is important. It is also possible to cleave the conjugate by using beta-glucuronidase enzyme. Although acid or alkali hydrolysis can be conducted by heating specimens for 30 min, usually several hours or more is needed for enzymatic hydrolysis but enzymatic hydrolysis method is gentler.

Derivatization is a chemical reaction where a polar group in a molecule (e.g., carboxyl or hydroxyl group) is converted chemically to a nonpolar group to make the molecule volatile so that it can be analyzed by GC/MS. In general, carboxylic acid (–COOH; e.g., THC-COOH and benzoylecgonine), hydroxyl (–OH, e.g., morphine), and primary or secondary amine (such as amphetamine and methamphetamine) containing molecules require derivatization.

Usually amphetamine and methamphetamine are analyzed as trifluoroacetyl or pentafluoropropyl derivatives although many other derivatization methods have been described in the literature. Benzoylecgonine is usually analyzed as trimethylsilyl or acetyl derivative. Codeine and morphine can also be analyzed as trimethylsilyl derivative, while PCP can be analyzed without derivatization.

More recently, LC-MS/MS are used for drug confirmation. One major advantage of LC-MS/MS method is that no derivatization is necessary because liquid chromatography can be used for analyzing both polar and nonpolar compounds. In contrast, GC/MS can analyze only nonpolar compounds which may be vaporized at higher temperature (200–250°C). For GC/MS analysis, electron ionization is commonly used although chemical ionization is also sometime used. However, special mass spectrometry method is needed for LC-MS/MS analysis. Commonly used mass spectrometric method is called electrospray ionization where after exiting liquid chromatography as pure analyte in mobile phase, a portion of that liquid is diverted through a small spray needle connected to the mass spectrometer. If the mass spectrometer is operated in positive ionization mode, then positively charged ions are drawn out of the needle contained within fine liquid droplets which are pulled toward the mass spectrometer. Simultaneously a warm inert gas also flows for evaporating solvents. If tandem mass spectrometry (triple quadrupole mass spectrometry) is used, then the first mass spectrometer (first quadrupole) acts as molecular prefilter allowing one ion to pass the mass spectrometer, for example, $m/z$ 304 for cocaine [M + hydrogen ion] in the positive ionization mode. This ion is called precursor ion or parent ion. The second quadrupole (not a mass spectrometer) acts as a collision chamber only where the precursor ion undergoes collision-induced dissociation due to interaction with inert gas producing fragment ions. These ions are called daughter ions or product ions. These product ions then enter the third quadrupole which is also a mass spectrometer, producing mass spectrum. Therefore, two mass spectrometers and a collision chamber are used in the overall design. Selection of a precursor ion in the first quadrupole, collison of the precursor ion in the second quadrupole, and selection of a product ion in the third quadrupole are known as selected ion monitoring. For example, in analysis of cocaine, the precursor ion is $m/z$ 304 and product ion is $m/z$ 182. When both ions are present, then it provides further confirmation of cocaine in the specimen. Although electrospray ionization is most common, other ionization methods such as atmospheric pressure chemical ionization methods are also available but used less commonly for drug confirmation.

A major advantage of LC-MS/MS for drug testing is that urine specimen can be screened for large amounts of drugs in one step. In one report, the authors developed protocol for screening of 100 drugs in urine using LC-MS/MS [12]. LC-MS/MS-based methods are also very valuable for confirming the presence of novel psychoactive substances (emerging designer drugs) where immunoassays are not available [13].

More recently, dilute and shoot protocols are available for rapid analysis of drugs using LC-MS/MS. This method enables diluting the urine specimen (after adding internal standards) with mobile phase solvent mixture followed by direct injection to LC-MS/MS system. Theoretically this is a very easy protocol for analysis (no liquid-liquid or solid-phase extraction to clean up the sample) at the expense of instrument wear and tear. Yen et al. reported a dilute and shoot protocol for rapid analysis of opioids, its metabolites, and antihistamines in urine using LC-MS/MS [14].

## Analytical true positive due to use of prescription drugs and other factors

As expected if a person is taking codeine for pain control, the opiate test should be positive. Similarly, positive benzodiazepine test result is expected if the person is taking a prescription medication of benzodiazepine class. Amphetamine and methamphetamine as well as their analogs are used in treating attention deficit disorders and as expected use of such drug could lead to positive amphetamine/methamphetamine drug testing result. Although not used frequently, cocaine is still used as a local anesthetic in ENT surgery and a patient may test positive for cocaine metabolite 1–2 days after the procedure.

Poppy seeds contain both codeine and morphine. Therefore, eating poppy seed containing food may cause false-positive opiate test results. In order to circumvent that problem, the cutoff concentration of opiate was increased to 2000 ng/mL from originally cutoff level of 300 ng/mL in Federal drug testing programs. However, some private employers still use 300 ng/mL cutoff concentration for opiate testing in their workplace drug testing protocols. Eating poppy seed containing food prior to drug testing (1–2 days) may cause positive opiate test result if 300 ng/mL cutoff concentration is used. However, a morphine level over 2000 ng/mL in the absence of substantial amount of codeine is highly suspicious of heroin abuse. Moreover, confirmation of 6-monoacteyl morphine is inconsistent with poppy seed use but proves heroin abuse.

Hemp oil is prepared from hemp seed which may contain trace amount of cannabinoid (marijuana), but ingesting hemp oil should not cause positive marijuana test result at 50 ng/mL cutoff level. Analytical true-positive test results due to the use of prescription medications are other factors summarized in Table 6. Important points regarding analytical true positives include:

- Prescription use of various drugs may cause analytical true-positive test results.

Table 6 Analytical positive drug test result (GC/MS confirmation) due to the use of prescription medication and other factors.

| Positive drug test | Prescription medication |
| --- | --- |
| Amphetamine | Adderall (contain amphetamine), lisdexamfetamine, desoxyn (contain methamphetamine), clobenzorex, ethyl amphetamine, mefenorex, benzphetamine, famprofazone |
| Cocaine | Cocaine use (as topical anesthetic) during ENT procedure<br>Drinking Inca tea or Coca de Mate tea (South American tea prepared from coca leaves) |
| Barbiturates | Short- and long-acting barbiturates |
| Benzodiazepines | Both short- and long-acting benzodiazepines |
| Opiates | Codeine, morphine, oxymorphone, hydrocodone, hydromorphone, oxycodone<br>Eating poppy seed containing food |
| Phencyclidine | None |
| Tetrahydrocannabinol | Marinol (synthetic marijuana), drinking hemp oil should not cause positive test at 50 ng/mL cutoff |
| Methadone | None |

- Positive opiate test result may occur in a person eating poppy seed containing food prior to drug test especially if the private employer still uses 300 ng/mL cutoff concentration for opiates.
- Drinking coca tea (health Inca tea, Inca tea, or mate de coca) may cause positive cocaine test result because these teas originating from South America are prepared from coca leaves.
- Use of cocaine as a local anesthetic during ENT surgery may cause positive cocaine test result 1–3 days after surgery.
- Ingestion of hemp oil or passive inhalation of marijuana should not cause positive marijuana test result.

## Issues of adulterated urine specimens in workplace drug testing

People try to beat drug test during workplace drug testing. Usually people drink detoxifying agents purchased from the Internet. Contrary to the claim that these agents can purge drugs out of circulation, these products contain only diuretics such as caffeine or hydrochlorothiazide, and if a person taking such product also drinks plenty of water (suggested in the package insert of these products), it produces diluted urine. Because specimen integrity check (pH, creatinine, specific gravity, and temperature) is routinely used in all urine specimen collected for workplace drug testing, low creatinine (below 20 mg/dL) indicates that urine is diluted and no further analysis is conducted. Specimen adulteration is considered as "refuse to testing" and the person is denied employment. Various household chemicals such a stable salt (sodium chloride), vinegar, liquid soap, liquid laundry bleach, sodium bicarbonate (baking soda), and lemon juice are added to urine after collection to beat drug test. Although some of these products may invalidate immunoassay screening step, all of these products can be detected indirectly by specimen integrity testing and specimen is rejected for further testing due to adulteration. However, Visine eye drop significantly interferes with enzyme multiplied immunoassay technique (EMIT) assays for drugs of abuse testing and other immunoassay, but the presence of Visine eye drops in an adulterated specimen cannot be detected by routine specimen integrity testing. EMIT marijuana assay is most affected by Visine.

More recently, several products are available through the Internet for use as in vitro adulterants. These products usually contain strong oxidizing agents such as potassium nitrite, pyridinium chlorochromate, glutaraldehyde, Stealth (hydrogen peroxide and peroxidase), and more recently zinc sulfate. These adulterants cannot be detected by routine specimen integrity testing, and such products can mask testing of marijuana metabolite not only in the immunoassays screening step but also in GC/MS confirmation step because THC-COOH is slowly oxidized by these products. However, spot tests and special urine dipstick analysis (AdultaCheck 6 or 10, Intect 7, etc.) are available for detecting these adulterants [15].

Sometimes patients in pain management sell narcotic analgesics prescribed to them in street for more money. In addition, they may go to various doctors (doctor shopping) to obtain more prescription for narcotics. Drug testing in urine is often conducted to ensure that they are compliant with therapy. These people want their urine drug test to be positive for the prescription drug and often add the tablet after grinding (to produce fine powder of the tablet for easy dissolving in the urine). However, during analysis, high amount of parent drug is recovered in the urine as expected, but no metabolite is detected indicating that these patients are attempting to cheat on their drug testing. Important points regarding urine adulteration:

- If household chemicals are used as urinary adulterants, specimen integrity testing (pH, creatinine, specific gravity, and temperature) should be able to identify such adulterants.
- Presence of Visine eye drops in adulterated urine cannot be determined by specimen integrity testing.
- The presence of Internet-based adulterants such as potassium nitrite, glutaraldehyde, pyridinium chlorochromate, and Stealth cannot be detected by specimen integrity testing. However, spot test and special urine dipsticks (AdultaCheck 6, Intect 7) can be used for their detection.
- Presence of an adulterant in urine specimen subjected to workplace drug testing is considered as "refusal to testing" and the person may be denied job. If an adulterant is present in a specimen, no further test is usually conducted including immunoassay screening for various drugs.

## Miscellaneous other issues in drugs of abuse testing

Lysergic acid diethylamide (LSD) was widely abused in 1960s and 1970s, and in some US cities, these drugs are reappearing in rave parties. There is a commercially available immunoassay for testing LSD. Rave party drug 3,4-methylenedioxymethampheatmine (MDMA) can be detected by some amphetamine immunoassays and also by using special immunoassays that detect both MDMA and amphetamine. However, other rave party drugs ketamine and gamma-hydroxybutyric acid cannot be detected by routine toxicology testing although reference toxicology laboratories offer testing for such drugs. Many designer drugs cannot be detected by routine toxicology analysis, including newly emerged designer drugs spice, K2 (synthetic marijuana) and bath salts (synthetic cathinone). Please see Chapter 17 for more detail.

As mentioned earlier, over 90% drug testing are conducted using urine specimens. Hair testing provides a larger window of detection (up to 6 months), while saliva (oral fluid) testing can identify impairment due to recent abuse of a drug. In European countries, oral fluid testing is gaining popularity for identifying impaired drivers. In addition, sweat testing can be used for continuous monitoring of a drug in a person undergoing drug rehabilitation.

## Summary/key points

- Marijuana is most popular abused drugs in the United States.
- Urine drug testing represents approximately 90% of all drug testing. Although initial immunoassay screening result using urine specimen may not be confirmed by gas chromatography/mass spectrometry (GC/MS) in medical drug testing, GC/MS confirmation (or confirmation by another method) is mandatory in legal drug testing. In addition, in legal drug testing, medical review officer (MRO) must review and certify the positive drug testing result, indicating there is no alternative explanation for a positive drug test. Moreover, chain of custody must be maintained.
- Currently, SAMHSA (substance abuse and mental health services administration) guidelines direct federal workplace drug testing. SAMHSA guidelines require testing for abused drugs, including amphetamines (amphetamine, methamphetamine, and 3, 4-methylenedioxy-methamphetamine; MDMA, Ecstasy), cocaine (tested as benzoylecgonine), opiates (morphine, codeine), heroin metabolite (6-maonoacetylmorphine), oxycodone/oxymorphone, hydrocodone/hydromorphone, marijuana (tested as marijuana metabolite), and phencyclidine (PCP).
- In urine, most drugs can be detected only for 2–3 days after abuse except for PCP (14 days) and chronic abuse of marijuana (up to 30 days).

- Amphetamine/methamphetamine immunoassay screening tests are subjected to interferences most commonly from over-the-counter cold medications containing ephedrine/pseudoephedrine.
- Although Vick's inhaler contains *l*-methamphetamine which has less cross-reactivity with amphetamine/methamphetamine immunoassay (*d*-methamphetamine is abused and measured by amphetamine/methamphetamine immunoassays), excessive use of Vick's inhaler may cause false-positive test result with both immunoassay and GC/MS confirmation. Chiral derivatization is essential to resolve this issue.
- Cocaine abuse is detected by the presence of benzoylecgonine in urine which is an inactive metabolite of cocaine.
- Abuse of cocaine and ethanol (alcohol) is dangerous because cocaethylene is formed due to interaction between ethanol and benzoylecgonine, and cocaethylene is an active metabolite with a long half-life. Deaths have been reported in individuals abusing both cocaine and alcohol although pure cocaine abuse may also be lethal.
- Cocaine may be used in ENT surgery as a local anesthetic and such use may cause positive urine specimen up to 2 days.
- Certain herbal tea (Mate de Coca and Health Inca tea) may contain cocaine, and drinking such tea may cause positive cocaine drug test.
- Heroin is metabolized to 6-monoacetylmorphine (also called 6-acetylmorphine) and then into morphine. Detection of 6-monoacetylmorphine (marker compound of heroin abuse) in urine is only possible if a person abuses heroin.
- Eating poppy seeds containing food may cause positive opiate tests result but, in a person, eating poppy seeds both codeine and morphine are present and can be confirmed by GC/MS. A morphine level above 2000 ng/mL with very little codeine (coming from crude opium used in preparation of heroin) or no codeine is indicative of heroin abuse.
- Opiate immunoassays usually utilize a morphine-specific antibody, and certain opioids cannot be detected by opiate immunoassays due to low cross-reactivities. These opioids include oxycodone and oxymorphone (keto opioids), methadone, propoxyphene, fentanyl, and its analogs.
- A common cause of positive PCP immunoassay screening result is the presence of dextromethorphan in the specimen. GC/MS confirmation should be negative.
- Marijuana is analyzed by detecting carboxylic acid derivative in urine. Use of synthetic marijuana (Mannitol) should cause positive marijuana tests result, but passive inhalation of marijuana should not cause positive test result because urine concentration of the metabolite should be well below the 50 ng/mL cutoff concentration.
- People try to beat drug tests by adulterating urine. Visine eye drop may interfere with various immunoassay screening tests, and the presence

Visine eye drop in urine cannot be determined by specimen integrity testing (pH, creatinine, specific gravity, and temperature). Internet-based adulterants such as potassium nitrite, pyridinium chlorochromate, stealth (peroxidase enzyme and hydrogen peroxide), glutaraldehyde, and zinc sulfate can invalidate drug tests, but their presence cannot be detected by specimen integrity tests. However, special urine dip stick tests (AdultaCheck 6 or 10, Intect 7, etc.), spot tests, and other laboratory-based tests are available for their detection except for adulteration with zinc sulfate. Such specimen integrity testing is essential in workplace drug testing.

# References

[1] Manchikanti L, Sanapati J, Benyamin RM, Atluri S, et al. Reframing the prevention strategies of the opioid crisis: focusing on prescription opioids, fentanyl and heroin epidemic. Pain Physician 2018;21:309–26.

[2] White AM, Hingson RW, Pan IJ, Yi HY. Hospitalization for alcohol and drug overdoses in young adults ages 18–24 in the United States, 1999–2008: results from the nationwide inpatient sample. J Stud Alcohol Drugs 2011;72:774–86.

[3] Yee LM, Wu D. False positive amphetamine toxicology screen results in three pregnant women using labetalol. Obstet Gynecol 2011;117:503–6.

[4] Bush D. The US mandatory guidelines for Federal workplace drug testing programs: current status and future considerations. Forensic Sci Int 2008;174:111–9.

[5] Solomon MD, Wright JA. False-positive for (+)-methamphetamine. Clin Chem 1977;23:1504.

[6] Wyman JF, Cody JT. Determination of l-methamphetamine: a case history. J Anal Toxicol 2005;29:759–61.

[7] Rollins DE, Jennison TA, Jones G. Investigation of interference by non-steroidal antiinflammatory drugs in urine tests for abused drugs. Clin Chem 1990;36:602–6.

[8] Boucher A, Vilette P, Crassard N, Bernard N, et al. Urinary toxicology screening: analytical interference between niflumic acid and cannabis. Arch Pediatr 2009;16:1457–60 [article in French].

[9] West R, Pesce A, West C, Crews B, et al. Comparison of clonazepam compliance by measurement of urinary concentration by immunoassay and LC-MS/MS in patient management. Pain Physician 2010;13:71–8.

[10] Von Seggern RL, Fitzgerald CP, Adelman LC, Adelman JU. Laboratory monitoring of OxyContin (oxycodone): clinical pitfalls. Headache 2004;44:44–7.

[11] Baker JE, Jenkins AJ. Screening for cocaine metabolite fails to detect an intoxication. Am J Forensic Med Pathol 2008;29:141–2.

[12] Remane D, Wetzel D, Peters FT. Development and validation of a liquid chromatography-tandem mass spectrometry (LC-MS/MS) procedure for screening of urine specimens for 100 analytes relevant in drug-facilitated crime (DFC). Anal Bioanal Chem 2014;406:4411–24.

[13] Olesti E, Pascual JA, Ventura M, Papaseit E, et al. LC-MS/MS method for the quantification of new psychoactive substances and evaluation of their urinary detection in humans for doping control analysis. Drug Test Anal 2020;12:785–97.

[14] Yen YT, Chang YJ, Lai PJ, Chang CL, et al. A study of opiate, opiate metabolites and antihistamines in urine after consumption of cold syrups by LC-MS/MS. Molecules 2020;25(4), E972.

[15] Dasgupta A, Chughtai O, Hannah C, Davis B, Wells A. Comparison of spot tests with AdultaCheck 6 and Intect 7 urine test strips for detecting the presence of adulterants in urine specimens. Clin Chim Acta 2004;34:19–25.

# CHAPTER 17

# Designer drugs, date rape drugs, LSD, volatiles, magic mushroom, and peyote cactus abuse

## Negative toxicology report

Usually 9–10 drug panel is commonly used for urine drug test. Therefore, if a person is abusing less commonly abused drugs, urine toxicology screen could be negative. Although abused less frequently, various designer drugs, some date rape drugs, magic mushroom, and peyote cactus cannot be detected by routine toxicological analysis. However, designer drugs such MDMA (3,4-methylenedioxymethamphetamine) and MDA (3,4-methylenedioxyamphetamine) could be detected by routine toxicology analysis due to sufficient cross-reactivity with amphetamine immunoassays. Moreover, specific immunoassay for detecting MDMA is also commercially available.

Designer drugs were initially synthesized by clandestine laboratories to avoid legal consequences of manufacturing and selling illicit drugs. Then in 1986, the United States Controlled Substances Act was amended in order to make manufacturing and selling of designer drugs illegal. The common types of designer drugs include amphetamine analogs, opiate analogs (including fentanyl derivatives), piperazine analogs, tryptamine-based hallucinogens, phencyclidine analogs, and gamma hydroxy butyric acid (GHB) analogs. GHB and its analogs are used in date rape situation (also called drug facilitated rape). These drugs cannot be detected by routine toxicology analysis.

## Abuse of amphetamine-like designer drugs including bath salts

Amphetamine is a sympathomimetic amine which is structurally related to phenethylamine. Designer drugs structurally related to amphetamine such as 3,4-methylenedioxymethamphetamine (MDMA, Street names; Ecstasy) and 3,4-methylenedioxy-amphetamine (MDA, Street name; Adam) are widely abused. In addition to MDMA other designer drugs structurally related to

361

amphetamine such as para-methoxyamphetamine (PMA) para-methoxy meth-amphetamine (PMMA, Street names; Killer, Dr. Killer), and 4-methylthioamphetamine (4-MTA, Street names; Golden eagle) have also been encountered. There are many other designer drugs that are psychostimulant and have amphetamine-like action. In addition, many of these drugs are more potent than amphetamine and there are reports of fatalities associated with use of such drugs. The common examples of these designer drugs include 3,4-methylenedioxy-N-ethylamphetamine (MDEA, Street names; Eve), 2,5-dimethoxy-4-methylamphetamine (DOM), 2,5-dimethoxy-4-methylthioamphetamine (DOT), and 2-CB (4-bromo-2,5-dimetoxyphenethy-lamine, Street names; Venus, Bromo, Ero, Neux). Another class of amphet-amine-like designer drugs are bath salts (synthetic cathinone) which are classified structurally as beta-keto-amphetamine. Various amphetamine-like designer drugs including bath salts are listed in Table 1.

**Table 1** Classification of designer psychostimulants which are amphetamine-like designer drugs.

| Structure | Example |
|---|---|
| Beta-keto amphetamines (bath salts) | 4-Methylmethcathinone (mephedrone), 4-methethcathinone, butylone, methylone (beta-keto-MDMA), ethylone (beta-keto-MDEA), 4-FMC (4-fluoromethcathinone), 4-methoxymethcathinone, MDPV (3,4-methylenedioxypyrovalerone: MDPV), etc. |
| 2,5-Dimethoxyamphetamine | DOB: 4-bromo-2,5-dimethoxyamphetamine<br>DOC: 4-chloro-2,5-dimethoxyamphetamine<br>DOI: 4-iodo-2,5-dimethoxyamphetamine<br>DOM: 2,5-dimethoxy-4-methylamphetamine<br>MDOB: 4-bromo-2,5-dimethoxymethamphetamine<br>TMA-2: 2,4,5-trimethoxyamphetamine |
| 2,5-Dimethoxy phenylamine (2C-series) | 2C-B: 4-bromo-2,5-dimethoxyphenethylamine<br>2C-I: 4-iodo-2,5-dimethoxyphenethylamine<br>2C-D: 2,5-dimethoxy-4-methylphenethylamine<br>2C-E: 4-ethyl-2,5-dimethoxyphenethylamine<br>2C-T2: 4-ethylthio-2,5-dimethoxyphenethylamine<br>2C-T4: 4-isopropylthio-2,5-dimethoxyphenethylamine<br>2C-T7: 2,5-dimethoxy-4-propylthiophenethylamine |
| Piperazines | BZP: 1-benzylpiperazine<br>MeOPP: 1-(4-methoxyphenyl) piperazine<br>TFMPP: trifluoromethyl-piperazine<br>mCPP: 1-(4-chlorophenyl)piperazine<br>pFPP: 1-(4-fluorophenyl) piperazine |
| 2-Aminoindanes | 5-IAI: 5-iodo2-aminoindane<br>MDAI: 5,6-methylenedioxy-2-aminoindane |
| Para-methoxy derivative | PMA: para-methoxyamphetamine<br>PMMA: para-methoxymethamphetamine |

Bath salts have been placed in Schedule 1 drug in September 2011 due to high abuse potential. Bath salts are synthetic drugs which are cathinone derivatives. Cathinone is a natural component of khat plants (*Catha edulis*), a flowering evergreen shrub cultivated as a bush or small tree which is native to Ethiopia, East Africa, and the Southern Arabian Peninsula. The leaves of khat tree have an aromatic odor and slight sweet taste. Its young bud and tender leaves contain amphetamine-like psychoactive substance cathinone (and related compounds) which upon chewing produces stimulation and euphoria. Chewing leaves of Khat is a part of Yemini culture where approximately 90% males and 50% females chew khat leaves on a regular basis. The World Health Organization considers khat as a drug of abuse. Khat is banned in European countries, United States, and Canada, while it is lawful in Yemen, Somalia, and Ethiopia [1]. Cathinone is present in varying amounts (78–343 mg) in 100 g of khat leaves. The stimulatory effects of khat appear approximately 30 min after initiation of chewing khat plant and the effect may last up to 3 h. Approximately 90% of active compound present in leaves are released during chewing. The main route of absorption of cathinone is through oral mucosa (approximately 60% of absorption) and the rest of cathinone is absorbed through the gastrointestinal tract. A typical session of Khat chewing is similar to the effect of 5 mg amphetamine [2]. Khat is known as "Herbal Ecstasy" because of its central nervous system stimulant effects similar to amphetamine. Chemical structure of cathinone is given in Fig. 1.

Bath salts are synthetic derivatives of cathinone and belong to the beta-keto amphetamine class structurally. In Europe and United States, commonly abused bath salts include 4-methylmethcathinone (mephedrone), 4-methethcathinone, butylone, methylone (beta-keto-MDMA), ethylone (beta-keto-MDEA), 4-FMC (4-fluoromethcathinone), 4-methoxymethcathinone, and MDPV (3,4-methylenedioxypyrovalerone: MDPV) [3]. To date, about 150 drugs which are synthetic cathinone have been identified on the clandestine drugs market, which are one of the largest groups of novel psychoactive substances (NPS) monitored by the United Nations Office on Drugs and Crime and the European Monitoring Center for Drugs and Drug Addiction. Synthetic cathinone (bath salts) in general mimic the psychostimulant properties of amphetamine, MDMA, or cocaine. These drugs act by increasing synaptic levels of monoamines (noradrenaline, dopamine, and serotonin) by interacting with

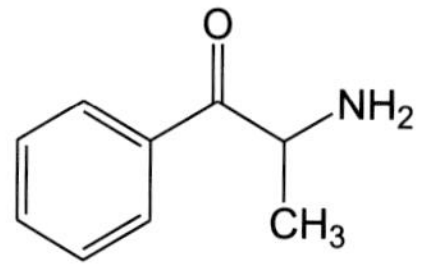

**FIG. 1**
Chemical structure of cathinone.

plasma membranes of various monoamine transporters including noradrenaline, dopamine, and serotonin transporters, thus inhibiting reuptake of respective neurotransmitters. However, their selectivity varies among various synthetic cathinone derivatives. Some drugs such as cathinone, mephedrone, and methcathinone can promote release of dopamine, while other drugs such as mephedrone, methylone, ethylone, and butylone can release serotonin [4]. Bath salts may cause serious life-threatening toxicity and even death. Euphoria after bath salt use lasts for 2–4 h and is more intense than euphoria after abuse of MDMA. Abuse of bath salts could be fatal. The deaths are attributed to hyperthermia, hypertension, cardiac arrest, and more in general to the classic serotonin syndrome. Only rarely the concentration of the parent drug causing fatality overcomes the value of 1 mg/L in postmortem biological fluids, indicating high potency of bath salts [5].

Bath salts cannot be detected by regular toxicological screen because these compounds do not cross-react with amphetamine immunoassays. However, a drug of abuse array V biochip array assay by Randox Corporation is capable of detecting synthetic cathinone if present in the urine specimen. This biochip assays have two antibodies specific for identification of various bath salts. Bath Salt I antibody is anti-methcathinone antibody with 100% cross-reactivity to mephedrone. The Bath Salt II antibody is anti-MDPV antibody capable of detecting MDPV as well as MDPBP (3–4-methylenedioxy-alpha-pyrrolidinobutiophenone). Other baths salts known to cross-react with these two antibodies include methylone, flephedrone, naphyyrone, and pentedrone. The proposed manufacturer's cut-off is 5 ng/mL for Bath Salt I assay and 39 ng/mL for Bath Salt II assay. However, bath salts can also be detected and confirmed using gas chromatography/mass spectrometry or liquid chromatography combined with tandem mass spectrometry (LC-MS/MS). The major advantages of LC-MS/MS for analysis of bath salts include high sensitivity and specificity as well as analysis of many bath salts in one run.

## Abuse of synthetic marijuana (Spice, K2)

The abuse of synthetic cannabinoid was first reported in Europe in early 2000s and in the United States around 2008. However, popularity of recreational use of synthetic cannabinoids is on rapid rise with synthetic cannabinoid-related intoxication calls to poison control centers in the United States increased by 240% between 2010 and 2011. In order to deal with severe health risks associated with use of synthetic cannabinoids, the US Drug Enforcement Administration (DEA) passed Synthetic Drug Abuse Prevention Act in 2012 which classified many synthetic cannabinoids in Schedule 1 drug indicating no known medical use but high abuse potential. Unfortunately, to circumvent this classification, many clandestine laboratories introduced newer versions of synthetic cannabinoids as evidenced by 330% increase in calls to poison control centers between January 2015 and May 2015 [6]. More than 177 novel synthetic cannabinoids are currently reported. Street names of synthetic cannabinoids include Spices, K2, and K2 blonde.

Synthetic cannabinoids were initially synthesized for research purposes. One of the first reported synthetic cannabinoids, JW-018, was originally synthesized by John William Huffman, a chemist at Clemson University to evaluate its therapeutic potential. It has been speculated that chemist in a clandestine laboratory copied his method and JWH-018 was introduced as a recreational drug in the underground market because JWH-018 like natural cannabinoid (tetrahydrocannabinol: THC) can interact with cannabinoid receptors in the brain.

Usually synthetic cannabinoids are sprayed on herbal mixtures and wrapped in foil packages with warning label "Not for Human Consumption" (street names: spice, K2, K2 blond, etc.). These products like cannabis are often smoked via a joint or water pipe but also can be consumed orally or intranasally. Synthetic cannabinoids are structurally different from natural tetrahydrocannabinol (THC) but are called synthetic cannabinoid because their mechanism of action is similar to THC. However, synthetic cannabinoids are more potent and may have longer half-life. The effects may appear within 10 min after inhalation and may last 2–6 h after use. Now there are more than 177 synthetic cannabinoids found through the WHO database and the Internet but most commonly detected synthetic cannabinoids can be broadly classified as aminoalkylindoles synthesized by John W. Hoffman (JWH-018, JWH-122) or Alexander Makriyannis (AM-2201), cyclohexylphenols originally synthesized at Pfizer (CP 47,497 and its analogs), benzoylindoles produced by Research Chemical Suppliers (RCS-4), and classical synthetic cannabinoid such as HU-210 produced at the Hebrew University [7]. Commonly encountered synthetic cannabinoids are listed in Table 2. Structure of THC and synthetic cannabinoid JWH-018, HU-210, and AM-2201 is given in Fig. 2.

**Table 2** Classification of synthetic cannabinoids based on structure.[a]

| Chemical class | Examples |
|---|---|
| Aminoalkylindoles | JWH-015, JWH-018, JWH-019, JWH-020, JWH-073, JWH-081, JWH-122, JWH-200, JWH-203, JWH-210, JWH-250, JWH-251, JWH-397, JWH-412, AM-1220 |
| Cyclohexylphenols | CP 47,497 and its homologs, CP 55,940 |
| Benzoylindoles | AM-694, AM-679, AM-2201, AM-2233, RCS-4 |
| Synthetic cannabinoids analog | HU-210, HU-211, HU-239, HU-243, HU-308, HU-320 |

[a]There are also few other synthetic cannabinoids such as UR-144, WN 55,212, etc. which do not belong to any particular class.

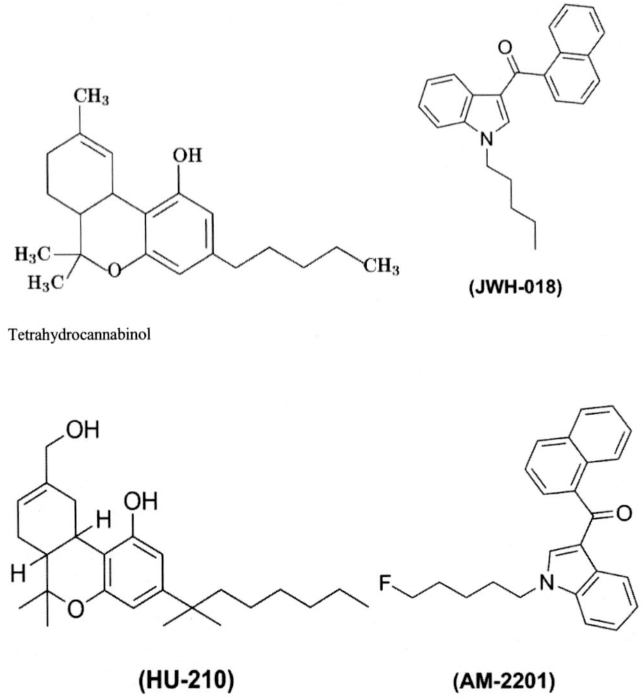

**FIG. 2**
Chemical structure of tetrahydrocannabinol (THC), and synthetic cannabinoids JWH-018, HU-210, and AM-2201.

Although structurally different from natural cannabinoid (THC: tetrahydrocannabinol), synthetic cannabinoids also bind with cannabinoid receptor 1 (CB1) and cannabinoid receptor 2 (CB2) which are found mostly in the central nervous system but also may be found in peripheral tissues including lungs, liver, and kidneys. However, within the brain, these receptors are mostly located in the cerebral cortex, hippocampus, basal ganglia, and cerebellum. In contrast to THC, which is a partial CB1 receptor agonist, synthetic cannabinoids are more potent full CB1 agonist with much higher affinity for CB1 receptors than THC. Some of the metabolites are also active with high affinity for CB1 receptors. By 2010, there has been 1057 synthetic marijuana-related toxicity in 18 states and District of Columbia. The users also report adverse effects such as rapid heartbeat, irritability, and paranoia. In addition, tremor, seizure, slurred speech, dilated pupils, hypokalemia, tachycardia, hypertension, and chest pain may also be present in a patient overdosed with synthetic cannabinoids. Long-term adverse effects are not known [8]. Additionally, a recent outbreak of coagulopathies and at least four associated deaths due to synthetic cannabinoids tainted with brodifacoum have been reported [9]. Because of higher affinity for cannabinoids receptors, synthetic cannabinoids are associated with higher rate of toxicity, hospital admission, and fatalities compared to abusers of natural cannabis. Shanks et al. reported three cases where fatalities were associated with synthetic cannabinoid (JWH-018 and JWH-073) use [10]. Fatalities associated with abuse of JWH-122, JWH-175, JWH-210, JWH-250, XLR-11, and AM-2201 have also been reported [11].

Synthetic cannabinoids do not cross-react with marijuana immunoassay. Therefore, urine toxicology screen is usually negative in a person suspected with synthetic cannabinoids overdose unless marijuana is also smoked along with synthetic cannabinoid. Many THC users also abuse synthetic cannabinoids. In this case, urine screen for marijuana metabolite should be positive. However, recently, ELISA assay (National Medical Services JWH-018 direct assay targeting JWH-018 metabolite: JWH-018 N (5-hydroxylpentyl)) for detecting certain synthetic cannabinoid in urine has been commercially available. The drug of abuse array V biochip array assay by Randox Corporation is also capable of detecting certain synthetic cannabinoids if present in the urine specimen. In general, synthetic cannabinoids are not routinely analyzed in most hospital-based clinical laboratories. Therefore, specimens should be shipped to a reference laboratory or a major academic medical center where such tests may be available. LC-MS/MS-based methods are most suitable for analysis of synthetic cannabinoids.

## Designer drugs that are opioid analogs

Heroin (diacetylmorphine) which was synthesized from morphine in 1874 can be considered as the first designer drug. Heroin is a class I scheduled drug in the United States. Heroin is metabolized to 6-monoacetylmorphine (also called 6-acetylmorphie) and then finally to morphine (see Chapter 16). The presence of 6-monoacetylmorphine in urine confirms heroin abuse.

Fentanyl is a widely used synthetic narcotic analgesic which is approximately 75–100 times more potent than morphine. Several analogs of fentanyl such as sufentanil, alfentanil, lofentanil, and remifentanil have been synthesized by pharmaceutical industries and are used clinically. Carfentanyl is used only in veterinary medicine. Currently in the United States, fentanyl is a Schedule II drugs which are used as an anesthetic. Injection of 50–100 μg fentanyl produces rapid analgesic effect and unconsciousness. Fentanyl is also available as lozenges and transdermal patches (Duragesic) for pain management. Oral transmucosal fentanyl citrate (Actiq) is a relatively new formulation where fentanyl is incorporated into a sweetened matrix to produce fentanyl lozenge. Therapeutic range of fentanyl is considered as 1–3 ng/mL in serum and toxicity of fentanyl is similar to opiate toxicity [12]. Many deaths have been reported due to

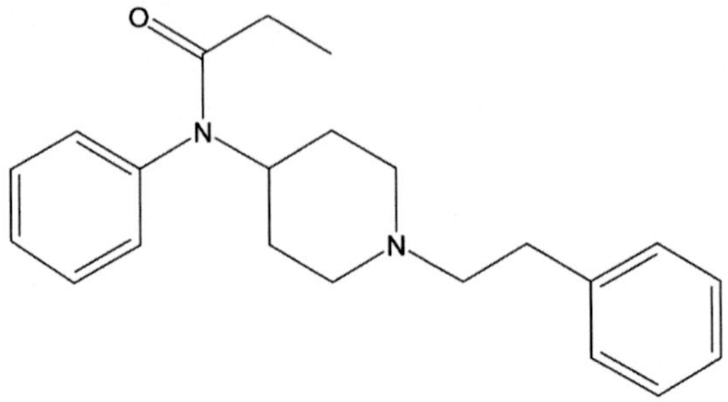

**FIG. 3**
Chemical structure of fentanyl.

fentanyl abuse. More recently, deaths due to abuse of heroin laced with fentanyl (fentanyl is cheaper than heroin) have been reported. The chemical structure of fentanyl is given in Fig. 3.

Fentanyl analog designer drugs China White (α-methylfentanyl) appeared in the underground market of California in 1979 causing over 100 deaths. In 1984, another illicit designer drug 3-methylfentanyl appeared as a street drug in California which was also related to fatal drug overdose. More recently, McIntyre et al. described a fatality associated with butyr-fentanyl. Using gas chromatography/mass spectrometry, the authors demonstrated that butyr-fentanyl concentration in the peripheral blood was 58 ng/mL while the concentration in vitreous was 40 ng/mL. The liver concentration was 320 ng/g [13]. Death due to abuse of acetyl fentanyl has also been reported. The acetyl fentanyl peripheral blood concentration was 260 ng/mL while the liver concentration was 1000 ng/kg in one person who died from acetyl fentanyl overdose [14].

Fentanyl and its analogs are metabolized through N-dealkylation. The major metabolite of fentanyl is nor-fentanyl. Similarly, major metabolite of alpha-methyl fentanyl is nor-fentanyl but 3-methyl fentanyl is converted into nor-3-methyl fentanyl. Neither fentanyl nor its analogs can be detected in routine urine drug testing by regular opiate screening assays because these compounds do cross-react with antibody used in opiate assays which is specific for morphine However, there is a specific ELISA assay for determination of fentanyl in urine. The concentrations of fentanyl and its analogs can be measured in serum, urine, and other biological matrix using gas chromatography/mass spectrometry or LC-MS/MS.

## Meperidine and phencyclidine-like designer drugs

Meperidine (Demerol) is a synthetic opioid which is used as a narcotic analgesic. It is available both as an injectable form and as oral use supplied as hydrochloride salt. Both meperidine and several of its analogs are abused. A common designer drug which is a meperidine analog is MPPP (1-methyl-4-phenyl-4-propionoxypiperdine). An impurity found in some illicitly synthesized MPPP preparation which is identified as 1-methyl-4-phenyl-1,2,3,6-tetrahydropyridine (MPTP) caused permanent Parkinsonism in a number of intravenous drug abusers. Meperidine and its analogs have negligible cross-reactivity with the opiate screening assay for drugs of abuse. However, specific immunoassay is available for detection of meperidine and its metabolites in urine. Chromatographic techniques are also available for analysis of meperidine and its analog.

There are a number of designer drugs that are analogs of phencyclidine. These drugs include N-(1-phenylcyclohexyl)-propanamine (PCPr), N-(1-phenylcyclohexyl)-3-methoxypropanamine (PCMPA), N-(1-phenylcyclohexyl)-2-methoxyethanamine (PCMEA), and N-(1-phenylcyclohexyl)-2-ethoxyethanamine (PCEEA). These drugs are extensively metabolized by liver enzymes.

## Rave party drugs and date rape drugs

In commonly abused drugs in rave parties are methamphetamine, MDMA (3,4-methylenedioxymethamphetamine), MDA (3,4-methyledioxyamphetamine), ketamine and also date rape drugs gamma-hydroxy butyric acid (GHB) as well as its analogs. Both MDMA and MDA are designer drugs which are synthesized and have structural similarity with amphetamine and methamphetamine. Rave party drugs are also known as club drugs. Abuse of MDMA and MDA may cause fatality.

Abuse of methamphetamine can be easily detected in urine using amphetamine immunoassays or immunoassays specifically designed to detect MDMA. However, detection of other rave party drugs such as GHB (street name; Blue, Nitro, GH) and its analogs, ketamine, and Rohypnol (flunitrazepam; street name: Roofies, Mexican valium, Roche) is difficult. Rohypnol and GHB are often used for drug-assisted sexual assault (date rape). Chemical structure of date rape drugs Rohypnol and GHB are given in Fig. 4. Ketamine is less frequently used both in rave parties and in date rape situation. Chemical structure of ketamine is also given in Fig. 4.

Flunitrazepam (Rohypnol), a benzodiazepine, is banned in the United States. However, this drug can be obtained illegally and usually benzodiazepine screening assay in urine at usual cut-off concentration of 200 ng/mL may not detect the presence of flunitrazepam due to its low concentration in urine. Forsman et al. using CEDIA benzodiazepine assay at a cut-off of 300 ng/mL failed to obtain a positive result in the urine of volunteers after they received a single

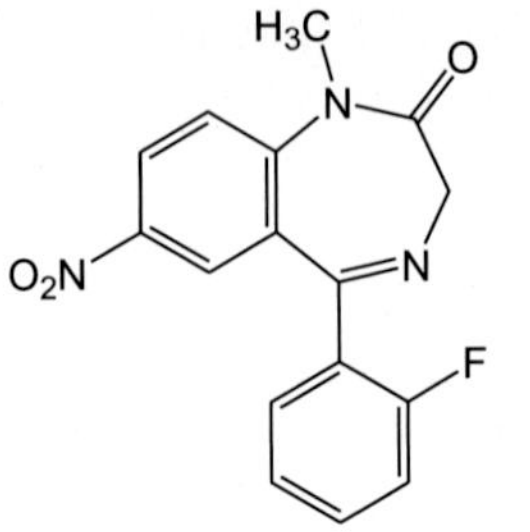

Rohypnol (Flunitrazepam)

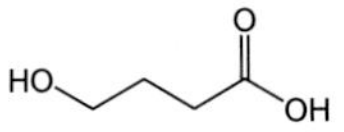

GHB (Gamma-hydroxybutyric acid)

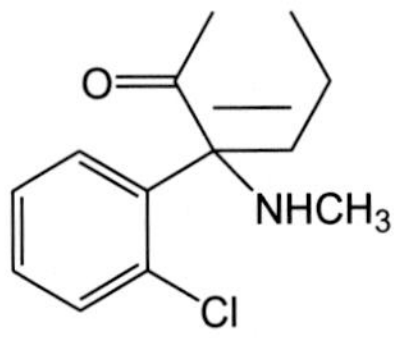

Ketamine

**FIG. 4**

Chemical structure of date rape drugs Rohypnol, GHB (most frequently used) and ketamine (less frequently used).

dose of 0.5 mg flunitrazepam. In addition, only 22 out of 102 urine specimens collected from volunteers after receiving the highest dose of flunitrazepam (2 mg) showed positive screening test result using the CEDIA benzodiazepine assay [15]. However, chromatographic methods such as gas chromatography/mass spectrometry or liquid chromatography combined with tandem mass spectrometry (LC-MS/MS) at a cut-off concentration of 40 ng/mL or lower should detect both flunitrazepam and its metabolite 7-amino flunitrazepam.

There is no commercially available immunoassay for ketamine, GHB, and its analogs. Therefore, only chromatographic methods (gas chromatography/mass spectrometry: GC/MS or liquid chromatography/tandem mass spectrometry: LC-MS/MS) are available for their analysis mostly in medical legal investigation. Ketamine is used through intravenous injection but it can also be added to a drink for abuse or drug-facilitated rape.

Although sale of GHB and its analogs for human consumption is against the law in the United States, several GHB analogs are commercially available as industrial solvents and for manufacture of plastic and other products, for example 1,4-butanediol. 1,4-Butanediol may be present in toys, especially toys manufactured outside the United States. Because 1,4-butanediol is endogenously converted into GHB, licking toys containing 1,4-butanediol may cause serious toxicity in toddlers. Ortmann et al. reported coma in a 20-month-old child due to ingestion of a plastic toy containing 1,4-butanediol [16].

Table 3 Rave party, date rape drugs, and GHB analogs.

| Drug class | Specific drugs |
| --- | --- |
| Rave party (club) drugs | 3,4-Methylenedioxyamphetamine (MDA) |
| | 3,4-Methylenedioxymethamphetamine (MDMA) |
| | Ketamine, methamphetamine |
| | Gamma-hydroxybutyric acid (GHB) |
| Date rape drugs | Rohypnol (Flunitrazepam) |
| | Gamma-hydroxybutyric acid (GHB) |
| GHB analogs | Gamma butyrolactone (BDL) |
| | 1,4-Butanediol (1,4-BD) |
| | Gamma hydroxy valeric acid (GHV) |
| | Gamma valerolactone (GVL) |

Gamma-valerolactone (GVL) is the most recent addition to designer drug list which is GHB analog. GVL is quickly metabolized in vivo by lactonase enzymes into 4-methyl gamma hydroxybutyrate (gamma-hydroxyvaleric acid) which is responsible for its pharmacological effects similar to GHB because this metabolite can bind GHB receptor. Rave party drugs, date rape drugs, and designer drugs related to GHB are listed in Table 3.

## LSD abuse

Lysergic acid diethylamide (LSD) is a semisynthetic substance derived from lysergic acid, a natural product found in parasitic rye fungus (*Claviceps purpurea*). LSD was first synthesized by Albert Hofmann, a scientist at the Sandoz AG Pharmaceutical Company (Basel, Switzerland) in 1938 while searching for pharmacologically active derivative of lysergic acid. He accidentally discovered dramatic psychoactive effect of LSD in 1943. Toward the end of 1960s, people used LSD for recreational and spiritual purposes leading to a "psychedelic movement." Although use of LSD has declined after that, it is still abused today. The optimum dosage abusers use is 100–200 µg. The chemical structure of LSD is given in Fig. 5.

After oral administration, LSD is absorbed and effects are observed 30–45 after ingestion. The oral bioavailability of LSD was determined to be 71%. After ingestion of LSD, maximum plasma concentrations could be reached 1.5 h (median value). The major metabolite detected in urine was 2-oxo-3-hydroxy-lysergicacid diethylamide.

Similar to active ingredients of magic mushroom and peyote cactus, LSD also binds to serotonin receptor. Studies have shown that serotonin 2A receptor (5-$HT_{2A}$) is critically involved in the formation of visual hallucinations and

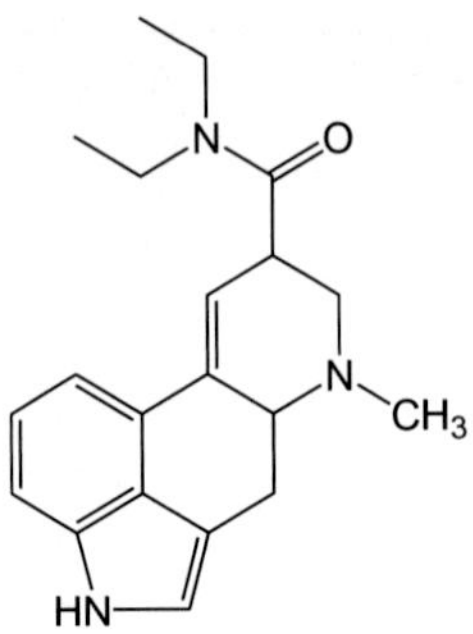

**FIG. 5**
Chemical structure of LSD.

cognitive impairment in LSD induced states. Moreover, activation of $5\text{-HT}_{2A}$ receptor by LSD leads to a hippocampal-prefrontal cortex-mediated breakdown of inhibitory processing, which might subsequently promote the formation of LSD-induced visual imageries [17].

Immunoassays are commercially available as screening tests for the presence of LSD in urine. For example, cloned enzyme donor immunoassay (CEDIA) LSD assay utilizes a monoclonal antibody capable of detecting LSD and the assay has a cut-off concentration of 0.5 ng/mL. CEDIA assay for LSD can be used either as a positive/negative screening mode or for obtaining semiquantitative values. Other immunoassays such as enzyme multiplied immunoassay technique (EMIT), kinetic interaction of microparticles in solution (KIMS), etc. are also commercially available. In addition, both GC/MS and LC-MS/MS can be used for confirmation of LSD and its metabolite in urine and other biological fluids.

## Abuse of volatiles (solvents)

The general populations are exposed to many volatile compounds which are used in common household products including cleaning agents, adhesives, paints, and cosmetic products. However, there are certain volatile compounds which are listed under drug class of abused inhalants. These compounds usually have low vapor pressure and high volatility at room temperature that enables use of these solvents as euphorigenic inhaled agents. These abused inhalants include aromatic hydrocarbons, aliphatic hydrocarbons, halocarbons, halogenated ethers, nitrous oxide, and alkyl nitrites.

Various readily available household and office products are abused including glue, adhesives, nail polish, nail polish remover, cigarette lighter fluid, butane gas, air fresheners, deodorant, hairspray, pain relieving spray, typewriter correction fluid, paint thinners, paint removers, and a variety of other agents. These household and office products contain toxic solvent such as toluene (paint, spray paint, adhesives, paint thinner, shoe polish), acetone (nail polish remover, typewriter correction fluid and markers), hexane (glue, rubber cement), chlorinated hydrocarbon (spot and grease removers), xylene (permanent markers), propane gas (gas to light the grill, spray paints), butane gas (lighter fluid, spray paint), and fluorocarbons (hair spray, analgesic spray, refrigerator coolant such as Freon). However, toluene is found in many household products such as glues and thinners which are widely abused due to psychoactive property of toluene.

Although solvent (inhalant) abuse is common among adolescents not only in the United States but also worldwide, this problem is often overlooked. In the United States, approximately 20% of adolescents have tried inhalants at least once by the time they had reached eighth grade. Analysis of data from the US Poison Control Centers from 1993 to 2008 showed 35,453 cases of toxicity due to abuse of inhalants. Prevalence was highest among children 12–17 years and peaked in 14 years of age. Inhalant abuse was more common in boys (73.5%) than in girls. Propellants, gasoline, and paint were the most frequent product categories. Butane, propane, and air freshener abuse are associated with the highest fatality rates [18].

Inhalant are abused either by breathing directly from a container or by soaking a rag with the solvent and then placing it over the nose and mouth. Moreover, abusers also pour the solvent in a plastic bag and then breathe fumes. Fatalities due to abuse of products containing toluene have been reported. Fatalities may occur from abusing diethyl ether, propane and butane.

In general, complete blood count, electrolyte, acid-base assessment, hepatic and renal chemistry profiles are ordered in victims with suspected volatile abuse. Blood alcohol should be ordered as many solvent abusers also abuse alcohol. Urine drug screen should also be ordered as illicit drugs are often detected in solvent abusers. Treatment is mostly supportive. Commonly abused inhalants and their composition are listed in Table 4.

## Magic mushroom abuse

Magic mushrooms (psychoactive fungi) that grow in the United States, Mexico, South America, and many other parts of the world contain psilocybin and psilocin which are hallucinogens and are Class I controlled substances. Magic mushrooms can be eaten raw, cooked with food, or dried and then consumed. These mushrooms can be mistaken for other nonhallucinogenic mushrooms or even poisonous mushrooms such as Amanita class. After ingestion, psilocybin, the major component of mushroom, is rapidly converted by

**Table 4** List of commonly abused inhalants and their composition.

| Product | Solvent found |
| --- | --- |
| Spray paint | Propane, butane, toluene, hydrocarbon, fluorocarbons |
| Hair spray | Propane, butane, fluorocarbons |
| Paint thinner | Toluene, methyl chloride, methanol |
| Shoe polish | Toluene |
| Nail polish remover | Acetone, toluene |
| Lighter fluid | Butane |
| Domestic fuel | Propane, butane, isooctane |
| Film cleaner, correction fluid | 1,1,1-Trichloroethane, acetone |
| Adhesive glue | Toluene, xylene, ethyl benzene, hexane |
| Air freshener | Fluorocarbons |
| Rubber cement, marker | Toluene, hexane, methyl chloride, acetone |
| Spot remover, degreasers | Chlorinated hydrocarbons |
| Gasoline | Combination of aliphatic and aromatic hydrocarbons and other volatile organic compounds |
| Refrigerator fluid | Trichlorofluoromethane (Freon) |
| Metal cleaner | n-Propyl bromide |

dephosphorylation into psilocin, which has psychoactive effects similar to lysergic acid diethylamide (LSD). Chemical structure of psilocybin and psilocin are given in Fig. 6. In general, duration of the trip after abusing magic mushroom may last between 2 and 6 h and effects ranged from intended feeling of relaxation, uncontrollable laughter, joy, euphoria, visual enhancement of colors, hallucination, and altered perception but some abusers may also experience negative effect such as depression or paranoia. Intoxication from use of magic mushroom is common. Some species of magic mushroom contain phenylethylamine which may cause cardiac toxicity. There are several reports of fatality due to ingestion of magic mushroom. One person died 6–8 h after ingestion of unknown quantity of magic mushroom. Postmortem toxicology analysis showed very high plasma concentration of psilocin (4000 ng/mL) [19]. Death of a 27-year-old Japanese male due to ingestion of magic mushroom has also been reported [20]. In addition, death from ingestion of Amanita mushroom due to mistaken identity as magic mushroom has been reported [21]. It is dangerous practice to eat wild mushrooms because there are other reports of severe toxicity due to ingestion of toxic wild mushroom as a result of misidentification [22]. Currently there is no immunoassay for determination of psilocybin and psilocin in body fluids. Therefore, chromatographic methods must be employed for their analysis, especially during a forensic investigation.

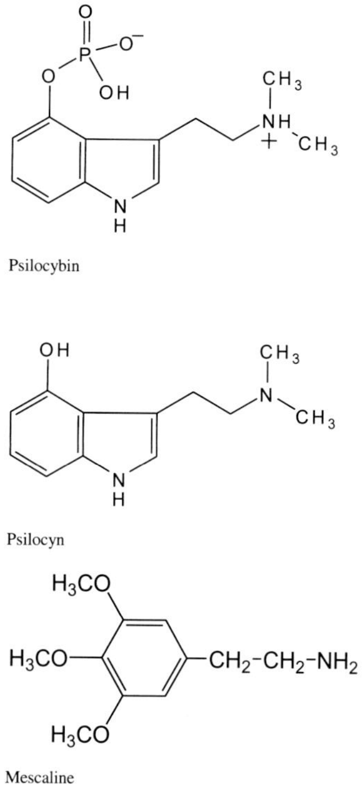

**FIG. 6**

Chemical structures of psilocybin, psilocin, and mescaline.

## Peyote cactus abuse

Peyote cactus (*Lophophora williamsii*) is a cactus that grows in the Southwestern part of the United States and Mexico. Peyote cactus is small and round without sharp spines and the top of the cactus which is above ground is called "crown." The crown consists of disk-shaped buttons which contain the psychoactive compound mescaline (3,4,5-trimethoxyphenethylamine). The chemical structure of mescaline is given in Fig. 6. Mescaline can be extracted from the peyote cactus or buttons can be chewed or soaked in water to produce psychoactive liquid. Mescaline is classified as a Class I controlled substance, but approximately 300,000 members of Native Americans Church can ingest peyote cactus legally as a religious sacrament during all night prayer in the Native American Church [23]. The mescaline content of peyote cactus is usually 0.4% in fresh cactus and 3%–6% in dried cactus. The highest psychedelic effect may be achieved within 2h of ingestion but the effect may last up to 8h [24]. The psychoactive effects of mescaline are similar to LSD (lysergic acid diethylamide) including deeply mystical feelings. Abuse of peyote cactus may cause serious

| Table 5 Active ingredients of magic mushroom and peyote cactus. | |
| --- | --- |
| **Substance abused** | **Active ingredient** |
| Magic mushroom | Psilocybin and psilocin, while psilocybin is also converted in vivo into psilocin |
| Peyote cactus | Mescaline |

toxicity requiring medical attention. Although rare, severe toxicity and even death may occur from mescaline overdose. One person who died under the influence of mescaline showed 9.7 μg/mL of mescaline concentration in blood, and 1163 μg/mL in urine [25]. Currently, there is no commercially available immunoassay for analysis of mescaline in body fluids and only chromatographic methods are available for its analysis. Although mescaline is metabolized into several different metabolites, a large amount of mescaline can be recovered unchanged in urine. Therefore, detection of mescaline in serum or urine can be used for establishing diagnosis of magic mushroom abuse. For death investigation, confirmation of the presence of mescaline in body fluid is essential in establishing the cause of death. Severe toxicity and even death may occur from mescaline overdose. Active ingredients of magic mushroom and peyote cactus are also listed in Table 5.

## Summary/key points

- Common designer drugs structurally related to amphetamine include 3,4-methylenedioxymethamphetamine (MDMA, Ecstasy), 3,4-methylenedioxy-amphetamine (MDA), para-methoxyamphetamine (PMA) para-methoxy methamphetamine (PMMA), and 4-methylthioamphetamine (4-MTA). Although the presence of MDMA and MDA in urine can be detected by amphetamine immunoassays or immunoassay especially designed for detecting MDMA, other amphetamine designer drugs have lower cross-reactivities and may not be detected by amphetamine immunoassays.
- Bath salts, also known as beta-keto amphetamines, are synthetic derivatives of cathinone (natural component of khat plant). Major bath salts are methylenedioxypyrovalerone (MDPV), and 4-methylmethcathinone (also known as mephedrone). These compounds have stimulant effects like amphetamine and cocaine. Abuse of bath salts may cause serious life-threatening toxicity and even death. Bath salts cannot be detected by regular toxicological screen because these compounds do not cross-react with amphetamine immunoassays. However, a drug of abuse array V biochip array assay by Randox

Corporation is capable of detecting some synthetic cathinone (bath salts) if present in the urine specimen.

- Since 2008, synthetic marijuana compounds sold as spice, K2, K2 blonde, or herbal high are gaining popularity among drug abusers. The first synthetic compound in this category JWH-018 was synthesized by Dr. John W. Huffman at Clemson University, but currently more than 100 compounds are available. Most common examples of spice are JWH-018, JWH-073, JWH-250, JWH-015, JWH-081, HU-210, HU-211 (synthesized in Hebrew University), and CP-47,497 (synthesized at Pfizer).

- Similar to natural cannabinoid (THC: tetrahydrocannabinol), synthetic cannabinoids bind with cannabinoid receptor 1 (CB1) and cannabinoid receptor 2 (CB2) which are found mostly in the central nervous system but also may be found in peripheral tissues including lungs, liver, and kidneys. However, within the brain, these receptors are mostly located in the cerebral cortex, hippocampus, basal ganglia, and cerebellum. In contrast to THC, which is a partial CB1 receptor agonist, synthetic cannabinoids are more potent full CB1 agonist with much higher affinity for CB1 receptors than THC. Some of the metabolites are also active with high affinity for CB1 receptors. Interestingly, synthetic cannabinoids are not structurally related to THC.

- Marijuana immunoassays cannot detect the presence of these illicit drugs if present in urine. However, few specialized immunoassays including Randox assay is available for detecting the presence of synthetic cannabinoids in urine.

- Rave party drugs MDMA (3,4-methylenedioxymethamphetamine) and MDA (3,4-methylediioxyamphetamine) can be detected by amphetamine/methamphetamine immunoassay or MDMA immunoassay. However, there is no immunoassay available for detection of other ketamine as well as date rape drugs gamma-hydroxy butyric acid (GHB) and its analogs.

- Fentanyl analog designer drug China White (α-methylfentanyl) and 3-methylfentanyl are abused, but neither fentanyl nor its analogs can be detected in routine urine drug testing by opiate immunoassays.

- LSD (lysergic acid diethylamide) is a semisynthetic substance derived from lysergic acid, a natural product found in parasitic rye fungus (*Claviceps purpurea*). LSD was first synthesized by Albert Hofmann in 1938. Toward the end of 1960s, people used LSD for recreational and spiritual purposes leading to a "psychedelic movement." Although use of LSD has declined after that, it is still abused today. The optimum dosage abusers use is 100–200 µg.

- Similar to active ingredients of magic mushroom and peyote cactus, LSD also binds to serotonin receptor. LSD can be analyzed by immunoassays, GC/MS, or LC-MS/MS.

- Various readily available household and office products which are volatiles are also abused most commonly by adolescents. These products include glue, adhesives, nail polish, nail polish remover, cigarette lighter fluid, butane gas, air fresheners, deodorant, hairspray, pain relieving spray, typewriter correction fluid, paint thinners, paint removers, and a variety of other agents. These household and office products contain toxic solvent such as toluene (paint, spray paint, adhesives, paint thinner, and shoe polish), acetone (nail polish remover, typewriter correction fluid, and markers), hexane (glue, rubber cement), chlorinated hydrocarbon (spot and grease removers), xylene (permanent markers), propane gas (gas to light the grill, spray paints), butane gas (lighter fluid, spray paint), and fluorocarbons (hair spray, analgesic spray, refrigerator coolant such as Freon). Solvent abuse may cause fatality.
- Magic mushrooms (psychoactive fungi) contain psilocybin and psilocin which are hallucinogens and are Class I controlled substances. After ingestion, psilocybin, often the major component of magic mushroom, is rapidly converted by dephosphorylation into psilocin which has psychoactive effects similar to LSD (lysergic acid diethylamide). Magic mushroom abuse cannot be detected in routine immunoassay screening for abused drugs in urine.
- Peyote cactus (*Lophophora williamsii*) is a small spineless cactus that grows in the Southwestern part of the United States and Mexico and contains the psychoactive compound mescaline, a Class I controlled substance with effects similar to LSD. Peyote cactus abuse cannot be detected in routine toxicology screen.

# References

[1] El-Menyar A, Mekkodathil A, Al-Thani H, Al-Motarreb A. Khat use: history and heart failure. Oman Med J 2015;30:77–82.

[2] Ishraq D, Jiri S. Khat habit and its health effect. A natural amphetamine. Biomed Pap Med Fac Univ Palacky Olomouc Czech Repub 2004;148:11–5.

[3] Fass JA, Fass AD, Garcia A. Synthetic cathinones (bath salts): legal status and patterns of abuse. Ann Pharamcother 2012;46:436–41.

[4] Gonçalves JL, Alves VL, Aguiar J, Teixeira HM, Câmara JS. Synthetic cathinones: an evolving class of new psychoactive substances. Crit Rev Toxicol 2019;49:549–66.

[5] Zaami S, Giorgetti R, Pichini S, Pantano F, et al. Synthetic cathinones related fatalities: an update. Eur Rev Med Pharmacol Sci 2018;22:268–74.

[6] Cooper ZD. Adverse effects of synthetic cannabinoids: management of acute toxicity and withdrawal. Curr Psychiatry Rep 2016;18:52.

[7] Seely CA, Lapoint J, Moran JH, Fattore L. Spice drugs are more than harmless herbal blends: a review of the pharmacology and toxicology of synthetic cannabinoids. Prog Neuropsychopharmacol Biol Psychiatry 2012;39:234–343.

[8] Wells D, Ott CA. The new marijuana. Ann Pharmacother 2011;45:414–7.

[9] Alipour A, Patel PB, Shabbir Z, Gabrielson S. Review of the many faces of synthetic cannabinoid toxicities. Ment Health Clin 2019;9:93–9.

[10] Shanks KG, Dahn T, Terrell AR. Detection of JWH-018 and JWH-073 by UPLC-MS-MS in postmortem whole blood. J Anal Toxicol 2012;36:145–52.

[11] Labay LM, Caruso JL, Gilson TP, Phipps RJ, et al. Synthetic cannabinoid drug use as a cause or contributory cause of death. Forensic Sci Int 2016;260:31–9.

[12] Mystakidou K, Katsouda E, Parpa E, Vlahos L, et al. Oral transmucosal fentanyl citrate: overview of pharmacological and clinical characteristics. Drug Deliv 2006;13:269–76.

[13] McIntyre IM, Trochta A, Gary RD, Wright J, et al. An acute butyr-fentanyl fatality: a case report with postmortem concentrations. J Anal Toxicol 2016;40:162–6.

[14] McIntyre IM, Trochta A, Gary RD, Malamatos M, et al. An acute acetyl fentanyl fatality: a case report with postmortem concentrations. J Anal Toxicol 2015;39:490–4.

[15] Forsman M, Nystrom I, Roman M, Berglund L, et al. Urinary detection times and excretion patterns of flunitrazepam and its metabolites after a single oral dose. J Anal Toxicol 2009;33:491–501.

[16] Ortmann LA, Jaeger MW, James LP, Schexnayder SM. Coma in a 20 month old child from an ingestion of a toy containing 1,4-butanediol, a precursor of gamma-hydroxybutyrate. Pediatr Emerg Care 2009;25:758–60.

[17] Schmidt A, Muller F, Lenz C, Dodler C, et al. Acute LSD effects on response inhibition neural networks. Psychol Med 2018;48:1464–73.

[18] Marsolek MR, White NC, Litovitz TL. Inhalant abuse monitoring trends by using poison control data 1993–2008. Pediatrics 2010;125:906–13.

[19] Lim TH, Wasywich CA, Ruygrok PN. A fatal case of magic mushroom ingestion in a heart transplant recipient. Intern Med J 2012;42:1268–9.

[20] Gonomori K, Yoshioka N. The examination of mushroom poisonings at Akita University. Leg Med (Tokyo) 2003;5(Suppl. 1):S83–6.

[21] Madhok M. Amanita bisporgera: ingestion and death from mistaken identity. Minn Med 2007;90:48–50.

[22] Madhok M, Scalzo AJ, Blume CM, Neuschwander-Tetri BA, et al. Amanita biosporigera ingestion: mistaken identity, dose related toxicity, and improvement despite severe hepatotoxicity. Pediatr Emerg Care 2006;22:177–80.

[23] Halpern JH, Sherwood AR, Hudson JI, Yugerlum-Todd D, Pope Jr HG. Psychological and cognitive effects of long term peyote use among Native Americans. Biol Psychiatry 2005;58:624–31.

[24] Nichols DE. Hallucinogens. Pharmacol Ther 2004;101:131–81.

[25] Reynolds PC, Jindrich EJ. A mescaline associated fatality. J Anal Toxicol 1985;9:183–4.

# CHAPTER 18

# Testing for ethyl alcohol (alcohol) and other volatiles

## Alcohol use and abuse

Alcohol (ethanol) has been used since prehistoric time (10,000 BCE) by mankind. The normal fermentation process which uses yeast cannot produce alcohol beverages with alcohol content over 14%. Therefore, hard liquors or spirits are produced using fermentation followed by distillation. Alcoholic beverages are full of calories and can be classified under two broad categories: beer and wine (produced by direct fermentation where alcohol content is usually <14%) and spirits (produced by fermentation followed by distillation. Alcohol content can be as high as 40% or more). Alcohol content of various alcoholic beverages varies widely, for example, beer contains approximately 4%–7% alcohol, while average alcohol content of vodka is 40%–50%. However, due to wide variation between serving size, one drink (often called one standard drink) contains approximately 0.6 oz of alcohol, which is equivalent to 14 g of pure alcohol. In the United States, a standard drink is defined as a bottle of beer (12 fluid ounces) containing 5% alcohol, 8.5 fluid ounces of malt liquor containing 7% alcohol, 5 fluid ounces glass of wine containing 12% alcohol, 3.5 fluid ounces of fortified wine like sherry or port containing about 17% alcohol, 2.5 fluid ounces of cordial or liqueur containing 24% alcohol, or one shot (1.5 fluid ounces) of a distilled spirit such as gin, rum, vodka, or whiskey. Alcohol content of various drinks are listed in Table 1.

Currently in the United States, the alcohol content of a drink is measured by the percentage of alcohol by the volume. The code of Federal Regulations requires that the label of alcoholic beverages must state the alcohol content by volume. Alcohol proof (also known as proof), also a measure of the content alcohol in alcoholic beverages by volume, was originated in England and was approximately 1.82 times the alcohol by volume in alcoholic beverages. However, the United Kingdom now uses the alcohol by volume standard instead of alcohol proof. In the United States, alcohol proof is defined as twice the percentage of alcohol in the beverage. Therefore, vodka containing 40% alcohol is considered as 80 proof in the United States. Alcoholic drinks primarily consist of

381

Clinical Chemistry, Immunology and Laboratory Quality Control. https://doi.org/10.1016/B978-0-12-815960-6.00014-5

**Table 1** Alcohol content of various alcoholic beverages.

| Beverage | One standard drink (ounce) | Alcohol content (%) |
| --- | --- | --- |
| Standard American beer | 12 | 4–6 |
| Table wine | 5 | 7–14 |
| Sparkling wine | 5 | 8–14 |
| Whiskey | 0.6 | 40–75 |
| Vodka | 0.6 | 40–50 |
| Gin | 0.6 | 40–49 |
| Rum | 0.6 | 40–80 |
| Tequila | 0.6 | 45–50 |

water, alcohol, and variable amounts of sugars and carbohydrates (residual sugar and starch left after fermentation), but negligible amounts of other nutrients such as proteins, vitamins, or minerals. However, distilled liquors such as cognac, vodka, whiskey, and rum contain no sugars. Red wine and dry white wines contain 2 to 10 g of sugar per liter, while sweet wines and port wines may contain up to 120 g of sugar per liter of wine. Beer and dry sherry contain 30 g of sugar per liter [1].

United States Department of Agriculture (USDA) and Department of Health and Human Services jointly published "Dietary Guidelines for Americans" every 5 years, suggesting Americans what constitutes a balanced diet. These guidelines also include suggestion for drinking in moderation. However, alcohol is not a component in USDA food pattern. If alcohol is consumed, the calories from alcohol must be accounted for when other foods are consumed so that daily calorie intake does not exceed the recommended limit (1600–2400 cal per day for women and 2000–3000 cal per day for men). The latest Dietary Guidelines for Americans 2015–2000, eighth edition, suggest that if alcohol is consumed it should be consumed in moderation following these guidelines:

- Up to one drink per day for women and up to two drinks per day for men and only by adults of legal drinking age (21 years or older).

One drink is defined by the guidelines as containing 14 g (0.6 fluid ounces) of pure alcohol.

People who mix alcohol and caffeine may drink more alcohol and become more intoxicated than they realize, increasing the risk of alcohol-related adverse events. Energy drinks are gaining popularity among young adults as well as among underage drinkers. Studies have indicated that energy drinks may increase craving for alcohol and binge drinking. When an energy drink which often contains caffeine is combined with alcohol, the desire to drink alcohol is more pronounced compared to drinking alcohol without consumption of energy drink. Moreover, pleasurable experience of drinking alcohol is also enhanced by consuming energy drinks at the same time [2].

National Institute on Alcohol Abuse and Alcoholism (NIAAA) considers high-risk drinking as following:

- Consuming four or more drinks in any day or eight or more drinks per week for women
- Consuming five or more drinks in any day or 15 or more drinks per week for men

Binge drinking is defined by NIAAA as consumption of five drinks in 2h for men and four drinks in 2h for women and such drinking pattern always produce a blood alcohol level of 0.08% or higher. However, another government agency SAMHSA (Substance Abuse and Mental Health Services Administration) defines binge drinking as consuming five or more alcoholic beverages on the same occasion in the past 30 days. SAMHSA also defines heavy drinking as consuming five or more drinks on the same occasion on each of 5 or more days in the past 30 days. Definition of moderate, high-risk, heavy, and binge drinking are listed in Table 2.

Alcohol abuse is a leading cause of mortality and morbidity internationally and is ranked by the World Health Organization (WHO) as one of the top five risk

**Table 2** Definition of moderate, high-risk, heavy, and binge drinking.

| Type of drinking | Definition/guideline |
|---|---|
| Moderate drinking (Dietary Guidelines for Americans 2015–2000, eighth edition) | Up to one drink per day for women and up to two drinks per day for men |
| High-risk drinking (NIAAA) | Consumption of four or more drinks on any day or eight or more drinks per week for women and five or more drinks on any day or 15 or more drinks per week for men |
| Heavy drinking (SAMHSA) guideline | Consuming five or more drinks in the same occasion on each of 5 or more days in the past 30 days in both male and female |
| Binge drinking (NIAAA) | Consuming five drinks in 2h period for men and four drinks in 2h for women. Such drinking practice produces a blood alcohol level of 0.08% or more |
| Binge drinking (SAMHSA) | Consuming five or more drinks in the same occasion at least 1 day in the past 30 days for both men and women |

NIAAA, National Institute on Alcohol Abuse and Alcoholism; SAMHSA, Substance Abuse and Mental Health Services Administration.

**Table 3** Physiological effect of various blood alcohol levels.

| Blood alcohol | Physiological effect |
|---|---|
| 0.02%–0.05% (20–50mg/dL) | Relaxation and general positive mood elevating effect of alcohol including increased social interactions |
| 0.08% (80mg/dL) | Legal limit of driving, minor impairment possible in a person who drinks rarely |
| 0.1%–0.15% (100–150mg/dL) | Euphoria, but sensory impairment and decreased cognitive ability and difficulty in driving a motor vehicle |
| 0.2% (200mg/dL) | Worsening of sensory motor impairment and inability to drive, decreased cognitive function, and visual impairment |
| 0.3% (300mg/dL) | Vomiting, incontinence, symptoms of alcohol intoxication |
| 0.4% (400mg/dL) | Stupor, coma, respiratory depression, hypothermia |
| 0.5% and more (500mg/dL) | Potentially lethal |

factors for disease burden. Physiological effects of various blood alcohol levels are listed in Table 3. Usually moderate drinking (1–2 standard drink) produces a blood alcohol level in the range of 0.05% which produces beneficial mood elevation effect of alcohol including increased social interaction, while consuming more alcohol in one occasion may produce undesirable effects.

## Health benefits of moderate drinking

Currently the best evidence of health benefit of drinking in moderation is the reduced risk of cardiovascular disease. However, drinking in moderation has many other health benefits as listed below:

- Reduced risk of coronary heart disease including myocardial infarction and angina pectoris
- Better survival chance after a heart attack
- Reduced risk of ischemic stroke
- Reduced risk of type 2 diabetes
- Reduced risk of forming gallstones and kidney stones
- Reduced risk of developing arthritis
- Reduced risk of developing age-related dementia and Alzheimer's disease
- Reduced risk of certain types of cancer
- Increased longevity
- Better perception of good health (consumption of wine only)
- Less chance of getting common cold (consumption of red wine only)

The relationship between alcohol consumption and coronary heart disease was examined in the original Framingham Heart Study. The alcohol consumption showed a U-shaped curve with reduced risk of developing such disease with moderate drinking, but high risk of developing such diseases with heavy drinking. It is beneficial to drink one drink per day or at least six drinks per week to reduce the risk of coronary heart disease and heart attack in men, but women may get health benefit of moderate drinking by just consuming one drink per week [3]. There are several hypotheses on how moderate drinking can reduce the risk of developing heart disease that are listed below:

- Increases the concentration of high-density lipoprotein cholesterol (HDL cholesterol)
- Reduces the concentration of low-density lipoprotein cholesterol
- Reduces narrowing of coronary arteries by reducing plaque formation
- Reduces risk of blood clotting
- Reduces level of fibrinogen

Studies have indicated that the level of increase in HDL cholesterol in blood may explain 50% of protective effect of alcohol over cardiovascular disease and the other 50% may be partly related to inhibition of platelet aggregation, thus, reducing blood clot formation in coronary arteries. It has been suggested that although alcohol can increase HDL cholesterol levels and inhibit platelet aggregation, the polyphenolic antioxidant compound found in abundant in red wine can further reduce platelet activity via other mechanisms than alcohol. Therefore, it appears that red wine is more protective against cardiovascular disease than other alcoholic beverages [4].

## Health hazards of heavy drinking

All benefits of drinking alcohol are lost with heavy drinking. The most common alcohol-related organ damage is fatty liver, which may lead to even cirrhosis of liver, a potentially fatal disease. A pregnant woman or a woman planning to be pregnant should not consume any alcohol due to risk of fetal alcohol syndrome in the newborn baby. In the United States, over 18 million people aged 18 and older suffer from alcohol abuse or dependency and only 7% of these people receive any form of treatment. The highest prevalence of alcohol dependency in the United States is observed among younger people with ages between 18 and 24. Hazards of heavy drinking include:

- alcoholic liver disease
- increased risk of cardiovascular disease
- brain damage
- increased risk of stroke
- damage to immune and endocrine system
- anxiety and mood disorder
- increased risk of various cancers

- poor outcome of pregnancy
- reduced life span and increased mortality

In addition, alcoholics are prone to deep depression and violent behavior. However, from clinical point of view alcoholic liver diseases and fetal alcohol syndrome have been studied in more detail than other alcohol-related problems. Alcohol-induced liver disease can be classified under three categories: fatty liver, alcoholic hepatitis, and liver cirrhosis. Heavy drinking for as little as a few days may produce fatty change in the liver (steatosis), which is reversed after abstinence. A person infected with hepatitis C should consult with his/her physician regarding safe consumption of alcohol because alcohol and hepatitis C act in synergy to cause liver damage. Alcohol is a small molecule, so it can easily pass through the placenta to the embryo and cause birth defects, which are collectively known as fetal alcohol spectrum disorders. If more severe signs of these birth defects are present in a newborn, the condition may be called "fetal alcohol syndrome." Drinking alcohol during pregnancy may even cause stillbirth. Poor outcomes associated with drinking alcohol during pregnancy include:

- stillbirth or death of the newborn shortly after birth
- preterm baby
- smaller birth weight and/or growth retardation of the baby
- neurological abnormality
- facial abnormalities
- intellectual impairment during development

## Metabolism of alcohol: Effect of gender

After consumption, alcohol is absorbed from the stomach and metabolized by the liver. A small amount of alcohol is also found in breath and is the basis of breath analysis of alcohol in suspected drivers driving with impairment. Factors that affect how the body handles alcohol include:

- age
- gender
- body weight
- amount of food consumed with alcohol
- race and ethnicity (genetic factors)

When alcohol is consumed, about 20% is absorbed from the stomach and the rest is absorbed from the small intestine. Food substantially slows down the absorption of alcohol, and sipping alcohol instead of drinking also slows the absorption. Peak blood alcohol concentration is also reduced if alcohol is consumed with food. A small amount of alcohol is metabolized by the enzyme present in the gastric mucosa, and also a small amount of alcohol is metabolized by the liver before it can enter the main bloodstream (first pass metabolism). Then the rest of the alcohol enters the systematic circulation. After drinking the same amount of alcohol, a man would have a lower peak blood alcohol level compared to a woman with the same body weight. This gender difference in the blood alcohol level is related to the different body water content between a male and a female. Other important points to remember regarding alcohol metabolism include:

- Alcohol follows zero-order kinetics during metabolism, which means no matter how high the blood alcohol level, only a certain amount of alcohol is removed from the body per hour. In contrast, most drugs follow first order kinetics which means higher the drug level in blood, faster is the metabolism.
- Women also metabolize alcohol slower than men. In women, alcohol concentration is reduced by 15–18 mg/dL (0.018%) per hour regardless of blood alcohol concentration. In men, this rate is 18–20 mg/dL per hour.
- Hormonal changes also play a role in the metabolism of alcohol in women although this finding has been disputed in the medical literature.

Human liver metabolizes alcohol using zero-order kinetics. Several enzyme systems are involved in the metabolism of ethanol, namely alcohol dehydrogenase (ADH), microsomal ethanol oxidizing system (MEOS), and catalase. These enzymes also metabolize other similar compounds such as methanol, isopropyl alcohol, and ethylene glycol. Most important enzyme for alcohol metabolism is alcohol dehydrogenase (ADH) which is found in hepatocytes and the enzyme catalyzes the following reaction:

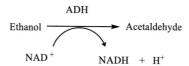

ADH activity is greatly influenced by the frequency of ethanol consumption. Adults who consume two to three alcoholic beverages per week metabolize ethanol at a rate much lower than alcoholics. For medium-sized male adults, the blood ethanol level declines at the rate of 18–20 mg/dL/h (0.018%/−0.020% hour). The average rate is slightly less in women than men. The major drug metabolizing family of enzymes found in the liver is the cytochrome P450 mixed function oxidase. Many members of this family of enzymes, most notably CYP3A4, CYP1A2, CYP2C19, and CYP2E1 isoenzymes, play vital roles in the metabolism of many drugs. For nonalcoholics, alcohol metabolism via CYP2E1 metabolic pathway is considered a minor, secondary route, but it becomes much more important in alcoholics. Because of the additional participation of CYP2E1, alcoholics can eliminate alcohol faster from their body compared to nonalcoholics.

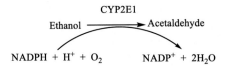

However, alcohol metabolism via CYP2E1 pathway produces free radicals which may be associated with liver damage in alcoholics. The acetaldehyde produced due to metabolism of alcohol regardless of pathway is subsequently converted to acetate (acetic acid) as the result of the action of mitochondrial aldehyde dehydrogenase (ALDH2). Acetaldehyde is fairly toxic compared to ethanol and must be metabolized fast.

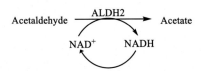

Acetate or acetic acid then enters the citric acid cycle which is a normal metabolic cycle of living cells and is converted into carbon dioxide and water. From the chemical point of view, the body oxidizes alcohol into carbon dioxide and water, and this process generates calories. Therefore, alcoholic drinks are high in calories. Metabolism of alcohol changes with advancing age because the activity of the enzymes involved in alcohol metabolism diminish with age. Water volume also reduces with advancing age. Therefore, an elderly person would have a higher blood alcohol level from consumption of the same amount of alcohol compared to a younger person of the same gender. Moreover, elderly persons consume more medications than younger people, and a medication may interact with alcohol.

## Metabolism of ethyl alcohol: Effect of genetic factors

The genes encoding aldehyde dehydrogenase enzyme are located on a single chromosome (chromosome 4), while *ALDH* genes are not localized in a single chromosome. Humans have 19 genes and 3 pseudogenes that encode the ALDH superfamily of isoenzymes, but only three of these genes, *ALDH1A1*, *ALDH1B1*, and *ALDH2*, are relevant to acetaldehyde metabolism. ALDH1A enzyme is usually found in cytosol, while ALDH1B1 and ALDH2 enzymes exert their function in mitochondria. ALDH2 plays most important role in the oxidation of acetaldehyde. The medication disulfiram (Antabuse) which is used in treating alcohol detoxification inhibits ALDH2 enzyme.

The most common polymorphism of *ALDH2* gene is *ALDH2*2* allele that encodes poorly active ALDH2 enzyme. The *ALDH2*2* allele is found among 45% of East Asians including Han Chinese, Japanese, and Koreans, but found rarely in other ethnic groups. It has been estimated that 540 million people worldwide (8% of world population) carry this allele. Although individuals who are *ALDH2*2* homozygotes (*ALDH2*2/*2*) show essentially inactive ALDH2 enzyme, but heterozygotes (*ALDH2*1/*2*) may show partial ALDH2 activity. Genetic and epidemiological studies have shown that *ALDH2*2* homozygous individuals are almost fully protected from developing alcohol use disorder, but heterozygous individuals have partial protection (approximately 60%). This is due to increased acetaldehyde levels in these individuals after consuming alcohol resulting in unpleasant reaction from drinking such as flushing (alcohol flush reaction or Asian flush), severe nausea, asthma attack, rapid heartbeat, and psychological distress [5,6]. However, superactive alcohol dehydrogenase enzyme (ADH) encoded by *ADH* gene may also cause acetaldehyde build up in blood causing unpleasant reaction. Therefore, three different scenarios may have protective effect on alcohol abuse:

- Polymorphism in *ALDH2* gene causing poor activity of aldehyde dehydrogenase enzyme.
- Polymorphism in *ADH* genes that lead to superactive alcohol dehydrogenase enzyme.
- Combination of superactive alcohol dehydrogenase activity and poor aldehyde dehydrogenase activity that offers total protection from consuming alcohol.

There are two superactive ADH enzymes due to polymorphisms of genes encoding ADH enzyme. One such polymorphism is *ADH1B*2* gene (Arg48His, rs1229984), which is found commonly among East Asians (Han Chinese, Japanese, Koreans, Filipinos, Malays, and aborigines of Australia and New Zealand) and also among approximately 25% of people of Jewish origin. This allele is also encountered in small frequency among Caucasians. Although *ADH1B*2* is more prevalent in Asians, if this allele is present in Caucasians it also provides protection from consuming alcohol [7]. Another polymorphism *ADH1B*3* (Arg370Cys, rs2066702) is found primarily in people of African descent and Native Americans. ADH isoenzymes encoded by *ADH1B*2* or *ADH1B*3* alleles are superactive ADH enzymes (due to significantly higher turnover rate of these enzymes) that results in 30- to 40-fold increase in metabolism of ethanol compared to normally functioning enzyme encoded by the wild-type *ADH1B*1* gene. As a result, acetaldehyde may build up in blood causing facial flushing and adverse side effects after alcohol use thus discouraging people carrying such alleles from drinking.

## Relation between whole blood alcohol and serum alcohol and legal limit of driving

Usually alcohol concentration in blood is measured in patients admitted to the emergency department with suspected drug and alcohol overdose. This is considered as medical blood alcohol determination because no chain of custody is maintained and alcohol concentration is confidential patient-related information, which cannot be disclosed to a third party. Medical alcohol determination is usually conducted in serum using automated analyzers and enzymatic assays for alcohol which can be easily automated. In addition, alcohol concentration in blood is measured in a driver suspected of driving with impairment (DWI). Legal alcohol testing is usually conducted using headspace gas chromatography and whole blood as the specimen.

- Legal limit of blood alcohol in all states of the United States is currently 0.08% whole blood alcohol (80 mg/dL).
- Serum alcohol concentration is higher than whole blood alcohol concentration due to higher amounts of water in serum (alcohol is freely water soluble).
- In order to convert serum alcohol level to whole blood alcohol level, serum alcohol level must be divided by 1.15. Therefore, serum alcohol concentration of 100 mg/dL (0.1%) is equivalent to 87 mg/dL (0.087%) whole blood concentration.

Rainey reported that ratio between serum and whole blood alcohol ranged from 0.88 to 1.59, but the median (most commonly occurred value) was 1.15. Therefore, dividing serum alcohol value by 1.15 would calculate whole blood alcohol concentration. The serum to whole blood alcohol ratio was independent of serum alcohol concentration and hematocrit [8].

One popular defense of DWI is endogenous production of alcohol. Although substantial alcohol may be produced endogenously in a decomposed body by the action of various microorganisms, the human body does not produce enough endogenous alcohol. In healthy individuals, who do not drink, usually endogenous alcohol levels are very low and often not detectable. Autobrewery syndrome (ABS), also known as gut fermentation syndrome, is a rarely diagnosed medical condition where ingested carbohydrates are converted to alcohol by fungi in the gastrointestinal tract. Patients with this condition become inebriated and suffer all the medical and social implications of alcoholism, including arrest for drunken driving. A previously active, healthy, 46-year-old man was treated with cephalexin 250 mg orally three times a day for 3 weeks for a complicated traumatic thumb injury. After that treatment, in one morning, he was arrested for presumed driving while intoxicated (DWI) with initial blood alcohol level of 200 mg/dL. The hospital personnel and police refused to believe him when he repeatedly denied alcohol ingestion. Further investigation confirmed the diagnosis of ABS when after a carbohydrate diet his blood alcohol was 57 mg/dL after 8 h. The carbohydrate challenge test was followed by upper and lower endoscopy to obtain intestinal secretions to detect fungal growth. Fungal cultures of secretions grew *Candida albicans* and *C. parapsilosis*. In addition, *Saccharomyces cerevisiae* (brewer's yeast) and *S. boulardii* were detected in his stool in addition to his normal stool bacterial flora, thus confirming diagnosis of ABS [9].

Although blood alcohol is usually directly determined in a driver suspected of driving under the influence of alcohol, blood alcohol level can also be predicted by using Widmark formula, which can be simplified as follows to calculate blood alcohol in percent:

$$C = (\text{Number of drinks} \times 3.1 / \text{Weight in pounds} \times r) - 0.015\,t$$

where $C =$ blood alcohol in percent (mg/dL), $r = 0.7$ for men and 0.6 for women.

Because most standard drinks contain approximately the same amounts of alcohol, it is only important to know how many drinks one person consumes. The type of drink does not matter and that makes the calculation easy. For example, if a 160 lb. man drinks five beers in a 2-h period, his blood alcohol at the end of drinking would be:

$$C = (5 \times 3.1 / 160 \times 0.7) - 0.015 \times 2$$
$$= 0.138 - 0.030$$
$$= 0.108\% \text{or blood alcohol was } 108\,\text{mg/dL}.$$

## Analysis of alcohol in body fluids: Limitations and pitfalls

Alcohol is most commonly measured in whole blood or serum. Alcohol concentration is also measured in urine, but less frequently. In hospital laboratories, ethyl alcohol is also analyzed using enzymatic method and automated analyzers. There are several different automated analyzers available from various diagnostic companies which are capable of analyzing alcohol in serum or plasma. Enzyme-based automated methods are generally not applicable for the analysis of whole blood although modified methods are available for the analysis of alcohol in urine specimens. Enzymatic automated analysis of alcohol is based on the following principle:

- Conversion of alcohol to acetaldehyde by alcohol dehydrogenase, and in this process, NAD is converted into NADH. NAD has no absorption at ultraviolet light at 340 nm wavelength, while NADH absorbs at 340 nm.

Therefore, an absorption peak is observed when alcohol is converted into acetaldehyde because NAD is also converted into NADH.

- The intensity of the peak is proportional to the amount of alcohol present in the specimen. If no alcohol is present, no peak is observed.

Usually methanol, isopropyl alcohol, ethylene glycol, and acetone have negligible effect on alcohol determination using enzymatic methods, but propanol, if present, may cause 15%–20% cross-reactivity with alcohol assay. Although isopropyl alcohol, which is used as rubbing alcohol, is common in households, propanol is used in much lesser frequency in household products. However, interference of lactate dehydrogenase and lactate in enzymatic method of alcohol determination is significant. Therefore, enzymatic alcohol assays are unsuitable for determination of alcohol in postmortem blood because high concentrations of lactate dehydrogenase (LDH) and lactate may be found in postmortem blood due to decomposition of carbohydrate by microorganisms. Postmortem blood alcohol must be determined by gas chromatography (GC) most commonly headspace GC. Lactate concentrations also tend to increase in trauma patients. Therefore, false-positive alcohol result may be observed in these patients if an enzymatic assay is used. Important points regarding interferences in enzymatic alcohol methods include:

- Enzymatic methods for alcohol determination are unsuitable for postmortem alcohol analysis due to high concentrations of LDH and lactate, and only gas chromatographic methods must be used. It is also important to note that negative urine alcohol but positive blood alcohol may indicate interference because LDH is absent in urine due to high molecular weight and cannot interfere with urine alcohol determination. However, for legal blood alcohol determination, GC method is always used.
- Alcohol may be produced by the activity of microorganisms after death. Therefore, elevated blood alcohol in postmortem specimen may not confirm alcohol intake prior to death and vitreous humor is a better source for determination of postmortem alcohol.
- Alternatively the presence of ethyl glucuronide and ethyl sulfate which are metabolites of alcohol in postmortem blood or urine confirms alcohol abuse prior to death. However, if postmortem blood alcohol level is positive but no ethyl glucuronide or ethyl sulfate can be detected in blood or urine, it is an indication of postmortem production of alcohol and the deceased did not consume alcohol prior to death.

In urine, alcohol can be determined up to 48 h after drinking depending on the amount of alcohol consumed, but usually no blood alcohol is detected not more than 24 h even after heavy drinking. Although dividing urine alcohol level by 1.3 may provide approximate blood alcohol level, this approach has many limitations. In addition, alcohol production in vitro after urine collection is a major problem for interpretation of urine alcohol level. Uncontrolled diabetes mellitus may cause glycosuria and if yeast infection is present, simultaneously, alcohol may be produced in vitro in the presence of yeast such as *Candida albicans* and glucose in urine. A 78-year-old man with a history of uncontrolled diabetes, multiple sclerosis, neurogenic bladder with recurrent urinary infection was brought to the emergency room for complain of reduced alertness. His blood alcohol was negative, but urine alcohol was positive. Considering his glycosuria as well as persistent urine yeast and bacterial colonization, his urine alcohol was considered to be false positive for alcohol ingestion (patient also denied consuming any alcohol). His urine alcohol was a by-product of local microorganism fermentation of carbohydrate [10]. In another case a 44-year-old Caucasian man tested positive for alcohol in urine although he reported having abstained from alcohol. A test for urine ethyl glucuronide and ethyl sulfate was used to validate the information in the patient history stating that the patient had not consumed alcohol for 1 year. Evaluation of the urinalysis results from the patient revealed fermentation in the context of glucosuria as the source of the urine ethanol [11]. Testing for ethyl glucuronide is recommended when positive urine alcohol result is inconsistent with patient denying alcohol consumption. Women with urinary tract infection may have similar problem. Storing urine at 4°C and using 1% sodium fluoride or potassium fluoride as a preservative can minimize the problem.

## Headspace gas chromatography for alcohol determination

The gold standard for the determination of blood alcohol is gas chromatography (GC) or gas chromatography/mass spectrometry (GC/MS). Usually headspace gas chromatography is used for the determination of alcohol in whole blood not only in forensic laboratories but also in larger medical centers and reference laboratories because this method is almost free from all interferences. The basic principle underlying all instrumentation for headspace gas chromatography is that an aliquot of the analyte from the vapor phase above a liquid such as whole blood containing alcohol and added internal standard (with known concentration) in a sealed vial or container must be reproducibly and effectively transferred to the inlet of a gas chromatograph. There are several means for accomplishing this, including gas tight syringe, transfer line, sample loop, and collection on a sorbent. The most important challenges are to ensure that the sample composition that reaches the gas chromatograph is truly representative of the composition of the headspace vapor in the vial and that the headspace vapor is representative of the composition of the original sample. Headspace gas chromatography with flame ionization detector is commercially available. However, it is important to note that the alcohol used for cleansing the venipuncture site does not jeopardize blood and plasma alcohol measurement with either headspace gas chromatography or enzymatic assay [12].

Instead of headspace gas chromatography, direct injection gas chromatography can also be used for the analysis of alcohol along with other volatiles. Smith et al. described determination of ethanol along with methanol, isopropanol, and acetone using capillary gas chromatography through direct sample injection. The author used 1-propanol (n-propanol) as the internal standard. Baseline separation was obtained between ethanol, other analytes, and the internal standard [13]. Although less commonly used for alcohol determination, several authors also used GC/MS for the analysis of ethanol and related volatiles in biological specimens. Xiao et al. described a GC/MS method for the analysis of whole blood ethanol using n-propanol as the internal standard. The mass spectrometer was operated in selected ion-monitoring mode (ionization source: electron ionization), monitoring $m/z$ 31 and 45 for ethanol and 31 and 59 for the internal standard. The method was linear for whole blood ethanol concentration between 4 and 126.3 mg/dL [14].

## Breath alcohol measurement

Breath alcohol is measured using breathalyzers which are used as a reliable estimate of blood alcohol concentrations since the 1970s. Evidentiary breathalyzers which are approved by the National Highway Traffic Safety Administration are reliable for breath alcohol measurement and values are admissible to court as evidence. Although interlock devices in the car also works on the principle of breathalyzers, such devices are not designed for legal alcohol testing. The principle of breath alcohol measurement is partition of alcohol between blood and alveolar air. The estimated ratio between breath alcohol and blood alcohol is 1:2100 and this ratio is utilized in various breathalyzers to calculate blood alcohol level based on the concentration of ethanol in exhaled air by multiplying breath alcohol concentration expressed in mg/L with 2100 to produce blood alcohol concentration in mg/L (or multiply by 210 to obtain blood alcohol concentration in mg/dL). Some breathalyzer software may automatically calculate blood alcohol value from observed breath alcohol level.

Breathalyzers are based on various technologies. While most breathalyzers are based on a single technology, some analyzers may involve two different technologies in order to achieve higher precision and accuracy. The earliest developed breathalyzer was based on reaction of alcohol if present in the exhaled air with a cocktail of chemicals containing sulfuric acid, potassium dichromate, silver nitrate, and water. However, this technology is subjected to interferences from a variety of substances. Some evidentiary breathalyzers are based on the principle of infrared spectroscopy (IR spectroscopy) for quantitative determination of alcohol in exhaled air. Another popular method is fuel cell technology. In addition, some evidentiary breathalyzers are based on both fuel cell and infrared spectroscopy technology giving them good sensitivity and specificity. Semiconductor alcohol sensors are utilized in inexpensive breathalyzers marketed to the general public. However, sensor response is nonspecific to alcohol and response may be nonlinear.

There are several issues that may produce misleading breathalyzer reading. Sometimes a driver stopped by the police may use mouthwash to hide any alcoholic breath. Because some mouthwashes contain alcohol, use of a mouthwash prior to taking a breath alcohol analysis may cause falsely elevated breath alcohol result. However, residual alcohol evaporates from the mouth rapidly and a mandatory 15 min waiting time in the police station when no food or drink is allowed eliminates possibility of a false-positive test result. Various methods used in such devices are listed in Table 4.

**Table 4** Examples of breathalyzers based on different technologies.

| Method | Technology | Examples of breathalyzers |
|--------|-----------|---------------------------|
| Colorimetry | Based on reaction of alcohol if present in the exhaled air with a cocktail of chemicals containing sulfuric acid, potassium dichromate, silver nitrate, and water. If alcohol is present, yellow color of the cocktail due to presence of potassium dichromate turns into green and the amount of alcohol in the exhaled air is estimated from the intensity of green color | Breathalyzer 900, Breathalyzer 1100 |
| Infrared | IR spectroscopy technology uses two different wavelengths (3.37 and 3.44 μm) specific for alcohol to determine its concentration | Intoxilyzer 5000, Intoxilyzer 8000, Data Master cdm |
| Fuel cell | The fuel cell is a porous disk coated with platinum oxide (also called platinum black) on both sides. The manufacturer mounts this fuel cell in the device along with the entire assembly so that when a person blows to the disposable mouthpiece the air can travel through the fuel cell. If any alcohol is present in the exhaled air, it is converted into acetic acid, hydrogen ion, and electrons on the top surface by the platinum oxide. Then hydrogen ions travel to the bottom surface and is converted into water by combining with oxygen present in the air. As a result, electric current is generated which is converted into alcohol concentration by the microprocessor present in the device | Alcotest Models 6510, 6810, 7410 etc. Alco-Sensor III and IV |
| Mixed | Infrared technology and fuel cell technology combined | Intox EC/IR |

## Biomarkers of alcohol abuse

Alcohol biomarkers could be indirect, direct of genetic biomarker. No single genetic mutation is related to alcohol and drug addiction. It is generally accepted that both environment (approximately 50% contribution) and various genetic factors (approximately 50% contribution) contribute to susceptibility of alcohol and drug addiction. Discussion of genetic mutations that may increase susceptibility to alcohol and drug addiction is beyond the scope of this book.

Indirect biomarkers include:

- liver enzymes (mostly commonly used is GGT: gamma glutamyl transferase)
- mean corpuscular volume (MCV)
- carbohydrate-deficient transferrin
- serum or urine beta hexosaminidase
- total serum or plasma sialic acid
- sialic acid index of apolipoprotein J
- 5-hydroxytryptophol

However, GGT, MCV (routinely measured during regular health check) and carbohydrate-deficient transferrin (first FDA-approved alcohol biomarker measured in serum/plasma, immunoassay available) are more often used in clinical practice.

Direct biomarkers include:

- ethyl glucuronide
- ethyl sulfate
- fatty acid ethyl esters
- phosphatidyl ethanol

However, most commonly used direct biomarker is ethyl glucuronide because immunoassays are commercially available for rapid determination of this marker measured commonly in urine.

Alcohol biomarkers are primarily used for screening patients for possible alcohol abuse and also identifying pregnant women who may be abusing alcohol, because fetal alcohol syndrome is a totally preventable disorder. Alcohol biomarkers are also used in emergency room settings, psychiatric clinics, and internal medicine settings because self-reporting of alcohol use is not always accurate as some patients are reluctant to admit a problem with alcohol. The addition of biomarkers may help identify individuals who need treatment for alcohol abuse.

Traditional biomarkers of alcohol use are indirect biomarkers, which are elevated in a person consuming moderate to heavy amounts of alcohol. These biomarkers are elevated due to toxicity of alcohol on a particular organ, for example, liver enzyme gamma glutamyl transferase (GGT) is elevated after heavy alcohol consumption. Mean corpuscular volume (MCV) as well as the first carbohydrate-deficient transferring, are also indirect biomarkers. In addition, serum and urine hexosaminidase and sialic acid are also indirect biomarkers of alcohol abuse, but used less often in clinical practice. In contrast, minor alcohol metabolites such as ethyl glucuronide, ethyl sulfate, or biomolecule derived from interaction of alcohol with other molecules such as fatty acid ethyl ester and phosphatidyl ethanol are direct biomarkers of alcohol consumption.

Because alcohol is produced by bacterial action after death, ethyl glucuronide and ethyl sulfate are postmortem markers of antemortem alcohol ingestion because neither ethyl glucuronide nor ethyl sulfate is formed after death. In one study involving 36 death investigations where postmortem ethanol production was suspected, ethyl glucuronide and ethyl sulfate were measured in both urine and blood of the deceased. In 19 out of the 39 deceased, the concentration of ethyl glucuronide in blood ranged from 0.1 to 23.2 ng/mL, while urinary ethyl glucuronide concentrations ranged from 1.9 to 182 ng/mL. For ethyl sulfate, the blood concentration ranged from 0.04 to 7.9 ng/mL, while urine concentrations ranged from 0.3 to 99 ng/mL. In 16 other individuals, no ethyl glucuronide or ethyl sulfate was detected. The authors concluded that in 36 cases, alcohol consumption before death was likely in 19 deceased only who showed positive ethyl glucuronide and ethyl sulfate concentrations in blood and urine [15].

Fatty acid ethyl esters are direct markers of alcohol abuse because they are formed due to chemical reaction between fatty acids and alcohol (ethanol). Fatty acids are integral part of structures of triglycerides, but a small amount of fatty acids, also known as free fatty acids are found in circulation. The chemical reaction between alcohol and fatty acid is known as esterification which is mediated by fatty acyl ethyl ester synthase (FAEE synthase), an enzyme found in abundance in liver and pancreas. Carboxylesterase lipase, another enzyme which liberates free fatty acids from complex lipids, also can induce the reaction between alcohol and fatty acids generating fatty acid ethyl esters. Fatty acid ethyl esters are found in circulation, but these compounds are also incorporated into hair follicle through sebum and can be used as a biomarker of alcohol abuse. There are four major fatty acid ethyl esters: ethyl myristate, ethyl palmitate, ethyl stearate, and ethyl oleate. These compounds are measured in blood or hair using gas chromatography/mass spectrometry. Then results are usually expressed as sum of all four fatty acid ethyl ester concentrations. Reference range and window of detection of alcohol abuse by using various markers are summarized in Table 5.

**Table 5** Characteristics of various alcohol biomarkers.

| Alcohol biomarker | Type of biomarker | Specimen | Detection window | Indicated use | Cut-off value |
|---|---|---|---|---|---|
| GGT | Indirect | Blood | 3–4 weeks | Chronic heavy drinking | 30 U/L |
| MCV | Indirect | Blood | 3–4 months | Chronic heavy drinking | 100 fl |
| % CDT | Indirect | Blood | 2–3 weeks | Moderate-to-heavy drinking | 2.4% |
| Serum beta hexosaminidase | Indirect | Blood | 7–10 days | Chronic heavy drinking | 35%[a] |
| Total plasma sialic acid | Indirect | Blood | 3 weeks | Differentiating heavy drinking from social drinking | 77.8 mg/dL in female 80 mg/dL in male |
| Ethyl glucuronide | Direct | Urine | 2–3 days | Moderate drinking | 100 ng/mL |
| Ethyl glucuronide | | Hair | Several months | Moderate-to-heavy drinking | 30 pg/mg |
| Ethyl sulfate | Direct | Urine | 1–2 days | Moderate drinking | 25 ng/mL |
| Fatty acid ethyl esters | | Hair | Only 24 h in serum but several months in hair. Therefore, hair is more frequently used | Showing abstinence or differentiating heavy drinking from social drinking depending on cut-off | 0.29 ng/mg of hair |
| Phosphatidyl ethanol | Direct | Blood | 2–3 weeks | Heavy drinking | 0.36 μmol/L |

[a]35% cut of value represent beta-Hex-B% (Hex-B represents heat stable beta hexosaminidase activity, therefore, Hex-B% is the ratio of heat stable beta hexosaminidase to total beta hexosaminidase activity multiplied by 100).

## Methanol abuse

Methanol (wood spirit) is found in many household chemicals (auto products, cleaning products, etc.), but methylated spirit is the most common household chemical that contains methanol. Methanol is well absorbed even through skin and may cause toxicity. Inhalation of methanol through carburetor cleaner is a major route of domestic exposure to methanol. Accidental ingestion of windshield washer fluid is also another common cause of methanol intoxication. Routine occupational exposure to methanol containing products is relatively safe. Like ethanol, exposure to methanol during pregnancy is dangerous.

A small amount of methanol is found in alcoholic beverages as a part of the natural fermentation process and this small amount does not cause any harm because the ethanol present in the drink protects the human body from methanol toxicity. However, illicit drinks prepared from methylated spirit cause severe and even fatal illness. Illegally prepared moonshine whiskey may contain much higher amounts of methanol and these are major sources of epidemic of methanol toxicity worldwide. Methanol is readily absorbed after ingestion or inhalation and enters into the blood stream. A small amount of methanol is excreted unchanged in urine and also through exhaled breath, but the majority of methanol is metabolized by the same enzyme in the liver that metabolizes ethanol—namely alcohol dehydrogenase. In this process formaldehyde is generated and is further metabolized by another liver enzyme acetaldehyde dehydrogenase to formic acid.

Methanol → Formaldehyde → Formic Acid

Methanol itself is relatively nontoxic and methanol toxicity is a classic example of "lethal synthesis," where metabolites of methanol in the body are the major cause of methanol toxicity. Formaldehyde, the end product of methanol metabolism, is the key factor in causing toxicity from methanol including blindness and death. Important points regarding methanol intoxication include:

- The lethal dose of methanol in humans is not fully established. Although it is assumed that ingestion of anywhere from 30 to 100 mL of methanol may cause death, fatality from methanol may occur even after ingestion of 15 mL of 40% methanol and blindness may result from consuming as little as 4 mL of methanol.
- If blood methanol concentration exceeds 20 mg/dL, treatment should be initiated. However, a clinician may treat a patient with much lower methanol concentration depending on clinical picture of the patient.

Best way to establish the diagnosis of methanol toxicity is direct measurement of methanol by gas chromatography. If that is not available, high anion gap and osmolar gap with suspected methanol ingestion can be used for the diagnosis of methanol poisoning. Methanol poisoning can be treated with infusion of ethanol (blood ethanol targeted as 100 mg/dL). The goal is to slow down production of formic acid, the toxic metabolite. In addition, 4-methylpyrazole (fomepizole), sodium bicarbonate, and even dialysis can be used for treating methanol poisoning.

## Abuse of ethylene glycol and other alcohols

Ethylene glycol is a colorless and relatively nonvolatile liquid which has a high boiling point and a sweet taste which is why children and pets tend to ingest it causing ethylene glycol toxicity. An adult may drink ethylene glycol as a substitute for ethanol or in a suicide attempt. Because of the low melting point and high boiling point, ethylene glycol is used as a major ingredient in automobile antifreeze. Ethylene glycol is used in deicing fluid, and in industries, ethylene glycol is widely used as a starting material for preparing various polyester products.

Because ethylene glycol is relatively nonvolatile, inhalation or exposure to ethylene glycol is not generally considered as an occupational health hazard. Absorption of ethylene glycol through the skin may cause serious toxicity especially if there are any skin lesions. The major route of exposure to ethylene glycol is ingestion of ethylene glycol containing fluids. Ethylene glycol is rapidly and completely absorbed from the intestinal tract after oral ingestion. Ethylene glycol itself is relatively nontoxic like methanol, but metabolites are toxic. Ethylene glycol is primarily metabolized in the liver (approximately 80%), while another 20% is excreted in the urine unchanged. Metabolism of ethylene glycol by the liver is a four-step process. Ethylene glycol is first metabolized to glycolaldehyde (also known as glycoaldehyde or glyoxal) by alcohol dehydrogenase and then glycolaldehyde is further metabolized by aldehyde dehydrogenase into glycolic acid. Finally, glycolic acid is transformed into oxalic acid through an intermediate glyoxylic acid. Oxalic acid then combines with calcium causing deposition of calcium oxalate in kidneys, resulting in severe renal failure.

$$Oxalic\ Acid + Ca^{2+} \rightarrow Calcium\ oxalate\ crystals\ causing\ nephrotoxicity$$

Major complications of ethylene glycol poisoning are metabolic acidosis and renal failure and these complications may even be fatal. The lethal dose of ethylene glycol is usually assumed as 100 mL, but there are reports with fatality from ethylene glycol poisoning even from ingestion of only 30 mL [16]. Blood levels of ethylene glycol are usually measured by headspace gas chromatography either singly or in combination with other volatile compounds such as methanol, acetone, and isopropyl alcohol. In addition, there are some enzymatic methods available for rapid determination of blood ethylene glycol levels using an automated analyzer in the clinical laboratory. Limitations of enzymatic methods for ethylene glycol include:

- Like the enzymatic method for alcohol, the enzymatic method for ethylene glycol determination produces a false-positive ethylene glycol level if lactate and lactate dehydrogenase are present in high concentration in the serum specimen.
- Interestingly in patients poisoned with ethylene glycol, falsely elevated lactate may be observed using blood gas analyzers, but chemistry analyzers usually do not show this false elevation [17].

Ethylene glycol poisoning is treated similarly to methanol poisoning using bicarbonate, ethanol, fomepizole, or hemodialysis. Propylene glycol, which is similar to ethylene glycol, is used as an industrial solvent and can also be used in antifreeze formulation. Propylene glycol is significantly less toxic than ethylene glycol and is preferred antifreeze used in motor homes and recreational vehicles. Propylene glycol is also used as a diluent for oral, topical, or intravenous pharmaceutical preparations so that active ingredients can be dissolved properly in the formulation. Isopropyl alcohol is also known as rubbing alcohol and is a 70% aqueous solution of isopropyl alcohol. Isopropyl alcohol is slowly metabolized by alcohol dehydrogenase to acetone. Acetone is also found in many domestic products, for example, nail polish remover. Although after methanol or ethanol (alcohol) poisoning both ketosis and acidosis may be observed (formic acid produced after methanol poisoning; acetic acid after alcohol poisoning) only ketosis (without acidosis) is observed with isopropyl alcohol poisoning because end product of isopropyl alcohol poisoning is acetone, a ketone but no organic acid is produced.

Neither isopropyl alcohol nor acetone can cause metabolic acidosis and poisoning from these compounds may be less life threatening than methanol or ethylene glycol poisoning, but there are reports of death from severe isopropyl alcohol poisoning. Overdose with methanol and ethanol may cause metabolic acidosis, but overdose with isopropyl alcohol causes ketosis without acidosis because isopropyl alcohol is converted into acetone.

## Summary/key points

- Alcohol content of various alcoholic beverages varies widely, but different amounts are consumed for different drinks. Therefore, one standard drink contains approximately 0.6 oz of pure alcohol or 14 g of pure alcohol.
- According to guidelines of moderate drinking, for men no more than two drinks a day and for women not more than one drink a day.
- Binge drinking is defined by NIAAA as consumption of five drinks in 2 h for men and four drinks in 2 h for women and such drinking pattern always produce a blood alcohol level of 0.08% or higher.
- Alcohol-induced liver disease can be classified under three categories: fatty liver, alcoholic hepatitis, and liver cirrhosis. Heavy drinking for as little as a few days may produce fatty change in the liver (steatosis) which is reversed after abstinence.
- Alcohol metabolism follows zero-order kinetics (blood alcohol is cleared at a constant rate regardless of blood alcohol level: 0.018%–0.020% male and 0.015%–0.018% for female) and metabolism depends on age, gender, body weight, amount of food consumed (less blood alcohol if consumed with food), and genetic makeup of the person. Usually it is

assumed that alcohol concentration in blood is reduced by 15–18 mg/dL per hour.

- Although CYP2E1 pathway for metabolism of alcohol into acetaldehyde is a minor pathway in most people consuming alcohol, in alcoholics this pathway metabolizes significant amount of alcohol due to saturation of alcohol dehydrogenase pathway. This metabolic pathway also produces free radicals which cause liver damage and increases oxidative stress.

- Three different combination of polymorphisms of gene encoding alcohol dehydrogenase and ALDH2 may have protective effect on alcohol abuse: polymorphism in *ALDH2* gene causing poor activity of aldehyde dehydrogenase enzyme, polymorphism in *ADH* genes that lead to superactive alcohol dehydrogenase enzyme, and combination of superactive alcohol dehydrogenase activity and poor aldehyde dehydrogenase activity that offers total protection from consuming alcohol.

- Legal limit of blood alcohol in all states of the United States is currently 0.08% whole blood alcohol (80 mg/dL). Serum alcohol concentration is higher than whole blood alcohol concentration. In order to convert serum alcohol level to whole blood alcohol level, serum alcohol level must be divided by 1.15. Therefore, serum alcohol concentration of 100 mg/dL (0.1%) is equivalent to 0.87 mg/dL (0.087%) whole blood concentration.

- Enzymatic methods for alcohol determination are commonly used in hospital laboratories, but high lactate dehydrogenase and lactate if present in specimen can falsely elevate serum alcohol level even if alcohol is absent. Headspace gas chromatography or gas chromatography with direct injection is considered as the gold standard because this method is not affected by high lactate and LDH as well as other interferences.

- Enzymatic assays are also unsuitable for alcohol determination in postmortem blood because high lactate dehydrogenase and lactate can be found in postmortem specimen due to cellular breakdown. In addition, alcohol may be produced by the activity of microorganisms after death. Therefore, elevated blood alcohol in postmortem specimen may not confirm alcohol intake prior to death. Therefore, vitreous humor is a better source for determination of postmortem alcohol. Alternatively the presence of ethyl glucuronide and ethyl sulfate which are metabolites of alcohol in postmortem blood or urine confirms alcohol abuse prior to death. However, if postmortem blood alcohol level is positive but no ethyl glucuronide or ethyl sulfate can be detected in blood or urine, it is an indication of postmortem production of alcohol and the deceased did not consume alcohol prior to death.

- Headspace gas chromatography as well as direct injection gas chromatography based method can analyze alcohol along with other volatiles (methanol, isopropyl alcohol, ethylene glycol, acetone, etc.) if present in specimen simultaneously in a single run.

- Endogenous production of alcohol is minimal except for very rare instances of patients with autobrewery syndrome, where if a patient takes carbohydrate-rich food, endogenous alcohol production in a high level is possible due to gut infection related to fungus.
- Although blood alcohol is more commonly measured, urine alcohol may also be measured in workplace drug testing. Again, gas chromatography is preferred although enzymatic methods can also be used. However, another problem of urine alcohol determination is in vitro production of alcohol if a female patient has poorly controlled diabetes and yeast infection because urine is contaminated with yeast and glucose present in urine can be converted into alcohol by the action of yeast.
- The principle of breath alcohol measurement is partition of alcohol between blood and alveolar air. The estimated ratio between breath alcohol and blood alcohol is 1:2100 and this ratio is utilized in various breathalyzers to calculate blood alcohol level based on the concentration of ethanol in exhaled air by multiplying breath alcohol concentration expressed in mg/L with 2100 to produce blood alcohol concentration in mg/L (or multiply by 210 to obtain blood alcohol concentration in mg/dL). Some breathalyzer software may automatically calculate blood alcohol value from observed breath alcohol level.
- The alcohol biomarkers include liver enzymes particularly GGT (gamma glutamyl transferase), mean corpuscular volume (MCV), carbohydrate-deficient transferrin, serum and urine hexosaminidase, sialic acid, acetaldehyde-protein adducts, ethyl glucuronide, ethyl sulfate, and fatty acid ethyl ester.
- Methanol is metabolized to toxic formaldehyde and formic acid which can cause metabolic acidosis, blindness, and death. The lethal dose of methanol in humans is not fully established. Although it is assumed that ingestion of anywhere from 30 to 100 mL of methanol may cause death, fatality from methanol may occur even after ingestion of 15 mL of 40% methanol and blindness may result from consuming as little as 4 mL of methanol.
- Ethylene glycol is metabolized finally into oxalic acid which combines with calcium forming calcium oxalate. Calcium oxalate crystals deposit in kidney causing renal failure and severe overdose of ethylene glycol may be fatal. Both methanol and ethylene glycol can be treated with ethanol infusion to slow down metabolism of methanol or ethylene glycol (ethanol is a preferred substrate for alcohol dehydrogenase which also metabolizes methanol and ethylene glycol). Hemodialysis is used for treating life-threatening methanol or ethylene glycol overdose. Overdose with methanol and ethanol may cause metabolic acidosis, but overdose with isopropyl alcohol causes ketosis without acidosis.

# References

[1] Liber CS. Relationship between nutrition, alcohol use and liver disease. Alcohol Res Health 2003;27:220–31.

[2] Marczinski CA. Can energy drinks increase the desire for more alcohol? Adv Nutr 2015;6:96–101.

[3] Piano MR. Alcohol's effects on the cardiovascular system. Alcohol Res 2017;38(2):219–41.

[4] Liberale L, Bonaventura A, Montecucco F, Dallegri F, Carbone F. Impact of red wine consumption on cardiovascular health. Curr Med Chem 2019;26:3542–66.

[5] Lai CL, Yao CT, Chau GY, Yang LF, Kuo TY, Chiang CP, et al. Dominance of the inactive Asian variant over activity and protein contents of mitochondrial aldehyde dehydrogenase 2 in human liver. Alcohol Clin Exp Res 2014;38:44–50.

[6] Yang M, Zhang Y, Ren J. ALDH2 polymorphism and ethanol consumption: a genetic-environmental interaction in carcinogenesis. Adv Exp Med Biol 2019;1193:229–36.

[7] Wall TL, Shea SH, Luczak SE, Cook TA, Carr LG. Genetic association of alcohol dehydrogenase with alcohol use disorders and endo phenotypes in white college students. J Abnorm Psychol 2005;114:456–65.

[8] Rainey P. Relation between serum and whole blood ethanol concentrations. Clin Chem 1993;39:2288–92.

[9] Malik F, Wickremesinghe P, Saverimuttu J. Case report and literature review of auto-brewery syndrome: probably an underdiagnosed medical condition. BMJ Open Gastroenterol 2019;6(1), e000325.

[10] Borgan SM, Sourial K, Robalino E, Camargo K, Ho M. A challenge is brewing: false-positive urine alcohol in an elderly diabetic male. Am J Med 2019;132(10):e738.

[11] Foley KFA. Positive urine alcohol with negative urine ethyl-glucuronide. Lab Med 2018;49:276–9.

[12] Lippi G, Simundic AM, Musile G, Danese E, et al. The alcohol used for cleansing the venipuncture site does not jeopardize blood and plasma alcohol measurement with head-space gas chromatography and an enzymatic assay. Biochem Med (Zagreb) 2017;27:398–403.

[13] Smith NB. Determination of volatile alcohols and acetone in serum by non-polar capillary gas chromatograph after direct sample injection. Clin Chem 1984;30:1672–4.

[14] Xiao HT, He L, Tong RS, Yu JY, Chen L, Zpou J, et al. Rapid and sensitive headspace gas chromatography-mass spectrometry method for the analysis of ethanol in the whole blood. J Clin Lab Anal 2014;28:386–90.

[15] Hoiseth G, Karinen R, Christophersen A, Morland J. Practical use of ethyl glucuronide and ethyl sulfate in postmortem cases as markers of antemortem alcohol ingestion. Int J Legal Med 2010;124:143–8.

[16] Walder AD, Tyler CKG. Ethylene glycol antifreeze poisoning: three case reports and a review of treatment. Anesthesia 1994;49:964–7.

[17] Sandberg Y, Rood PPM, Russcher H, Zwaans JJM, et al. Falsely elevated lactate in severe ethylene glycol intoxication. Neth J Med 2010;68:320–3.

# CHAPTER 19

# Common poisonings including heavy metal poisoning

## Poisoning from analgesics

Poisonings from analgesics are due to overdose or suicidal attempt using over-the-counter(OTC) drugs such as acetaminophen and aspirin (acetyl salicylate). Determination of serum acetaminophen and salicylate concentrations is useful in clinical laboratories for the diagnosis of poisoning from these drugs. Important points regarding acetaminophen toxicity include:

- Usually severe acetaminophen toxicity occurs in an adult after consuming 7–10 g of acetaminophen (15–20 tablets containing 500 mg acetaminophen in each).
- Acetaminophen is normally metabolized to glucuronide and sulfate conjugates by the liver enzymes but a small amount is metabolized by cytochrome mixed function oxidase family of enzymes to a toxic metabolite N-acetyl-p-benzoquinone imine which is detoxified after conjugating with glutathione present in the liver.
- Acetaminophen toxicity is due to formation of excess toxic metabolite: N-acetyl-p-benzoquinone imine during severe overdose because glutathione supply of the liver is depleted and this metabolite is no longer conjugated with glutathione.
- Antidote of acetaminophen poisoning is N-acetyl cysteine (Mucomyst). This antidote is a precursor of glutathione and also can detoxify the toxic metabolite of acetaminophen. This antidote must be administered as soon as possible but certainly within 8 h of acetaminophen overdose for maximum benefit.

Antidotes/treatments of various common poisonings are listed in Table 1. Although the therapeutic range of acetaminophen is 10–30 μg/mL and the toxic concentration is over 150 μg/mL (4 h after ingestion), alcoholics may experience severe liver toxicity after consuming moderate dosage of acetaminophen because alcohol is known to deplete glutathione supply of the liver. Strikingly

Clinical Chemistry, Immunology and Laboratory Quality Control. https://doi.org/10.1016/B978-0-12-815960-6.00025-X

**Table 1** Treatment/antidote for common poisonings.

| Poisoning | Treatment/antidote |
|---|---|
| Acetaminophen | N-Acetylcysteine (Mucomyst) |
| Acetyl salicylate | Activated charcoal, sodium bicarbonate to correct acid-base disorder, hemodialysis |
| Carbon monoxide | 100% oxygen or hyperbaric oxygen therapy |
| Cyanide | Hydroxocobalamin or any combination of amyl nitrate, sodium nitrate, ferrous sulfate, or dicobalt edentate |
| Tricyclic antidepressants | Sodium bicarbonate |
| Benzodiazepines | Flumazenil |
| Opiates | Naloxone |
| Methanol | Ethyl alcohol or fomepizole (4-methylpyrazole) |
| Ethylene glycol | Ethyl alcohol or fomepizole (4-methylpyrazole) |
| Organophosphorus/carbamate insecticides | Atropine, pralidoxime |
| Lead poisoning | Various chelating agents such as calcium sodium EDTA (ethylenediamine tetraacetic acid), D-penicillamine, DMSA (2,3-dimercaptosuccinic acid) |
| Mercury poisoning | BAL (British-Lewisite, Dimercaprol) |
|  | DMPS (2,3-dimercapto-1-propanesulfonate) |
| Arsenic poisoning | BAL, DMSA |

abnormal liver enzymes in alcoholics after consuming moderate dosage of acetaminophen has been well documented in the literature [1].

Aspirin (acetyl salicylate) poisoning is also common. After ingestion, acetyl salicylate is hydrolyzed by liver and blood esterase into salicylic acid (salicylate) which is the pharmacologically active drug. The therapeutic range is 150–250 μg/mL, although analgesic effect may be observed below 100 μg/mL level. Life-threatening toxicity is observed with salicylate levels at 500 μg/mL or above. Important points regarding salicylate toxicity include:

- Reye's syndrome is a rare but a potentially fatal (estimated fatality 30%–40%) condition that children and teenagers may develop during the treatment of fever due to viral infection with drugs containing acetylsalicylic (aspirin) acid and other salicylates. The high mortality rate from this disease is associated with the development of a rapidly progressing toxic encephalopathy and hepatic insufficiency. The etiology and pathogenesis of the Reye syndrome are not clear. Because the effect of acetylsalicylic acid is an important factor of development of Reye syndrome, salicylate use for treating cold in children and young adult is not recommended [2].

- Salicylate overdose results in direct stimulation of central respiratory system causing hyperventilation and respiratory alkalosis.
- Salicylate overdose also causes hyperthermia.
- Finally, salicylate overdose causes metabolic acidosis due to accumulation of organic acids as a result of inhibition of the Krebs cycle. Increased anion gaps and high serum osmolarity are indirect indications of salicylate overdose.

Activated charcoal can be given to an overdose patient to prevent further absorption. In addition, sodium bicarbonate may be used for correcting acid-base disorder. Finally, hemodialysis may be needed for a patient with salicylate level in near fatal range.

## Methyl salicylate poisoning

Methyl salicylate is a major component of oil of wintergreen which is prepared by distillation from wintergreen leaves. Methyl salicylate has an analgesic effect and is used in many OTC analgesic creams or gels designed only for topical use. The action of methyl salicylate is multimodal with analgesic, antiinflammatory, and rubefacient/counterirritant properties. The former arises from the rapid hydrolysis of the methyl ester yielding salicylic acid as the active agent after dermal absorption. Methyl salicylate also has a vasodilatory action resulting in an increased localized blood flow and, consequently, produces a rise in tissue temperature—its rubefacient action. Methyl salicylate if ingested is very poisonous. Methyl salicylate is a relatively common cause of poisoning of children, and ingestion of one teaspoon of oil of wintergreen may be fatal because it contains approximately 6 g of salicylate equivalent to ingesting approximately 20 aspirin tablets [3]. Poisoning from methyl salicylate can also occur due to abuse of topical cream containing methyl salicylate in adults because significant amount of salicylate can be detected in blood after excessive topical application of cream containing methyl salicylate. Bell and Duggin reported the case of a 40-year-old man who became acutely ill after receiving treatment from an unregistered naturopath after receiving an herbal skin cream treatment for psoriasis. The herbal cream contained methyl salicylate, and in this case, transcutaneous absorption of methyl salicylate was enhanced due to psoriasis. His salicylate level was highly elevated to 48.5 mg/dL at the time of presentation to the hospital and arterial blood gas showed metabolic acidosis superimposed on respiratory alkalosis consistent with the diagnosis of salicylate poisoning [4]. Death of a 17-year-old cross-country runner after excessive self-administration of a topical muscle pain relief treatment containing methyl salicylate has been reported. The American Association of Poison Control Centers Toxic Exposure Surveillance System has reported that about 77% of the enquiries relating to exposures/incidents of methyl salicylate poisoning involve children under 6 years of age [5]. Popular topical ointment Bengay contains 15% methyl salicylate while Ben-gay Muscle Pain/Ultra Strength contains 30% methyl salicylate. Many Chinese medicines and medicated oils contain high amounts of methyl salicylate.

## Carbon monoxide poisoning

Carbon monoxide poisoning may occur accidently from faulty ventilated home heating unit or inhaling exhaust of car with gasoline engine in a suicide attempt. In general, carbon monoxide is produced due to incomplete combustion and also produced during burning wood, charcoal grills, propane grills, and many gasoline-operated instruments. In addition, carbon monoxide is also present in cigarette smoke. When inhaled, carbon monoxide tightly binds with the hemoglobin producing carboxyhemoglobin. Because binding affinity of carbon monoxide is 250 times more than the binding affinity of oxygen, in the presence of carbon monoxide, hemoglobin preferentially binds with carbon monoxide causing severe hypoxia. Because carbon monoxide is odorless, it cannot be detected and sometimes victims are not aware that they are exposed to lethal carbon monoxide environment.

Blood level of carbon monoxide is usually determined spectrophotometrically by measuring carboxyhemoglobin level using Co-oximeter. Commercially available instruments perform absorption measurement of blood specimens at various wavelengths to determine the concentration of oxyhemoglobin, deoxyhemoglobin, carboxyhemoglobin, and methemoglobin. More recently, noninvasive pulse Co-oximeters are available for the measurement of various hemoglobin components for screening in the emergency department to identify patients with carbon monoxide poisoning [6]. Important factors regarding carboxyhemoglobin levels in blood include:

- Usually nonsmokers living in rural area have carboxyhemoglobin level in blood <0.5% while urban nonsmokers may have levels up to 2%.
- Smokers may have carboxyhemoglobin level of 5%–6%.
- Minor symptoms of carbon monoxide poisoning such as shortness of breath may be experienced with carboxyhemoglobin concentration of 10% and higher while at 30% carboxyhemoglobin level, full-blown symptoms of carbon monoxide poisoning such as severe headache, fatigue, nausea, and vomiting as well as difficulty in judgment are observed.
- A carboxyhemoglobin level of 60%–70% and above may cause respiratory failure and even death. A value over 80% may cause quick death.

- Fetal hemoglobin has slightly different spectrophotometric properties than adult hemoglobin. Therefore, falsely high carboxyhemoglobin value up to 7% has been reported in neonates using Co-oximeter. A previously healthy 3-month-old girl presented to the pediatric emergency department with smoke inhalation from a malfunctioning furnace was alert and afebrile, with Glasgow Coma Scale of score of 15 and normal vital with mildly elevated carboxyhemoglobin level. Her carboxyhemoglobin was still 11.2% after receiving 100% oxygen for 6 h. Further investigation revealed that her falsely elevated carboxyhemoglobin level was due to interference of fetal hemoglobin in spectrophotometric measurement of carboxyhemoglobin [7].

Treatment of a person poisoned with carbon monoxide is to administer 100% oxygen. Half-life of carboxyhemoglobin is reduced significantly when a victim breathes 100% oxygen rather than room air. If the victim is placed in a hyperbaric chamber (oxygen treatment at two to three times atmospheric pressure), the half-life of carboxyhemoglobin is approximately 15 min. However, this treatment is usually reserved for a victim experiencing severe carbon monoxide poisoning.

## Cyanide poisoning

Most lethal form of cyanide poisoning is inhaling hydrocyanic acid (HCN), also known as prussic acid. If inorganic cyanide such as potassium cyanide is ingested for committing suicide, it is converted into HCN causing toxicity and even fatality. Cyanide poisoning is relatively uncommon. The most common source is smoke inhalation from a fire in the house because burning common household substances such as plastic, silk, and rubber may produce cyanide smoke containing HCN. Therefore, firefighters are also at high risk of cyanide poisoning. However, during fires, both carbon monoxide and cyanide poisoning may occur. Lundquist et al. studied blood cyanide and carboxyhemoglobin levels in 19 victims who were found dead in building after fire. The results indicated that 50% of the victims had been exposed to toxic levels of HCN but 90% to the toxic levels of carbon monoxide [8].

Some stone fruits (fruits that contain a pit or solid core) such as apricots, cherries, peaches, pears, plums, and prunes contain cyanogenic glycoside. Massive ingestion of pits of these fruits may be dangerous but eating flesh of these fruits is not a concern. Cyanogenic glycoside is also present in cassava roots and fresh bamboo shoots. Therefore these foods should be cooked before consuming.

After exposure to HCN, cyanide ion tightly binds with hemoglobin producing cyanhemoglobin causing severe hypoxia. Although body can transform cyanide into relatively nontoxic thiocyanate, this process is slow and not effective to avoid life-threatening cyanide poisoning. Cyanide poisoning produces non-specific symptoms. Therefore, determining blood cyanide level is useful for diagnosis. Usually, spectroscopic methods are used for cyanide determination and normal blood cyanide level is less than $0.2\,\mu g/mL$. A cyanide level above $2\,\mu g/mL$ can produce severe toxicity and level above $5\,\mu g/mL$ may be lethal if not treated immediately.

There are several antidotes for treating cyanide poisoning including inhalation of amyl nitrate, and administration of sodium thiosulfate, sodium nitrate, ferrous sulfate, and dicobalt edentate as well as hydroxocobalamin. Oxygen therapy is also useful. If cyanide poisoning victims receive prompt medical care, life can be saved. Borron et al. concluded in their study that 67% of the patients with confirmed cyanide poisoning after smoke inhalation survived after successful administration of hydroxocobalamin antidote [9].

## Overdose with tricyclic antidepressants

Although overdose with tricyclic antidepressant was a serious problem in the past, the number of cases related to tricyclic overdose is on decline because these antidepressants are prescribed less frequently today due to availability of newer psychoactive drugs with improved safety margin especially SSRI (selective serotonin reuptake inhibitors) drugs. Some of the tricyclic antidepressants are metabolized to their active metabolites. Interestingly, active metabolites are also available as drugs for therapy (Table 2). Two major manifestations of tricyclic antidepressants include:

- Central nervous system anticholinergic effects, such as dry mouth and skin, flushing and urinary retention
- Cardiovascular effects

Mortality from tricyclic antidepressant overdose is usually due to cardiovascular toxicity including cardiac arrhythmia. Usually in the electrocardiogram, QRS prolongation of more than 100 ms is a sign of severe toxicity which are usually associated with total tricyclic antidepressant level of 1000 ng/mL or more. Heart rates above 100 or more are also associated with severe toxicity [10]. Tricyclic

| Table 2 Tricyclic antidepressants which have active metabolite. | |
|---|---|
| **Tricyclic antidepressant** | **Active metabolite** |
| Imipramine | Desipramine |
| Amitriptyline | Nortriptyline |
| Doxepin | Nordoxepin |

antidepressant overdose can be treated by alkalization using sodium bicarbonate. Other therapies may also be initiated depending on the clinical condition of the patient as determined by the physician.

## Benzodiazepine and opiate overdose

Benzodiazepines are one of the most commonly observed prescription medications along with opioid prescription medications. Flumazenil, a specific benzodiazepine antagonist, is useful in reversing the sedation and respiratory depression which are characteristics of benzodiazepine overdose. However, some controversies exist in using flumazenil in treating mixed benzodiazepine/tricyclic antidepressant overdose due to possible precipitation of seizure activity [11].

Naloxone (Narcan) is a synthetic derivative of oxymorphone which can antagonize pharmacological effects of opiates. Therefore, naloxone is an excellent antidote for opiate overdose not only with morphine and heroin, but also with other opioids such as hydrocodone, oxycodone, and oxymorphone. This antidote is administered by intravenous injection (poorly absorbed after oral administration) and onset of action is very rapid (within 2–5 min). However, it is important to note that opioid addiction is treated with methadone.

## Alcohol poisoning

Poisoning with alcohol (ethanol), methanol, and ethylene glycol is also common. In addition to standard care such as gastric irrigation, both methanol and ethylene glycol poisoning can be treated with intravenous administration of ethyl alcohol to achieve a blood ethanol level of 0.1% (100 mg/dL) so that the metabolism of methanol and ethylene glycol to their toxic metabolite (formic acid for methanol and oxalic acid for ethylene glycol) can be minimized. This is possible because methanol and ethylene glycol are metabolized by alcohol dehydrogenase enzyme but ethanol is the preferred substrate for alcohol dehydrogenase. Fomepizole (4-methylpyrazole, administered intravenously) is also a competitive inhibitor of alcohol dehydrogenase which can be used in treating methanol and ethylene glycol poisoning instead of ethyl alcohol. However, both fomepizole and ethyl alcohol cannot be used simultaneously because fomepizole also significantly increases the elimination of ethyl alcohol. Depending on the blood level of methanol and ethylene glycol, dialysis may also be initiated. Treatment of alcohol poisoning is mostly standard but ethanol is also dialyzable.

It is important to note that alcohol abuse is treated in a detoxification center with disulfiram (Antabuse) which inhibits alcohol dehydrogenase and a person supposed to be alcohol-free experiences uncomfortable physical reaction (nausea, flushing, vomiting, and headaches) due to the accumulation of acetaldehyde after drinking even one drink. Other medications used in treating alcohol abuse include naltrexone that blocks the good feeling after drinking and acamprosate that helps in combating alcohol craving. However, in contrast to disulfiram, naltrexone, and acamprosate do not produce uncomfortable feeling after drinking alcohol in a person supposed to be sober.

## Poisoning from organophosphorus and carbamate insecticides

Pesticides are widely used in agriculture to control insects, fungi, weeds, and microorganisms in order to increase the crop yield, but chemicals in pesticides may cause toxicity in humans if the chemical comes in contact with skin, eyes, inhaled or is swallowed (more commonly in a suicide attempt). Pesticides can be absorbed through the skin after exposure, but it can be absorbed more readily from the eyes or through lungs after breathing contaminated air since many pesticides form droplets or fumes. Pesticides can be grouped according to chemical classes. The most commonly used chemical classes include organophosphates (also called organophosphorus), organochlorines, and carbamates. Thiocarbamate and dithiocarbamate are subclasses of carbamate pesticides.

Organophosphate and to some extent carbamate pesticides inhibit cholinesterase. Two major human cholinesterase enzymes are acetylcholinesterase, found primarily in nerve tissues and erythrocytes, and pseudocholinesterase, which is found in both serum and the liver. Poisoning with carbamate is usually less severe because carbamate binds to acetylcholinesterase reversibly, while binding with organophosphorus is irreversible. Nevertheless, both organophosphorus and carbamate toxicity may cause serious life-threatening situations including fatality.

Currently, red cell (erythrocyte) acetylcholinesterase activity or serum plasma cholinesterase (also known as pseudocholinesterase or butyryl-cholinesterase) activity can be measured for the diagnosis as well as for monitoring progress of therapy in patients poisoned with organophosphate or carbamate insecticides. The antidote for organophosphorus and carbamate poisoning is atropine to block the muscarinic action of excess acetylcholine. Pralidoxime is also given to reactivate cholinesterase.

## Lead poisoning

Lead is a heavy metal and a divalent cation that has been used in human civilization for a long time. In 1976 lead-based paints in toys were banned in the

United States, and lead-based household paints were banned in 1978, and in 1986, lead-based gasoline was phased out. Today, major sources of lead exposures are as follows:

- Deteriorated lead-based paint in older housing.
- Old water pipes with lead-based soldering.
- Soil contaminated with lead.
- Artists who use stained glass and glazed ceramic.
- Moonshine liquors.
- Many Asian herbal remedies and Indian Ayurvedic medicine are contaminated with heavy metals such as lead, arsenic, and mercury. Lead poisoning after taking Indian Ayurvedic medicine has been reported.
- An unusual source of lead could be the edible parts of birds killed by lead shots.
- A rare source is firearm injury and retention of a part of lead bullet in one's body.

Lead enters the body either through inhalation or through ingestion. Lead is distributed in the body in three main compartments: blood, soft tissue, and bone. Approximately, 99% of lead in blood is bound to red blood cells and 1% is available freely in plasma. The skeleton is the main lead depot in the human body and may represent approximately 90%–95% body burden of lead in adults. Important issues regarding lead toxicity include:

- Lead inhibits delta-aminolevulinic acid dehydrogenase, one of the enzymes that catalyze the synthesis of heme from porphyrin thus increasing the erythrocyte concentration of protoporphyrin. Lead also inhibits enzyme ferrochelatase leading to inhibition of incorporation of iron in protoporphyrin. As a result, zinc protoporphyrin (ZPP) is produced and remains in circulation in red blood cells for 120 days with elevated concentration. A hematofluorometer may be used to directly measure ZPP fluorescence in whole blood without extraction. The instrument measures the ratio of ZPP fluorescence to heme absorption. The normal zinc protoporphyrin (ZPP) level is $<35\,\mu g/dL$. Elevated zinc protoporphyrin is a biological indicator of both lead poisoning and iron deficiency.
- Lead exposure may cause microcytic anemia and acute lead toxicity may cause hemolytic anemia. Lead is a nephrotoxic agent.
- Neurotoxicity of lead produces more adverse effects in children than adults.

Although the presence of erythrocytic zinc protoporphyrin can be used for the diagnosis of lead poisoning, zinc protoporphyrin levels are also elevated in anemia. The zinc protoporphyrin to heme ratio can in this case also be used for the diagnosis of lead poisoning. In addition, determination of blood lead level is also useful for assessing lead exposure. Heparinized whole blood is the appropriate specimen for this test since most of the lead in blood is bound to erythrocytes. Lead testing on a dried filter paper blood spot is also used routinely by some laboratories for screening of lead poisoning especially in small children and neonates. The lead level in blood can be measured by atomic absorption spectrophotometry. However, other techniques such as anodic stripping voltammetry and inductively coupled plasma mass spectrometry (ICP-MS) can also be used. Determination of lead in 24 h urine is also useful to monitor the therapy for lead toxicity. For the collection of 24 h urine sample, the patient should void directly into a lead-free container (a borosilicate glass or polyethylene containers are lead-free).

The Center for Disease Control recommends that blood lead levels of children should be less than 10 µg/dL. In many cases of lead poisoning, removing the person from the sources of exposure is sufficient to reduce the blood lead level. The WHO (World Health Organization) defines lead levels over 30 µg/dL indicative of significance exposure to lead. Chelating therapy is usually performed for blood lead concentrations over 60 µg/dL. Calcium sodium EDTA is an effective chelating agent, but D-penicillamine is also effective in treating lead poisoning. More recently, 2,3-dimercaptosuccinic acid (DMSA) has been introduced as an orally given chelating agent to replace EDTA.

## Mercury poisoning

Mercury is a heavy metal with known toxicity. Most human exposure to mercury is from outgassing of mercury from dental amalgam or eating fish and seafood contaminated with mercury. Occupational exposure to mercury vapor also causes mercury toxicity. The target organ for inhaled mercury vapor is the brain. Mercury exposure may also occur from inorganic mercury (mercurous and mercuric salt) as well as from organic mercury (mostly methyl mercury). Mercury in all forms is toxic because it alters the tertiary and quaternary structure of proteins by binding with sulfhydryl groups. Consequently, mercury can potentially impair the function of any organ, but neurological functions are mostly affected.

Exposure of methyl mercury from eating fish, shellfish (both fresh water and sea), or sea mammals is dangerous for pregnant women and children. Over 3000 lakes in the United States have been closed for fishing due to mercury contamination [12]. Methyl mercury may be present in higher amounts in predatory fish. Pregnant women, women planning to be pregnant, and children should not eat fish (shark, swordfish, king mackerel, golden bass, and snapper) where high amounts of mercury could accumulate. Other people can eat up to 7 oz of these fish per week. Salmon, cod, flounder, catfish, and other seafood such as crabs and scallops may also contain mercury in lower amounts. Determination of blood, urine, and even hair level of mercury is useful for the determination of exposure to mercury. WHO experts have determined that 24 h urine mercury level over 50 μg is indicative of excessive exposure to mercury.

Mercury poisoning is treated with chelating therapy. BAL (British-Lewisite, dimercaprol), a chelating agent developed to treat arsenic poisoning is also effective in chelating mercury. However, this chelating agent is toxic and more recently safer chelating agents such as DMPS (2,3-dimercapto-1-propanesulfonate) have been developed for clinical use.

## Arsenic poisoning

Soluble inorganic arsenic (as arsenic salts) is acutely toxic, and ingestion of a large dosage may cause gastrointestinal, neurological, and cardiovascular toxicity and eventually may be fatal. Long-term chronic exposure to arsenic may occur from drinking water from wells where groundwater is contaminated with arsenic. This may be a particular problem in many developing countries and such long-term exposures may cause cancer in the skin, lungs, bladder, and kidney. Hypertension and cardiovascular disorders are also common after chronic exposure to arsenic. In addition, early chronic exposure to arsenic may cause skin changes such as hyperkeratosis and pigmentation change. The WHO guidelines indicate that the maximum limit of arsenic in drinking water should be 10 μg/L. However, arsenic levels are >200 μg/L in some drinking water wells in Bangladesh [13].

Another source of arsenic poisoning is the use of some herbal remedies imported from Asia and Indian Ayurvedic medicines. In some Indian Ayurvedic medicine, arsenic and other heavy metals are used as components. Some kelp supplements may contain high amount of arsenic. Serum arsenic concentrations are elevated for a short time only after exposure. Therefore, hair analysis of arsenic is useful to investigate chronic exposure to arsenic. Hair arsenic levels greater than 1 μg/g in dry hair are indicative of excessive exposure to arsenic. Severe arsenic toxicity can be treated with chelation therapy. Originally, BAL was used widely as a chelating agent but it has been mostly replaced by safer chelating agents such as DMSA.

## Poisoning from other metals

Metal toxicity may also occur from exposure to other heavy metals, such as aluminum, antimony, beryllium, cadmium, chromium, cobalt, copper, iron, manganese, nickel, selenium, silicon, and thallium. However, toxicities from these heavy metals are observed less frequently than lead, mercury, and arsenic poisoning. Aluminum toxicity has been documented after excessive use of aluminum-based OTC antacids secondary to aluminum-containing phosphate binder agents, but such toxicity is rare. Patients undergoing dialysis are also exposed to aluminum, but toxicity is very rare. Therefore, routine plasma monitoring of dialysis patients is not necessary, and it should only be conducted if excessive aluminum exposure is suspected.

The classic accumulation of copper in the human body is due to the genetic disease known as "Wilson Disease." Iron supplements are used for treating anemia, but excess ingestion of iron supplements may cause iron overload and toxicity. Silicon toxicity was a public health concern when asbestos-containing products were used as insulating material in houses. Asbestos dust inhalation is particularly harmful. However, asbestos is no longer used for most instances. Silicone-based implants (containing elastomers of silicon) may be another source of silicon exposure, especially if there is a rupture of an implant. Silicone lymphadenopathy (presence of silicone in the lymph node) is a recognized complication of breast augmentation. It is thought to occur when silicone droplets migrate from breast implants to lymph nodes. A 35-year-old Italian woman, 10 years after bilateral cosmetic breast augmentation, showed extensive granulomatous inflammation, numerous histiocytes, and multinucleated giant cells containing star-shaped structures known as "asteroid bodies" in her axillary lymph node fine-needle biopsy [14].

## Summary/key points

- Usually, severe acetaminophen toxicity occurs in an adult after consuming 7–10 g of acetaminophen (15–20 tablets containing 500 mg acetaminophen in each). Acetaminophen is normally metabolized to glucuronide and sulfate conjugates by the liver enzymes but a small amount is metabolized by cytochrome mixed function oxidase family of enzymes to a toxic metabolite N-acetyl-p-benzoquinone imine which is detoxified after conjugating with glutathione present in the liver. In severe acetaminophen overdose, glutathione supply in the liver is exhausted and this toxic metabolite causes liver damage. Antidote of acetaminophen poisoning is N-acetyl cysteine (Mucomyst) which is a precursor of glutathione and also can detoxify the toxic metabolite of acetaminophen. This antidote must be administered as soon as possible but certainly within 8 h of acetaminophen overdose for maximum benefit.
- Acetylsalicylate (aspirin) can cause potentially life-threatening toxicity known as "Reye's syndrome" (30%–40% fatality) in children and adolescents when treated for symptoms for certain viral infection such as varicella and influenza. Therefore, salicylate use is contraindicated for these groups of patients.

- Salicylate overdose results in direct stimulation of central respiratory system causing hyperventilation and respiratory alkalosis and hypothermia followed by metabolic acidosis.
- Methyl salicylate should be used only as a topical analgesic. If ingested, it is very toxic.
- Carbon monoxide poisoning causes increase in carboxyhemoglobin level. However, smokers may have carboxyhemoglobin level of 5%–6%. A carboxyhemoglobin level of 60%–70% and above may cause respiratory failure and even death. Fetal hemoglobin has slightly different spectrophotometric properties than adult hemoglobin. Therefore falsely high carboxyhemoglobin value up to 7% has been reported in neonates using Co-oximeter. Treatment for carbon monoxide poisoning is 100% oxygen or oxygen therapy using a hyperbaric chamber.
- Cyanide poisoning can be life-threatening. Hydrogen cyanide (HCN) is the most lethal form. The most common source is smoke inhalation from a fire in the house because burning common household substances, such as plastic, silk, and rubber may produce cyanide smoke containing HCN. After exposure to HCN, cyanide ion tightly binds with hemoglobin producing cyanhemoglobin which results in severe hypoxia. Usually, spectroscopic methods are used for cyanide determination and normal blood cyanide level is less than $0.2 \mu g/mL$. A cyanide level above $2 \mu g/mL$ can produce severe toxicity and level above $5 \mu g/mL$ may be lethal if not treated immediately. There are several antidotes for treating cyanide poisoning including inhalation of amyl nitrate, and administration of sodium thiosulfate, sodium nitrate, ferrous sulfate, and dicobalt edentate as well as hydroxocobalamin.
- Overdose with tricyclic antidepressants causes central nervous system anticholinergic toxic effects, such as dry mouth and skin, flushing, and urinary retention as well as cardiovascular toxicity. Mortality from tricyclic antidepressant overdose is usually due to cardiovascular toxicity including cardiac arrhythmia. Usually in the electrocardiogram, QRS prolongation of more than 100 ms is a sign of severe toxicity which are usually associated with total tricyclic antidepressant level of 1000 ng/mL or more. Heart rates above 100 or more are also associated with severe toxicity. Tricyclic antidepressant overdose can be treated by alkalization using sodium bicarbonate.
- Benzodiazepines overdose can be treated with flumazenil while naloxone is a good antidote for treating opioid overdose.
- Methanol and ethylene glycol overdose can be treated with either ethyl alcohol or fomepizole (4-methylpyrazole).
- Both organophosphorus and carbamate toxicity may cause serious life-threatening situations including fatality, but in general, carbamate insecticides are less toxic. Red cell (erythrocyte) acetylcholinesterase

activity or serum plasma cholinesterase (also known as pseudocholinesterase or butyryl-cholinesterase) activity can be measured for diagnosis as well as for monitoring the progress of therapy in patients poisoned with organophosphate or carbamate insecticides.

- The Center for Disease Control recommends that blood lead levels of children should be less than 10 µg/dL. Blood lead measurement is a good way for diagnosis of lead toxicity. Lead inhibits delta-aminolevulinic acid dehydrogenase, one of the enzymes that catalyze synthesis of heme from porphyrin, thus increasing the erythrocyte concentration of protoporphyrin. Lead also inhibits enzyme ferrochelatase leading to inhibition of incorporation of iron in protoporphyrin. As a result, zinc protoporphyrin (ZPP) is produced and remains in circulation for 120 days. ZPP is measured in whole blood where normal level is <35 µg/dL. Elevated ZPP may be due to lead poisoning or anemia. Lead exposure may cause microcytic anemia and acute lead toxicity may cause hemolytic anemia. Lead is a nephrotoxic agent. In addition, neurotoxicity of lead produces more adverse effects in children than in adults.

- Normal blood mercury level should be lower than 10 µg/L (1 µg/dL) but blood level over 50 µg/L (5 µg/dL) is indicative of severe methyl mercury toxicity. Higher levels may be observed after poisoning of inorganic mercury. Eating fish growing in contaminated water may cause methyl mercury exposure. A 24 h urine mercury level over 50 µg is indicative of excessive exposure to mercury as determined by WHO experts.

- Arsenic poisoning may occur from drinking well water contaminated with arsenic. Herbal supplements especially some Indian Ayurvedic medicines may be contaminated with heavy metals including lead, arsenic, and mercury.

## References

[1] Seeff LB, Cuccherini BA, Zimmerman HJ, Adler E, et al. Acetaminophen hepatotoxicity in alcoholics: a therapeutic misadventure. Ann Intern Med 1986;104:399–404.

[2] Chornomydz I, Boyarchuk O, Chornomydz A. Reye (Ray's) syndrome: a problem everyone should remember. Georgian Med News 2017;272:110–8.

[3] Davis JE. Are one or two dangerous? Methyl salicylate exposure in toddlers. J Emerg Med 2007;32:63–9.

[4] Bell AJ, Duggin G. Acute methyl salicylate toxicity complicating herbal skin treatment for psoriasis. Emerg Med (Fremantle) 2002;14:188–90.

[5] Anderson A, McConville A, Fanthorpe L, Davis J. Salicylate poisoning potential of topical pain relief agents: from age old remedies to engineered smart patches. Medicines (Basel) 2017;4(3). pii: E48.

[6] Suner S, Partridge R, Sucov A, Valente J, et al. Non-invasive pulse CO-oximetry screening in the emergency department identifies occult carbon monoxide toxicity. J Emerg Med 2008;34:441–50.

[7] Mehrorta S, Edmonds M, Lim RK. False elevation of carboxyhemoglobin: a case report. Pediatr Emerg Care 2011;27:138–40.

[8] Lundquist P, Rammer L, Sorbo B. The role of hydrogen cyanide and carbon monoxide in fire causalities: a prospective study. Forensic Sci Int 1989;43:9–14.

[9] Borron SW, Baud FJ, Barriot P, Imbert M, et al. Prospective study of hydroxocobalamin for acute cyanide poisoning in smoke inhaler. Ann Emerg Med 2007;49:794–801.

[10] Lavoie FW, Gansert GG, Weiss RE. Value of initial ECG findings and plasma drug levels in cyclic antidepressant overdose. Ann Emerg Med 1990;19:696–700.

[11] Krisanda TJ. Flumazenil: an antidote for benzodiazepine toxicity. Am Fam Physician 1993;47:891–5.

[12] Bernhoff RA. Mercury toxicity and treatment: a review of literature. J Environ Public Health 2012;2012:460508.

[13] Bolt HM. Arsenic: an ancient toxicant of continuous public health impact, from Iceman to Otzi until now. Arch Toxicol 2012;86:825–30.

[14] Malzone MG, Campanile AC, Gioioso A, Fucito A, et al. Silicone lymphadenopathy: presentation of a further case containing asteroid bodies on fine-needle cytology sample. Diagn Cytopathol 2015;43:57–9.

# CHAPTER 20

# Pharmacogenomics

## Introduction to pharmacogenomics

The goal of pharmacogenomics (also known as *pharmacogenetics*) is to understand the effect of polymorphisms of genes encoding drug metabolism enzymes, transporters, and receptors of drugs on the outcome of drug therapy. The first pharmacogenomics discovery was made over 50 years ago when it was demonstrated that patients with a genetic polymorphism leading to deficiency of glucose-6-phosphate dehydrogenase developed hemolysis after treatment with primaquine [1]. With the completion of the Human Genome Project and availability of pharmacogenomics testing, currently there are over 100 drugs for which pharmacogenomics testing may benefit patients, but in reality, pharmacogenomics may be most beneficial in patients receiving warfarin therapy, chemotherapy with certain anticancer drugs and opioids [2]. Examples of drugs where pharmacogenomics testing are useful are listed in Table 1.

There are variabilities at the DNA level between individuals which govern many characteristics of the person including his or her ability to respond to a particular drug therapy. The effect of genetic polymorphism of butylcholinesterase on the metabolism of neuromuscular blocking agent succinylcholine and mivacurium used during general anesthesia has been well documented. Many current clinical applications of pharmacogenomics involve drug metabolism, but with continued research into pharmacogenomics involving pharmacodynamics. Targeted cancer therapies utilize medications designed to impact specific somatic mutations in neoplasms. Many of the current pharmacogenomic clinical applications focus on the cytochrome P450 (CYP) mixed function oxidase enzymes. The CYP enzyme represents the most important family of enzymes responsible for drug metabolism, comprises a large group of heme-containing enzymes. These enzymes are found in abundance in the liver and other organs. The major CYP isoforms responsible for the metabolism of drugs

Clinical Chemistry, Immunology and Laboratory Quality Control. https://doi.org/10.1016/B978-0-12-815960-6.00020-0

**Table 1** Commonly used pharmacogenomic biomarkers for selected drugs.

| Drug | Biomarker | Function of biomarker | Comments |
|---|---|---|---|
| Warfarin | CYP2C9 | Drug metabolism | For warfarin pharmacogenomic both CYP2C9 and VKORC1 should be tested |
| Clopidogrel (Plavix) | CYP2C19 | Drug metabolism | Asian population should be tested for CYP2C19 poor metabolizer because clopidogrel may not effective |
| Codeine | CYP2D6 | Drug metabolism | Children who are ultrarapid metabolizer and breast-feeding infant where mother is ultrarapid metabolizer are at danger of morphine toxicity and even fatality |
| Tramadol | CYP2D6 | Drug metabolism | Poor metabolizers may not get adequate pain control where ultrarapid metabolizers are at higher risk of toxicity |
| Tamoxifen | CYP2D6 | Drug metabolism | Tamoxifen is a prodrug which must be converted into endoxifen. Poor metabolizers show poor response to tamoxifen therapy |
| Tricyclic antidepressants, citalopram, fluoxetine, aripiprazole, iloperidone | CYP2D6 | Drug metabolism | Poor metabolizers clear drug slowly |
| Tacrolimus | CYP3A5 | Drug metabolism | Polymorphism may affect dose selection |
| Azathioprine, 6-mercaptopurine | Thiopurine methyltransferase (TPMT) | Drug metabolism | Profound bone marrow toxicity following standard doses of azathioprine and 6MP due to low TPMT enzymatic activity (TPMT poor metabolizers) |
| Irinotecan | UGT1A1 | Drug metabolism | UGT1A1*28 allele is associated with irinotecan toxicity |
| Carbamazepine | HLA-B*1502 | Genetic link to hypersensitivity | HLA-B*1502, which is more abundant among Asians, is associated with severe skin rash including Stevens-Johnson syndrome following treatment with carbamazepine |
| Abacavir | HLA-B*5701 | Genetic link to hypersensitivity | Strongly associated with hypersensitivity toward the anti-HIV drug abacavir |
| Allopurinol | HLA-B*5801 | Genetic link to hypersensitivity | Associated with hypersensitivity |

include CYP1A2, CYP2B6, CYP2C9, CYP2C19, CYP2D6, CYP2E1, and CYP3A4/CYP3A5. Major points regarding CYP enzymes include:

- CYP3A4 is the predominant isoform of the CYP family (almost 30%) usually responsible for metabolism of many drugs.
- Genetic polymorphisms of CYP2D6, CYP2C9, and CYP2C19 have been well studied and account for some wide interindividual response to various drugs.

## Basic pharmacogenomics

The most common type of genetic variation is single-nucleotide polymorphism (SNP), a situation in which some individuals have one nucleotide at a given position while other individuals have another nucleotide (e.g., cytosine vs guanine, C/G). SNPs occurring in the coding region of genes can change the amino acid sequence that results when DNA is transcribed into RNA and the RNA is then translated into protein. Less common types of genetic variation include insertions or deletions (collectively known as *indels*), alterations of gene splicing, partial or total gene deletions, gene duplication or multiplication, and variation in gene promoter affecting gene expression [3].

## Nomenclature

Although there is a systematic nomenclature for precisely describing genetic variation (e.g., following the HUGO Gene Nomenclature standards), genetic variants important in pharmacogenomics are still referenced by a historical naming system based on when variants were first characterized and reported. By this historical convention, the normal allele (wild type) is defined as *1 (e.g., CYP2C9*1). In order of historical discovery, variant alleles were designated *2, *3, *4, and so on. Unfortunately, this nomenclature does not give any clue to the nature of the genetic variation. For instance, *2 and *5 could represent fairly benign genetic variants (e.g., point mutations with minimal impact on protein function), whereas *6 could signify a variant that results in complete absence of enzyme activity such as gene deletion or a mutation resulting in a frameshift or protein truncation.

## Various metabolizer types

Based on the response to a drug, individuals can be classified under different class:

- Extensive metabolizers (EM): These individuals are carriers of two normal copies of the gene (wild type) that encodes drug metabolizing enzyme.
- Intermediate metabolizers (IM) have enzymatic activity roughly 50% that of extensive metabolizers. The most common genetic reason underlying intermediate activity is one copy of the normal gene and one variant copy associated with low activity (i.e., heterozygous for a low activity variant allele).
- Poor metabolizers (PM) represent individuals with little or no enzymatic activity, due to either lack of expression of the enzyme (partial or complete gene deletions; intron splice variants; frameshift mutations) or

mutations that reduce enzymatic activity (e.g., a variant impacting the catalytic site) in both alleles.

- Ultrarapid metabolizers (UM) have enzymatic activity significantly greater than that of the average population (extensive metabolizers). This often results when an individual has more than the normal two copies of a gene. Gene duplication or multiplication is not seen with many genes but can occur with CYP2D6.

For many drugs, CYP or other drug-metabolizing enzymes inactivate the drug and thus play an important role in clearance. Poor metabolizers may thus experience drug toxicity with standard doses due to slower drug clearance. Ultrarapid metabolizers may see little clinical effect with standard drug doses due to accelerated clearance of the drug or may be more prone to drug toxicity if the metabolite is active. Special circumstances arise when pharmacogenomics impacts prodrugs. For example, clopidogrel (antiplatelet agent; Plavix) is a prodrug which is converted into the active metabolite by CYP2C19. Poor metabolizers (CYP2C19*2 and CYP2C19*3) may not get therapeutic benefits after treating with clopidogrel. These alleles are common among Asian population and therapy with clopidogrel after cardiovascular event may not be effective [4].

Codeine is metabolized into morphine by CYP2D6 and morphine is a much stronger analgesic than codeine. Poor metabolizers of CYP2D6 may not get adequate pain control if treated with codeine. Tamoxifen, a selective estrogen receptor modulator used in breast cancer treatment, is also functionally a prodrug dependent on CYP2D6 for conversion to metabolites with higher affinity for estrogen receptors. CYP2D6 poor metabolizers will show less response to standard doses of tamoxifen.

## Other polymorphic enzymes that metabolize drugs

Other polymorphically expressed drug-metabolizing enzymes are N-acetyltransferase (NAT 1 and NAT 2) and thiopurine-S-methyltransferase (TPMT). The slow acetylator phenotype of the NAT1/2 polymorphism results in isoniazid-induced peripheral neuropathy and sulfonamide-induced hypersensitivity reactions. TPMT catalyzes inactivation of various anticancer and antiinflammatory drugs. In addition, polymorphism of uridine-5 phosphate glucuronosyltransferase (UDP-glucuronosyltransferase) may also play vital role in metabolism of certain drugs, for example, irinotecan, an anticancer drug. This enzyme is responsible for conjugation of glucuronic acid with the drug molecule in the phase II metabolism, thus inactivating the drug. This enzyme is mostly found in the liver but also may be present in other organs. There are two main families of UDP-glucuronosyltransferase: UGT1 and UGT2. Polymorphisms of UGT1A1 and UGT2B7 play important roles in phase II metabolism of certain drugs.

## Polymorphism of transporter proteins and receptors

Most drug responses are determined by the interplay of several gene products that influence pharmacokinetics and pharmacodynamics, i.e., drug metabolizing enzymes, drug transporters, and drug targets. With the complete sequencing of the human genome, it has been estimated that approximately 500–1200 genes code for drug transporters and today the best characterized drug transporter is the multidrug-resistant transporter P-glycoprotein/MDR1, the gene product of MDR1 (multiple drug resistant protein). Compared to drug metabolizing enzymes, much less is known about the genetic polymorphisms of drug targets and receptors but molecular research has revealed that many of the genes that encode drug targets demonstrate genetic polymorphism.

## Pharmacogenomics and warfarin therapy

Warfarin is a synthetic compound available as a racemic mixture of 50% R-warfarin and 50% S-warfarin. Warfarin remains the commonly prescribed oral anticoagulant, although the growing use of other oral anticoagulants (e.g., direct thrombin and factor X inhibitors) has reduced warfarin usage. Standard monitoring of warfarin therapy utilizes prothrombin time (PT) or international normalized ratio (INR which is calculated from PT and is a way of standardizing PT value expressed in seconds regardless of testing method), with INR target ranges for specific indications (e.g., atrial fibrillation and mechanical heart valve). Warfarin-related bleeding complications is a major complicating factor of warfarin therapy. The goal of warfarin pharmacogenomics is to help predict both the initial and maintenance doses of warfarin. One limitation of pharmacogenomics testing is that it does not eliminate the need to monitor warfarin therapy by INR.

Two major genes are involved in warfarin pharmacogenomics: CYP2C9 and VKORC1 (VKORC1; vitamin K epoxide reductase complex). CYP2C9 metabolizes warfarin to an inactive metabolite. VKORC1 is the molecular target inhibited by warfarin, thus reducing the production of vitamin K-dependent clotting factors. CYP2C9-poor metabolizers are at risk for warfarin toxicity at standard doses because of reduced clearance of the drug and excess concentrations of warfarin. The two most common mutations associated with poor metabolizers are CYP2C9*2 and *3. VKORC1 has several genetic variants that influence warfarin pharmacodynamics. Collectively, genetic variations in CYP2C9 and VKORC1 account for approximately one-third of the observed variation in warfarin dosage that leads to stable anticoagulation. Additional proteins involved in warfarin effect include CYP4F2 and gamma-glutamyl carboxylase (GGCX) with genetic variation in these genes also having an effect on warfarin. Various algorithms are available to predict optimal warfarin dosage [5]. One commonly used algorithm (www.warfarindosage.org) is freely available and open-access.

## Pharmacogenomics of selected anticancer drugs

There are increasing pharmacogenomic applications involving cancer therapy. Some examples are related to drug metabolism, while others involve somatic mutations in the cancer itself. Azathioprine and 6-mercaptopurine (6MP) are drugs used in the treatment of cancers (e.g., acute lymphoblastic leukemia) and autoimmune disorders such as inflammatory bowel disease and rheumatoid arthritis. Azathioprine is the prodrug of 6MP. One route for inactivation and clearance of 6MP is by the enzyme thiopurine methyltransferase (TPMT). Although azathioprine and 6MP both can produce toxic effects on the bone marrow if used in high doses, approximately 1 in 300 Caucasians (less in most other populations) experiences very profound bone marrow toxicity following standard doses of azathioprine and 6MP due to low TPMT enzymatic activity (TPMT poor metabolizers). Several clinical laboratory tests are available to determine whether individuals will have difficulty metabolizing 6MP and azathioprine [6]. In 2004–2005, the package inserts for azathioprine and 6MP were revised to include warnings on toxicity related to genetic variation of TPMT. TPMT-poor metabolizers can still receive 6MP or azathioprine, but need markedly reduced doses.

Irinotecan, a chemotherapeutic agent used in the treatment of colorectal cancer and some solid tumor, is inactiavted by glucuronidation mediated by UDP-glucuronosyltransferase 1A1 (UGT1A1), an enzyme also responsible for conjugation of bilirubin. A rare mutation in UGT1A1 can result in the Crigler-Najjar syndromes a potentially fatal disease in childhood unless liver transplantation is performed. A milder mutation, designated UGT1A1*28, is the most common cause of Gilbert syndrome, a usually benign condition often diagnosed incidentally in the primary care setting following detection of increased unconjugated bilirubin during routine checkup or in the workup of jaundice. These individuals are, however, at high risk for severe toxicity following irinotecan therapy. With standard doses, such individuals may develop life-threatening neutropenia or diarrhea poorly responsive to therapy [7]. Genetic testing for UGT1A1*28 is FDA-approved and, similar to 6MP and azathioprine, the package insert for irinotecan now includes specific information on UGT1A1 genetic variation. However, genetic screening for UGT1A1 has not been adopted universally.

Tamoxifen is used for treating estrogen receptor positive breast cancer. Tamoxifen is a prodrug which is converted into the active metabolite endoxifen (4-hydroxy-N-desmethyl-tamoxifen) mostly by CYP2D6. Poor metabolizers have lower levels of endoxifen in blood and respond poorly to tamoxifen therapy [8]. Approximately 5%–14% of Caucasians, 0%–5% Africans, and 0%–1% of Asians lack CYP2D6 activity, and these individuals are poor metabolizers because they carry two defective alleles [9]. Lower serum endoxifen levels are observed in poor metabolizers compared to extensive metabolizers after a standard dose.

Pharmacogenomics also find application in oncology in targeted therapies for certain cancers. One of the best examples is the use of trastuzumab (Herceptin) for breast cancers overexpressing the HER2 protein. In the pathology workup of breast cancer, determination of HER2 expression status determines whether trastuzumab is a therapeutic option. A more recent trend in oncology is the emergence of *paired diagnostics*, where a drug and its companion diagnostic test are approved simultaneously. An example is vemurafenib (Zelboraf), a drug that targets metastatic melanoma with a specific mutation in *BRAF*. Vemurafenib and the companion test to determine *BRAF* mutation status were approved simultaneously by the FDA.

## Pharmacogenomics of selected opioid drugs

In general, the majority of therapeutic drugs used in pain management including codeine, dihydrocodeine, fentanyl, hydrocodone, methadone, morphine, oxycodone, and tramadol as well as tricyclic antidepressants are metabolized by polymorphic CYP450 enzymes such as CYP2D6, CYP3A4, and/or uridine diphosphate glucuronosyltransferase (UGT2B7). The wide range of genetic polymorphisms of CYP2D6 leads to four distinct groups of metabolizers including ultrarapid metabolizers containing multiple copies of the CYP2D6 gene, extensive metabolizers with a single wild-type copy of the CYP2D6 gene, intermediate metabolizers showing decreased enzymatic activity and poor metabolizers (PMs) with almost no detectable activity. Differences in drug metabolism due to polymorphism of these genes can lead therapeutic failure or toxicity depending on the individual drug. Codeine is metabolized by the liver into the more active drug morphine. Approximately, 5%–10% of Caucasians and 1%–4% of most other ethnic groups have decreased CYP2D6 activity and these patients may not get adequate pain control following codeine therapy. In contrast, codeine toxicity may occur in CYP2D6 ultra-rapid metabolizing children who are administered codeine (e.g., for analgesia for dental procedures) or in breastfeeding infants whose mothers are CYP2D6 ultra-rapid metabolizers taking codeine for pain control after delivery [10]. In either case, the excess of morphine can cause respiratory depression and other symptoms of opiate overdose. There is a report of death of 13-day-old infant due to high morphine level in blood originated from breast milk of the mother who was taking codeine and also an ultrarapid CYP2D6 metabolizer [11]. The FDA issued separate warnings on use of codeine in breastfeeding mothers and for postsurgical pain management for children in 2007 and 2012, respectively.

## Pharmacogenomics of selected psychoactive drugs

Pharmacogenomics play an important role in determining serum or plasma levels of selected psychoactive drugs. The tricyclic antidepressant amitriptyline is metabolized by CYP2C19 to the active metabolite nortriptyline. Then CYP2D6 is needed for deactivation of nortriptyline. Adverse drug reactions tend to be associated with nortriptyline concentrations, and CYP2D6 poor metabolizers are more likely to suffer from adverse effects due to buildup of nortriptyline concentration. Smith and Curry reported the case of a comatose woman who intentionally overdosed with amitriptyline. She demonstrated rising total tricyclic antidepressant concentration over the next 6 days after admission to the hospital. The level started declining on day seven. Genotyping showed the patient to be homozygous for CYP2D6*4 allele, the most common cause of CYP2D6 enzymatic deficiency among Caucasians [12]. Paroxetine, a selective serotonin reuptake inhibitor (SSRI), is metabolized by CYP2D6 and also is an inhibitor of CYP2D6. Poor metabolizers are at higher risk of adverse effects from paroxetine [13].

## Pharmacogenomics of miscellaneous other drugs

Organ transplant recipients receive immunosuppressants in order to prevent organ rejection. These drugs are metabolized by cytochrome P-450 family of enzymes including CYP3A4 and CYP3A5. Although polymorphisms of CYP3A4 do not alter the enzyme activity significantly, polymorphisms of CYP3A5 may be clinically more significant, as enzyme activities may vary significantly between different alleles. Pharmacogenomics testing can be used to identify polymorphisms of CYP3A5 to predict optimal initial dosage of tacrolimus [14].

Recent developments in understanding severe immunological adverse reactions following therapy with certain drugs indicate that HLA-B (major histocompatibility class I type B gene complex) is a biomarker to strongly predict onset of serious skin rash such as Stevens-Johnson syndrome. HLA-B57 and HLA-B58 are major histocompatibility class-I allotypes that are predictive of clinically important immune phenotype. It has been demonstrated that HLA-B*1502, which is more abundant among Asians, is associated with severe skin rash including Stevens-Johnson syndrome following treatment with carbamazepine. The FDA has released a warning suggesting that HLA-B testing in Asian be performed before carbamazepine therapy is given. Although a strong correlation between carbamazepine induced Stevens-Johnson syndrome has been found among Han Chinese patients, HLA-B*1511, a member of HLA-B75 found among Japanese, is also a risk factor for such adverse reactions from carbamazepine therapy [15]. HLA-B*5701 is strongly associated with hypersensitivity toward the anti-HIV drug abacavir. In addition, HLA-B*5801 is associated with hypersensitivity to allopurinol [16].

## Methods for pharmacogenomics testing

A variety of methods have been used to identify DNA polymorphisms affecting genes involved in enzymes responsible for drug metabolism or drug target receptor genes. These techniques either focus on analysis of a selected number of genes (examples: classic Sanger sequencing, pyrosequencing, real-time polymerase chain reaction, and melting curve analysis) or highly multiplexed analysis of a large number of genes (various microarray, microchip, and microbead techniques). The first test for which the FDA has granted market approval using a DNA microarray is the AmpliChip CYP450 (Roche), which genotypes cytochrome P450 (CYP2D6 and CYP2C19). The assay utilizes a patient's blood or a specimen from a buccal swab. In addition to DNA microarray, some currently available pharmacogenomics tests include a test that detects variations in the UGT1A1 gene, which produces the enzyme UDP-glucuronosyltransferase, and another test that detects genetic variants of the CYP2C9 and VKORC1 (vitamin K epoxide reductase) enzymes. Under CLIA rules, home brew pharmacogenomics testing is considered high-complexity testing and must follow strict guideline qualifications of the laboratory director and testing personnel.

## Summary/key points

- Single-nucleotide polymorphism (SNPs) accounts for over 90% of genetic variation in human genome including genes that code enzymes responsible for drug metabolisms.
- Individuals who have two normal genes are extensive metabolizers (EM) who metabolize a drug normally but individuals with two nonfunctional genes are poor metabolizers (PM). Intermediate metabolizers (IM) may have one active allele and one nonactive allele for the same gene while ultrarapid metabolizers (UM) may have multiple copies of active genes and these individuals metabolize a particular drug so fast that the drug may not have any pharmacological effect.
- CYP3A4 is the predominant isoform of the CYP family (almost 30%) usually responsible for metabolism of many drugs. However, genetic polymorphisms of CYP2D6, CYP2C9, and CYP2C19 account for some wide interindividual response to various drugs and pharmacogenetics testing usually focuses on polymorphism of these isoenzymes.
- N-acetyltransferase (NAT 1 and NAT 2) and thiopurine-S-methyltransferase (TPMT) are also important drug-metabolizing enzymes that show significant genetic polymorphism. In addition,

polymorphism of uridine-5 phosphate (glucuronosyltransferase UDP-glucuronosyltransfearse, enzyme responsible for conjugation in Phase II part of drug metabolism) may also play a vital role in the metabolism of certain drugs, for example, irinotecan, an anticancer drug.

- Warfarin pharmacogenomics has been well established where genetic polymorphism of CYP2C9, the enzyme that metabolizes warfarin, and polymorphism of warfarin's target receptor (VKORC1; vitamin K epoxide reductase complex) play an important role in pharmacological action of warfarin.

- Thiopurine anticancer drugs such as 6-mercaptopurine (6-MP), thioguanine, and azathioprine are metabolized by thiopurine S-methyltransferase (TPMT) and polymorphism of TRMT determines pharmacological response of these anticancer drugs.

- Irinotecan is metabolized to an active metabolite 7-ethyl-10-hydroxycamptothecin (SN-38) which is then detoxified by UDP-glucuronosyltransferase (UGT1A1). Therefore, decreased activities of UGT1A1 caused by polymorphism in genes controlling enzymatic activity in particular UGT1A1*28 may cause severe toxicity in an individual after treating with a standard dose due to accumulation of active metabolite SN-38.

- Polymorphism of CYP2D6 is important for pharmacological activities of certain drugs that are metabolized by this enzyme. Codeine is converted into morphine by CYP2D6. Poor metabolizers of CYP2D6may not get adequate pain control after taking codeine. Tamoxifen is a prodrug which is converted into the active metabolite endoxifen (4-hydroxy-N-desmethyl-tamoxifen) mostly by CYP2D6. Poor metabolizers have lower levels of endoxifen in blood and respond poorly to tamoxifen therapy.

- It has been demonstrated that HLA-B*1502, which is more abundant among Asians, is associated with severe skin rash including Stevens-Johnson syndrome following treatment with carbamazepine.

- The first test for which the FDA has granted market approval using a DNA microarray is the AmpliChip CYP450 (Roche), which genotypes cytochrome P450 (CYP2D6 and CYP2C19). Now many pharmacogenomics test kits are commercially available.

## References

[1] Beutler E. Drug induced hemolytic anemia. Pharmacol Rev 1969;21:73–103.

[2] Wang L, McLeod HL, Weinshilboum RM. Genomics and drug response. N Engl J Med 2012;364:1144–53.

[3] St Sauver JL, Bielinski SJ, Olson JE, Bell EJ, et al. Integrating pharmacogenomics into clinical practice: promise vs reality. Am J Med 2016;129:1093–9. e1091.

[4] Xi Z, Fang F, Wang J, AlHelal J, et al. CYP2C19 genotype and adverse cardiovascular outcomes after stent implantation in clopidogrel-treated Asian populations: a systematic review and meta-analysis. Platelets 2019;30:229–40.

[5] (a) Klein TE, Altman RB, Eriksson N, Gage BF, et al. Estimation of the warfarin dose with clinical and pharmacogenetic data. N Engl J Med 2009;360:753–64. (b) Zhou S. Clinical pharmacogenomics of thiopurine S-methyltransferase. Curr Clin Pharamcol 2006;1:119–28.

[6] Asadov C, Aliyeva G, Mustafayeva K. Thiopurine S-methyltransferase as a pharmacogenetic biomarker: significance of testing and review of major methods. Cardiovasc Hematol Agents Med Chem 2017;15:23–30.

[7] de Man FM, Goey AKL, van Schaik RHN, Mathijssen RHJ, et al. Individualization of irinotecan treatment: a review of pharmacokinetics, pharmacodynamics, and pharmacogenetics. Clin Pharmacokinet 2018;57:1229–54.

[8] Ingle JN. Pharmacogenomics of tamoxifen and aromatase inhibitors. Cancer 2008;112 (3 Suppl):695–9.

[9] Frueh FW, Amur S, Mummaneni P, Epstein RS, et al. Pharamacogenomics biomarkers information in drug labels approved by the United States food and drug administartion: prevaslance and related drug use. Pharamcotherapy 2008;28:992–8.

[10] Willmann S, Edginton AN, Coboeken K, Ahr G, et al. Risk to the breast-fed neonate from codeine treatment to the mother: a quantitative mechanistic modeling study. Clin Pharmacol Ther 2009;86:634–43.

[11] Koren G, Cairns J, Chitayat D, Gaedigk A, et al. Pharmacogenomics of morphine poisoning in a breastfed neonate of a codeine prescribing mother. Lancet 2006;368:704.

[12] Smith JC, Curry SC. Prolonged toxicity after amitriptyline overdose in a patient deficient in CYP2D6 activity. J Med Toxicol 2011;7:220–3.

[13] Chen R, Shen K, Hu P. Polymorphism in CYP2D6 affects the pharmacokinetics and dose escalation of paroxetine controlled-release tablet in healthy Chinese subjects. Int J Clin Pharmacol Ther 2017;55:853–60.

[14] Campagne O, Mager DE, Brazeau D, Venuto RC, Tornatore KM. Tacrolimus population pharmacokinetics and multiple CYP3A5 genotypes in black and white renal transplant recipients. J Clin Pharmacol 2018;58:1184–95.

[15] Kaniwa N, Saito Y, Aihara M, Matsunaga K, et al. HLA-B*1511 is a risk factor for carbamazepine induced Stevens-Johnson syndrome and toxic epidermal necrolysis in Japanese patients. Epilepsia 2010;51:2461–5.

[16] Yan D, Zhang Y. A response letter to allopurinol-induced toxic epidermal necrolysis and association with HLA-B*5801 in white patients. Pharmacogenet Genomics 2018;28:268–9.

# CHAPTER 21

# Point-of-care testing

## Introduction to point-of-care testing

Point-of-care testing (POCT) also known as near patient testing is analysis of patient's specimen near or at the site of patient care outside the clinical laboratory. Point-of-care diagnostics provide clinicians access to rapid and actionable diagnostic results at or near the site of patient care to facilitate faster treatment decision-making. POCT devices are used in emergency department, physician's office, pain management clinic, or even roadside for identifying drunk drivers using breath alcohol analyzer. POCT can also be used in a clinical laboratory, for example, rapid pregnancy test or blood gas analysis. In addition, POCT devices may be used by a patient for home monitoring, for example, blood glucose and hemoglobin A1c. Moreover, sending specimen to a clinical laboratory for analysis of same analyte that can be performed using POCT device may cause significant delays in turnaround time. For example, turnaround time for cardiac troponins is recommended to be less than 1 h when specimens are sent to clinical laboratory. Therefore POCT of cardiac troponin I is useful in emergency department for rapid analysis and quick decision-making; whether patient should be admitted or discharged. It is important to note that POCT does not replace clinical laboratory where higher precision and accuracy could be achieved compared with POCT, but complement clinical laboratory when availability of rapid results can benefit patient care [1]. Moreover, POCT is not available for many laboratory tests and when available, results obtained by such device must be correlated with results obtained in clinical laboratory (gold standard).

Example of POCT includes but not limited to:

- Glucometers for measuring blood glucose, devices for measuring hemoglobin A1c.
- POCT analyzers to perform tests such as creatinine, electrolytes, hemoglobin, hematocrit, blood gases, ionized calcium, lactate, troponin, and B-type natriuretic peptide (BNP).

**433**

- Device for measuring INR, D-dimer, intraoperative PTH, etc.
- Devices for detecting some infectious diseases.
- Dipsticks for urine chemistry as well as specialized dipsticks to check urine adulteration for workplace drug testing.
- Refractometers for measuring specific gravity.

Another advantage of POCT is that many tests are "waived test." Waived testing is designated by Clinical Laboratory Improvement Amendments (CLIA) of the Food and Drug Administration (FDA) as simple tests that carry a low risk for an incorrect result. Sites performing only waived testing must have a CLIA certificate of waiver and follow the manufacturer's instructions but do not need to follow rigorous guidelines for more complex testing such as moderate or high complexity testing which are usually conducted in centralized clinical laboratory. However, some POCT tests are moderately complex and are not waived tests. Nevertheless, major advantages of POCT test include:

- POCT devices require no sample processing, can deliver test results at patient bedside or in a physician office within 10–20 min.
- POCT devices can be connected to Laboratory Information Systems (LIS) through wireless system.
- Many tests are "waived tests" requiring little supervision. Most devices have internal quality control protocol. Therefore unlike clinical laboratory tests which must be conducted by trained laboratory professionals, nurses and other health care professionals can run POCT without need for experienced medical technologist.

## POCT devices: Methodologies

Most POCT devices use the same reaction modes as used in the central laboratory assays. However, unlike the central laboratory assays where serum or plasma is mainly analyzed after centrifugation of whole blood, POCT devices are capable of analysis of whole blood. As a result, POCT device may utilize a chromatographic paper, which serves the same role as reaction flow cells, hematocrit is filtered, and the serum is drawn by capillary action, reacting with immobilized reagents and generating the signal, which is calibrated to give the analyte concentration. These signals can be electrical (e.g., glucose) or colorimetric (e.g., gold nanoparticles) and overall assay could be qualitative (positive/negative), semiquantitative or quantitative [2].

Lateral flow assay is a well-established platform for POCT capable of analyzing proteins, nucleic acids, drugs of abuse, toxins, and bacterial and viral pathogens. The entire immunoassay process, including washing and signal-reporting,

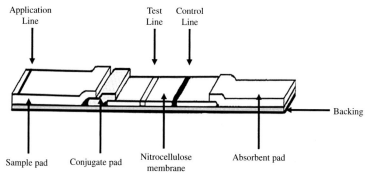

**FIG. 1**

Diagram of a lateral flow immunoassay strip. *Adapted from Luzzi V. Point of care devices for drugs of abuse testing: limitations and pitfalls. In: Dasgupta A, editor. Critical issues in alcohol and drugs of abuse testing. Academic Press; 2019 [chapter 11]. p. 142 [Reprinted with permission].*

can be integrated into one capillary paper-based strip starting with few drops of sample. As a result, lateral flow assay is one of the most successfully clinically translated POCT platform with a global market size of USD 5.55 billion in 2017 and an expected market size of USD 8.24 billion by 2022 [3].

Lateral flow format is a solid-phase competitive immunoassay often available as strips, cup, or a cassette that can be read manually or using a reader. A diagram of a lateral flow immunoassay strip is shown in Fig. 1. Close to the sample application line, there are a labeled conjugate, and a labeled control antigen pads. Up in the strip, further away from the application line, there is a line for test and above it, the control line. The test line contains detection antibodies against the target analyte or labeled conjugate. The control line contains antibodies against the control antigens. As the sample migrates from the application line to the labeled conjugate and control antigen pads, it carries its contents to the test and control lines. If the concentration of the target analyte in the sample is below the cutoff, a color line will appear because the labeled conjugate will bind to the immobilized detection antibodies. In contrast, if the target analyte is above the cutoff, it will compete with the labeled conjugate and inhibit color formation. Results from strips, cups, and cassettes can be evaluated visually by inspecting the presence or absence of color which usually develops in less than 30 min. Colloidal gold nanoparticles or other colored labels conjugated to detection antibodies are used in lateral flow assay. Direct visualization of colorimetric lateral flow assay readout is very useful for clinicians to make an immediate medical decision. However, subjective judgment variation in visual interpretation may exist among end-users, caused by the differences of illumination setting and personal visual ability such as color blindness. Thus, it could lead to controversial readouts, especially when the colorimetric signal is close to threshold. In order to circumvent this problem, quantitative colorimetric readout of test strips, various optical strip readers, and image processing algorithms are available. However, the sensitivity and quantification ability of POCT are intrinsically limited by the colorimetric signal readout. Quantitative lateral flow assays utilizing fluorescence, magnetic or Raman reporters instead of colorimetric labels have also been developed to circumvent this problem [3].

The POCT testing can be done for a single analyte or as a panel. Some strips and cassettes can also be read by semiautomated or automated endpoint readers. Among the chemistry assays, precision, accuracy, and interferences present the biggest challenges for POCT. Overall, all POCT devices have poorer precision and sensitivity than same assays conducted in clinical laboratory. Although lateral flow assays are widely used in POCT, other methodologies are also used such as ion selective electrodes and chemical sensors.

## Glucose measurement using POCT

POCT devices for glucose measurement mostly use the enzyme glucose oxidase, employing the following reaction:

Glucose + $O_2$ (Glucose oxidase)➔Gluconic acid + $H_2O_2$

Electrical signals are derived from the decomposition of $H_2O_2$. Pitkin and Rice reviewed analytical performances of glucometers and commented that the trade-off in moving to these POCT devices has been a reduction in accuracy, especially in the hypoglycemic range. Furthermore, many of these devices were originally developed, marketed, and received Food and Drug Administration regulatory clearance as home use devices for patients with diabetes. Without further review, many of these POCT devices have found their way into the hospital environment and are used frequently for measurement during intense insulin therapy, where accurate measurements are critical. Unfortunately, the reduction in accuracy of glucose measurement, especially in the hypoglycemic range, compared with the glucose test results obtained by analyzers in clinical laboratories is a major limitations of such devices [4]. In another study, the authors concluded that all five POCT devices have comparable precision and accuracy, but the major difference was in interferences—from hematocrit, maltose, ascorbic acid, acetaminophen, galactose, dopamine, and uric acid [5]. Sartor et al. found that POCT glucose values were falsely elevated in burn patients undergoing vitamin C therapy; the authors observed a bias of 10–200 mg/dL in seven such patients, probably caused by ascorbic acid interference with the POC glucose assay [6].

## Measuring multiple analytes using POCT

Many POCT devices are available commercially where one device is capable of measuring many analytes for patient management using very small amount of specimen and disposable test cartridges (some cartridge can measure multiple analytes where others measure a single analyte). For example, the iSTAT, a hand-held POCT device manufactured by Abbott laboratories is capable of measuring many analytes. For analysis, few drops of whole blood is applied to a specific cartridge and then the cartridge is inserted into the device for analysis. Within a short period of time, results are displayed on the screen and in some model, results can be transmitted to a computer system wirelessly. There is no issue of carryover, because a cartridge is discarded after single use. The iSTAT device is capable of analyzing sodium, potassium, chloride, ionized calcium, glucose, BUN (blood urea nitrogen), creatinine (Chem 8 + cartridge and other cartridges), lactate (CG4 + cartridge), hematocrit, and hemoglobin (Chem 8 + cartridge and other cartridges), blood gas parameters, PT/INR, ACT (activated coagulation time) Kaolin, ACT Celite, beta-hCG, troponin I, CK-MB and BNP (using individual cartridge specific for such analyte). POCT devices are also available from various other diagnostic companies, for example, Roche Diagnostics.

POCT for creatinine and bilirubin have been available for several decades. The POCT creatinine assays use enzymatic methods for quantitative measurement of creatinine with redox outputs that are displayed electronically. Radiological contrast agents have been reported to interfere in POCT creatinine assays, resulting in false-positive results. POCT testing of lactate, magnesium, ionized calcium, sodium, potassium, chloride, and bicarbonate utilizes specific ion-sensitive electrodes. However, discordance between the POCT results (using whole blood) and clinical laboratory analyzer-based results (using serum or plasma) is often seen—probably because of hematocrit interference. Dashevsky et al. however, observed good agreement between POCT and clinical laboratory-based test results for electrolytes (sodium, potassium), blood urea nitrogen, creatinine, and hematocrit in the emergency center of a large hospital [7].

## Cardiac markers

With the need of rapid diagnosis of acute myocardial infarction in emergency department, POCT is available for cardiac troponin I, cardiac troponin T, CK-MB, and also for BNP (B-type natriuretic peptide) and NT-proBNP. Unfortunately, POCT troponin assays are not as sensitive as troponin assays performed in clinical laboratories. Ter Avest et al. observed that bedside AQT90-flex cTnT results were systematically lower than laboratory-based modular hs-cTnT (Roche Diagnostics) results. The authors concluded that AQT90-flex POCT cTnT assay is not yet sensitive and reliable enough to be used to exclude acute myocardial infarction in the emergency department with a single blood draw at the time of presentation [8]. Furthermore, the same interferants for clinical laboratory-based troponin tests (heterophilic interference, biotin, etc.) also affect the POCT troponin tests. Similarly, POCT NT-proBNP assay (Roche Diagnostics) was found to be interfered by biotin (above 10 mg/L), human antimouse antibodies (HAMA), and cardiac drugs (bisoprolol and BNP). However, The NT-proBNP results obtained by using POCT showed good analytical performance and excellent agreement with the laboratory method and therefore suitable for clinical use [9].

## Pregnancy test

There are many POCT for pregnancy tests (beta-hCG in urine) available. These assays are qualitative competition immunoassays. Hook effect is a problem with POCT urine pregnancy test which occurs when excess hCG in the urine saturates the detection antibody and prevents formation of the necessary antibody-antigen-antibody sandwich, causing a false-negative result. This situation could be observed in individuals with conditions such as gestational trophoblastic disease. Variant hook effect occurs when one antibody with an affinity for beta core fragment of hCG (βcf-hCG) becomes saturated while the other antibody does not recognize βcf-hCG, causing a false-negative result. Evaluation of 11 common qualitative POCT urine pregnancy test kits showed that nine are susceptible to false-negative results due to βcf-hCG [10]. One case study reported false-positive POCT urine pregnancy tests because of the presence of leukocytes in the specimen [11].

## Coagulation POCT

POCT Coagulation meters are used to report prothrombin time and international normalized ratio (INR), activated partial thromboplastin time (aPTT), and activated coagulation time (ACT) to monitor anticoagulation therapy (warfarin or heparin). Moiz et al. measured a total of 200 specimens for INR using POCT coagulometer (CoaguChek XS Pro, Roche Diagnostics) and laboratory analyzer (Sysmex CS2000i). The authors observed excellent correlation between both results (correlation coefficient = 0.973). The overall mean difference was 0.21 INR. However, the mean difference was found to be increased as INR results increased; the mean difference was 0.09 in the subtherapeutic range (≤1.9 INR), 0.29 in the therapeutic range (2.0–3.0 INR) while 0.4 in the supratherapeutic range (>3.0 INR). The authors concluded that positive bias observed in INR measurement using POCT-INR is evident in the supratherapeutic range which could affect the dosing decision. Therefore, high INR values obtained by using POCT should be confirmed with the laboratory-based INR measurement [12].

## Infectious disease

Several commercial POCT devices are capable of testing blood for infectious diseases: for example, *Alere* HIV 1 testing, which could be performed on a finger stick. The *Murex* from Abbott is a 10-min POCT that can use whole blood, serum, or plasma. The *Trinity Biotech* POC for HIV 1 uses recombinant antigens and gold nanoparticles to enhance the signal. Some of them (e.g., *OraQuick* from OraSure) can even use saliva as sample, but saliva is not the best specimen for HIV testing.

Other POCT devices for infectious diseases have capabilities to test for streptococci, methicillin-resistant *Staphylococcus aureus* (MRSA), Candida, influenza virus, sexually transmitted diseases, Tuberculosis (*Mycobacterium tuberculosis*, especially in developing countries), respiratory pathogens, and urinary tract infecting pathogens. In addition, POCT is available for procalcitonin (a sepsis marker) and has been launched in the market.

## POCT for drugs of abuse

POCT devices are widely used in both emergency department and physician's office for drug screening. In contrast to many POCT devices used for other analytes where results are quantitative, POCT drugs-of-abuse testing only produces qualitative results (positive/negative) if a drug is present at a concentration above the cutoff as recommended by SAMHSA (see Chapter 16). Therefore, POCT devices for drugs-of-abuse testing are for screening purpose only and if drug confirmation is needed, urine specimen must be sent to a laboratory for confirmation using either gas chromatography/mass spectrometry (GC/MS) or liquid chromatography combined with tandem mass spectrometry (LC-MS/MS). Although laboratory-based immunoassay screenings step suffers from interferences, reliability and effectiveness of drugs-of-abuse POCT devices have been challenged by several studies. The main issues are due to the specificity of the antibodies used in manufacturing POCT devices, the inconsistency of cutoffs among reagent lots of the same device, and the difficulty of interpreting the result depending on the read time used on the device [13].

Several POCT devices are available for drugs-of-abuse testing using mostly urine, but some devices can also screen drug levels in oral fluids. These include Triage Drugs of Abuse Panels and Triage TOX Drug Screens (Biosite, Inc., San Diego, California), OnTrak TesTcard 9 (Varian, Inc., Palo Alto, California), Rapid Drug Screen (American Bio Medica, Kinderhook, New York), Intercept (OraSure Technologies, Inc., Bethlehem, Pennsylvania), Profile-II (MEDTOX Diagnostics, Burlington, North Carolina), and Status DS (LifeSign LLC, Somerset, New Jersey). These POCT devices also utilize solid-phase immunoassay format with a visible endpoint in a disposable cartridge. Device formats include dipsticks, cup devices, cards, and plastic cassettes. For example, the Triage TOX Drug Screen device utilizes competitive fluorescence immunoassay for the presence of amphetamine, methamphetamine, barbiturates, benzodiazepines, cocaine (as benzoylecgonine), methadone, phencyclidine, opiates, tetrahydrocannabinol (as THC-COOH), and tricyclic antidepressants in urine. Results are available within 10 min. The interpretation of the results are nonsubjective because Triage Reader is capable of reading such results. The device has built-in quality controls and is capable of electronic record keeping [14].

Many pain management clinics use urine cups that contain POCT urine drug testing strips to assess the presence of drugs in urine specimens. Urine POCT strips utilize the immunochromatography principle to qualitatively assess the presence or absence of around 10–15 commonly abused drugs. However, due to low analytical sensitivity and specificity, the likelihood of false-negative and false-positive results is high. Therefore, positive results from POCT urine drug testing cups are presumptive positive and should be confirmed by GC/MS or LC-MS/MS if needed. In one report, the authors observed that oxycodone has the highest false-negative results of 25% when POCT was used for monitoring pain management patients compared with laboratory-based LC-MS/MS assay indicating limitations of POCT in monitoring pain management patients [15].

Interestingly, if urine specimens are collected in urine drug testing cups, such as Discover Plus (American Screening, Shreveport, LA), U-screen (US Screening Source Inc., Pewee Valley, KY), and First Sign (Hem Sure, Irwindale, CA) and later sent to a reference laboratory for drug confirmation, concentrations of several drug could be reduced. The reduction in the concentration of drugs is not specific to POCT urine drug testing strips as urine chemistry dipsticks show the same effect. However, drug concentrations are not reduced if urine is collected in regular urine collection cup. Therefore, urine specimens that are exposed to POCT urine drug testing strips for longer than 24 h should not be used for measurements of drugs with a confirmatory method (LC-MS/MS) of urine drug testing. It is recommended that before the shipment of urine specimen, the strips should be removed from the POCT urine drug testing cups. As an alternative solution, an aliquot of the specimen could be transferred to a regular urine cup to be transported to a clinical toxicology laboratory before exposing it to POCT urine drug testing strips [16].

Result interpretation is important for POCT drugs-of-abuse testing because the absence of a line indicates that a drug is present at or above the defined threshold and even a faint line should be interpreted as negative. Furthermore, the window to read the result is typically 5–10 min, and if an operator extends the read time, false results can be obtained. The Triage TOX Drug Screens offer the advantage of a one-step method and an instrument-read cartridge that avoids the need to visually interpret multiple bands and permits results to be interfaced directly to the laboratory information system [17].

While oral fluids have been used as alternative samples to urine, limitations apply on using oral fluid in drug-of-abuse testing when using POCT devices. Oral fluid is usually collected on a device that contains a buffer as preservative. The ratio between the buffer and the oral fluid is not necessarily constant and may preclude accurate drug concentration measurements. A systematic metaanalysis of previously published studies evaluating the reliability of detecting five commonly monitored drugs in oral fluid using POCT devices also warns about the variability in specificity and sensitivity of POCT [18].

## Important issues with POCT devices

One major advantage POCT devices is in getting results quickly without the need of sending the specimens to a central laboratory. The perception is that laboratory results correlate really well with POCT device results, but there are limitations to this assumption. For example, abnormal high INR obtained by POCT device must be verified by using a clinical laboratory. POCT device complexity as defined by CLIA can be waived on nonwaived. However, it is important that clinical laboratory professionals must be in charge of oversighting POCT testing in a hospital setting especially during validation of POCT analyzers which is commonly performed by comparison of results obtained by a POCT device and by a laboratory-based analyzer. Moreover, laboratory professionals should also train users who may not be having training of a medical technologist.

While many point-of-care tests are designed to be relatively simple and low risk to use, they are not error-proof. Individuals using point-of-care tests, even healthcare practitioners, must carefully follow test directions and be familiar with the test system. Some point-of-care tests such as those used to adjust doses of medications have the potential to lead to serious health consequences if not performed properly. Many large hospitals have point-of-care coordination teams to ensure that testing procedures are properly followed.

It is important that the convenience of point-of-care testing does not tempt users to apply them beyond their intended purpose or to misinterpret results. For example, glucose meters and point-of-care hemoglobin A1c tests are designed only for monitoring diabetes and should not be used for diagnosis or screening. Point-of-care testing has many benefits but only with strict training of correct usage, proper quality control and understanding limitation of such devices. In general, results obtained in clinical laboratory may be more accurate than results obtained by POCT devices. Therefore, when an unexpected result is obtained using a POCT device, such results must be confirmed by clinical laboratory-based test before making any diagnostic decision.

## Key points

- Point-of-care testing (POCT) also known as near patient testing is the analysis of patient's specimen near or at the site of patient care outside the clinical laboratory. Point-of-care diagnostics provide clinicians access to rapid and actionable diagnostic results at or near the site of patient care to facilitate quick treatment decision-making. The major advantages are portability, low cost, and user friendliness.
- POCT require no sample processing, and can deliver test results at patient bedside or in a physician office within 10–20 min.
- POCT devices can be connected to Laboratory Information Systems (LIS).
- Many POCT tests are "waived tests" requiring little supervision. Most devices have internal quality control protocol. Therefore, nurses and other health care professionals can run such test without need for experienced medical technologist.
- Most POCT devices use the same reaction modes as used in the central laboratory assays. However, unlike the central laboratory assays where serum or plasma is mainly analyzed after centrifugation of whole blood, POCT devices analyze whole blood.
- Lateral flow format is a solid-phase competitive immunoassay often available as strips, cup or a cassette that can be read manually or using a reader. This format is commonly used in POCT devices. Few drops of blood are applied to a test cartridge and then the cartridge is inserted in the device. Within a short period of time, results are displayed. Some POCT devices can send the result wirelessly to a central computer so that results could be included in patient's chart.
- Many POCT devices are capable of measuring many analytes for patient management using very small amount of specimen. For example, for iSTAT, a handheld POCT device manufactured by Abbott Laboratories is capable of analyzing sodium, potassium, chloride, ionized calcium, glucose, BUN, creatinine (Chem 8+ cartridge and other cartridges), lactate (CG4+ cartridge), hematocrit and hemoglobin (Chem 8+ cartridge and other cartridges), blood gas parameters, PT/INR, ACT (activated coagulation time) Kaolin, ACT Celite, beta-hCG, troponin I, CK-MB, and BNP (using individual cartridge specific for such analyte). POCT devices are also available from various other diagnostic companies, for example, Roche Diagnostics.
- There are many POCT for pregnancy tests (beta-hCG in urine) available. These assays are qualitative competition immunoassays. Hook effect is a problem with POCT urine pregnancy test which occurs when excess hCG in the urine saturates the detection antibody and prevents formation of the necessary antibody-antigen-antibody sandwich, causing a false-negative result. Variant hook effect occurs when one antibody with an

affinity for beta-core fragment of hCG (βcf-hCG) becomes saturated while the other antibody does not recognize βcf-hCG, causing a false-negative result.

- POCT Coagulation meters are used to report prothrombin time and international normalized ratio (INR), activated partial thromboplastin time (aPTT), and activated coagulation time (ACT) to monitor anticoagulation therapy (warfarin or heparin). However, abnormal INR especially high INR observed using a POCT device should be verified using a clinical laboratory due to bias between laboratory best INR test and POCT device if INR is elevated.
- Several commercial POCT devices are capable of testing blood for infectious diseases: for example, *Alere* HIV 1 testing, which could be performed on a finger stick. The *Murex* from Abbott is a 10-min POCT that can use whole blood, serum, or plasma. The *Trinity Biotech* POC for HIV 1 uses recombinant antigens and gold nanoparticles to enhance the signal. Some of them (e.g., *OraQuick* from OraSure) can even use saliva as sample, but saliva is not the best specimen for HIV testing. Other POCT devices for infectious diseases have capabilities to test for streptococci, methicillin-resistant *Staphylococcus aureus* (MRSA), Candida, influenza virus, sexually transmitted diseases, Tuberculosis (*Mycobacterium tuberculosis*, especially in developing countries), respiratory pathogens, and urinary tract infecting pathogens. In addition, POCT is available for procalcitonin (a sepsis marker) and has been launched in the market.
- POCT devices are widely used in both emergency department and physician's office for drug screening. In contrast to many POCT devices used for many analytes where results are quantitative, POCT drugs-of-abuse testing only produces qualitative results (positive/negative) if a drug is present at a concentration above the cutoff as recommended by SAMHSA. Therefore, POCT devices for drugs-of-abuse testing are for screening purpose only and if drug confirmation is needed, urine specimen must be sent to a laboratory for confirmation.

## References

[1] Florkowski C, Don-Wauchope A, Gimenez N, Rodriguez-Capote K. Point-of-care testing (POCT) and evidence-based laboratory medicine (EBLM) – does it leverage any advantage in clinical decision making? Crit Rev Clin Lab Sci 2017;54:471–94.

[2] Luppa PB, Bietenbeck A, Beaudoin C, Giannetti A. Clinically relevant analytical techniques, organizational concepts for application and future perspectives of point-of-care testing. Biotechnol Adv 2016;34:139–60.

[3] Li Z, Chen H, Wang P. Lateral flow assay ruler for quantitative and rapid point-of-care testing. Analyst 2019;144:3314–22.

[4] Pitkin AD, Rice MJ. Challenges to glycemic measurement in the perioperative and critically ill patient: a review. J Diabetes Sci Technol 2009;3:1270–81.

[5] Lv H, Zhang GJ, Kang XX, et al. Factors interfering with the accuracy of five blood glucose meters used in Chinese hospitals. J Clin Lab Anal 2013;27:354–66.

[6] Sartor Z, Kesey J, Dissanaike S. The effects of intravenous vitamin C on point-of-care glucose monitoring. J Burn Care Res 2015;36:50–6.

[7] Dashevsky M, Bernstein SL, Barsky CL, Taylor RA. Agreement between serum assays performed in ED point-of-care and hospital central laboratories. West J Emerg Med 2017;18:403–9.

[8] Ter Avest E, Visser A, Reitsma B, Breedveld R, Wolthuis A. Point-of-care troponin T is inferior to high-sensitivity troponin T for ruling out acute myocardial infarction in the emergency department. Eur J Emerg Med 2016;23:95–101.

[9] Zugck C, Nelles M, Katus HA, Collinson PO, et al. Multicenter evaluation of a new point-of-care test for the determination of NT-proBNP in whole blood. Clin Chem Lab Med 2006;44:1269–77.

[10] Latifi N, Kriegel G, Herskovits AZ. Point-of-care urine pregnancy tests. JAMA 2019; 2019:15833.

[11] Jao HF, Er TK, Hsiao JK, Chiang CH. False-positive urine pregnancy test due to leukocyte interference. Ann Lab Med 2012;32:167–8.

[12] P Moiz B, Rashid A, Hasan M, Jafri L, Raheem A. Prospective comparison of point-of-care device and standard analyzer for monitoring of international normalized ratio in outpatient oral anticoagulant clinic. Clin Appl Thromb Hemost 2018;24:1153–8.

[13] Mastrovitch TA, Bithoney WG, DeBari VA, Nina AG. Point-of-care testing for drugs of abuse in an urban emergency department. Ann Clin Lab Sci 2002;32:383–6.

[14] Attema-de Jonge ME, Peeters SY, Franssen EJ. Performance of three point-of-care urinalysis test devices for drugs of abuse and therapeutic drugs applied in the emergency department. J Emerg Med 2012;42:682–91.

[15] Hall B, Daly S. How accurate are point-of-care urine drug screens in patients taking chronic opioid therapy? J Fam Pract 2018;67:177.

[16] Haidari M, Mansani S, Ponds D, Romero L, et al. Storage of urine specimens in point of care (POC) urine drug testing cups reduces concentrations of many drugs. Clin Chim Acta 2019;499:81–6.

[17] Melanson SE. Drug-of-abuse testing at the point of care. Clin Lab Med 2009;29:503–9.

[18] Scherer JN, Fiorentin TR, Borille BT, Pasa G, et al. Reliability of point-of-collection testing devices for drugs of abuse in oral fluid: a systematic review and meta-analysis. J Pharm Biomed Anal 2017;143:77–85.

# CHAPTER 22

# Biotin interferences with immunoassays

## Biotin

Biotin (vitamin B7) previously known as vitamin H is an essential vitamin which acts as a coenzyme for carboxylase reactions in gluconeogenesis, fatty acid synthesis, and amino acid catabolism. Biotin is present in many foods including some vegetables, fruits (banana for example), salmon, chicken, eggs, whole-grain cereals, whole wheat bread, dairy products, and nuts. The daily recommended allowance of biotin is only 30 microgram which can be easily achieved by eating a balanced diet. Biotin deficiency is rare, mostly due to genetic predisposition.

Although no scientific data support taking biotin supplement for healthy hair, nail and skin, approximately 15% to 20% people in the US consume biotin-containing supplements anyway. Most people take 0.5 to 1 mg biotin supplement but some people take 5 or 10 mg biotin daily. High-dose biotin (up to 300 mg per day) is also used as an experimental therapy to reduce symptoms of multiple sclerosis.

## Serum biotin levels

In general, biotin concentration is very low in people not taking any supplement. For subjects not taking supplement, biotin concentrations usually vary from 0.4 to 1.2 ng/mL [1]. However, taking biotin supplement results in higher serum biotin levels which is dose-dependent. Mardach et al. reported that when five normal adults consumed 1 mg of biotin daily, the mean biotin level of 8.6 ng/mL in blood was observed 1 to 3 h after the last biotin dose [2]. In another study, the authors reported that biotin concentrations varied from range: 10 to 73 ng/mL, 1 h postdose after taking 5 mg biotin supplement each day. However, the median serum biotin concentration was 91 ng/mL (range: 53 to 141 ng/mL) after taking 10 mg supplement and 184 ng/mL (range: 80 to 355 ng/mL) after ingesting 20 mg biotin supplement each day [3]. As expected,

**445**

serum biotin levels would be much higher after taking biotin in pharmaceutical dosages such as 100 mg or 300 mg daily. The mean peak biotin concentration 1 to 3 h after taking 100 mg biotin supplement was 495 ng/mL. The plasma biotin concentrations varied from 700 to 900 ng/mL when subjects ingested a single dose of 300 mg biotin supplement and blood was drawn 2 h after taking the supplement. After 24 h, the biotin plasma levels were reduced to around 75 ng/mL [4]. Serum levels of biotin in people taking no supplement as well as taking biotin supplements are summarized in Table 1.

## Biotin interference

Many automated immunoassays incorporate biotinylated antibodies and streptavidin-coated magnetic beads as a means of immobilizing antigen-antibody complexes to the solid phase. Biotin is a small molecule which when conjugated with active macromolecule such as an assay antibody, it rarely interferes with the function of the labeled molecule. Streptavidin is a glycoprotein which has a high affinity for biotin and the binding is not only highly specific but also resistant to changes in temperature, pH, and also can withstand the presence of denaturing agents and organic solvents. By attaching streptavidin protein to a moiety, such as a fluorophore, an enzyme, or a gold nanoparticle, the streptavidin-biotin interaction can be utilized as a detection method in a variety of immunoassays including enzyme-linked immunosorbent assay (ELISA), chemiluminescent assay (CLIA) as well as in Western Blotting and flow cytometry. This design offers many advantages including signal amplification and increased sensitivity.

**Table 1** Biotin levels in human taking no supplement and after taking biotin supplement.

| Biotin supplement | Serum/plasma biotin concentration |
| --- | --- |
| No supplement | 0.4–1.2 ng/mL |
| 300 micrograms | 0.59–2.77 ng/mL in men 2 h after taking supplement 0.42–3.11 ng/mL in women 2 h after taking supplement |
| 1 mg supplement daily for 7 days | Mean level 8.4 ng/mL 1–3 h after the last biotin dose |
| 5 mg | 10–73 ng/mL, 1 h postdose |
| 10 mg | 53–141 ng/mL, 1 h postdose |
| 20 mg | 80–355 ng/mL, 1 h postdose |
| 100 mg | Peak level 495 ng/mL |
| 300 mg | 700–900 ng/mL, 2 h postdose |

Exogenous biotin if present in high amount in the specimen, it competes with biotinylated reagents for binding sites on streptavidin reagent. In competitive immunoassay, concentration of the analyte is inversely proportional to the signal intensity of the washed solid phase. Biotin reduces signal intensity in competitive immunoassay by occupying streptavidin-binding sites thus falsely increases true analyte concentration. In sandwich immunoassay format, the concentration of the analyte is directly proportional to the signal intensity after washing of solid phase. Exogenous biotin competes with the binding of labeled complex to the solid phase and reduces the signal intensity, thus producing falsely lower analyte concentration. In general, competitive immunoassays are used for detecting relatively small molecules while sandwich assays are more effective in detecting large molecules. Therefore the key points are:

- Biotin causes positive interference in competitive immunoassays.
- Biotin causes negative interferences in sandwich immunoassays.

However, mere presence of biotin/streptavidin in assay design does not necessarily imply that the assay will be subjected to biotin interference. In some assays, the biotin-labeled reagents are combined during manufacturing process (pre-bound reagents). Such assays are expected to be less susceptible or unaffected in the presence of high biotin concentrations in the serum or plasma [5]. Currently, Roche Diagnostics, Siemens Healthineers Laboratory Diagnosis, Ortho Clinical Diagnosis, Beckman Coulter and Immunodiagnostics System manufacture and market biotin-based immunoassays. Commonly affected tests due to biotin interferences are listed in Table 2.

Biotin interference in immunoassays that utilizes biotin in assay design is dose-dependent. Biotin has no impact on clinical laboratory test results in patients taking no supplement or multivitamins (usually contains 30 microgram biotin) or biotin supplement up to 5 mg per day. A biotin dose of 10 mg (more than 300 times than recommended daily allowance of 30 microgram per day) may affect certain assays but not all biotin-based assays. However, if a person is taking a pharmaceutical dosage of biotin (300 mg per day), it will affect almost all biotin-based assays because biotin interference is very significant at 500 ng/mL or higher which can be easily achieved after taking 300 mg biotin supplement. As biotin half-life varies widely depending on the amount of biotin dosage ingested, a particular patient may show different level of interferences depending on several factors:

- How much biotin is ingested?
- How much time has elapsed between ingestion of biotin and time of blood draw?
- Which biotin-based immunoassay is used?

**Table 2** Commonly reported biotin interferences in biotin-based immunoassays.

| Type of test | Competitive immunoassays showing positive interferences (Falsely elevated values) | Sandwich immunoassays showing negative interferences (falsely lower values) |
|---|---|---|
| Thyroid function test | Free T4, total T4, free T3, total T3, anti-TPO (antithyroid peroxidase), anti-TSHR (antithyrotropin receptor antibody) | Thyroid stimulating hormone (TSH), Thyroglobulin |
| Hormones | Testosterone, dehydroepiandrosterone-sulfate (DHEAS), estradiol, progesterone, cortisol | Parathyroid hormone (PTH), luteinizing hormone (LH), follicle stimulating hormone (FSH), growth hormone, prolactin, insulin, C-peptide, adrenocorticotropic hormone (ACTH) |
| Cardiac markers | | Troponin I, Troponin T, CK-MB |
| Serology testing | | HIV antigen/antibody combo, hepatitis C virus antibody, hepatitis B surface antigen, hepatitis B surface antibody, hepatitis B antigen, hepatitis B-core antibody IgM, hepatitis A antibody, IgM, hepatitis A antibody total |
| Other analytes | 25-Hydroxy vitamin D, vitamin $B_{12}$, folate | NT-Pro-BNP, hCG, ferritin, myoglobin, sex hormone-binding globulin |
| Tumor marker | | Alpha-fetoprotein (AFP), cancer antigen 125 (CA-125), carbohydrate antigen 19-9 (CA 19-9), prostate-specific antigen (PSA), free PSA |

## FDA safety communication on biotin

In 2018 the Federal Drug Administration (FDA) posted a safety communication, alerting the public, health care providers, laboratory personnel, and laboratory test developers that biotin can significantly interfere with certain laboratory test results which may go undetected in patients who are ingesting high levels of biotin as dietary supplements. The FDA has observed an increase in the number of reported adverse events including one death related to biotin interference with laboratory test results. The FDA reported that one patient taking high levels of biotin died from falsely low troponin results when a troponin test known to have biotin interference was used. However, the FDA did not specify which diagnostic company marketed the troponin assay [6].

## Biotin interferences in thyroid function tests

Clinically significant interference of biotin in thyroid function tests is probably the most well-documented interference of biotin in clinical laboratory test results. Biotin shows positive interference with thyroxine (T4), free thyroxine (FT4), triiodothyronine (T3), and free T3 (FT3) assays because these are relatively small molecules and immunoassays used for their analysis are competitive immunoassays. In contrast, thyroid-stimulating hormone (TSH) is a large molecule (molecular weight: 28 kDa) which requires sandwich immunoassay format for analysis. As a result, biotin falsely lowers TSH value which may lead to false diagnosis of hyperthyroidism. Graves' disease, a common cause of hyperthyroidism, has also been misdiagnosed in a normal patient due to elevated FT4, FT3, and falsely lower TSH value [7]. Piketty et al. reviewed the issue of false biochemical diagnosis of hyperthyroidism due to biotin interference in streptavidin-biotin-based immunoassays and commented that lack of clinical correlation with laboratory test results, for example, test results indicating hyperthyroidism in a clinically euthyroid patient should indicate the possibility of assay interference including biotin interference [8].

## Biotin falsely lower parathyroid hormone level

Parathyroid hormone (PTH) maintains serum calcium homeostasis by directly altering calcium metabolism in bone and kidney and indirectly in the intestine. Testing of PTH is important because primary hyperthyroidism is the leading cause of hypercalcemia in outpatient setting and the third most common endocrine disorder. Immunoassays for PTH are usually sandwich immunoassays. The PTH levels in a 60-year-old woman ranged from 90 to 336 pg/mL (reference range: 10 to 60 pg/mL) but when the patient was being worked up for surgery, her PTH level was undetectable but serum calcium level was elevated. At that time, it was revealed that she was taking 1.5 mg biotin supplement daily for hair growth. After the biotin was stopped, the PTH test was repeated a month later and the PTH value was 197 pg/mL. The Roche PTH assay utilizes biotinylated antibody and is susceptible to biotin interference. The authors also described a second case report where PTH levels were undetectable when the patient was taking 5 mg of biotin supplement daily for neuropathic pain [9].

## Biotin interferes with cardiac function tests

Biotin falsely lowers troponin T result using Roche analyzer. For example, troponin T value was reduced from 39 to 3 ng/L in the presence of 500 ng/mL biotin. The high vulnerability of the troponin T assay to biotin interference is clinically concerning because high dose biotin therapy can lower troponin T results to below 99th percentile cut-off and in the early investigation of a potential myocardial infarction, this might lead inappropriate discharge of a patient and potentially missing a diagnosis of myocardial infarction. Therefore, a false-negative troponin I in the context of atypical chest pain and nonspecific ECG change may have dire clinical consequences because values may be falsely lower due to biotin interference [10]. Djebrani et al. investigated biotin interference in troponin I assay on Dimension EXL analyzer that utilizes biotin-streptavidin binding in assay design. The authors reported a significant negative interference of biotin on cardiac troponin I determinations for concentrations of 100 ng/mL and above [11]. Sancar commented that interference of biotin in cardiac troponin measurement may cause significant clinical problem [12]. Because CK-MB and pro-BNP tests are also sandwich immunoassays, biotin-based immunoassays for CK-MB and pro-BNP also show negative interference.

## Biotin interferences in multiple immunoassays

Li et al. investigated the effect of biotin on performance of hormone and non-hormone assays using six healthy adults (2 women and 4 men; mean age: 38 years) who ingested 10 mg biotin for 7 days. The analytes tested by the authors include TSH, total T4, total T3, FT4, FT3, PTH, prolactin, N-terminal pro-brain natriuretic peptide (NT-pro-BNP), 25-hydroxyvitamin D, prostate-specific enzyme (PSA) and ferritin using various assays (23 biotinylated assays and 14 nonbiotinylated assays) and analyzers (Vitros, Centaur, Vista, Cobas e602 and Architect) obtained from different diagnostic companies (Ortho Clinical Diagnosis, Siemens, Roche Diagnostics and Abbott Laboratories). Biotin ingestion was associated with falsely lower NT-pro-BNP results (using Vitros 5600 analyzer) in all participants. In contrast, biotin ingestion falsely increased 25-hydroxy vitamin D assay using Cobas e602 analyzer. Interestingly, biotin ingestion has no effect on prolactin using biotin-based immunoassays on Cobas e602 and Vista analyzer. As expected, biotin showed no interference in nonbiotinylated assays. The authors commented that differential biotin interference tolerance among assays is likely due to the amounts of biotinylated antibodies or analogs used, the availability of streptavidin-binding sites and other factors in the assay reagents (e.g., streptavidin-coated magnetic particles) [13].

Biotin can also interfere with urine-based testing because approximately 50% biotin is secreted in urine unchanged. In general, 20% of ingested biotin dose is excreted in urine within 4 h. In one study, the authors investigated potential interference of biotin in qualitative point of testing for hCG (human chorionic gonadotropin) both in vitro and in vivo and showed that only QuickVue point of care device for detecting hCG was producing invalid result because control line was very faint in the presence of significant amount of biotin but other devices (Alere 20, Alere 25, Icon 20, Osom, QuPID and SureVue) were not affected. When volunteers ingested 10 mg biotin supplement each day for 7 days, invalid hCG test using QuickVue was observed with urine of some subjects due to very faint control line [14].

## Diagnostic companies using biotin-based assays: A review

Homes et al. reviewed current manufacturer's instructions for the use of 374 immunoassays performed by eight of the most popular immunoassay analyzers in the US and commented that 221 of these assays are biotin-based assays. The authors identified 44 biotin-based assays marketed by Roche as high risk for biotin interference followed by 28 assays marketed by Ortho Clinical Diagnosis, 13 assays marketed by Siemens, and 6 assays marketed by Beckman Coulter. Interestingly, Abbott Laboratories have four biotin-based immunoassays but none of them are affected by biotin because biotin is present in bound form. Assays marketed by DiaSorin are not biotin-based, hence free from biotin interference [15].

## Approaches to overcome biotin interference

The possibility of interference in an assay may be suspected from any or combination of various observations:

- Lack of correlation between test results and clinical presentation of the patient is a strong suspicion of assay interference. For example, observation of low TSH and elevated FT3 and FT4 indicating hyperthyroidism in a euthyroid patient is an indication of assay interference including biotin interference.
- Generation of implausible laboratory tests results: extreme deviation from values expected in pathological condition is an indication of assay interference.
- Nonlinearity after dilution is an indication of assay interference including biotin interference.
- Discordant results by different assays for the same analyte are an indication of assay interference.

One approach to eliminate biotin interference in clinical laboratory test is to use an alternative assay platform that does not use biotin in assay design. Currently, most assays marketed by Abbott Laboratories do not use biotin in assay design and only four assays that are biotin-based are not affected by high levels of biotin. Another approach is dilution study:

- Higher concentration of the analyte in diluted specimen compared to original specimen using sandwich immunoassay
- Lower than expected concentration of the analyte in diluted specimen compared to original specimen using competitive immunoassays

**Table 3** Approaches to overcome biotin interference in immunoassays.

| Approach to overcome biotin interference | Comments |
| --- | --- |
| Use immunoassays that do not use biotin in assay design | This is probably the best approach. Currently all assays marketed by DiaSorin do not use biotin/streptavidin interaction. Moreover, only four assays on Abbott's Architect platform use biotin but they are not subjected to biotin interference |
| Serial dilution | This is the simplest approach for laboratories using biotin-based immunoassays. Nonlinearity during dilution is an indication for biotin interference |
| Waiting for biotin clearance | Biotin has a short half-life and after taking 5 mg biotin supplement waiting for 8 h may be sufficient to eliminate biotin interference. American Thyroid Association recommends waiting for at least 2 days after cessation of high dose biotin therapy before blood draw for thyroid testing. After taking 100–300 mg biotin waiting for at least 3 days is advised but for individuals with renal failure, biotin which is cleared renally may cause interference up to 15 days after last dose. However, for acutely ill patients no such waiting period is possible |
| Remove biotin by pretreatment of specimen with streptavidin-coated microparticle | Incubation of serum with streptavidin-coated microparticle depletes up to 1000 ng/mL of biotin from specimen and such pretreatment eliminates biotin interference in both competitive and sandwich immunoassays that utilize biotin in assay design |

Another approach to overcome biotin interference in biotin-based assays is to wait sufficient time for biotin to be eliminated from the body. Roche Diagnostics recommend waiting for 8 h before blood collection for patients taking 5 mg of biotin supplement. The American Thyroid Association Guidelines recommend discontinuation of biotin at least 2 days before conducting thyroid function tests [16]. However, such approach is not possible in emergency situation where cardiac troponin tests (negative interference of biotin) cannot be delayed in a patient suspected of myocardial infarction. Currently, best approach to overcome biotin interference in biotin-based immunoassays is pretreatment of specimen with streptavidin-coated microparticle to deplete biotin present in specimens. However, such approach is laborious [17].

Approaches to avoid biotin interferences in immunoassays are listed in Table 3.

## Key points

- Biotin (vitamin B7) previously known as vitamin H is an essential vitamin which acts as coenzyme for carboxylase reactions in gluconeogenesis, fatty acid synthesis, and amino acid catabolism. The daily recommended allowance of biotin is only 30 microgram which can be easily achieved by eating a balanced diet. Biotin deficiency is rare, mostly due to genetic predisposition.

- Although no scientific data support taking biotin supplement for healthy hair, nail and skin, approximately 15% to 20% people in the US consume biotin-containing supplements anyway. Most people take 0.5 to 1 mg biotin supplement but some people take 5 or 10 mg biotin daily. High dose biotin (up to 300 mg per day) is also used as an experimental therapy to reduce symptoms of multiple sclerosis.
- Many automated immunoassays incorporate biotinylated antibodies and streptavidin-coated magnetic beads as a means of immobilizing antigen-antibody complexes to the solid phase. This design offers many advantages including signal amplification and increased sensitivity. However, streptavidin/biotin-based immunoassays are vulnerable to biotin interferences if biotin is present in excess concentration.
- Biotin causes positive interference in competitive immunoassays.
- Biotin causes negative interferences in sandwich immunoassays.
- However, in some assays, the streptavidin (or antibiotin antibody) and the biotin-labeled reagents are combined during manufacturing process (prebound reagents). Such assays are expected to be less susceptible or unaffected in the presence of high biotin concentrations in the serum or plasma.
- Currently, Roche Diagnostics, Siemens Healthineers Laboratory Diagnosis, Ortho Clinical Diagnosis, Beckman Coulter and Immunodiagnostics System manufacture and market biotin-based immunoassays. In general, immunoassays marketed by Abbott Laboratories are not biotin-based. Therefore, Abbott assays as expected are free from biotin interferences.
- Most commonly reported clinical misdiagnosis due to biotin interferences in immunoassays is wrong diagnosis of Graves' disease/hyperthyroidism. Biotin if present in sufficient concentration falsely increases FT4, FT3, and total T4 values but falsely lower TSH value just mimicking clinical profile of hyperthyroidism.
- Negative interference of biotin in troponin I and troponin T measurement may cause missed diagnosis of myocardial infarction. FDA reported the death of a person due to falsely lower troponin value to biotin interference. Biotin also causes negative interference with CK-MB and NT-pro-BNP.
- Biotin causes positive interferences with some hormone assays which are small molecules including testosterone, estradiol, progesterone, and cortisol. However, biotin causes negative interference with other hormones which are large molecules, such as parathyroid hormone, LH, FSH, growth hormone, prolactin, insulin, C-peptide, and ACTH.
- Biotin causes negative interferences with tumor markers measurement (alpha-fetoprotein, CA-125, CA 19-9, PSA, free PSA, etc.) using sandwich assays. In addition, biotin also causes negative interference with serological tests for hepatitis and HIV.

- If biotin interference is suspected then sending the test to another laboratory which does not use biotin-based immunoassay is the best way to resolve biotin interference.
- Nonlinearity during dilution also confirms assay interference but may not be related to biotin only.
- Another approach to overcome biotin interference in biotin-based assays is to wait for sufficient time for biotin to be eliminated from the body. Roche Diagnostics recommend waiting for 8 h before blood collection for patients taking 5 mg of biotin supplement. The American Thyroid Association Guidelines recommend discontinuation of biotin at least 2 days before conducting thyroid function tests. However, such approach is not possible in emergency situation where cardiac troponin tests cannot be delayed in a patient suspected of myocardial infarction.
- Currently, best approach to overcome biotin interference in biotin-based immunoassays is pretreatment of specimen with streptavidin-coated microparticle to deplete biotin present in specimens. However, such approach is laborious.

# References

[1] Peyro Saint Paul L, Debruyne D, Bernard D, Mock DM, et al. Pharmacokinetics and pharmacodynamics of MD1003 (high-dose biotin) in the treatment of progressive multiple sclerosis. Expert Opin Drug Metab Toxicol 2016;12:327–44.

[2] Mardach R, Zempleni J, Wolf B, Cannon MJ, et al. Biotin dependency due to a defect in biotin transport. J Clin Invest 2002;109:1617–23.

[3] Grimsey P, Frey N, Bending G, Zitzler J, et al. Population pharmacokinetics of exogenous biotin and the relationship between biotin serum levels and in vitro immunoassays interference. Int J Pharmacokinet 2017;2:247–56.

[4] Peyro Saint Paul L, Debruyne D, Bernard D, Mock DM, Defer GL. Pharmacokinetics and pharmacodynamics of MD1003 (high-dose biotin) in the treatment of progressive multiple sclerosis. Expert Opin Drug Metab Toxicol 2016;12:327–44.

[5] Samarasinghe S, Meah F, Singh V, Basit A, et al. Biotin interference with routine clinical immunoassays: understanding the causes and mitigate risk. Endocr Pract 2017;23:989–98.

[6] Biotin (vitamin B7). Safety communication—Biotin may interfere with lab tests (issued by the FDA), https://www.fda.gov/safety/medwatch/safetyinformation/safetyalertsforhumanmedicalproducts/ucm586641.htm.

[7] Trambas CM, Sikaris KA, Lu ZX. More on biotin treatment mimicking Graves' disease. N Engl J Med 2016;375:1698.

[8] Piketty ML, Polak M, Flechtner I, Gonzales-Briceño L, et al. False biochemical diagnosis of hyperthyroidism in streptavidin-biotin-based immunoassays: the problem of biotin intake and related interferences. Clin Chem Lab Med 2017;55:780–8.

[9] Waghray A, Milas M, Nyalakonda K, Siperstein AE. Falsely low parathyroid hormone secondary to biotin interference: a case series. Endocr Pract 2013;19:451–5.

[10] Trambas C, Lu Z, Yen T, Sikaris. Characterization of the scope and magnitude of biotin interference in susceptible Roche Elecsys competitive and sandwich immunoassay. Ann Clin Biochem 2018;55:205–15.

[11] Djebrani M, Jabbour F, Hulot FD, De Mazancourt P. Biotin analytical interference on Siemens Immunoanalysis on Dimension EXL®: impact on cardiac troponin I and total vitamin D (D 2 +D 3) assays. Ann Biol Clin (Paris) 2019;77:693–6.

[12] Sancar F. Update on biotin interference with troponin tests. JAMA 2019;322:2277.

[13] Li D, Radulescu A, Sherstha RT, Root M, et al. Association of biotin ingestion with performance of hormone and non-hormone assays in healthy adults. JAMA 2017;318:1150–60.

[14] Williams GR, Cervinski MA, Nerenz RD. Assessment of biotin interference with qualitative point of care hCG test devices. Clin Biochem 2018;53:168–70.

[15] Holmes EW, Samarasinghe S, Emanuele MA, Meah F. Biotin interference in clinical immunoassays: a cause for concern. Arch Pathol Lab Med 2017;141:1459–60.

[16] Ross DS, Burch HB, Cooper DS, Greenlee MC, et al. 2016 American Thyroid Association guidelines for diagnosis and management of hyperthyroidism and other causes of thyrotoxicosis. Thyroid 2016;26:1343–421.

[17] Trambas C, Lu Z, Yen T, Sikaris K. Depletion of biotin using streptavidin-coated microparticles: a validated solution to the problem of biotin interference in streptavidin-biotin immunoassays. Ann Clin Biochem 2018;55:216–26.

# CHAPTER 23

# Hemoglobinopathies and thalassemia

## Introduction

Hemoglobinopathies are inherited structural disorders of hemoglobin. Thalassemia is reduced rate of globin chain synthesis. There are over a thousand different types of hemoglobinopathies. Most are clinically insignificant. The wide variation in clinical manifestation of these disorders is attributable to both genetic and environmental factors [1]. Some hemoglobinopathies such as sickle cell disease can be fatal if not treated. Hematopoietic stem cell transplantation, the only established cure, is becoming increasingly safer and cost-effective [2].

## Hemoglobin structure and synthesis

Hemoglobin, the oxygen-carrying pigment of erythrocytes, consists of a heme portion (protoporphyrin IX, a heterocyclic organic compound containing four pyrrole rings, chelated with iron) and four globin chains. Six distinct species of normal hemoglobin are found in human, three in normal adults, and three in fetal life. The globulins associated with hemoglobin molecule (both embryonic stage and after birth) include alpha chain ($\alpha$-chain), beta chain ($\beta$-chain), gamma chain ($\gamma$-chain), delta chain ($\delta$-chain), epsilon chain ($\varepsilon$-chain), and zeta chain ($\zeta$-chain). In embryonic stage, hemoglobin Grower and hemoglobin Portland are found but these are replaced by hemoglobin F (Hb F: two alpha chains and two gamma chains) in fetal life. Interestingly, Hb F has higher oxygen affinity than adult hemoglobin and is capable of transporting oxygen in peripheral tissues in hypoxic fetal environment. In the third trimester, genes responsible for beta- and gamma-globulin synthesis are activated and as a result, adult hemoglobin such as hemoglobin A (Hb A: two alpha chains and two beta chains) and hemoglobin $A_2$ (Hb $A_2$: two alpha chains and two delta chains) may also be found in neonates, but Hb F is still the major component. Newborn babies and infants up to 6 months of age do not depend on Hb A synthesis, although switch from Hb F to Hb A occurs around 3 months of

457

Clinical Chemistry, Immunology and Laboratory Quality Control. https://doi.org/10.1016/B978-0-12-815960-6.00005-4

**Table 1** Embryonic, fetal, and adult hemoglobin.

| Period of life | Hemoglobin species | Globulin chains | % Present in adult |
|---|---|---|---|
| Embryonic | Gower-1 | Two zeta, two epsilon | |
| | Gower-2 | Two alpha, two epsilon | |
| | Portland-1 | Two zeta, two gamma | |
| | Portland-2 | Two zeta, two beta | |
| Fetal | Hemoglobin F | Two alpha, two gamma | |
| Adult | Hemoglobin A | Two alpha, two beta | 95%–97% |
| | Hemoglobin A2 | Two alpha, two delta | <3.5% |
| | Hemoglobin F | Two alpha, two gamma | <1% |

age. Therefore, disorders due to beta-chain defect such as sickle cell disease tend to manifest clinically after 6 months of age, although diseases due to alpha-chain defect are manifested in utero or following birth. Embryonic, fetal, and adult hemoglobins are summarized in Table 1. The different types of naturally occurring embryonic, fetal, and adult hemoglobins vary in their tetramer-dimer subunit interface strength (stability) in the liganded (carboxyhemoglobin or oxyhemoglobin) state [3].

The normal hemoglobin (HbA) in adults contains two alpha chains and two beta chains. Each alpha chain contains 141 amino acids in length and each beta chain contains 146 amino acids. Hemoglobin $A_2$ (HbA$_2$) contains two alpha chains and two delta chains. The gene for the alpha chain is located in chromosome 16 (two genes in each chromosome, a total of four genes), while genes for beta (one gene in each chromosome, a total of two genes) gamma and delta chains are located on chromosome 11. Adults have mostly HbA and small amounts of HbA$_2$ (less than 3.5%) and Hb F (less than 1%). A small amount of fetal hemoglobin persists in adults due to a small clone of cells called F cells. When hemoglobin is circulating with erythrocytes, glycosylation of the globin chains may take place. These are referred to as X1c (X being any hemoglobin, for example, HbA1c; glycosylated hemoglobin, a biomarker for diabetes). When hemoglobin molecule is aging, glutathione is bound to cysteine at the 93rd position of the beta chain. This is HbA3 or HbA1d which is found in old red blood cells. In addition, like HbA1c, other glycosylated forms such as HbC1c, HbC1d, Hb S1c, and HbS1d may also exist in circulation.

Heme is synthesized in a complex way involving enzymes in both mitochondrion and cytosol. In the first step, glycine and succinyl CoA combines in mitochondria to form delta aminolevulinic acid which is transported into cytoplasm and is converted into porphobilinogen by the action of enzyme aminolevulinic acid dehydrogenase. Then porphobilinogen is converted finally into coproporphyrinogen III through several steps involving multiple enzymes.

Then coproporphyrinogen III is transported into mitochondria and is converted into protoporphyrinogen III by coproporphyrinogen III oxidase enzyme. Then protoporphyrinogen III is converted into protoporphyrin IX by protoporphyrinogen III oxidase enzyme and finally protoporphyrin IX is converted into heme (incorporating iron in the protoporphyrin IX molecule) by ferrochelatase enzyme. Eventually, heme is transported into cytosol and combines with globulin to form the hemoglobin molecule.

## Introduction to hemoglobinopathies

Hemoglobinopathies, which result from quantitative defects in hemoglobin (alpha- and beta-thalassemia) and hemoglobin structural variants, are prevalent in malaria-endemic regions (the Mediterranean, Asia, and sub-Saharan Africa) owing to natural selection. They have also become increasingly prevalent in nonendemic Europe, North America, and Australia owing to population migrations. As a result, hemoglobinopathies become a global health problem [4]. However, out of over 1000 hemoglobinopathies reported, most of such disorders are asymptomatic, although in several other cases hemoglobinopathies are associated with significant clinical disorder including fatality.

Hemoglobinopathies are transmitted in autosomal recessive fashion. Therefore, carriers who have one mutated gene and one normal gene are usually healthy or slightly anemic. When both parents are carriers, then children have 25% chance of being normal (one normal gene from father and another from mother), 25% chance of being severely affected by the disease (homozygous: one muted gene from mother and another from father) as well as 50% chance of being a carrier (one muted gene and one normal gene). Hemoglobinopathies are caused by inherent mutation of genes coded for globin synthesis. Point mutation of the gene in coding region (exons), which causes production of defective globin, results in formation of abnormal hemoglobin (hemoglobin variants) [5]. In general, hemoglobinopathies can be divided into three major categories:

- Quantitative disorders of hemoglobin synthesis: Production of structurally normal but decreased amounts of globin chains (alpha- and beta-thalassemia).
- Disorder (qualitative) in hemoglobin structure: Production of structurally abnormal globulin chains such as hemoglobin S, C, O, or E. Sickle cell syndrome (Hb S) is the most common example of such disease.
- Failure to switch globin chain synthesis after birth: for example, hereditary persistence of fetal hemoglobin (Hb F), a relatively benign condition. It may coexist with thalassemia or sickle cell disease and will result in decreased severity of such diseases (protective effect).

The World Health Organization (WHO) estimated that at least 5.2% of world population (and over 7% of pregnant women) carries significant hemoglobin variant. Approximately 1.1% of couples worldwide are at a risk of having children with a hemoglobin disorder. Annually over 330,000 affected conceptions or birth worldwide are affected by hemoglobinopathies. Approximately 275,000 have a sickle cell disorder and require early diagnosis and prophylaxis. In addition, about 56,000 have a major thalassemia, including at least 30,000 requiring regular transfusion to survive and 5500 who die prematurely due to α-thalassemia major. Most affected babies born in high-income countries survive with a chronic hemoglobin disorder but most born in low-income countries die before the age of 5 years. As a result, hemoglobin disorders contribute to 3.5% of mortality in children under age of 5 years worldwide, but unfortunately in in Africa, the mortality rate is 6.4% [6].

The clinically significant hemoglobinopathies include α- and β-thalassemia, sickle cell disease, HbE disease, and HbC disease. Currently, population screening programs for hemoglobinopathies are available in all continents of the globe. The main aim of screening is to reduce the number of affected births and, in the case of sickle cell disease, reduce childhood morbidity and mortality [4]. In West Africa, approximately 25% of individuals are heterozygous for hemoglobin S (Hb S) gene which is related to sickle cell diseases. In addition, high frequencies of Hb S gene alleles are also found in people of Caribbean, South and Central Africa, Mediterranean, Arabian Peninsula, and East India. Hemoglobin C (Hb C) is found mostly in people living or originating from West Africa. Hemoglobin E (Hb E) is widely distributed between East India and Southeast Asia with highest prevalence in Thailand, Laos, and Cambodia but may be sporadically observed in parts of China and Indonesia.

Hemoglobinopathies can be due to alpha chain defect, beta chain defect, gamma chain defect, or delta chain defect. Commonest beta chain defect hemoglobinopathies are:

- Hb S (seen most often in African Americans)
- Hb C (seen most often in African Americans)
- Hb E (seen most often in S E Asians)

Commonest alpha chain defect:

- Hb G (seen most often in African Americans)

Commonest delta chain defect:

- Hb A2′ (seen most often in African Americans)

Thalassemia is an inherited autosomal recessive blood disorder characterized by abnormal hemoglobin which may cause anemia. Therefore, thalassemia syndrome is not due to structural defect in globin chain but due to lack of sufficient synthesis of globin chain. This is due to mutations which disrupt gene expression. When there is reduced amount of alpha chain synthesis, the condition is termed alpha-thalassemia. When beta chain synthesis is reduced, the condition is termed beta-thalassemia. Similarly, delta and delta beta-thalassemia may also occur. Of these, alpha and beta-thalassemia are seen most often. In general, β-thalassemia is observed in Mediterranean, Arabian Peninsula, Turkey, Iran, West and Central Africa, India, and Southeast Asian countries, while α-thalassemia is commonly observed in parts of Africa, Mediterranean, Middle East, and throughout Southeast Asia [7].

## Alpha-thalassemia

There are two genetic loci for alpha gene resulting in four genes (alleles) for alpha hemoglobin (α/α, α/α) on chromosome 16. Two alleles are inherited from each parent. Alpha-thalassemia occurs when there is a defect or deletion in one or more of four genes responsible for alpha-globin production. When there is underproduction of the alpha-globin chain, then the body tries to compensate by increased production of beta, gamma, and delta globin chains. Four beta-globin chains may form a novel hemoglobin called Hb H. Four gamma-globin chains may form another novel hemoglobin called Hb Bart's.

Alpha-thalassemia can be divided into four categories:

- The silent carriers: Characterized by only one defective or deleted gene but three functional genes. These individuals have no health problem. An unusual case of silent carrier is individuals carrying one defective constant spring mutation but three functional genes. These individuals also have no health issue.
- Alpha thalassemia trait: Characterized by two deleted or defective genes and two functional genes. These individuals may have mild anemia.
- Alpha-thalassemia major (Hemoglobin H Disease): Characterized by three deleted or defective genes and only one functional gene. These patients have persistent anemia and significant health problems. In a newborn with alpha-thalassemia major, gamma-globin chain production is still active and thus Hb Bart's presence is expected. However, when the child becomes older, gamma-globin chain production ceases and is overtaken by beta-globin chain production causing development of Hb H.
- Hydrops Fetalis: Characterized by no functional alpha gene and these individuals have hemoglobin Bart. This condition is not compatible with life unless intrauterine transfusion is initiated.

When an alpha gene is functional, it is denoted as "α" and if not functional or deleted, denoted as "–". There is not much difference in impaired alpha-globin synthesis between a deleted gene and a nonfunctioning defective gene. With deletion or defect of one gene (-/α, α/α), little clinical effect is observed because three alpha genes are enough to allow normal hemoglobin production. These patients are sometimes referred to as "silent carriers" since there is no clinical symptom but mean corpuscular volume (MCV) and man corpuscular hemoglobin (MCH) may be slightly decreased. These individuals are diagnosed by deduction only when they have children with thalassemia trait or hemoglobin H disorder. An unusual case of silent carrier state is when an individual carrying one hemoglobin Constant Spring mutation but three functional genes. Hemoglobin Constant Spring (hemoglobin variant isolated from a family of ethnic Chinese background from Constant Spring district of Jamaica) is a hemoglobin variant where mutation of alpha-globin gene produces an abnormally long alpha chain (172 amino acids instead of normal 141 amino acids). Hemoglobin Constant Spring is due to nondeletion mutation of alpha gene which results in production of unstable alpha-globin. Moreover, this alpha-globin is produced in very low quantity (approximately 1% of normal expression level) and is found in people living or originating in Southeast Asia.

When two genes are defective or deleted, alpha-thalassemia trait is present. There are two forms of alpha-thalassemia trait. Alpha-thalassemia 1 (-/-, α/α) results from the *cis* deletion of both alpha genes on the same chromosome. This mutation is found in Southeast Asian populations. Alpha-thalassemia 2 (-/α, -/α) results from the trans-deletion of alpha genes on two different chromosomes. This mutation is found in the African and African American populations (prevalence of disease 28% in African Americans). Only in the case of cis-deletion, zeta globin is expressed in carriers. In alpha-thalassemia trait, two functioning alpha genes are present and as a result, erythropoiesis is almost normal in these individuals, but a mild microcytic hypochromic anemia (low MCV and MCH) may be observed. This form of the disease may mimic iron deficiency anemia. Therefore, distinguishing alpha-thalassemia from iron deficiency anemia is essential.

If three genes are affected (-/-, -/α), the disease is called hemoglobin H disease, which is a severe form of alpha-thalassemia causing severe anemia requiring blood transfusion. Because only one alpha gene is responsible for production of alpha-globin in Hb H disease, high beta (β)-globin to α-globin ratio (2 to 5-fold increase in β-globin production) may result in formation of a tetramer containing only β and this form of hemoglobin is called Hb H (four β chains). This form of hemoglobin cannot deliver oxygen in peripheral tissues because hemoglobin H has a very high affinity for oxygen. A microcytic hypochromic anemia with target cells and Heinz bodies (which represents precipitated Hb H) is present in the peripheral blood smear of these patients. Moreover, red cells that contain hemoglobin H are sensitive to oxidative stress, and may be more susceptible to hemolysis, especially when oxidants such as sulfonamides are administered. More mature erythrocytes also contain increasing amounts of precipitated hemoglobin H (Heinz bodies). These are removed from the circulation prematurely, which may also cause hemolysis. Therefore, clinically, these patients experience a varying severity of chronic hemolytic anemia. Due to the subsequent increase in erythropoiesis, erythroid hyperplasia may result and cause bone structure abnormalities with marrow hyperplasia, bone thinning, maxillary hyperplasia, and pathologic fractures. When hemoglobin Constant Spring is associated with Hb H disease, a more severe form of anemia is observed requiring frequent transfusion [8]. However, when a child inherits one hemoglobin Constant Spring gene from father and one from mother, then hemoglobin Constant Spring disease is present which is less severe than hemoglobin H-hemoglobin Constant Spring disease, but severity is comparable to hemoglobin H disease. Patients with hemoglobin H and related diseases require transfusion and chelation therapy to remove excess iron.

When four genes are defective or deleted, (-/-, -/-), the result is Hemoglobin Bart's disease where alpha-globin is absent because no gene is present to promote alpha-globin synthesis and as a result four gamma ($\gamma$) chains form a tetramer. As in Hb H, the hemoglobin in Hb Bart's is unstable, which impairs the ability of the red cells to release oxygen to the surrounding tissues. The fetus usually cannot survive gestation causing stillbirth with hydrops fetalis. However, more recently, with support through intrauterine transfusion and neonatal intensive care unit, survival may be possible but survivors have severe transfusion-dependent anemia. Farashi and Harteveld summarized clinical presentation, diagnostic criteria, and molecular basis of alpha-thalassemia [9].

## Beta-thalassemia

Beta-thalassemia is due to deficit or absence of beta-globin resulting in excess production of alpha-, gamma-, and delta-globin chains. HbA$_2$ has two alpha chains and two delta chains. As a result, in beta-thalassemia, HbA$_2$ levels are high, typically greater than 3.5%. HbF has two alpha chains and two gamma chains. Thus, in beta-thalassemia, HbF levels are also high. Synthesis of beta-globin may vary from near complete presence to complete absence causing various severities. Beta-thalassemia is due to mutation of genes (one gene each on chromosome 11) and more than 200 point mutations have been reported. However, deletion of both genes is rare. Beta-thalassemia can be broadly divided into three categories:

- Beta-Thalassemia Trait: Characterized by one defective gene and one normal gene and individuals may experience mild anemia but are not transfusion dependent. Here the key feature is presence of Hb A$_2$, greater than 3.5%. Analysis of the complete blood count (CBC) values should demonstrate microcytic hypochromic red cells with mild anemia. The RDW is not significantly increased. The RBC count is disproportionately increased when compared to the hemoglobin level.

- Beta-Thalassemia Intermedia: Characterized by two defective genes but some beta-globin production is still observed in these individuals. However, some individuals may have significant health problems requiring intermittent transfusion.
- Beta-Thalassemia Major (Cooley's anemia): Characterized by two defective genes but almost no function of either gene leading to no synthesis of beta-globin. These individuals have severe form of disease requiring lifelong transfusion and may have shortened life span. These individuals have significantly elevated levels of HbF. In some cases of beta-thalassemia major, $HbA_2$ may not be elevated.

If a defective gene is incapable of producing any beta-globin, it is characterized as "$\beta^0$" causing more severe form of beta-thalassemia. However, if the mutated gene can retain some function, it is characterized as "$\beta^+$". In the case of one gene defect, beta-thalassemia minor (trait; patients are $\beta^0/\beta$ or $\beta^+/\beta$) is observed and individuals are either normal or mildly anemic. These patients have increased $HbA_2$. In addition, HbF may also be elevated. MCV and MCH are low but these patients are not transfusion dependent. If both genes are affected resulting in severely impaired production of beta-globin ($\beta^0/\beta^0$ or $\beta^+/\beta^0$), the disease is severe and is called beta-thalassemia major (also known as Cooley's anemia). However, due to the presence of fetal hemoglobin, symptoms of beta-thalassemia major are not observed prior to 6 months of age. Patients with beta-thalassemia major have elevated $HbA_2$ and HbF (although in some individual $HbA_2$ may be normal). If production of beta-globin is moderately hampered then the disease is called beta-thalassemia intermediate ($\beta^0/\beta$ or $\beta^+/\beta^+$). These individuals have less severe disease than beta-thalassemia major. In patients with beta-thalassemia major, excess alpha-globulin chain precipitates leading to hemolytic anemia. These patients require lifelong transfusion and chelation therapy. Interestingly, having $\beta^0$ or $\beta^+$ does not predict the severity of disease because patients with both types have been diagnosed with beta-thalassemia major or intermedia. Major features of alpha and beta-thalassemia are summarized in Table 2.

## Delta-thalassemia

Delta-thalassemia, a rare disease, is due to mutation of genes responsible for synthesis of delta chain and is characterized by decreased or total absence of production of delta and beta-globin. A mutation that prevents formation of delta chain is called delta$^0$ and if some delta chain is formed, the mutation

**Table 2** Major features of alpha and beta-thalassemia.

| Disease | Number of deleted gene | Comments |
|---|---|---|
| Alpha-thalassemia silent carrier | One of four gene deletions | Asymptomatic<br>May have low MCV, MCH |
| Alpha-thalassemia trait | Two of four gene deletions | Asymptomatic or mild microcytic hypochromic anemia |
| Hemoglobin H disease | Three of four gene deletions | Microcytic hypochromic Anemia. Hb H found in adults and Hb Bart's found in neonates. Hemoglobin H may coexist with hemoglobin Constant Spring, a more severe disease than Hb H. |
| Hydrops fetalis | Four of four gene deletions | Hemoglobin Bart's disease<br>Most severe form may cause stillbirth/Hydrops fetalis |
| Beta-thalassemia trait | One gene defect | Asymptomatic |
| Beta-thalassemia intermedia | Both genes defective | Variable degree of severity as some beta-globin is still produced |
| Beta-thalassemia major | Both genes defective | Severe impairment or no beta-globin synthesis. Severe disease with anemia, splenomegaly, requiring lifelong transfusion. |

is termed as delta+. If an individual inherits two delta$^0$ mutations, no delta chain is produced and no HbA$_2$ can be detected in blood (normal level < 3.5%). However, if an individual inherits two delta+ mutations, decreased HbA$_2$ is observed. Tang et al. described cases of a Chinese family with δ-thalassemia, in which the daughter is a homozygous for δ-thalassemia with complete absence of Hb A$_2$ and the mother is a heterozygote with low level of Hb A$_2$. The father, however, is a heterozygote with a normal Hb A$_2$ value due to coinheritance of a β-thalassemia. Although no abnormal clinical or hematological findings were noted in the individuals with δ-thalassemia, it is important to note that diagnosis of β-thalassemia can be missed during routine screening when β-thalassemia and δ-thalassemia coexist in a subject because in beta-thalassemia, HbA$_2$ is increased, but the presence of delta-thalassemia may reduce HbA$_2$ concentration, masking diagnosis of beta-thalassemia trait [10].

In delta-thalassemia, as a compensatory mechanism, gamma chain synthesis is increased resulting in significant amount of fetal hemoglobin (Hb F) in blood which is homogenously distributed in red blood cells. This condition is found in many ethnic groups but especially observed in individuals with ancestry from Greece or Italy. Heterozygous individuals are asymptomatic with normal Hb A$_2$ but homozygous individuals rarely experience mild symptoms.

## Sickle cell disease

Sickle cell disease is the most common monogenic disease. The prevalence of the disease is high throughout large areas in sub-Saharan Africa, the Mediterranean basin, the Middle East, and India because of the remarkable level of protection that the sickle cell trait provides against severe malaria. The "malaria hypothesis" formulated by Haldane in 1949 and by Allison in 1954 is an example of natural selection and balanced polymorphism, a process that is ongoing. Because of slave trading and contemporary population movements, the distribution of sickle cell disease is currently worldwide with an estimated 100,000 persons having this disease in the United States [11].

The terminology "sickle cell disease" (SCD) includes all manifestations of abnormal hemoglobin S (HbS), which includes sickle cell trait (Hb AS), homozygous sickle cell disease (Hb SS), and a range of mixed heterozygous hemoglobinopathies such as hemoglobin SC disease, hemoglobin SD disease, hemoglobin SO Arab disease, and hemoglobin S combined with beta-thalassemia. Individuals with sickle cell trait are typically asymptomatic and do not require medical intervention. Sickle cell disease is a commonly observed hemoglobin disorder in the United States affecting 1 in every 500 birth of African American birth and 1 in every 36,000 Hispanic American births. Sickle cell disease is a dangerous disorder and symptoms of sickle disease starts before age one with chronic hemolytic anemia, developmental disorder, crisis including extreme pain (sickle cell crisis), high susceptibility to various infections, spleen crisis, acute thoracic syndrome, and increased risk of stroke. Optimally treated individuals may have a life span of 50 to 60 years [12].

In sickle cell disease, normal round shape of red blood cell (RBC) is changed into a crescent shape and hence the name "sickle cell". In the heterozygous form (Hb AS), sickle cell disease protects from infection of *Plasmodium falciparum* malaria but not in the more severe form of homozygous sickle cell disease (Hb SS). The genetic defect-producing sickle hemoglobin is a single nucleotide substitution at codon 6 of the beta-globin gene on chromosome 11 that results in a point mutation in beta-globin chain of hemoglobin (substitution of valine for glutamic acid at sixth position). Hemoglobin S is formed when two normal alpha-globin combines with two mutant beta-globin. Because of this hydrophobic amino acid substitution, Hb S polymerizes upon deoxygenation and multiple polymers bundle to rod-like structure resulting in deformed RBC. Various possible diagnoses of patients with Hb S hemoglobinopathy include sickle cell trait (Hb AS), sickle cell disease (Hb SS), and sickle cell disease status post RBC transfusion/ exchange. Patients with sickle cell trait may also have concomitant alpha-thalassemia and diagnosis of HbS/beta-thalassemia (0/+/++) is also occasionally made. Double heterozygous states of Hb SC, Hb SD, and Hb SO Arab are important sickling states which should not be missed [13].

Hemoglobin C is formed due to substitution of glutamic acid residue with a lysine residue at the 6th position of beta-globin. Individuals who are heterozygous with Hb C disease are asymptomatic with no apparent disease but homozygous individuals have almost all hemoglobin (>95%) as Hb C and experience chronic hemolytic anemia and pain crisis. However, individual who are heterozygous with both hemoglobin C and hemoglobin S (Hb SC disease) have weaker symptoms than sickle cell disease because Hb C does not polymerize as readily as Hb S.

Patients with Hb SS disease may have increased Hb F. The distribution of Hb F among the haplotypes of Hb SS are Hb F 5%–7% in Bantu, Benin or Cameroon, Hb F 7%–10% in Senegal and Hb F 10%–25% in Arab-Indian. Hydroxyurea also causes increase in Hb F. This is usually accompanied by macrocytosis. Hb F can also be increased in Hb S/HPFH (HPFH: hereditary persistence of fetal hemoglobin). Hb $A_2$ values are typically increased in sickle cell disease and more so on high-performance liquid chromatographic (HPLC) analysis. This is because the post translational modification form of Hb S, Hb S1d produces a peak in the A2 window. This elevated value of Hb $A_2$ may produce diagnostic confusion with Hb SS disease and Hb S/beta-thalassemia. It is important to remember that microcytosis is not a feature of Hb SS disease and patients with Hb S/ beta-thalassemia typically exhibit microcytosis.

Hb SS patients and Hb S/beta 0 thalassemia patients do not have any Hb A, unless the patient has been transfused or has undergone red cell exchange. Glycated Hb S has the same retention time (approximately 2.5 min) as Hb A in HPLC. This will produce a small peak in the A window and raise the possibility of Hb S/beta+ thalassemia. Hb S /alpha-thalassemia is considered when the percentage of Hb S is lower than expected. Classical cases are 60% of Hb A and approximately 35%–40% of Hb S. Cases of Hb S/alpha-thalassemia should have lower values of Hb S, typically below 30% with microcytosis. Similar picture will also be present in patients with sickle cell trait and iron deficiency. Various features of sickle cell disease are summarized in Table 3.

## Hereditary persistence of fetal hemoglobin (HPFH)

In individuals with hereditary persistence of fetal hemoglobin, significant amounts of fetal hemoglobin (Hb F) can be detected well into adulthood. In normal adults, Hb F represents less than 1% of total hemoglobin, but in HPFH, the percentage of HB F can be significantly elevated, but $HbA_2$ is also normal. HPFH is divided into two major groups: deletional and nondeletional. Deletional HPFH is caused by variable length deletion in beta-globin gene cluster leading to decreased or absent beta-globin synthesis and compensatory increase in gamma-globin synthesis with a pancellular or homogenous

**Table 3** Major features of sickle cell disease.

| Disease | Hemoglobin variants | Features |
|---|---|---|
| Sickle cell trait | Hb AS | Hb A > HbS; HbA 50%–60%; Hb S: 30–40% <br> No apparent illness |
| Sickle cell disease | Hb SS | HbS (majority), Hb $A_2$: <3.5%, HbF (high), no HbA <br> Severe disease with chronic hemolytic anemia |
| Sickle cell-$\beta^0$-thalessemia | Hb S$\beta^0$ | HbS(majoroity), Hb $A_2$: >3.5%, Hb F(high), no HbA <br> Low MCV and low MCH; severe disease |
| Sickle cell-$\beta^+$-thalassemia | Hb S$\beta^+$/++ | HbS (majority), Hb $A_2$: >3.5%, Hb F (high), <br> Hb A; 5%–40%, <br> Variable mild-to-moderate sickle cell disease |
| Hemoglobin SC disease | Hb SC | Hb S: 50%, Hb C: 50% <br> Moderate sickling disease but chronic hemolytic anemia may be present |
| Hemoglobin S/HPFH | | Hb S: 60%, Hb $A_2$: <3.5%, Hb F: 30%–40% <br> Behaves as sickle cell trait |

distribution of Hb F in red blood cells. Nondeletional HPFH is a broad category of related disorders with increased Hb F typically distributed heterocellularly. Heterocellular distribution is also seen in beta-thalassemia and delta beta-thalassemia.

Both homozygous and heterozygous of HPFH are asymptomatic with no clinical or significant hematological change, although individuals with homozygous HPFH may show up to 100% Hb F while in heterozygous typically shows 20%–28% HbF. If HPFH is associated with sickle cell, it can reduce the severity of disease. Compound heterozygotes for sickle hemoglobin (Hb S) and HPFH have high level of Hb F but these individuals experience few if any sickle cell disease-related complication [14]. If HPFH is associated with thalassemia, individuals also experience less severe disease.

## Other hemoglobin variants

Hemoglobin D (hemoglobin D Punjab, also known as hemoglobin D Los Angeles) is formed due to substitution of glutamine for glutamic acid. Hb D Punjab is one of the most commonly observed abnormality worldwide not only found in Punjab region of India but also in Italy, Belgium, Austria, and Turkey. Hemoglobin D disease can occur in four different forms including heterozygous Hb D trait, Hb D-thalassemia, Hb SD disease, and very rarely, homozygous Hb D disease. Heterozygous Hb D disease is a benign condition with no apparent illness but when Hb D is associated with Hb S or beta-thalassemia, clinical conditions such as sickling disease and moderate hemolytic anemia may be observed. Heterozygous Hb D is rare and usually presents with mild hemolytic anemia and mild-to-moderate splenomegaly [15].

Hemoglobin E is due to point mutation of beta-globin which results in substitution of lysine for glutamic acid in position 26. Individuals with HbE also produce less beta-globin resulting in a thalassemia-like phenotype. Hb E is unstable and can form Heinz bodies under oxidative stress. Hb E trait is associated with moderately severe microcytosis but usually no anemia is present. However, individuals with Hb E homozygous present with modest anemia similar to thalassemia trait. However, when beta-thalassemia is combined with Hb E, for example, in Hb E/ $\beta^0$ thalassemia, patients may have significant anemia requiring transfusion similar to patients with beta-thalassemia intermedia.

Hemoglobin O-Arab (Hb-O-Arab; also known as Hb Egypt) is a rare abnormal hemoglobin variant where at position 121 of the beta-globin, normal glutamic acid is replaced by lysine. Hb-O-Arab is found in people from Balkans, Middle East, and Africa. Patients who are heterozygous for Hb-O-Arab may experience mild anemia and microcytosis similar to patients with beta-thalassemia minor. The homozygous form is extremely rare. Patients with Hb S/Hb O Arab may experience severe clinical symptoms similar to individuals with Hb S/S. Patients with Hb-O Arab/beta-thalassemia may also experience severe anemia with hemoglobin level between 6 and 8 g/dL and splenomegaly [16].

Hemoglobin Lepore is an unusual hemoglobin molecule which is composed of two alpha chains and two delta beta chains as a result of fusion gene of delta and beta genes. The delta beta chains have the first 87 amino acids of the delta chain and 32 amino acids of the beta chain. There are three common variants of hemoglobin Lepore: Hb Lepore Washington, also known as Hb Lepore Boston, Hb Lepore Baltimore, and Hb Lepore Hollandia. Hemoglobin Lepore is seen in individuals from of Mediterranean descent. Individuals with HbA/ Hb Lepore are asymptomatic with Hb Lepore representing 5%–15% of hemoglobin and slightly elevated Hb F (2%–3%). Homozygous Lepore individuals suffer from severe anemia similar to patients with beta-thalassemia intermedia with Hb Lepore representing 8%–30% of hemoglobin and remainder hemoglobin is Hb F. Patients with Hb Lepore/ beta-thalassemia experiences severe disease similar to patients with beta-thalassemia major.

Hemoglobin G-Philadelphia (Hb G) is the commonest alpha chain defect observed 1 in 5000 of African Americans and is associated with alpha-thalassemia 2 deletions. Therefore, these individuals have only 3 functioning alpha gene and Hb G represents 1/3 of total hemoglobin. Hb S is the commonest beta chain defect often observed in African American population while Hb G is the commonest alpha chain defect and again occurs most often in the African American people. If an individual has one parent who has sickle cell trait (HbAS) and the other parent is Hb G trait (HbAG), then the individual will have the one normal alpha chain and one normal beta chain but will also have one abnormal alpha chain representing as G and one abnormal beta chain representing as S. Therefore, this individual can then form following hemoglobin:

- HbA (two normal alpha chains and two normal beta chains)
- HbG (two abnormal alpha G chains and two normal beta chains)
- HbS (two normal alpha chains and two abnormal beta S chains)
- HbS/G (two abnormal alpha G chains and two abnormal beta S chains)

In addition, Hb G2 (alpha2, delta2) which is the counterpart of HbA2 is also present.

Increase in fetal hemoglobin percentage is associated with multiple pathologic states. These include beta-thalassemia, delta beta-thalassemia, and hereditary persistence of fetal hemoglobin (HPFH). While beta-thalassemia is associated with high $HbA_2$ and the latter two states are associated with normal $HbA_2$ values. Hematologic malignancies are associated with increased hemoglobin F and include acute erythroid leukemia (AML, M6) and juvenile myelomonocytic leukemia (JMML). Aplastic anemia is also associated with an increase in percentage of Hb F. In elucidating the actual cause of high Hb F, it is important to consider the actual percentage of Hb F, $HbA_2$ values as well as correlation with CBC and peripheral smear. It is also important to note that drugs (hydroxyurea, sodium valproate, erythropoietin) and stress erythropoiesis may also result in high Hb F. Hydroxyurea is used in sickle cell disease patients to increase the amount of Hb F, presence of which may help to reduce the clinical effects of the disease. Measuring the level of Hb F may be useful in determining the appropriate dose of hydroxyurea. In 15%–20% of cases of pregnancy, HbF may be raised to values as much as 5%.

Other rarely reported hemoglobinopathies involve hemoglobin I, hemoglobin J, hemoglobin Hope, as well as unstable hemoglobin such as hemoglobin Koln, hemoglobin Hasharon, and hemoglobin Zurich (for these unstable hemoglobin, isopropanol test is positive). Certain rarely reported hemoglobin variants are hemoglobin Malmo, hemoglobin Andrew, hemoglobin Minneapolis, hemoglobin British Columbia, and hemoglobin Kempsey. Patients with these rare hemoglobin variants experience erythrocytosis. Hemoglobin I is due to a single alpha-globin substitution (substitution of lysine at position 16 for glutamic acid). Hb I is clinically insignificant unless in a rare occasion it is associated with alpha-thalassemia where approximately 70% hemoglobin is Hb I. Hemoglobin J is characterized as a fast-moving band in hemoglobin electrophoresis (band found close to anode, farthest point from application of sample) and more than 50 variants have been reported including Hb J Cape Town, and Hb J Chicago. However, heterozygous hemoglobinopathy involving Hb J is

**Table 4** Various other common hemoglobinopathies.

| Diagnosis | Hemoglobin/ Hematological | Comments |
|---|---|---|
| Hb C trait (Hb AC) | Hb A:60%; Hb C:40% Normal/Microcytic | Hb C implies ancestry from West Africa, clinically insignificant |
| Hb CC disease | No Hb A, Hb C almost 100% Mild microcytic | Mild chronic hemolytic anemia |
| Hb C trait/ Alpha-thalassemia | Hb A: major hemoglobin Hb C < 30% | |
| Hb C/beta-thalassemia | Microcytic, hypochromic | Moderate-to-severe anemia with splenomegaly |
| Hb E trait (Hb AE) | Hb A major, Hb E: 30%–35% Normal/microcytic | No clinical significance, found in Cambodia, Laos and Thailand (Hb E triangle, where Hb E trait is 50%–60% of population) and South East Asia |
| Hb E disease | No Hb A; mostly Hb E Microcytic hypochromic red cells +/− anemia | Usually asymptomatic |
| Hb E trait with alpha-thalassemia | Majority is A: Hb E < 25% | |
| Hb O trait (Hb AO) | Majority is A: Hb O: 30%–40% Normal CBC | Clinically insignificant but Hb S/O is a sickling disorder |
| Hb D trait (Hb AD) | Hb A > Hb D Normal CBC | Clinically insignificant; Hb S/D is a sickling disorder |
| Hb G trait (HbAG) | Hb A > Hb G Normal CBC | Clinically insignificant |

clinically insignificant. In hemoglobin Hope, aspartic acid is substituted for glycine at position 136 of the beta chain. Important other hemoglobinopathies are summarized in Table 4.

# Laboratory investigation of hemoglobinopathies

Multiple methodologies exist to detect hemoglobinopathies and thalassemias. Several methods that are routinely employed include gel electrophoresis, high-performance liquid chromatography (HPLC), capillary electrophoresis, and isoelectric focusing. If any one method detects an abnormality, a second method must be used to confirm the abnormality. In addition, relevant clinical history, review of the complete blood count (CBC), and peripheral smear provide important correlation in the pursuit of an accurate diagnosis.

## Gel electrophoresis

Alkaline gel hemoglobin electrophoresis where hemoglobin molecules assume a negative charge and migrate toward the anode is the first step for identifying abnormal hemoglobin. If abnormality is observed, it can be followed by acid gel electrophoresis. In an acid pH, some hemoglobins assume a negative charge and migrate toward the anode, whereas others are positively charged and migrate toward the cathode. Electrophoresis can be carried out on filter paper, a cellulose acetate membrane, a starch gel, a citrate agar gel, or an agarose gel. Separation of different hemoglobin is largely but not solely dependent on the charge of the hemoglobin molecule. Change in the amino acid composition of the globin chains results in alteration of the charge of the hemoglobin molecule, resulting in change of the speed of migration. In the alkaline gel, following main lanes are observed from the bottom of the gel moving toward the top:

- C lane
- S lane
- F lane and
- A lane

A faint band which usually represents the enzyme carbonic anhydrase may be seen before the main C lane. Any bands above the A lane represent fast hemoglobins. These can be HbJ, HbI, HbN, Hb Bart's, and HbH. Hb H and HbI which are able to move the furthest and are located in the H lane. The location of the H lane is the same distance from J as A is from J in the opposite direction. Patterns of various band in gel electrophoresis are summarized in Table 5.

In the acid gel, the main lanes from the top to bottom are:

- C lane
- S lane
- A lane and
- F lane

## Interpretation of alkaline and acid gels

We start with the alkaline gel, moving on to the acid gel. Normal individuals will have a prominent band in the A lane in both the alkaline and acid gels.

Table 5 Migration of various hemoglobin bands in alkaline gel and acid gel electrophoresis.

| Region | Hemoglobin present |
|---|---|
| Alkaline gel electrophoresis | |
| Top band (farthest from origin: H Lane) | Hb H, Hb I |
| | Hb Bart's and Hb N are between Hb J and Hb H |
| J Lane | Hb J |
| A Lane | Hb A |
| F Lane | Hb F |
| S Lane | Hb S, Hb D, Hb G, Hb Lepore |
| C Lane | Hb C, Hb E, Hb O, Hb A2, Hb S/G hybrid |
| Carbonic anhydrase band (faint) | Hb G2, Hb A2', Hb CS |
| Acid gel electrophoresis | |
| Top band (fastest from origin) C Lane | Hb C |
| S Lane | Hb S |
| | Hb S/G hybrid; Hb O and Hb H are between S and A lane |
| A lane | Hb A, Hb E, Hb A2, Hb D, Hb G, Hb Lepore, Hb J |
| | Hb I, Hb N, Hb H |
| F Lane | Hb F, Hb Hope, Hb Bart's |

Very faint bands may be seen in the F lanes. If we see a prominent band in the C lane in the alkaline gel, then we need to think of HbC or HbE or HbO. If there is a prominent band in the A lane in the same patient, then we may be dealing with C trait or E trait or O trait. If there is no band in the A lane, then we are probably dealing with homozygous forms of the hemoglobinopathies. Now, we need to look at the acid gel. If there is prominent band in the C lane, then the band seen in the C lane of the alkaline gel is truly due to HbC. If in the acid gel there is no prominent band in the C lane or S lane or F lane, but there is only a band in the A lane, then the abnormal Hb is most likely HbE. If on the acid gel there is a band between the S and A lane, then the abnormal Hb is most likely HbO.

After we have formed a tentative diagnosis form the gels, a second method, either HPLC or capillary electrophoresis, must be performed to verify the abnormality. Again, as a different example, we see a band in the S lane in the alkaline gel. Differential now includes Hb S, HbD, Hb G, and Hb Lepore. If this is true HbS, then we expect to see a band in the S lane in the acid gel. HbD, HbG, and Hb Lepore migrate in the A lane in the acid gel. Hb Lepore is usually underproduced. Bands due to Hb Lepore are thus faint. Other under-produced hemoglobins are:

- Hb Constant Spring
- HbA2′

How should we differentiate HbD, HbG, and Lepore? By capillary electrophoresis, they are all found in the same zone, zone 6. HPLC is best to differentiate these three abnormal hemoglobins. Hb Lepore has a retention time of approximately 3.7 min, the same window as HbA2 and HbE. Also, please note that Hb Lepore is underproduced. Individuals with Hb Lepore also have minimally increased Hb F, which provides additional clue. Both HbD and HbG have the same retention time, approximately 3.9 to 4.2 min. However, individuals with HbG have an additional peak on HPLC at 4.6 to 4.7 min due to HbG2. This is not evident with HbD.

For diagnosis of Hb S/G hybrid on alkaline gel electrophoresis, one band is expected in the A lane, one band in the S lane (due to Hb S and Hb G), one band in the C lane (due to S/G hybrid), and one band in the carbonic anhydrase area (due to HbG2). Therefore, a total of four bands should be observed. If the band in the carbonic anhydrase is not prominent, at least three bands should be seen. On the acid gel electrophoresis, one band is expected in the A lane (due to Hb A, Hb G, and Hb G2) and one band in the S lane (due to Hb S and HbS/G hybrid). In capillary electrophoresis, a band should be seen in zone 5 (Hb S) and a band in zone 6 (Hb G). It is important to emphasize that for hemoglobinopathies, finding of gel electrophoresis must be confirmed by a second method, HPLC or capillary electrophoresis.

## High-performance liquid chromatography (HPLC)

HPLC systems utilize a weak cation exchange column system and a sample of RBC lysate in buffer is injected into the system followed by application of a mobile phase so that various hemoglobins can partition (interact) between stationary phase and mobile phase. The time required for different hemoglobin molecules to elute is referred to as retention time. The eluted hemoglobin molecules are detected by absorbance at 405 nm. HPLC permits the provisional identification of many more variant types of hemoglobins that cannot be distinguished by conventional gel electrophoresis. When HPLC is used, a recognized problem is carryover of specimen from one to the next specimen. For example, if the first specimen belongs to a patient with sickle cell disease (Hb SS), then a small peak may be seen at the "S" window in the next specimen. This can lead to diagnostic confusion as well as the sample to be re-run. Approximate retention times of common hemoglobins in a typical HPLC analysis are summarized in Table 6.

**Table 6** Approximate retention times of various hemoglobins in HPLC analysis.

| Approximate retention time | Hemoglobin |
|---|---|
| 0.7 min (Peak 1) | Acetylated Hb F, Hb H, Hb Bart's, Bilirubin |
| 1.1 min | Hb F |
| 1.3 min (Peak 2) | Hb A1c, Hb Hope |
| 1.7 min (Peak 3) | Aged Hb A (HbA1d), Hb J, HbN, Hb I |
| 2.5 min | Hb A, HbS1c |
| 3.7 min | Hb A2, Hb E, Hb Lepore, Hb S1d |
| 3.9–4.2 min | Hb D, Hb G |
| 4.5 min | Hb S, HbA2', Hb C1c<br>Hb O Arab has a broad range from 4.5 to 5 min |
| 4.6–4.7 | Hb G2 |
| 4.9 min | Hb C (preceding the main peak is a small peak, Hb C1d), Hb S/G hybrid, Hb CS (3 peaks: 2%–3%) |

## Capillary electrophoresis

In capillary electrophoresis, a thin capillary tube made of fused silica is used. When an electric filed is applied, the buffer solution within the capillary generates an electro-endosmotic flow that moves toward the cathode. Separation of individual hemoglobins takes place due to differences in overall charges. Different hemoglobins are represented in different zones. Capillary zone electrophoresis has an advantage over HPLC, in that hemoglobin adducts (glycated hemoglobins and the aging adduct Hb X1d) do not separate from the main hemoglobin peak in capillary electrophoresis, and make interpretation easier. For example, in capillary electrophoresis of hemoglobin using Sebia, electrophoresis is conducted in alkaline buffer using high voltage and individual hemoglobin peaks (separated in 15 zones) are detected at 415 nm. Common hemoglobin zones in capillary electrophoresis are given in Table 7.

Other less commonly used methodologies include isoelectric focusing, DNA analysis and mass spectrometry. It is important to note that hemoglobinopathies may interfere with measurement of glycosylated hemoglobin (HbA1c) providing an unreliable result. When an HbA1c result is inconsistent with a patient's clinical picture, possibility of hemoglobinopathy must be considered. Depending on the methodology used for measurement of HbA1c, such as HPLC, immunoassay etc., results may be falsely elevated or lowered. Patients with Hb C trait particularly show variable results. In such case, a test which is not affected by hemoglobinopathy such as fructosamine measurement (represent average blood glucose: 2–3 weeks) may be used [17].

**Table 7** Various zones in which common hemoglobins appear in capillary electrophoresis.

| Zone | Hemoglobin |
|------|-----------|
| Zone 1 | HbA2' |
| Zone 2 | Hb C, Hb CS |
| Zone 3 | Hb $A_2$, Hb O-Arab |
| Zone 4 | Hb E, Hb Koln |
| Zone 5 | Hb S |
| Zone 6 | Hb D-Punjab/Los Angeles/Iran Hb G-Philadelphia |
| Zone 7 | Hb F |
| Zone 8 | Acetylated Hb F |
| Zone 9 | Hb A |
| Zone 10 | Hb Hope |
| Zone 11 | Denatured Hb A |
| Zone 12 | Hb Bart's |
| Zone 13 | |
| Zone 14 | |
| Zone 15 | Hb H |

## Reporting normal hemoglobin pattern

Normal hemoglobin electrophoresis pattern:

When hemoglobin electrophoresis reveals a normal pattern, i.e., HbF is <1%, HbA2 is <3.5% and the remainder is HbA, check the CBC values. If the MCV and MCH are not low, the case can be signed out as normal hemoglobin electrophoresis.

## Diagnostics tips for thalassemia, sickle cell disease, and other hemoglobinopathies

For diagnosis of alpha-thalassemia, routine blood analysis (CBC) is the first step. Mean corpuscular volume (MCV), mean corpuscular hemoglobin (MCH), and red cell distribution width (RDH) provide important clues not only in diagnosis of thalassemias but also in other hemoglobin disorders. Thalassemias are characterized by hypochromatic and microcytic anemia and it is important to differentiate thalassemia from iron deficiency anemia because iron supplement has no benefit in patients with thalassemia. Often silent carriers of alpha-thalassemia are diagnosed incidentally when their CBC shows a mild microcytic anemia. However, serum iron and serum ferritin levels are normal in a silent carrier of alpha-thalassemia but reduced in a patient with iron deficiency anemia. In addition, microcytic anemia with normal RDW also indicate thalassemia trait. In hemoglobin H disease, MCV is further reduced, although in iron deficiency anemia MCV is rarely less than 80 fl. In addition, MCH is also reduced. For children, MCV of <80 fl may be common and Mentzer index (MCV/red blood cell count) is useful in differentiating thalassemia from iron deficiency anemia. In iron deficiency anemia, this ratio is usually greater than 13, but in thalassemia, this value is less than 13. However, for accurate diagnosis of alpha-thalassemia, genetic testing is essential. Hemoglobin electrophoresis is not usually helpful for diagnosis of alpha-thalassemia except in infant where presence of Hb Bart's or Hb H indicates alpha-thalassemia, but hemoglobin electrophoresis is usually normal in an individual with alpha-thalassemia trait. However, in an individual with Hb H disease, the presence of hemoglobin H in electrophoresis along with Hb Bart's provides useful diagnostic clue. In hydrops fetalis, newborn often dies or born with gross abnormalities. Circulating erythrocytes are markedly hypochromic and presence of anisopoikilocytosis. In addition, many nucleated erythroblasts are present in peripheral blood smear. Most of the hemoglobin observed in electrophoresis is Hb Bart's. Genetic testing of parents is essential for counseling of parents who may give birth to a baby with hydrops fetalis.

A patient with beta-thalassemia major disease can be identified during infancy, but after six months of age, these patients present with irritability, growth retardation, abnormal swelling, and jaundice. Individuals with microcytic anemia but milder symptoms that start later in life are suffering from beta-thalassemia intermedia. Hemoglobin electrophoresis of individuals with beta-thalassemia trait usually has reduced or absent Hb A, elevated levels of Hb $A_2$, and elevated levels of Hb F. Therefore, for the diagnosis of beta-thalassemia trait, the proportion of $HbA_2$ relative to the other hemoglobins is an important indicator. In certain cases, $HbA_2$ variants may also be present. In such cases, the total $HbA_2$ ($HbA_2$ and $HbA_2$ variant) needs to be considered for the diagnosis of beta-thalassemia. $HbA_2'$ is the most common of the known Hb $A_2$ variants reported in 1% to 2% of African Americans detected in heterozygous and homozygous states and in combination with other Hb variants and thalassemia. The major clinical significance of $HbA_2'$ is that for the diagnosis or exclusion of beta-thalassemia minor, the sum of $HbA_2$ and Hb $A_2'$ must be considered. $HbA_2'$, when present, accounts for a small percentage (1%–2%) in heterozygotes and is difficult to detect by gel electrophoresis. It is however, easily detected by capillary electrophoresis and HPLC. In HPLC, Hb $A_2'$ elutes in the "S" window. In Hb AS trait and HB SS disease, Hb $A_2'$ could be masked by the presence of Hb S. In Hb AC trait and Hb CC disease, glycosylated Hb C will also elute in the "S" window. In these conditions, Hb $A_2'$ will remain undetected. Conversely, sickle cell patients on chronic transfusion protocol or recent, efficient RBC exchange may result in a very small percentage of Hb S which the pathologist may interpret as Hb A2'. It has been documented that the HbA$_2$ concentration may be raised in HIV during treatment. Severe iron deficiency anemia can reduce HbA$_2$ levels and this may obscure diagnosis of beta-thalassemia trait. Hematological features of alpha and beta-thalassemia are given in Table 8.

Hb F quantification is useful in the diagnosis of beta-thalassemia and other hemoglobinopathies. But quantification of Hb F may be an issue when HPLC is used. Fast variants (e.g., Hb H or Hb Bart's) may not be quantified as they may elute off the column before the instrument begins to integrate in many systems designed for adult samples. This will affect the quantity of Hb F. If an alpha-globin variant separates from Hb A, then there should be an Hb F variant that will often separate from normal Hb F, but it may not separate from other hemoglobin adducts present and then the total Hb F will not be adequately quantified. Hb F variants may also be due to mutation of the gamma-globin chain, and again this may result in a separate peak and incorrect quantification. Some beta chain variants and / adducts will not separate from Hb F and this will lead to incorrect quantification. If Hb F appears to be greater than 10% on

**Table 8** Hematological features of alpha- and beta-thalassemia.

| Disease | CBC | Hemoglobin electrophoresis |
|---------|-----|---------------------------|
| Alpha-thalassemia[a] | | |
| Silent carrier | Hb: Normal, MCH <27 pg | Normal |
| Trait | Hb: Normal, MCH <26 pg, MCV <75 fl | Normal |
| Hb H disease | Hb: 8–10 g/dL, MCH <22 pg, MCV Low | Hb H: 10%–20% |
| Hydrops fetalis | Hb <6 g/dL, MCH <20 pg | Hb Bart's: 80%–90% Hb H < 1% |
| Beta-thalassemia[a] | | |
| Minor | Hb: Normal or low, MCV: 55–75 fl[b] MCH: 19–25 pg | Hb A$_2$ > 3.5% |
| Intermedia | Hb: 6–10 g/dL, MCV: 55–70 fl | Hb A$_2$: Variable |
| Mild or compound Heterozygous | MCH: 15–23 pg | Hb F up to 100% |
| Major | Hb <7 g/dL, MCV: 50–60 fl MCH: 14–20 pg | Hb A$_2$: Variable Hb F: High |

[a]*Mentzer Index for Children is <13 for both alpha and beta-thalassemia.*

[b]*MCV: Abnormal: Adult <80 fl, Children (7–12): <76, Children (6 months–6): <70.*

HPLC, its nature should be confirmed by an alternative method to exclude mis-identification of Hb N or Hb J as Hb F. Characterization of patient with high Hb F includes evaluation of the following:

- Consider if Hb F is physiologically appropriate for age.
- Beta-thalassemia: trait, intermedia (20%–40%) or major (60%–98%). Here Hb $A_2$ will also be raised. Patients should have microcytic hypochromic anemia with normal RDW and disproportionately high RBC count. Peripheral smear should exhibit target cells.
- Delta-Beta-Thalassemia: Here $HbA_2$ is normal but increased Hb F due to increase in gamma chains. However, the increase in gamma chains does not entirely compensate for the decreased beta chains. Moreover, alpha chain is present in excess. Trait shows microcytosis without anemia. Homozygous patients have severity of disease compared to thalassemia intermedia.

Hemoglobin electrophoresis is useful in diagnosis of sickle cell disease by identifying Hb S. Diagnostic approach of sickle cell disease is summarized in Table 9. However, solubility test can also aid in diagnosis of sickle cell disease. When a blood sample containing Hb S is added to a test solution containing saponin (to lyse cells) and sodium hydrosulfite (to deoxygenate the solution), a cloudy turbid suspension is formed if Hb S is present. If no Hb S is present, the solution remains clear. False-negative result may be observed if Hb S is <10% as often observed in infants younger than 3 months of age [13].

Table 9 Diagnostic approach of sickle cell hemoglobinopathy.

| Hemoglobin pattern | Diagnosis/Comments |
|---|---|
| A patient has Hb A and Hb S | Hb AS trait or Hb SS disease (post transfusion) or Hb S/beta+-thalassemia or a normal person transfused from a donor with Hb AS |
| | trait. Transfusion history is essential for diagnosis. For a patient with Hb AS trait, Hb A is majority and Hb S is 30%–40%; if donor was Hb S trait then S% is usually between 0.8% and 14% of the total hemoglobin. In Hb S/beta$^+$ thalassemia Hb $A_2$ is expected to be high and there should be microcytosis and hypochromia of the red cells. Hb A% is typically 5%–25% depending on severity of genetic defect |
| A patient has Hb S but no Hb A | Hb SS disease; Hb S/beta$^0$ thalassemia, Hb A2 is elevated with low MCV and MCH. |
| A patient has Hb S and high Hb F | Hb S/HPFH and Hb SS disease while patient is on hydroxyurea |
| | High MCV favors hydroxyurea; medication history will be required |

For diagnosis of Hb S/G hybrid on alkaline gel electrophoresis, one band is expected in the A lane, one band in the S lane (due to Hb S and Hb G), one band in the C lane (due to S/G hybrid), and one band in the carbonic anhydrase area (due to HbG2). Therefore, a total of four bands should be observed. If the band in the carbonic anhydrase is not prominent, at least three bands should be seen. On the acid gel electrophoresis, one band is expected in the A lane (due to Hb A, Hb G and Hb G2) and one band in the S lane (due to Hb S and HbS/G hybrid). In electrophoresis, a band should be seen in zone 5 (Hb S) and a band in zone 6 (Hb G). It is important to emphasize that for hemoglobinopathies, finding of gel electrophoresis must be confirmed by a second method, HPLC or capillary electrophoresis.

In the presence of Hb S, if a higher value of Hb F is observed, then HbS/HPFH can be suspected. In this case, CBC should be normal and Hb F should be between 25% and 35%. However, with Hb S/beta-thalassemia, Hb F could also be high. In HPFH and Hb S/HPFH, distribution of Hb F in red cells is normocellular, but in delta beta-thalassemia and Hb SS with high HbF, it is heterocellular. Kleihauer Betke tests or flow cytometry with anti F antibody will illustrate the difference. Interpretations of various other hemoglobinopathies are given in Table 10. Alternatively, if an initial band is present in C lane in alkaline gel, logical approach for diagnosis of hemoglobinopathies is given in Fig. 1. If an initial band is present in E lane in alkaline gel, diagnostic approaches for hemoglobinopathies are given in Fig. 2.

## Apparent hemoglobinopathy after blood transfusion

Blood transfusion history is essential in interpreting abnormal hemoglobin pattern because small peaks of abnormal hemoglobin may appear from blood transfusion. Apparent hemoglobinopathy after blood transfusion is rarely reported but it may cause diagnostics dilemma resulting in repeated unnecessary testing. Kozarski et al. reported 52 incidences of apparent hemoglobinopathies out of which 46 were Hb C, 4 were Hb S, and 2 were Hb O-Arab. The proportion of abnormal hemoglobin ranged from 0.8% to 14% (median: 5.6%). The authors recommended identifying and notifying the donor in such event [18].

## Universal newborn screening

Universal newborn screening for hemoglobinopathies is now required in all 50 states and District of Columbia. In addition, American College of Obstetricians and Gynecologist provides guideline for screening of couples who may be at risk of having children with hemoglobinopathy. Persons of Northern Europe,

**Table 10** Diagnostic approach to common hemoglobinopathies.

| Diagnosis | Features |
| --- | --- |
| Diagnosis of Hb C | Band in the C lane in the alkaline gel: possibilities are C, E, or O |
| | Band in the C lane in the acid gel |
| | HPLC shows a peak around 5 min with a small peak just before this main peak (HbC1d). A small peak may also be observed at 4.5 min (HbC1c) |
| | Or |
| | Capillary electrophoresis shows a peak in zone 2 |
| Diagnosis of Hb E | Band in the C lane in the alkaline gel: possibilities are C, E, or O |
| | Band in A lane in acid gel |
| | HPLC shows a peak at 3.5 min and is greater than 10% |
| | Or |
| | Capillary electrophoresis shows a peak in zone 4 |
| Diagnosis of Hb O | Band in the C lane in the alkaline gel: possibilities are C, E, or O |
| | Band between A and S lane in acid gel |
| | HPLC shows a peak between 4.5 and 5 min |
| | Or |
| | Capillary electrophoresis shows a peak in zone 3(O-Arab) |
| Diagnosis of Hb S | Band in the S lane in the alkaline gel; Possibilities S, D, G, Lepore |
| | Band in the S lane in the acid gel |
| | HPLC shows a peak at 4.5 min |
| | Or |
| | Capillary electrophoresis shows a peak in zone 5 |
| Diagnosis: Hb D | Band in the S lane in the alkaline gel; Possibilities S, D, G, Lepore |
| | Band in the A lane in acid gel |
| | HPLC shows a peak at 3.9 to 4.2 min; no additional peak |
| | Or |
| | Capillary electrophoresis shows a peak in zone 6 |
| Diagnosis: Hb G | Band in the S lane in the alkaline gel; Possibilities S, D, G, Lepore |
| | Band in the A lane in acid gel |
| | HPLC shows a peak at 3.9 to 4.2 min and a small additional peak (G2) |
| | Or |
| | Capillary electrophoresis shows a peak in zone 6 |
| Diagnosis of Hb Lepore[a] | Band (faint) in the S lane in the alkaline gel; Possibilities S, D, G Lepore |
| | Band in the A lane in acid gel |
| | HPLC shows a peak at 3.7 min ($A_2$ peak); quantity is lower than D or G or E. Small increase in HbF% |
| | Or |
| | Capillary electrophoresis shows a peak in zone 6 |

[a]Hb Lepore band in the alkaline gel is faint.

Japanese, Native Americans, or Korean descents are at low risk for hemoglobinopathies but people with ancestors from Southeast Asia, Africa, or Mediterranean are at higher risk. A complete blood count should be done to accurately measure hemoglobin. If all parameters are normal and the couple belongs to a low-risk group, no further testing may be necessary. For higher risk couples,

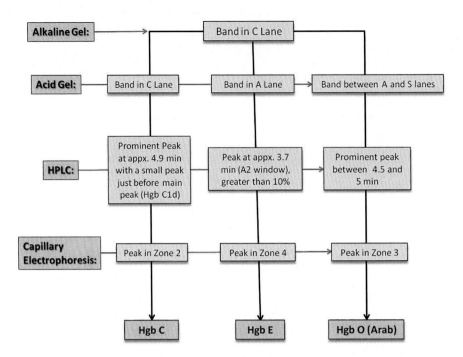

**FIG. 1**

Interpretation of hemoglobinopathy when a band is present in C lane in alkaline gel. *(Figure courtesy of Andres Quesda, M.D.)*

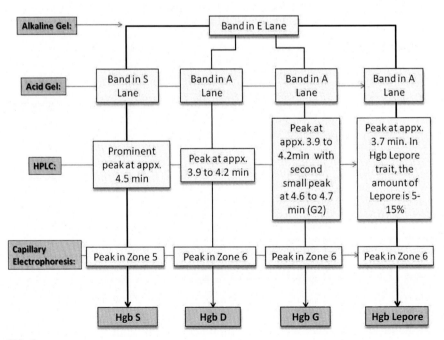

**FIG. 2**

Interpretation of hemoglobinopathy when a band is present in E lane in alkaline gel. *(Figure courtesy of Andres Quesda, M.D.)*

hemoglobin analysis by electrophoresis or other method is recommended. Solubility test for sickle cell may be helpful. Genetic screening can help physicians to identify couple at risk of having children with hemoglobinopathy. Molecular protocols for hemoglobinopathies started in 1970s using Southern blotting and restriction fragment length polymorphism analysis for prenatal sickle cell disease. With development of polymerase chain reaction (PCR), molecular testing for hemoglobinopathies now requires much less DNA for analysis [19]. However, currently, molecular testing for diagnosis of hemoglobinopathies, which certainly establishes a firm diagnosis, especially for alpha-thalassemia trait (direct gene analysis), is available in large academic medical centers and reference laboratories only.

## Key points

- The normal hemoglobin (HbA) in adults contains two alpha chains and two beta chains. Each alpha chain contains 141 amino acids in length and each beta chain contains 146 amino acids. Hemoglobin $A_2$ (Hb$A_2$) contains two alpha chains and two delta chains. The gene for the alpha chain is located in chromosome 16 (two genes in each chromosome, a total of four genes), while genes for beta (one gene in each chromosome, a total of two genes), gamma, and delta chains are located on chromosome 11.
- When hemoglobin is circulating with erythrocytes, glycosylation of the globin chains may take place. These are referred to as X1c (X being any hemoglobin, for example, HbA1c). When hemoglobin molecule is aging, glutathione is bound to cysteine at the 93rd position of the beta chain. This is HbA3 or HbA1d. Just like Hb A1c, there can exist HbC1c, HbC1d, Hb S1c, and HbS1d.
- Heme is synthesized in a complex way involving enzymes in both mitochondrion and cytosol.
- Hemoglobinopathies can be divided into three major categories: Quantitative disorders of hemoglobin synthesis: Production of structurally normal but decreased amounts of globin chains (Thalassemia syndrome), Disorder (qualitative) in hemoglobin structure: Production of structurally abnormal globulin chains such as Hemoglobin S, C, O, or E. Sickle cell syndrome is the most common example of such disease and failure to switch globin chain synthesis after birth: Hereditary persistence of fetal hemoglobin (Hb F), relatively benign condition, may coexist with thalassemia or sickle cell disease but decreased severity of such diseases (protective effect).
- Hemoglobinopathies are transmitted in autosomal recessive fashion.

- Disorders due to beta-chain defect such as sickle cell disease tend to manifest clinically after 6 months of age, although diseases due to alpha-chain defect are manifested in utero or following birth.
- The hemoglobin variants of most clinical significance are hemoglobin S, C, and E. In West Africa, approximately 25% of individuals are heterozygous for hemoglobin S (Hb S) gene which is related to sickle cell diseases. In addition, high frequencies of Hb S gene alleles are also found in people of Caribbean, South and Central Africa, Mediterranean, Arabian Peninsula, and East India. Hemoglobin C (Hb C) is found mostly in people living or originating from West Africa. Hemoglobin E (Hb E) is widely distributed between East India and Southeast Asia with highest prevalence in Thailand, Laos, and Cambodia, but may be sporadically observed in parts of China and Indonesia.
- Thalassemia syndrome is not due to structural defect in globin chain but due to lack of sufficient synthesis of globin chain and is also a genetically inherited disease. Thalassemic syndrome can be divide into alpha ($\alpha$)-thalassemia and beta ($\beta$)-thalassemia. In general, $\beta$-thalassemia is observed in Mediterranean, Arabian Peninsula, Turkey, Iran, West and Central Africa, India, and Southeast Asian countries while $\alpha$-thalassemia is commonly observed in parts of Africa, Mediterranean, Middle East, and throughout Southeast Asia.

Alpha-thalassemia occurs when there is a defect or deletion in one or more of four genes responsible for alpha-globin production. Alpha-thalassemia can be divided into four categories:

- The Silent Carriers: Characterized by only one defective or deleted gene but three functional genes. These individuals have no health problem. An unusual case of silent carrier is individuals carrying one defective Constant Spring mutation but three functional genes. These individuals also have no health problem.
- Alpha-Thalassemia Trait: Characterized by two deleted or defective genes and two functional genes. These individuals may have mild anemia.
- Alpha-thalassemia major (Hemoglobin H Disease): Characterized by three deleted or defective genes and only one functional gene. These patients have persistent anemia and significant health problems. When hemoglobin H disease is combined with hemoglobin Constant Spring, the severity of disease is more than hemoglobin H disease. However, if the child inherits one hemoglobin Constant Spring from mother and one from father, then the child has homozygous hemoglobin Constant Spring and severity of disease is similar to hemoglobin H disease.
- Hydrops Fetalis: Characterized by no functional alpha gene and these individuals have hemoglobin Bart. This condition is not compatible with life unless intrauterine transfusion is initiated.

- Hemoglobin Constant Spring (hemoglobin variant isolated from a family of ethnic Chinese background from Constant Spring district of Jamaica) is hemoglobin variant where mutation of alpha-globin gene produces an abnormally long alpha chain (172 amino acids instead of normal 146 amino acids). Hemoglobin Constant Spring is due to nondeletion mutation of alpha gene which results in production of unstable alpha-globin. Moreover, this alpha-globin is produced in very low quantity (approximately 1% of normal expression level) and is found in people living or originating in Southeast Asia.
- Beta-thalassemia can be broadly divided into three categories:
- Beta-Thalassemia Trait: Characterized by one defective gene and one normal gene and individuals may experience mild anemia but are not transfusion dependent.
- Beta-Thalassemia Intermedia: Characterized by two defective genes but some beta-globin production is still observed in these individuals. However, some individuals may have significant health problems requiring intermittent transfusion.
- Beta-Thalassemia Major (Cooley's anemia): Characterized by two defective genes but almost no function of either gene leading to no synthesis of beta-globin. These individuals have severe form of disease requiring lifelong transfusion and may have shortened life span.
- Patients with beta-thalassemia major have elevated $HbA_2$ and HbF (although in some individual HbF may be normal).
- In the heterozygous form (Hb AS), sickle cell trait protects from infection of *Plasmodium falciparum* malaria but not in the more severe form of homozygous sickle cell disease (Hb SS). The genetic defect-producing sickle hemoglobin is a single nucleotide substitution at codon 6 of the beta-globin gene on chromosome 11 that results in a point mutation in beta-globin chain of hemoglobin (substitution of valine for glutamic acid at sixth position).
- Double heterozygous states of Hb SC, Hb SD, and Hb SO Arab are important sickling states which should not be missed.
- Hemoglobin C is formed due to substitution of glutamic acid residue with a lysine residue at the 6th position of beta-globin. Hemoglobin E is due to point mutation of beta-globin which results in substitution of lysine for glutamic acid in position 26.
- Hemoglobin Lepore is an unusual hemoglobin molecule which is composed of two alpha chains and two delta beta chains as a result of fusion gene of delta and beta genes. The delta beta chains have the first 87 amino acids of the delta chain and 32 amino acids of the beta chain.

Individuals with HbA/ Hb Lepore are asymptomatic with Hb Lepore representing 5%–15% of hemoglobin, slightly elevated Hb F (2%–3%), and with MCV as well as MCH. However, homozygous Lepore individuals suffer from severe anemia similar to patients with beta-thalassemia intermedia with Hb Lepore representing 8%–30% of hemoglobin and remainder hemoglobin is Hb F.

- Hemoglobin G-Philadelphia (Hb G) is the commonest alpha chain defect observed 1 in 5000 of African Americans and is associated with alpha-thalassemia 2 deletions.
- It is possible that an African American individual may have Hb S/Hb G where hemoglobin molecule contains one normal alpha chain, one alpha G chain, one normal beta chain, and one beta S chain. This can result in detection of various types of hemoglobin in the blood including Hb A (alpha2, beta 2), Hb S (alpha2, beta S2), Hb G (alpha G2, beta 2), and HbS/G (alpha G2, beta S2). In addition, Hb G2 (alpha2, delta2), which is the counterpart of $HbA_2$, is also present.
- Increase in fetal hemoglobin percentage is associated with multiple pathologic states. These include beta-thalassemia, delta beta-thalassemia, and hereditary persistence of fetal hemoglobin (HPFH). While beta-thalassemia is associated with high $HbA_2$, the latter two states are associated with normal $HbA_2$ values.
- Hematologic malignancies are associated with increased hemoglobin F and include acute erythroid leukemia (AML, M6) and juvenile myelomonocytic leukemia (JMML). Aplastic anemia is also associated with an increase in percentage of Hb F. In elucidating the actual cause of high Hb F, it is important to consider the actual percentage of Hb F, $HbA_2$ values, as well as correlation with complete blood count (CBC) and peripheral smear findings. It is also important to note that drugs (hydroxyurea, sodium valproate, and erythropoietin) and stress erythropoiesis may also result in high Hb F. Hydroxyurea is used in sickle cell disease patients to increase the amount of Hb F, the presence of which may help to reduce the clinical effects of the disease. Measuring the level of Hb F may be useful in determining the appropriate dose of hydroxyurea. In 15%–20% of cases of pregnancy, HbF may be raised to values as much as 5%.

## References

[1] Rahimi Z. Genetic, epidemiology, hematological and clinical features of hemoglobinopathies in Iran. Biomed Res Int 2013;2013:803487.

[2] Faulkner LB, Uderzo C, Masera G. International cooperation for the cure and prevention of severe hemoglobinopathies. J Pediatr Hematol Oncol 2013;35:419–23.

[3] Manning LR, Russell JR, Padovan JC, Chait BT, et al. Human embryonic, fetal and adult hemoglobins have different subunit interface strength. Correlation with lifespan in the red cell. Protein Sci 2007;16, 164101658.

[4] Goonasekera HW, Paththinige CS, Dissanayake VHW. Population screening for hemoglobinopathies. Annu Rev Genomics Hum Genet 2018;19:355–80.

[5] Giordano PC. Strategies for basic laboratory diagnostics of the hemoglobinopathies in multi-ethnic societies: interpretation of results and pitfalls. Int J Lab Hematol 2013;35:465–79.

[6] Modell B, Darlison M. Global epidemiology of hemoglobin disorders and derived service indicators. Bull World Health Organ 2008;86:480–7.

[7] Rappaport VJ, Velazquez M, Willaims K. Hemoglobinopathies in pregnancy. Obset Gynecol Clin N Am 2004;31:287–317.

[8] Sriiam S, Leecharoenkiat A, Lithanatudom P, Wannatung T, et al. Proteomic analysis of hemoglobin H Constant Spring (HB H-CS) erythroblasts. Blood Cells Mol Dis 2012;48:77–85.

[9] Farashi S, Harteveld CL. Molecular basis of α-thalassemia. Blood Cells Mol Dis 2018;70:43–53.

[10] Tang HS, Wang DG, Huang LY, Li DZ. δ-Thalassemia with complete absence of Hb A 2 in a Chinese family. Hemoglobin 2018;42:135–7.

[11] Piel FB, Steinberg MH, Rees DC. Sickle cell disease. N Engl J Med 2017;376:1561–73.

[12] Pinto VM, Balocco M, Quintino S, Forni GL. Sickle cell disease: a review for the internist. Intern Emerg Med 2019;14:1051–64.

[13] Lubin B, Witkowska E, Kleman K. Laboratory diagnosis of hemoglobinopathies. Clin Biochem 1991;24:363–74.

[14] Ngo DA, Aygun B, Akinsheye I, Hankins JS. Fetal hemoglobin levels and hematological characteristics of compound heterozygous for hemoglobin S and deletional hereditary persistence of fetal hemoglobin. Br J Haematol 2012;156:259–64.

[15] Pandey S, Mishra RM, Pandey S, Shah V, et al. Molecular characterization of hemoglobin D Punjab traits and clinical-hematological profile of patients. Sao Paulo Med J 2012;130:248–51.

[16] Dror S. Clinical and hematological features of homozygous hemoglobin O-Arab [beta 121 Glu – LYS]. Pediatr Blood Cancer 2013;60:506–7.

[17] Smaldone A. Glycemic control and hemoglobinopathy: when A1C may not be reliable. Diabetes Spectrum 2008;21:46–9.

[18] Kozarski TB, Howanitz PJ, Howanitz JH, Lilic N, et al. Blood transfusion leading to apparent hemoglobin C, S and O-Arab hemoglobinopathies. Arch Pathol Lab Med 2006;130:1830–3.

[19] Mankhemthong K, Phusua A, Suanta S, Srisittipoj P, et al. Molecular characteristics of thalassemia and hemoglobin variants in prenatal diagnosis program in northern Thailand. Int J Hematol 2019;110:474–81.

# CHAPTER 24

# Protein electrophoresis and immunofixation

## Introduction

Serum protein electrophoresis, urine protein electrophoresis, serum immunofixation, and urine immunofixation are all performed primarily to investigate suspicion of monoclonal gammopathy. Monoclonal gammopathy is present when a monoclonal protein (M protein or paraprotein) is identified in a patient's serum, urine, or both. Monoclonal gammopathy could be monoclonal gammopathy of unknown significance, smoldering multiple myeloma, or multiple myeloma. Multiple myeloma is a neoplasm of clonal plasma cells which originate from the postgerminal lymphoid B-cell lineage and develop after lineage commitment in the bone marrow of progenitor cells. Multiple myeloma accounts for approximately 10% of hematologic malignancies and is the second most common hematologic malignancy in the world. The median age of onset is 66 years, and only 2% of patients are less than 40 years of age at diagnosis [1].

Multiple myeloma evolves from a premalignant condition known as monoclonal gammopathy of undetermined significance (MGUS; term coined in 1978), which is present in 3%–4% of the general population over the age of 50 years. Studies have shown that MGUS almost always precedes multiple myeloma with a risk of progression of approximately 1% per year [2]. The most common heavy chain subtype in MGUS is IgG, which is found in approximately 70% of patients, followed by IgM (15%), IgA (12%), and biclonal gammopathy (3%). MGUS is more common in men than women. Patients with IgM MGUS have an increased risk of developing non-Hodgkin lymphoma, chronic lymphocytic leukemia, immunoglobulin light chain (AL) amyloidosis, and Waldenstrom macroglobulinemia at a rate of 1.5% per year. However, patients with IgG or IgA MGUS may progress mainly to multiple myeloma. Interestingly, patients with an IgA MGUS have a greater probability for progression to multiple myeloma. Smoldering multiple myeloma (term coined in 1980) is an intermediate stage between MGUS and multiple myeloma, and is associated with a higher risk of progression of approximately 10% per year [3].

Clinical Chemistry, Immunology and Laboratory Quality Control. https://doi.org/10.1016/B978-0-12-815960-6.00021-2

Although MGUS with heavy chain (IgG, IgA, IgM) has much higher prevalence, light chain MGUS, the premalignant precursor of light chain multiple myeloma, has a prevalence of only 0.8% in population 50 years of age or older. Light chain MGUS may progress to light chain multiple myeloma or AL amyloidosis [4].

MGUS and smoldering multiple myeloma are asymptomatic plasma cell disorders, while multiple myeloma is a malignant neoplasm usually of a single clone of plasma cells which often proliferate diffusely throughout the bone marrow. However, these cells may also form a tumor known as plasmacytoma. In general, in MUGS, concentration of monoclonal protein (IgG, IgA, or IgM) should be <3 g/dL and monoclonal plasma cells in the bone marrow should be <10%. Moreover, patients must be asymptomatic. In smoldering myeloma, serum monoclonal protein concentration should be ≥3 g/dL and monoclonal plasma cells in bone marrow should be 10%–60%, but patient must be asymptomatic. In contrast, multiple myeloma is defined as the presence of end organ (classic "CRAB" criteria) damage in parallel with the presence of a paraprotein spike or monoclonal plasma cells or biopsy-proven bony or extramedullary plasmacytoma. The CRAB criteria includes: (C) hypercalcemia, calcium >11 mg/dL; (R) renal insufficiency, GFR <40 mL/min and/or elevated creatinine; (A) anemia, hemoglobin 2.0 g/dL lower than the limit of normal; and (B) bone lesions [5].

The paraprotein can be an intact immunoglobulin (one of the five immunoglobulin class: IgG, IgA, IgM, IgD, or IgE), only kappa or lambda light chain (κ or λ; light chain myeloma, light chain deposition disease, or amyloid light chain amyloidosis), or rarely found only as heavy chains (heavy chain disease). In general, IgG is the most commonly encountered paraprotein, while IgD is encountered infrequently and IgE is reported rarely. In one study based on 630 patients with paraproteins, IgG kappa was observed in 230 patients, IgG lambda in 99 patients, IgA kappa in 45 patients, IgA lambda in 19 patients, IgM kappa in 91 patients, IgM lambda in 28 patients followed by kappa light chain in 36 patients and lambda light chain in 20 patients. Biclonal gammopathy (discussed later) was observed in 62 patients [6].

Paraproteins can be detected in the serum and can also be excreted into the urine. Sometimes, if the paraprotein is only light chain (light chain disease) then this is detected only in urine but not in the serum. The serum may paradoxically exhibit only hypogammaglobulinemia. It is important to note that the presence of paraprotein in serum, urine, or both indicates monoclonal gammopathy and not necessarily the presence of multiple myeloma in a patient. Multiple myeloma is one of the causes of monoclonal gammopathy. Transient monoclonal gammopathy may be observed in an immunocompromised patient suffering from infection due to an opportunistic pathogen such as cytomegalovirus [7]. Monoclonal gammopathy is usually observed in patients over 50 years of age and is rare in children. Gerritsen et al. studied 4000 pediatric patients over a 10-year period and observed monoclonal gammopathy only in 155 children, but such gammopathies were found most frequently in patients suffering from primary and secondary immunodeficiency, hematological malignancies, autoimmune disease, and severe aplastic anemia. Follow-up analysis revealed that most of these monoclonal gammopathies were transient [8].

Agarose gel electrophoresis and capillary electrophoresis are two principal methods employed in screening for paraproteins. Both methods are applicable for both serum and urine specimens. Once a paraprotein is detected, confirmation and isotyping of paraprotein are essential which is usually achieved by immunofixation. For urine immunofixation, the best practice is to utilize a 24-h urine specimen which has been concentrated; such technique allows for detection of even a faint band.

Despite significant advancements in the understanding of molecular mechanisms, diagnostic methods, prognostication, and the treatment options in multiple myeloma over the last decade, this disease remains a heterogeneous disease with differing outcomes. In the last few years, serum monoclonal proteins including the serum-free light chain assays, imaging, and cytogenetics have been used to predict the outcomes of multiple myeloma patients receiving different types of therapies. In addition, liquid biopsies using circulating tumor cells, tumor DNA, and novel immune biomarkers are potentially being investigated. These novel potential biomarkers not only accurately detect the mutational status of different cancers compared to standard methods, but also serve as prognostic and predictive biomarkers for disease relapse and response to therapy [9]. However, detailed discussion on these novel biomarkers is beyond the scope of this book.

## Serum protein electrophoresis

Serum protein electrophoresis (SPEP) is an inexpensive, easy to perform screening procedure for initial identification of monoclonal bands (paraproteins). Monoclonal bands are usually seen in the gamma zone but may be seen in proximity of the beta band or rarely in the alpha-2 region. Blood can be collected in a tube with clot activator, and after separation from blood components, serum is then applied to a buffer-saturated support medium, such as agarose, which is then placed on an electrophoretic apparatus that consists of buffer reservoirs and positive and negative electrodes. The application of electric current to the supporting medium separates the proteins according to their net charge at a standard alkaline pH of 8.6. After a predetermined time of exposure to an electric field, the special paper is removed, dried, placed on a fixative to prevent further diffusion of specimen components followed by staining to visualize various protein bands. Coomassie brilliant blue is a common staining agent to visualize band in serum protein electrophoresis. Then using a densitometer, each fraction is quantitated. The serum protein components are separated into five major fractions:

- albumin
- alpha-1 globulin (alpha-1 zone)
- alpha-2 globulin (alpha-2 zone)
- beta globulins (beta zone often splits into beta-1 and beta-2 bands)
- gamma globulins (gamma zone)

Albumin and globulins are two major fractions of electrophoresis pattern. Albumin, the largest band, lies closest to the positive electrode (anode) due to its negative charge, while globulins tend to move toward negative electrode (cathode). Albumin has a molecular weight of approximately 67 kDa (67,000 Da). Reduced intensity of this band is observed in inflammation, liver dysfunction, uremia, nephrotic syndrome, and other conditions that lead to hypoalbuminemia, such as critically ill patients and pregnancy. A smear observed in front of the albumin band may be due to hyperbilirubinemia or due to presence of certain drugs. A band in front of the albumin band may be due to prealbumin (a carrier for thyroxine and vitamin A), which is commonly seen in cerebrospinal fluid specimens or serum specimens in patients with malnutrition. Two, rather than one, albumin bands may represent bisalbuminemia. This is a familial abnormality with no clinical significance.

Congenital analbuminemia (CAA) is an inherited, autosomal recessive disorder with an incidence of 1:1,000,000 live births. Affected individuals have a highly significant decreased concentration, or complete absence, of serum albumin, a trait confirmed by serum protein electrophoresis and immunochemistry technique. Moreover, analysis of the albumin gene is necessary for the molecular diagnosis. This disease may cause serious consequences in the prenatal period, and may lead to death mainly from fluid retention and infection in lower respiratory tract during neonatal period and early childhood. In contrast, CAA is better tolerated during adulthood. Clinically, in addition to the low level of albumin, the patients usually have hyperlipidemia, but they usually also have mild edema, reduced blood pressure, and fatigue. The fairly mild symptoms in adulthood are due to compensatory increment of other plasma proteins [10]. Pseudo-analbuminemia due to the presence of a slow moving albumin variant appearing in the alpha-1 region of serum protein electrophoresis has also been reported [11].

Moving toward the negative electrode (cathode), the alpha zone is the next band after albumin followed by beta and gamma zones. The alpha zone can be subdivided into two zones: the alpha-1 band and alpha-2 band. The alpha-1 band mostly consists of alpha-1 antitrypsin (90%), alpha-1-antichymotrypsin (faint band), and thyroid-binding globulin. Alpha-1 antitrypsin is an acute phase reactant and its concentration is increased in inflammation and other conditions. The alpha-1 antitrypsin band is decreased in patients with alpha-1 antitrypsin deficiency or decreased production of globulin in patients with severe liver disease. At the leading edge of this band, a haze due to high-density lipoprotein (HDL) may be observed, though different stains are used (Sudan Red 7B or Oil Red O) for lipoprotein analysis using electrophoresis. The alpha-2 band consists of alpha-2 macroglobulin, haptoglobin, and ceruloplasmin. Because both haptoglobin and ceruloplasmin are acute phase reactants, this band is increased in inflammatory states. Alpha-2 macroglobulin is increased in nephrotic syndrome and cirrhosis of liver.

The beta zone may consist of two bands, beta-1 and beta-2. Beta-1 is mostly composed of transferrin and low-density lipoprotein. An increased beta-1 band is observed in iron deficiency anemia due to increased level of transferrin. This band may also be elevated in pregnant women. Very low-density lipoprotein usually appears in the prebeta zone. The beta-2 band is mostly composed of complement proteins. If two bands are observed in the beta-2 region, it implies either electrophoresis of plasma specimen (fibrinogen band) instead of serum specimen or IgA paraprotein.

Much of the clinical interest of serum protein electrophoresis is focused on the gamma zone because immunoglobulins mostly migrate to this region. Usually the C-reactive protein band is found between the beta and gamma regions. Serum protein electrophoresis is most commonly ordered when multiple myeloma is suspected and observation of a monoclonal band (paraprotein) indicates that monoclonal gammopathy may be present in the patient. If paraprotein is observed in serum protein electrophoresis, the following steps are performed:

- The monoclonal band is measured quantitatively using densitometric scan of the gel.
- Serum and/or urine immunofixation is conducted to confirm the presence of the paraprotein as well as determining the isotype of the paraprotein.
- A serum light chain assay is conducted or recommended to the ordering clinician.

Monoclonal gammopathy can be due to various underlying diseases including multiple myeloma. When two distinct paraprotein bands are observed, it is biclonal gammopathy. Biclonal gammopathies can result from proliferation of two clones of plasma cells with each producing an unrelated monoclonal protein or from the production of two monoclonal proteins by a single clone of plasma cells. Approximately 1.5% to 3% of multiple myeloma cases present with biclonal paraproteinemia [12]. Mullikin et al. reported that 23 out of 393 patients with biclonal gammopathy of unknown significance in their study had progression to either multiple myeloma, smoldering myeloma, light chain amyloidosis, or Waldenstrom macroglobulinemia with the dominant clone being the principal player through the course of the disease in 21 patients [13]. Several cases of isotype switching have been reported in myeloma patients with biclonal spikes.

A patient may also have nonsecretory myeloma, as in the case of a plasma cell neoplasm, in which the clonal cells neither produce nor secrete paraproteins. When a monoclonal band is identified using serum protein electrophoresis, serum immunofixation and 24-h urine immunofixation are typically recommended. There are certain situations where a band may be apparent, but in reality, it is not a monoclonal band. Examples include:

- Fibrinogen is seen as a discrete band when electrophoresis is performed on plasma instead of serum specimen. This fibrinogen band is seen between the beta and gamma regions. If the electrophoresis is repeated after the addition of thrombin, this band should disappear. In addition, immunofixation study should be negative.
- Intravascular hemolysis results in release of free hemoglobin in circulation which binds to haptoglobin. The hemoglobin-haptoglobin complex may appear as a large band in the alpha-2 area. Serum immunofixation studies should be negative in such cases.
- In patients with iron deficiency anemia, concentrations of transferrin may be high which may result in a band in the beta region. Again, immunofixation should be negative.
- Patients with nephrotic syndrome usually show low albumin and total protein, but this condition may also produce increased alpha-2 and beta fractions. Bands in either of these regions may mimic a monoclonal band.
- When performing gel electrophoresis, a band may be visible at the point of application. Typically, this band is present in all samples performed at the same time using same agarose gel support material.

Common problems associated with interpretation of serum protein electrophoresis are summarized in Table 1. A low concentration of a paraprotein may not be detected by serum electrophoresis. There are also certain situations where a false-negative interpretation could be made on serum electrophoresis. These situations include:

- A clear band is not seen in case of alpha heavy chain disease (HCD). This is presumably due to the tendency of these chains to polymerize. Heavy chain diseases are rare B-cell lymphoproliferative neoplasm characterized

---

**Table 1** Common problems associated with serum protein electrophoresis.

Serum protein electrophoresis performed using plasma instead of serum produces an additional distinct band between beta and gamma zones due to fibrinogen but such band is absent in subsequent immunofixation study.

A band may be seen at the point of application. Typically, this band is present in all samples performed at the same time.

If concentration of transferrin is high (for example, iron deficiency), a strong band in the beta region is observed.

In nephrotic syndrome, prominent bands may be seen in alpha-2 and beta regions which are not due to monoclonal proteins.

Hemoglobin-haptoglobin complexes (seen in intravascular hemolysis) may produce a band in the alpha-2 region.

Paraproteins may form dimers, pentamers, polymers, or aggregates with each other resulting in a broad smear rather than a distinct band.

In light chain myeloma, light chains are rapidly excreted in the urine and no corresponding band may be present in serum protein electrophoresis.

---

by the production of a monoclonal component consisting of monoclonal immunoglobulin heavy chain without associated light chain.

- In mu HCD a localized band is found in only 40% of cases. Panhypogammaglobulinemia is a prominent feature in such patients.
- In occasional cases of gamma HCD, a localized band may not be seen.
- When a paraprotein forms dimers, pentamers, polymers, or aggregates with each other or when forming complexes with other plasma components, a broad smear may be visible instead of a distinct band.
- Some patients may produce only light chains which are rapidly excreted in the urine and no distinct band may be present in the serum protein electrophoresis. Urine protein electrophoresis is more appropriate for the diagnosis of light chain disease. When light chains cause nephropathy and result in renal insufficiency, excretion of the light chains are hampered and a band may be seen in serum electrophoresis.
- In some patients with Ig D myeloma, the paraprotein band may be very faint.

Hypogammaglobulinemia may be congenital or acquired. Among the acquired causes are multiple myeloma and primary amyloidosis. Panhypogammaglobulinemia can occur in about 10% of cases of multiple myeloma. Most of these patients have a Bence Jones protein in the urine, but lack intact immunoglobulins in the serum. Bence Jones proteins are monoclonal free kappa or lambda light chains in the urine. Detection of Bence Jones protein may be suggestive of multiple myeloma or Waldenstrom macroglobulinemia. Panhypogammaglobulinemia can also be seen in 20% of cases of primary amyloidosis. It is important to recommend urine immunofixation studies when panhypogammaglobulinemia is present in serum protein electrophoresis.

**Table 2** Abnormal serum protein electrophoresis pattern due to various diseases other than monoclonal gammopathy.

| Disease | Abnormal pattern | | | | |
|---|---|---|---|---|---|
| | Albumin | Alpha-1 | Alpha-2 | Beta | Gamma |
| Acute inflammation | Reduced | Increased | Increased | No change | No change |
| Chronic inflammation | Reduced | Increased | Increased | No change | Increased |
| Nephrotic syndrome[a] | Reduced | No change | Increased | Increased[b] | No change |
| Liver disease/cirrhosis[a] | Reduced | No change | No change | Beta-gamma bridging | Increased |
| Polyclonal gammopathy | No change | No change | No change | Increased | No change |

[a]Total protein is reduced.

[b]Increased beta zone due to secondary hyperlipoproteinemia.

Although monoclonal gammopathy is the major reason for serum protein electrophoresis, polyclonal gammopathy may also be observed in some patients. Monoclonal gammopathies are associated with a clonal process that is malignant or potentially malignant. However, polyclonal gammopathy, in which there is a nonspecific increase in gamma globulins, may not be associated with malignancies. Many conditions may lead to polyclonal gammopathies. Serum protein electrophoresis may also exhibit changes, which imply specific underlying clinical conditions other than monoclonal gammopathy. Common features of serum protein electrophoresis in various disease states other than monoclonal gammopathy include:

- Inflammation: Increased intensity of alpha-1 and alpha-2 regions, sharp leading edge of alpha-1 may be observed, but with chronic inflammation the albumin band may be decreased with increased gamma zone due to polyclonal gammopathy.
- Nephrotic syndrome: In nephrotic syndrome, the albumin band is decreased due to hypoalbuminemia. In addition, the alpha-2 band may be more distinct.
- Cirrhosis or chronic liver disease: A low albumin band due to significant hypoalbuminemia with a prominent beta-2 band and beta-gamma bridging are characteristics features of liver cirrhosis or chronic liver disease. In addition, polyclonal hypergammaglobulinemia is observed.

Various clinical conditions other than monoclonal gammopathy leading to abnormal patterns in serum protein electrophoresis are listed in Table 2.

## Urine electrophoresis

Urine protein electrophoresis is analogous to the serum protein electrophoresis and is used to detect monoclonal proteins in the urine. Ideally it should be performed on a 24-h urine sample (concentrated 50–100 times). Molecules <15 kDa are filtered through a glomerular filtration process and are excreted freely into urine. In contrast, only selected molecules with molecular weight between 16 and 69 kDa can be filtered by the kidney and may appear in the urine. Albumin is approximately 67 kDa. Therefore, trace albumin in urine is physiological.

Molecular weight of the protein, concentration of the protein in the blood, charge, and hydrostatic pressure all regulate passage of a protein through the glomerular filtration process.

Proteins that pass through glomerular filtration include albumin, alpha-1 acid glycoprotein (orosomucoid), alpha-1 microglobulin, beta-2 microglobulin, retinol-binding protein, and trace amounts of gamma globulins. However, 90% of these are reabsorbed and only a small amount may be excreted in the urine.

Normally total urinary protein is <150 mg/24 h and consists of mostly albumin and Tamm-Horsfall protein (secreted from ascending limb of loop of Henle). The extent of proteinuria can be assessed by quantifying the amount of protein-uria as well as expressing as protein to creatinine ratio. Normal protein to creatinine ratio is <0.5 in children 6 months to 2 years of age, <0.25 in children above 2 years, and <0.2 in adults.

Proteinuria with minor injury (typically only albumin is lost in urine) can be related to vigorous physical exercise, congestive heart failure, pregnancy, alcohol abuse, or hyperthermia. Overflow proteinuria can be seen in patients with myeloma or massive hemolysis of crush injury (myoglobin in urine). In addition, beta-2 microglobulin, eosinophil-derived neurotoxin, and lysozyme can produce bands in urine electrophoresis. Therefore, immunofixation studies are required to document true paraproteins and ruling out the presence of other proteins in urine electrophoresis.

Proteinuria can be classified as glomerular, tubular, or combined proteinuria. Glomerular proteinuria can be subclassified as selective glomerular proteinuria (urine will have albumin and transferrin bands) or nonselective glomerular proteinuria (urine demonstrate presence of all different types of proteins). In glomerular proteinuria, the dominant protein present is always albumin. In tubular proteinuria, albumin is a minor component. The presence of alpha-1 microglobulin and beta-2 microglobulin are indicators of tubular damage. Please see Chapter 11 for more detail.

## Immunofixation studies

Immunofixation combines the resolution of serum protein electrophoresis and the specificity of an antigen-antibody precipitation reaction. This step is usually performed after serum protein electrophoresis to investigate the nature of paraprotein. In immunofixation, serum specimen from a patient suspected of monoclonal gammopathy is applied to replicate lanes (six lanes total, one lane is untreated) followed by conventional serum protein electrophoresis as described earlier in the chapter. However, after electrophoresis, instead of staining, one lane is untreated (to produce regular serum protein electrophoresis), but each of other five lanes are treated with a specific antibody. Therefore, one lane is treated with antibody specific to IgG, next lane with IgA-specific antibody, following lane with IgM-specific antibody. The two remaining lanes are each treated with specific antibody against each light chain (kappa and lambda). If a particular paraprotein is present, it will produce insoluble precipitate due to formation of antigen-antibody complex. After washing to remove unreacted proteins the entire gel is stained with Coomassie brilliant blue or amido black. Then gel is interpreted manually. Although serum is the most common sample source, urine or cerebrospinal fluid can also be examined to detect and/or characterize paraproteins or their fragments.

A normal serum protein electrophoresis does not exclude diagnosis of myeloma because approximately 11% myeloma patients may have normal serum protein electrophoresis. Therefore, serum and urine immunofixation studies should be performed regardless of serum electrophoresis results if clinical suspicion is high. It is also important to note that paraprotein in serum protein electrophoresis may not be a true band unless it is identified by using serum or urine immunofixation since these tests are more sensitive than serum protein electrophoresis. In addition, immunofixation technique can also determine the particular isotype of the monoclonal protein. However, immunofixation technique cannot estimate the quantity of the paraprotein. In contrast, serum protein electrophoresis is capable of estimating the concentration of a paraprotein.

In multiple myeloma, sometimes only free light chains are produced. The concentration of the light chains in serum may be so low that these light chains remain undetected using serum protein electrophoresis and even serum immunofixation. In such cases immunofixation on a 24-h urine sample is useful. Another available test is detection of serum-free light chains by immunoassay, which allows for calculating the ratio of kappa to lambda free light chains. This test is more sensitive than urine immunofixation.

One source of possible error in urine immunofixation study is the "step ladder" pattern. Here multiple bands are seen in the kappa (more often) or lambda lanes and are indicative of polyclonal spillage rather than monoclonal spillage into the urine. During urine immunofixation, five or six faint, regular diffuse bands with hazy background staining between bands may be seen. This is referred to as the step ladder pattern and is a feature of polyclonal hypergammaglobulinemia with spillage into the urine.

# Capillary zone electrophoresis

Capillary zone electrophoresis is a faster and alternative method of gel electrophoresis. Gel electrophoresis is often carried out in a buffer at pH 8.6, resulting in most proteins having an overall negative charge. Detection of proteins is accomplished through visualization using stains, such as Coomassie brilliant blue, and visible bands are quantified by densitometry.

Capillary electrophoresis also known as capillary zone electrophoresis is performed in a capillary tube instead of a conventional gel and does not require staining. Capillary electrophoresis device consists of a high voltage power supply, a sample introduction system, a capillary tube, a detector, and an output device. Capillary tubes are usually composed of fused silica with an internal diameter of only 20–100 μm. Specimens are applied to capillary at the anode and proteins in the presence of an alkaline liquid buffer move from anode to cathode and are detected usually with 200 nm wavelength (UV detection). Capillary electrophoresis is popular in clinical laboratories because of fast analysis time and utilization of small sample volume compared to agarose gel electrophoresis. Serum protein separation pattern using capillary electrophoresis is similar to conventional gel electrophoresis [14]. If a paraprotein is present, it produces a distinct peak in the gamma region. However, subtle changes in the gamma zones (distortion) may also represent underlying monoclonal gammopathy. Interpretation can be subjective and a relatively high percentage of cases may be referred for ancillary studies such as immunofixation depending on preference of the pathologist interpreting the result. However, disregarding a subtle change in capillary zone electrophoresis may potentially results in missing a case which is a limitation of capillary electrophoresis compared to convention gel electrophoresis. Capillary electrophoresis pattern of a normal serum is given in Fig. 1.

The detection of proteins in capillary electrophoresis is often based on UV detection at 200 nm. The radio-opaque agents used in imaging study absorb at the same wavelength and may appear as a monoclonal band. However, this interference is not present in gel electrophoresis or immunofixation. Certain antibiotics such as piperacillin-tazobactam, ceftriaxone, 5-flurocystosine, and sulfamethoxazole administration also may cause interference in capillary electrophoresis due to absorbance at UV wavelength, producing a monoclonal band in the α or β regions. Immunofixation can be utilized to rule out interference due to antibiotics [15]. Monoclonal immunoglobulins also directly interfere with serum albumin measurement by the capillary zone electrophoresis method leading to a systematic overestimation of serum albumin concentrations proportional to the serum monoclonal immunoglobulin level [16].

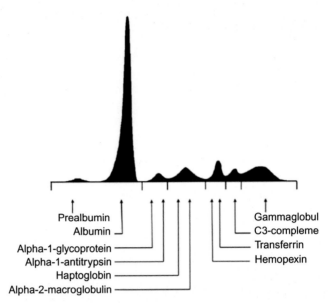

**FIG. 1**

Major proteins present in capillary serum electrophoresis. *(From Regeniter A, Siede WH. Peaks and tails: evaluation of irregularities in capillary serum protein electrophoresis. Clin Biochem 2018;51:48–55. Copyright Elsevier. Reprinted with permission.)*

## Free light chain assay

Patients with monoclonal gammopathy may have negative serum protein electrophoresis and serum immunofixation studies. This may be due to very low levels of paraproteins and light chain gammopathy in which the light chains are very rapidly cleared from the serum by the kidneys. Because of this, urine electrophoresis and urine immunofixation is part of the work up for cases where monoclonal gammopathy is a clinical consideration. Urine electrophoresis and urine immunofixation studies are also performed to document the amount (if any) of potentially nephrotoxic light chains being excreted in the urine in a case of monoclonal gammopathy.

Quantitative serum assays for kappa and lambda free light chain has increased the sensitivity of serum testing strategies for identifying monoclonal gammopathies, especially the light chain diseases. Cases which may appear as nonsecretory myeloma can actually be cases of light chain myeloma. Free light chain assays allow disease monitoring as well as providing prognostic information for monoclonal gammopathy of undetermined significance (MGUS), smoldering myeloma.

The rapid clearance of light chains by the kidney is reduced in renal failure. Levels may be 20 to 30 times higher than normal in end-stage renal disease.

In addition, the kappa/lambda ratio may be as high as 3:1 in renal failure (normal 0.26–1.65). Therefore, patients with renal failure may be misdiagnosed as kappa light chain monoclonal gammopathy. If a patient has lambda light chain monoclonal gammopathy, with the relative increase in kappa light chain in renal failure, the ratio may become normal. Thus, a case of lambda light chain monoclonal gammopathy may be missed.

## Mass spectrometry in diagnosis of multiple myeloma

Established guidelines from the International Myeloma Working Group recommend diagnostic screening for patients suspected of plasma cell proliferative disease using protein electrophoresis, free light chain measurements, and immunofixation electrophoresis of serum and urine in certain cases. However, mass spectrometry can also be applied for the analysis of paraproteins because it is technically feasible to detect low levels of monoclonal proteins in peripheral blood using mass spectrometry. These methods are based on the fact that a paraprotein has a specific amino acid sequence, and therefore, a specific mass. This mass can be tracked over time and can serve as a surrogate marker in the presence of clonal plasma cells. New mass spectrometry-based methods are capable of detecting the paraprotein in blood at low levels, with detection limits approximately 100 times lower than that of serum immunofixation. Another major advantage of mass spectrometry-based methods is differentiation between paraproteins and monoclonal antibody drugs which may be used in the treatment of newly diagnosed multiple myeloma patients. These antibody drugs interfere with conventional serum protein electrophoresis and immunofixation.

Several mass spectrometry-based methods to detect the paraprotein in serum have been described to date. The first step is enriching immunoglobulins from serum followed by processing the purified immunoglobulins into smaller components, and finally measuring the mass of the component by mass spectrometry. Although the general steps of the methods are the same, the analytical details vary significantly. These include top-down analysis in which an intact protein is measured, middle-down analysis in which a protein is first digested into large pieces, and bottom-up analysis in which proteins are cleaved into small peptides. These mass spectrometry methods also differ in terms of the techniques and reagents used to purify immunoglobulins from serum and the mass spectrometers that are used to analyze the proteins and protein fragments [17].

Various methods have been reported for the analysis of paraprotein using mass spectrometry. For example, immunoglobulin enrichment coupled with matrix-assisted laser desorption ionization time-of-flight mass spectrometry (MALDI-TOF) can be used to identify paraprotein [18]. Murray and Willrich described two mass spectrometric methods for the analysis of paraproteins. One method is based on a tryptic digest of immunoglobulins and using selective reaction monitoring mass spectrometric methods to detect unique peptides from the variable region of specific immunoglobulin (also referred to as the "clonotypic" peptide approach), and a second method is based on disassembling immunoglobulins by chemical reduction and measuring the mass distribution of immunoglobulin light chain (also known as miRAMM). Interestingly, miRAMM method has several advantages for adaptation to routine testing including the ability to adapt to MALDI-TOF mass spectrometers, elimination of the chromatography step allowing for improved throughput suitable for large volume reference laboratory, and simplicity of MALDI-TOF data for the analysis of paraprotein [19]. Detailed discussion on this topic is beyond the scope of this book.

## Paraprotein interferences in clinical laboratory tests

Interference of paraprotein can produce both false-positive and false-negative test results depending on the analyte and the analyzer. However, the magnitude of interference may not correlate with the amount of paraprotein present in serum. The most common interferences include falsely low high-density lipoprotein (HDL) cholesterol, falsely high bilirubin, and altered values of inorganic phosphate. Other tests in which altered results may occur include LDL (low-density lipoprotein cholesterol) cholesterol, C-reactive protein, creatinine, glucose, urea nitrogen, inorganic calcium, and blood count. Roy reported a case of a 65-year-old man with IgG kappa (3.5 g/dL) multiple myeloma who showed undetectable HDL-cholesterol. However, his HDL-cholesterol level 2 years ago was 42 mg/dL, indicating that his undetectable HDL-cholesterol was due to interference of paraprotein [20]. A 51-year-old man diagnosed with IgG lambda multiple myeloma showed an elevated bilirubin level of 19.6 mg/dL without jaundice, but his direct bilirubin level in serum was within normal range. This indicated that his hyperbilirubinemia may be artifactual. The patient had a monoclonal gammopathy of 10 g/dL and his total serum protein was 14.3 g/dL. Therapy with lenalidomide and dexamethasone resulted in a reduction of monoclonal gammopathy and during last follow up by authors prior to publishing this report, his monoclonal gammopathy was 4.8 g/dL and his total bilirubin was 0.35 g/dL, a value which was within normal limits indicating that prior bilirubin of 19.6 mg/dL was due to interference of paraprotein in bilirubin assay [21].

## Other plasma cell disorders

Plasma cell neoplasm is characterized by the clonal expansion of terminally differentiated B lymphoid cells that have undergone somatic hypermutation that results in production of monoclonal immunoglobulin protein. Plasma cell neoplasm encompass a spectrum of diseases and include MUGS, smoldering myeloma, idiopathic Bence Jones proteinuria, monoclonal gammopathy of renal significance (MGRS), POEMS syndrome, amyloidosis, and light and heavy chain disposition disease. MGUS, smoldering myeloma, and multiple myeloma have been discussed earlier. In this section other plasma cell disorders and plasma cell morphology are summarized.

Monoclonal gammopathy of renal significance (MGRS) represents a group of disorders in which a monoclonal immunoglobulin secreted by a nonmalignant or premalignant B-cell or plasma cell clone results in renal damage. These disorders do not meet diagnostic criteria for overt, symptomatic multiple myeloma or a lymphoproliferative disorder. The term MGRS was proposed in 2012 by the International Kidney and Monoclonal Gammopathy Research Group to collectively describe patients who would otherwise meet the criteria for monoclonal gammopathy of undetermined significance (MGUS) but demonstrate renal injury attributable to the underlying monoclonal protein.

In MGRS, the renal lesions are primarily caused by the abnormal deposition or activity of monoclonal proteins in the kidney. Deposition of the monoclonal proteins may occur within the glomeruli, tubules, vessels, or interstitium of the kidney depending on the specific biochemical characteristics of the pathogenic light and/or heavy chains involved.

MGRS-associated kidney diseases encompass a wide spectrum of renal pathology and include such lesions as immunoglobulin-associated amyloidosis, the monoclonal immunoglobulin deposition diseases (MIDDs; light chain deposition disease, heavy chain deposition disease, and light and heavy chain deposition disease), proliferative glomerulonephritis with monoclonal immunoglobulin deposits (PGNMID), C3 glomerulopathy with monoclonal gammopathy, light chain proximal tubulopathy (Fanconi syndrome), among others.

Patients with MGRS frequently develop progressive kidney disease and end-stage renal disease (ESRD). MGRS-related kidney diseases recur in most patients after kidney transplantation and can lead to rapid allograft loss.

Polyneuropathy, organomegaly, endocrinopathy, M-protein, skin changes (POEMS) syndrome, also known as osteosclerotic myeloma, is a paraneoplastic syndrome due to an underlying plasma cell neoplasm. The major criteria for the syndrome are polyradiculoneuropathy, clonal plasma cell disorder, sclerotic bone lesions, elevated vascular endothelial growth factor, and the presence of Castleman disease. Minor features include organomegaly, endocrinopathy, characteristic skin changes, papilledema, extravascular volume overload, and thrombocytosis. Diagnoses are often delayed because the syndrome is rare and can be mistaken for other neurologic disorders, most commonly chronic inflammatory demyelinating polyradiculoneuropathy [22].

Monoclonal immunoglobulin deposition disease: In this disease, tissue deposition of monoclonal Ig, Ig chain, or fragment takes place. In Ig-related primary amyloidosis, Congo red staining of amyloid protein in tissue can be used for diagnostic purpose. Under electron microscopy (EM), amyloid deposits appears to be composed of linear, nonbranching, aggregated fibrils that are 7.5–10 nm thick, of indefinite length arranged in a loose meshwork. Monoclonal immunoglobulin deposition disease can be related to systemic light chain or heavy chain deposition disease.

Plasma cell myeloma: In classical or symptomatic plasma cell myeloma, paraprotein is found in serum or urine (typically >3 g/L of IgG or >25 g/L of IgA or >1 g/24 h of urine light chain), bone marrow shows clonal plasma cells (usually >10% of nucleated cells in bone marrow), or there is evidence of plasmacytoma. In addition, there is evidence of organ or tissue damage (CRAB). Other variants include: (a) smoldering: laboratory criteria met, but no CRAB (the plasma cells in the bone marrow are >10% but <60%); (b) nonsecretory (85% nonsecretors have Ig in the cytoplasm but are not secreting them and 15% nonproducers); (c) plasma cell leukemia (>2000 plasma cells/mm$^3$ or 20% by diff).

## Morphology of plasma cells in myeloma

In myeloma, plasma cells are found as interstitial clusters or focal nodules or in diffuse sheets in the bone marrow. The cells may appear mature or immature, plasmablastic or pleomorphic. The following descriptive terms have been used but none are diagnostic for myeloma:

- Mott cell: plasma cell with grape-like cluster of cytoplasmic inclusion
- Russell bodies: plasma cells cherry red cytoplasmic inclusions
- Dutcher bodies: plasma cell with intranuclear inclusions
- Flame-shaped cells: plasma cells vermillion staining glycogen rich IgA
- Thesaurocytes: plasma cells with ground glass cytoplasm

## Cerebrospinal fluid electrophoresis

Qualitative assessment of cerebrospinal fluid by electrophoresis (CSF electrophoresis) for oligoclonal bands is an important diagnostic for multiple sclerosis. Multiple sclerosis is an inflammatory demyelinating disease of the central nervous system. Cerebrospinal fluid is used in the diagnosis of multiple sclerosis to identify intrathecal IgG synthesis as reflected qualitatively by the presence of oligoclonal bands in CSF electrophoresis and quantitatively by IgG index or IgG synthesis ratio. The immunoglobulin increase in cerebrospinal fluid in multiple sclerosis is predominantly IgG, though the synthesis of IgM and IgA may also be increased.

Oligoclonal bands are defined as at least two bands seen in the CSF lane with no corresponding band present in the serum lane. Thus, it is crucial to perform CSF and serum electrophoresis simultaneously. Oligoclonal bands may be found in 90% or more patients clinically diagnosed with multiple sclerosis. However, oligoclonal bands may be seen in patients with CNS infections such as Lyme disease and in patients with autoimmune diseases, brain tumors, and lympho-proliferative disorders. Thus, it is important to note that detection of oligoclonal bands is not always due to multiple sclerosis. The characteristic feature of CSF electrophoresis pattern is the presence of a prealbumin band and a band in the beta-2 region due to desialylated transferrin (also known as beta-1 transferrin). These bands are not present in the serum lane.

An abnormality of cerebrospinal fluid IgG production may be expressed as percentage of the total protein, as a percentage of albumin, or by the use of the Ig G index. IgG index is defined as:

$$IgG\ index = \frac{[CSF\ IgG/CSF\ Albumin]}{[Serum\ IgG/Serum\ Albumin]}$$

Normal value for IgG index in adult is 0.23–0.64. However, it is important to correlate clinical findings with IgG index and the findings of the electrophoresis. In multiple sclerosis, cerebrospinal fluid is grossly normal and the pressure is also normal. The total leukocyte count is normal in the majority of patients. If the white blood cell count is elevated it rarely exceeds 50 cells/μL. Lymphocytes are the predominant cells found, and the total protein concentration of cerebrospinal fluid is also within normal range. If there is a systemic immune reaction or a monoclonal gammopathy also present, then bands should be present in both the serum and CSF lanes and should correspond with each other. These are not oligoclonal bands. Although detection of oligoclonal bands is one of the major reasons to order cerebrospinal fluid electrophoresis, other characteristics features of cerebrospinal fluid electrophoresis include:

- Prealbumin band (transthyretin band) is seen anodal (positive electrode) to albumin (a band above the albumin band). This band is not present in serum electrophoresis.
- Albumin band in CSF electrophoresis is slightly anodal to the albumin band present in the corresponding serum electrophoresis.
- The alpha-2 band is significantly denser in the serum protein electrophoresis than CSF electrophoresis because alpha-2 microglobulin and haptoglobin do not cross the blood-brain barrier due to relatively

large molecular weight. However, if the brain barrier is damaged or if the cerebrospinal fluid is collected from a traumatic tap, this band may be denser in CSF electrophoresis.

- In contrast to serum protein electrophoresis where two bands are present in the beta region, CSF electrophoresis usually shows three bands in the beta region including the beta-1 band, another band for C3, and beta-2 band due to the presence of desialated transferrin (also known as tau protein).

Another reason for ordering CSF electrophoresis is to establish or rule out leakage of cerebrospinal fluid through the nose. The presence of prealbumin and tau protein helps to identify the sample as cerebrospinal fluid leaking through the nose. In addition, immunofixation studies can also be performed with antibodies directed against the tau protein.

## Summary/key points

- Serum protein electrophoresis, urine electrophoresis, serum immunofixation, and urine immunofixation are all performed primarily to investigate suspicion of monoclonal gammopathy in a patient. Monoclonal gammopathy is present in a patient when a monoclonal protein (also called paraprotein or M protein) is identified in a patient's serum, urine, or both. The paraprotein can be an intact immunoglobulin, only light chains (light chain myeloma, light chain deposition disease, or amyloid light chain amyloidosis), or rarely found only as heavy chains (heavy chain disease).
- Multiple myeloma evolves from a premalignant condition known as monoclonal gammopathy of undetermined significance (MGUS). Smoldering multiple myeloma is an intermediate stage between MGUS and multiple myeloma. MGUS and smoldering multiple myeloma are asymptomatic plasma cell disorders, while multiple myeloma is a malignant neoplasm usually of a single clone of plasma cells.
- In general, in MUGS, concentration of monoclonal protein (IgG, IgA, or IgM) should be <3 g/dL and monoclonal plasma cells in the bone marrow should be <10%. Moreover, patients must be asymptomatic. In smoldering myeloma, serum monoclonal protein concentration should be ≥3 g/dL, monoclonal plasma cells in bone marrow should be 10%–60%, but patient must be asymptomatic. In contrast, multiple myeloma is defined as the presence of end organ (classic "CRAB" criteria) damage in parallel with the presence of a paraprotein spike or monoclonal plasma cells (plasma cells ≥10% or biopsy-proven bony or extramedullary plasmacytoma. The CRAB criteria includes: (C) hypercalcemia, calcium > 11 mg/dL; (R) renal insufficiency, GFR <40 mL/min and or

elevated creatinine; (A) anemia, hemoglobin 2.0 g/dL lower limit of normal; and (B) bone lesions.

- The paraprotein can be an intact immunoglobulin (one of the five immunoglobulin class: IgG, IgA, IgM, IgD, or IgE), only light chains (κ or λ; light chain myeloma, light chain deposition disease, or amyloid light chain amyloidosis), or rarely found only as heavy chains (heavy chain disease). In general, IgG is the most commonly encountered paraprotein, while IgD is encountered infrequently and IgE is reported rarely.

- Agarose gel electrophoresis and capillary electrophoresis are two principal methods employed in screening for paraproteins. Both methods are applicable for both serum and urine specimens. However, capillary electrophoresis is more automated and faster. Coomassie brilliant blue is a common staining agent to visualize band in serum protein electrophoresis, but in capillary electrophoresis, bands are detected at 200 nm using automated reader. Paraproteins are seen usually in the gamma region of the electrophoresis, but also may be present in beta or rarely in the alpha-2 region.

- Once a paraprotein is detected, confirmation and the isotyping of paraprotein are essential which is usually achieved by immunofixation. In about 3% of cases two paraproteins may be detected. This is referred to as biclonal gammopathy. A patient may also have nonsecretory myeloma, as in the case of a plasma cell neoplasm, in which the clonal cells neither produce nor secrete M proteins. The most commonly observed paraprotein is IgG followed by IgA, light chain, and rarely IgD or IgE. A normal serum protein electrophoresis does not exclude diagnosis of myeloma because approximately 11% myeloma patients may have normal serum protein electrophoresis. Therefore, serum and urine immunofixation studies should be performed regardless of serum electrophoresis results if clinical suspicion is high.

- Components of serum protein electrophoresis include albumin, alpha-1 globulins (alpha-1 zone), alpha-2 globulins (alpha-2 zone), beta globulins (beta zone often splits into beta-1 and beta-2 bands), and gamma globulins (gamma zone).

- Reduced intensity of the albumin band is observed in inflammation, liver dysfunction, uremia, nephrotic syndrome, and other conditions that lead to hypoalbuminemia. A smear observed in front of the albumin band may be due to hyperbilirubinemia or due to the presence of certain drugs. A band in front of the albumin band may be due to prealbumin, (a carrier for thyroxine and vitamin A) which is commonly seen in cerebrospinal fluid specimens or serum specimens in patients with malnutrition. Two, rather than one, albumin bands may represent bisalbuminemia. This is a familial abnormality with no clinical significance.

- The alpha-1 band mostly consists of alpha-1 antitrypsin (AT) (90%), alpha-1-antichymotrypsin, and thyroid-binding globulin. Alpha-1 antitrypsin is an acute phase reactant and its concentration is increased in inflammation and other conditions. The alpha-1 antitrypsin band is decreased in patients with alpha-1 antitrypsin deficiency or decreased production of globulin in patients with severe liver disease. At the leading edge of this band, a haze due to high-density lipoprotein (HDL) may be observed. The alpha-2 band consists of alpha-2 macroglobulin, haptoglobin, and ceruloplasmin. Because both haptoglobin and ceruloplasmin are acute phase reactants, this band is increased in inflammatory states. Alpha-2 macroglobulin is increased in nephrotic syndrome and cirrhosis of liver.
- The beta zone may consist of two bands, beta-1 and beta-2. Beta-1 is mostly composed of transferrin and low-density lipoprotein. An increased beta-1 band is observed in iron deficiency anemia due to increased level of free transferrin. This band may also be elevated in pregnant women. Very low-density lipoprotein usually appears in the prebeta zone. The beta-2 band is mostly composed of complement proteins. If two bands are observed in the beta-2 region, it implies either electrophoresis of plasma specimen (fibrinogen band) instead of serum specimen or IgA paraprotein.
- There are certain situations where a band may be apparent, but in reality, it is not a monoclonal band. For example, fibrinogen is seen as a discrete band between beta and gamma regions when electrophoresis is performed on plasma instead of serum specimen. If the electrophoresis is repeated after the addition of thrombin, this band should disappear. In addition, immunofixation study should be negative. Intravascular hemolysis results in release of free hemoglobin in circulation which binds to haptoglobin. The hemoglobin-haptoglobin complex may appear as a large band in the alpha-2 area. Serum immunofixation studies should be negative in such cases.
- In patients with iron deficiency anemia, concentrations of transferrin may be high which may result in a band in the beta region. Again, immunofixation should be negative.
- Patients with nephrotic syndrome usually show low albumin and total protein, but this condition may also produce increased alpha-2 and beta fractions. Bands in either of these regions may mimic a monoclonal band.
- Hypogammaglobulinemia may be congenital or acquired. Among the acquired causes are multiple myeloma and primary amyloidosis. Panhypogammaglobulinemia can occur in about 10% of cases of multiple myeloma. Most of these patients have a Bence Jones protein in the urine, but lack intact immunoglobulins in the serum.

- Common features of serum protein electrophoresis in various disease states other than monoclonal gammopathy include:
  - Inflammation: Increased intensity of alpha-1 and alpha-2 with sharp leading edge of alpha-1 may be observed, but with chronic inflammation the albumin band may be decreased with increased gamma zone due to polyclonal gammopathy.
  - Nephrotic syndrome: In nephrotic syndrome, the albumin band is decreased due to hypoalbuminemia. In addition, the alpha-2 band may be more distinct.
  - Cirrhosis or chronic liver disease: A low albumin band due to significant hypoalbuminemia with a prominent beta-2 band and beta-gamma bridging are characteristics features of liver cirrhosis or chronic liver disease. In addition, polyclonal hypergammaglobulinemia is observed.
  - Proteinuria can be classified as glomerular, tubular, or combined proteinuria. Glomerular proteinuria can be subclassified as selective glomerular proteinuria (urine will have albumin and transferrin bands) or nonselective glomerular proteinuria (urine will have presence of all different types of proteins). In glomerular proteinuria, the dominant protein present is always albumin. In tubular proteinuria, albumin is a minor component. The presence of alpha-1 microglobulin and beta-2 microglobulin are indicators of tubular damage.
- One source of possible error in urine immunofixation study is the "step ladder" pattern. Here multiple bands are seen in the kappa (more often) or lambda lanes and are indicative of polyclonal spillage rather than monoclonal spillage into the urine. During urine immunofixation five or six faint, regular diffuse bands with hazy background staining between bands may be seen. This is referred to as the step ladder pattern and is a feature of polyclonal hypergammaglobulinemia with spillage into the urine.
- The rapid clearance of light chains by the kidney is reduced in renal failure. Levels may be 20 to 30 times higher than normal in end-stage renal disease. In addition, the kappa/lambda ratio may be as high as 3:1 in renal failure (normal 0.26–1.65). Therefore, patients with renal failure may be misdiagnosed as kappa light chain monoclonal gammopathy. If a patient has lambda light chain monoclonal gammopathy, with the relative increase in kappa light chain in renal failure, the ratio may be come normal. Thus, a case of lambda light chain monoclonal gammopathy may be missed.
- Plasma cell neoplasm encompass a spectrum of diseases and include MUGS, smoldering myeloma, idiopathic Bence Jones proteinuria, monoclonal gammopathy of renal significance (MGRS), POEMS syndrome, amyloidosis, light and heavy chain disposition disease.

- Monoclonal gammopathy of renal significance (MGRS) represents a group of disorders in which a monoclonal immunoglobulin secreted by a nonmalignant or premalignant B cell or plasma cell clone results in renal damage. These disorders do not meet diagnostic criteria for overt, symptomatic multiple myeloma or a lymphoproliferative disorder.
- Polyneuropathy, organomegaly, endocrinopathy, M-protein, skin changes (POEMS) syndrome, also known as osteosclerotic myeloma is a paraneoplastic syndrome due to an underlying plasma cell neoplasm.
- Characteristics features of cerebrospinal fluid (CSF) electrophoresis include prealbumin band (transthyretin band) which is seen anodal (positive electrode) to albumin (a band above the albumin band). This band is not present in serum electrophoresis but may be present in capillary zone electrophoresis of serum proteins.
- Albumin band in CSF electrophoresis is slightly anodal to the albumin band present in the corresponding serum electrophoresis.
- The alpha-2 band is significantly denser in the serum protein electrophoresis than CSF electrophoresis because alpha-2 microglobulin and haptoglobin do not cross the blood-brain barrier due to relatively large molecular weight. However, if the brain barrier is damaged or if the cerebrospinal fluid is collected from a traumatic tap, this band may be denser in CSF electrophoresis.
- In contrast to serum protein electrophoresis where two bands are present in the beta region, CSF electrophoresis usually shows three bands in the beta region including the beta-1 band, another band for C3, and beta-2 band due to the presence of desialated transferrin (also known as tau protein).
- Cerebrospinal fluid is used in the diagnosis of multiple sclerosis to identify intrathecal IgG synthesis as reflected qualitatively by the presence of oligoclonal bands in CSF electrophoresis and quantitatively by IgG index or IgG synthesis ratio.
- Oligoclonal bands are defined as at least two bands seen in the CSF lane with no corresponding band present in the serum lane. Thus, it is crucial to perform CSF and serum electrophoresis simultaneously.
- Another reason for ordering CSF electrophoresis is to establish or rule out leakage of cerebrospinal fluid through the nose. The presence of prealbumin and tau protein helps to identify the sample as cerebrospinal fluid leaking through the nose. In addition, immunofixation studies can also be performed with antibodies directed against the tau protein.

## References

[1] Kyle RA, Therneau TM, Rajkumar SV, Offord JR, et al. A long-term study of prognosis of monoclonal gammopathy of undetermined significance. N Engl J Med 2002;346:564–9.

[2] Rajkumar SV, Kumar S. Multiple myeloma: diagnosis and treatment. Mayo Clin Proc 2016;91:101–19.

[3] Mouhieddine TH, Weeks LD, Ghobrial IM. Monoclonal gammopathy of undetermined significance. Blood 2019;133:2484–94.

[4] Dispenzieri A, Katzmann JA, Kyle RA, Larson DR, et al. Prevalence and risk of progression of light-chain monoclonal gammopathy of undetermined significance (LC-MGUS): a newly defined entity. Lancet 2010;375(9727):1721–8.

[5] Rajkumar SV. Multiple myeloma: every year a new standard? Hematol Oncol 2019;37(Suppl 1):62–5.

[6] Roberts-Thomson PJ, Nikoloutsopoulos T, Smith AJ. IgM paraproteinaemia: disease associations and laboratory features. Pathology 2002;34:356–61.

[7] Vodopick H, Chaskes SJ, Solomon A, Stewart JA. Transient monoclonal gammopathy associated with cytomegalovirus infection. Blood 1974;44:189–95.

[8] Gerritsen E, Vossen J, van Tol M, Jol-van der Zijde C, et al. Monoclonal gammopathies in children. J Clin Immunol 1999;9:296–305.

[9] Levin A, Hari P, Dhakal B. Novel biomarkers in multiple myeloma. Transl Res 2018;201:49–59.

[10] Minchiotti L, Caridi G, Campagnoli M, Lugani F, et al. Diagnosis, phenotype, and molecular genetics of congenital analbuminemia. Front Genet 2019;10:336.

[11] Gras J, Padros R, Marti I, Gomez-Acha JA. Pseudo-analbuminemia due to the presence of a slow albumin variant moving into the alpha1 zone. Clin Chim Acta 1980;104:125–8.

[12] Kancharla P, Patel E, Hennrick K, Ibrahim S, Goldfinger M. A rare presentation of biclonal gammopathy in multiple myeloma with simultaneous extramedullary involvement: a case report. Case Rep Oncol 2019;12:537–42.

[13] Mullikin TC, Rajkumar SV, Dispenzieri A, Buadi FK, Lacy MQ, Lin Y, et al. Clinical characteristics and outcomes in biclonal gammopathies. Am J Hematol 2016;91:473–5.

[14] Regeniter A, Siede WH. Peaks and tails: evaluation of irregularities in capillary serum protein electrophoresis. Clin Biochem 2018;51:48–55.

[15] Bossuyt X. Interferences in clinical capillary zone electrophoresis of serum proteins. Electrophoresis 2004;25:1485–7.

[16] Padelli M, Labouret T, Labarre M, Le Reun E, et al. Systematic overestimation of human serum albumin by capillary zone electrophoresis method due to monoclonal immunoglobulin interferences. Clin Chim Acta 2019;491:74–80.

[17] Chapman JR, Thoren KL. Tracking of low disease burden in multiple myeloma: using mass spectrometry assays in peripheral blood. Best Pract Res Clin Haematol 2020;33, 101142.

[18] Milani P, Murray DL, Barnidge DR, Kohlhagen MC, et al. The utility of MASS-FIX to detect and monitor monoclonal proteins in the clinic. Am J Hematol 2017;92:772–9.

[19] Murray DL, Willrich MAV. Evolution of myeloma testing in clinical chemistry with mass spectrometry. J Appl Lab Med 2019;4:474–6.

[20] Roy V. Artifactual laboratory abnormalities in patients with paraproteinemia. South Med J 2009;102:167–70.

[21] Cascavilla N, Falcone A, Sanpaolo G, D'Arena G. Increased serum bilirubin level without jaundice in patients with monoclonal gammopathy. Leuk Lymphoma 2009;50:1392–4.

[22] Dispenzieri A. POEMS syndrome: 2019 update on diagnosis, risk stratification, and management. Am J Hematol 2019;94:812–27.

# CHAPTER 25

# Human immunodeficiency virus (HIV) and hepatitis testing

## Human immunodeficiency virus testing

Human immunodeficiency virus (HIV) is a slowly replicating retrovirus that causes acquired immunodeficiency syndrome (AIDS). HIV infection is sexually transmitted but is also transmitted due to sharing infected needle, blood transfer, and from mother to her newborn. In body fluid, HIV is present as free viral particles and also as a virus within infected immune cells. HIV infects vital cells involved in human immune function such as helper T cells especially CD4 + T cells, macrophages, and dendritic cells. HIV virus is transmitted as enveloped RNA virus and upon entry into target cell, viral RNA genome is converted into double-stranded DNA by a virally encoded reverse transcriptase that is transported along with viral genome in virus particle during infection. Two major types of HIV virus have been characterized: HIV-1 and HIV-2. HIV-1 was initially discovered and is the cause of majority of HIV infection worldwide. The HIV-2 virus is not widely spread worldwide, with only 1–3 million infected people, mainly in West Africa. HIV-2 infection is characterized by low and frequently undetectable viral loads and a slower course to AIDS. Although HIV-2 is endemic in West Africa, HIV-2 infection in North America has recently been described but HIV-2 progresses to symptomatic disease at a much slower rate than HIV-1. While treatment of HIV-1 infection is well characterized, there is far less experience among physicians in treating persons infected with HIV-2 virus and there are also controversies when to initiate therapy with a goal of reaching immune restoration while minimizing drug toxicity [1]. Several groups are classified within HIV-1 (Table 1). HIV-1 M group is the main type of HIV virus seen in clinical practice and M type can be divided into several subtypes [2].

## Window period in HIV infection

The diagnosis of HIV infection is most commonly achieved by detecting antibody against HIV in body fluid using a screening test followed by a

Clinical Chemistry, Immunology and Laboratory Quality Control. https://doi.org/10.1016/B978-0-12-815960-6.00015-7

**Table 1** Various HIV-1 viral types and groups.

| Type | Group | Comments |
|---|---|---|
| HIV-1 | | Related to viruses found in chimpanzees and gorillas |
| HIV-1 | M | M denotes "major" and this is the most common HIV responsible for AIDS pandemic. HIV-1 group M is subdivided into A–D, F–H, and J–K. HIV-1 group M, subtype B remains the predominant viral infection throughout North American and many developed countries but HIV-1 group M subtype C has highest prevalence worldwide among infected individuals |
| HIV-1 | N | N denotes for "non-M, non-O" and seen in Cameroon |
| HIV-1 | O | O denotes for "outlier" and common in Cameroon but usually seen outside West-central Africa |
| HIV-1 | P | P denotes for "pending the identification of further human cases": the virus was isolated from a Cameroonian woman residing in France |

confirmatory test. In 1985, US Food and Drug Administration (FDA) approved the first ELISA (enzyme-linked immunosorbent assay) to detect the presence of antibodies against HIV in serum. The major purposes of HIV testing include the following:

- Diagnosis of HIV infection in a suspected individual
- Testing for an individual who wishes to know whether infected by HIV
- Since 1985, screening of blood products for HIV was initiated which provided protection to countless individuals from transmission of HIV through blood or blood products
- Testing of potential donors before using organs or tissue for transplantation
- Epidemiological surveillance for health care officers to determine specific need in a community

HIV testing should be done in individuals who have clinical signs of acute or chronic infection.

Patients who have a high risk exposure to HIV: Annual or more frequent screening may also be considered for high-risk individuals.

After HIV infection, viral RNA can be measured within 10–12 days and viral p24 antigen can be measured afterward. The time needed for the appearance of HIV-specific antibodies in serum is known as serological "window period." In this period, only viral RNA and possible p24 antigen can be detected, but if screening is performed using a method that detects HIV-specific antibodies, the test could be negative. In general, IgM may be the first antibody against HIV that may be detected in circulation followed by IgG antibodies which appear approximately 3–4 weeks after infection. In general, within 1–2 months, HIV antibodies are usually present in almost all of the infected individuals although for few individuals, it may take up to 6 months for antibodies to appear in circulation. To reduce the spread of HIV infection, Center for Disease Control and Prevention (CDC) recommends HIV testing as a part of routine health care to all patients living in an area with high prevalence of HIV infection (>1%), high-risk patients who reside in low HIV-infected area, pregnant women as well as anyone requesting HIV testing. Patients with HIV infection benefits from early detection because infection in the late stage may have advanced immune suppression and may not get full benefits of antiretroviral therapy [3]. In 2006 CDC recommended expanded HIV screening in emergency departments.

False-negative HIV-1 antibody testing is most commonly attributed to a "window period" prior to the development of HIV-1-specific antibodies. In several case reports, patients treated with HAART (highly active antiretroviral therapy) very early in the course of disease may not have HIV-1 antibody possibly due to drug-induced viral load suppression. In very rare cases, a patient infected with HIV as evidenced by high viral load may not have HIV-specific antibody for a long time or persistent lack of antibody, but these patients present with severe immunodeficiency [4].

Safety of blood products used for transfusion is important and in the United States, blood and plasma are tested for antibodies to HIV-1 and HIV-2. In the United States, donated blood and plasma are tested for antibodies to human immunodeficiency virus (HIV-1 and HIV-2) by screening with an enzyme immunoassay (EIA) as well as an HIV-1 p24 antigen assay. Nevertheless, it is estimated that from 1 in 450,000 to 1 in 660,000 US blood donations may transmit HIV because of blood donations during the infectious window period, prior to seroconversion. Currently, available HIV antibody tests have an average infectious window period of 25 days. The p24 antigen can detect HIV infection 2–3 weeks after exposure. However, tests based on nucleic acid amplification (NAT) can detect HIV RNA much earlier. The introduction of nucleic acid amplification technology (NAT) for screening individual donations has remarkably improved the safety of blood products. It has been estimated that the risk of HIV infection in the United States has been reduced from 25 per 10 million donations with antibody testing only to 13–14 per 10 million donations with the addition of NAT testing [5]. Testing for HIV can be broadly divided into screening tests and confirmatory tests. Screening tests include standard testing, rapid HIV testing, and combination of HIV antibody and antigen testing. Confirmatory testing is performed by Western blot and recombinant immunoblot or line immunoassay (LIA).

## Standard HIV testing

Standard screening is performed by various immunoassays. The test is based on the detection of IgG antibody against HIV-1 antigens in the serum. HIV antigens include p24, gp 120, and gp 41 and antibodies to gp 41 and p24 are the first detectable serologic markers following HIV infection. IgG antibodies appear as early as 3 weeks but most likely within 12 weeks following HIV infection in the majority of patients and generally persist for life. Assays for IgM antibodies are not used as they are relatively insensitive. HIV viruses are categorized into several groups (Table 1) and M is considered as the pandemic strain and accounting for the vast majority of infection. The two important issues regarding HIV screening tests are the ability of the test to detect non-M strains and the timing of the test postexposure. If the patient has not yet seroconverted, the antibody is absent and the individual may have a false-negative test. There are also rare patients with HIV infection who become seronegative although these patients show earlier seropositive results after exposure to HIV virus. Some other causes of false-negative result include:

- Fulminant HIV infection
- Immunosuppression or immune dysfunction
- Delay in seroconversion following early initiation of antiretroviral therapy

The chance of a false-positive serologic test for HIV is extremely low. Testing of viral RNA is an approach that should be used to resolve such issues. Enzyme immunoassays (EIA) are widely used in clinical laboratories for HIV testing and various versions of EIA are used depending on assay platform. Since introduction in 1985, ELISA method has been evolved and now third-generation ELISA assays (sandwich format) can detect IgG and IgM antibodies and antibodies to all of the M subtypes as well as N and O group. Antibody testing method capable of detecting O group is important for the screening of blood products. More recently, fourth-generation assays have been introduced which simultaneously detect HIV antibodies and p24 antigen reducing the window period to only 13–15 days. These assays are often referred as HIV antigen/antibody combo assays and the first such assay was Architect HIV combo assay approved by the FDA in 2010. The sensitivity and specificities of these assays can reach over 99%. Enzyme-linked immunofluorescent assays (ELFA) are the modified version of ELISA technique that utilizes solid phases with a greater surface contact area to reduce incubation time. These assays use an enzyme and fluorescent substances which are converted into fluorescent products by the cation of enzymes and can be measured with fluorescent detectors. Assays based on this method can be adopted in automated analyzers. Chemiluminescence methods where chemiluminescent compounds are used to label antigen or antibody can also be adopted in automated high-throughput analyzers. Examples of some automated HIV testing are summarized in Table 2. Although

**Table 2** Examples of automated HIV tests.

| Analyzer | Diagnostic company | Comment |
|---|---|---|
| Architect | Abbott Laboratories | 4th Generation Chemiluminescent Sandwich Assay: CMIA (chemiluminescence Microparticle Immunoassay) Detecting both Antibodies and antigen |
| Centaur XP and CP | Siemens | 3rd Generation Chemiluminescent Sandwich Assay detecting antibodies OR 4th Generation Chemiluminescent Sandwich Assay Detecting both Antibodies and Antigen |
| Cobas | Roche | 4th Generation Chemiluminescent Sandwich Assay: ECLIA (Electrochemiluminescence) Detecting both Antibodies and Antigens |
| Liasion XL | Diasorin | 4th Generation Chemiluminescent Sandwich Assay Detecting both Antibodies and Antigens |
| Access 2 Unicel Dxl | Beckman | 4th Generation Chemiluminescent Sandwich Assay Detecting both Antibodies and Antigens |

various standard assays for HIV testing have excellent sensitivity and specificity, false-positive tests results may still be observed. Vardinon et al. studied 520 patients undergoing hemodialysis and observed 23 (4.4%) positive test results with enzyme immunoassay but results were indeterminant using confirmation Western blot. Five years of follow-up showed no seroconversion indicating positive test results using enzyme immunoassay as false-positive test result [6]. Therefore, patients undergoing hemodialysis may show false-positive test result with enzyme immunoassay indicating the need of confirmatory testing.

## Testing algorithm

Commonly used testing algorithms use these sequences:

- Individuals are tested using a fourth-generation antigen/antibody combination HIV-1/2 immunoassay. This assay detects HIV-1 and HIV-2 antibodies as well as the HIV P24 antigen. If the individual is negative, no further testing is necessary.
- If the fourth-generation assay is positive, an HIV-1/HIV-2 antibody differentiation assay is done. The test confirms the result of the first assay as well as tells us whether the patient is infected with HIV-1 or HIV-2 or both.
- A HIV RNA level is indicated if the fourth-generation combination assay is positive and the differentiation assay is negative or indeterminate. This will detect the early acute infection.

In an alternate approach, initial testing is done by ELISA to detect the presence of antibodies. If the test is positive, then a confirmatory Western blot is done. This approach may miss early infection, as antibodies may not have developed. In addition, HIV-2 infections may be misclassified as HIV-1.

## Rapid HIV antibody testing

Rapid HIV antibody tests can provide test result in less than 30 min and can be adopted in point-of-care settings. Since 2002, the FDA approved six rapid HIV tests (Table 3). Most of these tests can utilize whole blood thus avoiding the need of centrifuging specimen to obtain serum. However, OraQuick Advanced HIV1/2 assay can use whole blood, serum, or oral fluid. More recently, the FDA has approved OraQuick in Home HIV test, a rapid home use HIV kit using oral fluid where the test result can be obtained in 20–40 min. Most rapid HIV tests are based on the principle of enzyme immunoassays which are utilized in clinical laboratories using automated or semiautomated analyzers and most tests detect HIV antibodies by incorporating HIV envelop region antigens in the test methodology. However, both false-positive and false-negative results may occur with rapid HIV tests and it is important to confirm initial findings with a laboratory-based HIV assay. These tests are used for rapid screening only. Delaney et al. evaluated six rapid HIV antibody tests and observed sensitivities over 95% and specificity over 99% for all rapid tests. However, false-negative and false-positive results were observed in all rapid HIV assays [7]. Facente et al. observed that false-positive rate of oral fluid HIV test increases near expiration date of the kit [8].

## Confirmatory HIV test

Commonly used confirmatory test for diagnosis of HIV infection include Western blot and line immunoassay (LIA). Confirmatory test is conducted if the screening assay is positive. In Western blot, HIV denatured proteins are blotted on strips of nitrocellulose membrane which is then incubated with serum obtained from the patient. If serum contains antibodies against these viral

**Table 3** Examples of some rapid HIV tests available.

| Rapid test | Specimen | Methodology |
|---|---|---|
| OraQuick Advance HIV-1/2 | Whole blood, oral fluid, plasma | Lateral flow |
| MultiSpot HIV-1/2 | Serum, plasma | Flow through |
| Uni-Gold Recombigen HIV-1 | Whole blood, serum, plasma | Lateral flow |
| Reveal G3 HIV-1 | Serum, plasma | Flow through |
| Clearview complete HIV-1/2 | Whole blood, serum, plasma | Lateral flow |
| Clearview STAT-PAK HIV-1/2 | Whole blood, serum, plasma | Lateral flow |

proteins (most commercial assays use antigen from both HIV-1 and HIV-2), they will bind to these proteins and antigen-antibody reaction is visualized using an enzyme labeled secondary antibody and a matching substrate. A colorimetric reaction leads to formation of bands representing antigen-antibody complex indicating positive result. Results of Western blot can be positive, negative, or indeterminant. The disadvantages of Western blot are high cost and subjective interpretation. However, LIA methods based on recombinant proteins and or synthetic peptides capable of detecting antibodies to HIV-1 and HIV-2 are more specific, producing fewer indeterminant results than Western blot.

## HIV viral load test and related assays

Routine annual HIV viral load monitoring is recommended by the World Health Organization for monitoring patients on antiretroviral treatment. HIV viral load test determines the number of copies of RNA of HIV virus present in per milliliter of serum or plasma and is expressed as copies/mL or in log scale. HIV viral load test indicates viral replication and is often conducted to monitor the progress of antiretroviral therapy. In addition, CD4 lymphocyte counts are also measured to evaluate the immune system. The success of antiretroviral therapy is measured both clinically and by suppression of viremia below 50 copies/mL in two successive measurement, although this cut-off value has been contested. The nucleic acid tests (NAT) are commercially available tests that can identify HIV nucleic acid (either RNA or proviral DNA) and these testes are based on the principle of polymerase chain reaction (PCR), real-time PCR, nucleic acid sequence-based amplification or ligase chain reaction. NAT assays are useful in special population such as in the window period of infection when antibody against HIV is absent in serum, and newborns of HIV-infected mother where maternal antibody against HIV is present in the serum of the newborn. Amplification of proviral DNA allows detection of immune cells that harbor quiescent provirus as well as cells infected with actively replicating HIV. This test is useful in the diagnosis of HIV infection in infants and children up to 18 months of age born to HIV-infected mothers. However, false-negative DNA PCR test result may occur in children treated with antiretroviral therapy and may lead to inappropriate discontinuation of antiretroviral therapy [9].

A variety of molecular platforms exist for accurate, high-throughput HIV viral load testing, but they require sophisticated laboratory infrastructure and highly skilled laboratory technicians. The logistical requirements to reliably transport samples to centralized laboratory facilities, which include scheduling specimen pick-up and maintaining sample integrity, can also present challenges. Fortunately, point-of-care devices are also available for determining HIV viral load. A point-of-care HIV viral load assay must provide information on the level of viremia consistent with treatment failure, the threshold of which is currently defined by the WHO as 1000 RNA copies/mL. In addition, any point-of-care HIV viral load assay must be easy to use by local health care workers with a minimal amount of training. Examples of such point-of-care devices include the m-PIMA HIV-1/2 viral load test (a quantitative nucleic acid amplification test for viral load measurement of HIV type 1 groups M/N and O, and HIV-2 in plasma samples under 70 min by Abbott Laboratories, Xpert HIV-1 viral load test by Cepheid etc.) [10]. Xpert HIV-1 viral load test requires 1 mL of plasma and the instrument automates the test process including RNA extraction, purification, reverse transcription, and cDNA real-time quantitation using one fully integrated cartridge. The test result is available in 90 min.

## Introduction to hepatitis testing

Hepatitis infection is a worldwide problem. Hepatitis could be A, B, C, D, or E. However, infections from the hepatitis A virus, hepatitis B virus, or hepatitis C virus are most common. Liver injury can result from nearly any hepatitis virus infection with systemic involvement, however, including cytomegalovirus and herpes simplex virus.

Hepatitis A is a problem with many developing countries because the hepatitis A virus may spread from contaminated water (fecal-oral route), after consuming contaminated food or handling contaminated objects. However, hepatitis A infection is rare in developed countries including the United States due to availability of hepatitis A vaccine in 1995 and better hygiene. Hepatitis A is a small, nonenveloped, RNA virus of the genus *Hepatovirus*, which was first isolated in 1979, and humans are its only natural host. Hepatitis A virus can tolerate a pH as low as 1, high heat, as well as capable of surviving on surfaces outside of the human body for months [11]. Laboratory testing of hepatitis A is straightforward where IgM antihepatitis A antibody denotes recent infection and IgG antihepatitis A antibody appears in the convalescent phase of acute hepatitis.

Hepatitis B is a double-stranded, enveloped DNA virus of the Hepadnaviridae family. It is the smallest DNA virus known, which replicates in hepatocytes and causes dysfunction of the liver. Hepatitis B is transmitted via the percutaneous or permucosal route. Transmission can occur during sexual intercourse, through sharing of needles, when blood from an affected individual enters another person via an open wound or mucosal surface, and vertically from mother to child during childbirth. It is not spread through casual contact. The incubation period ranges from 45 to 180 days but the hepatitis B surface antigen (HBsAg) can be detected in blood as early as 30–60 days after exposure [11].

Hepatitis C is an RNA virus transmitted by exposure to hepatitis C-infected blood or blood-containing body fluid (such as sharing contaminated needle during drug abuse), sexual transmission, occupational injury, perinatal transmission, receiving piercings or tattoos from unregulated body art studios, and transfusion of unscreened blood before 1992. Patients infected with hepatitis C are often asymptomatic but 75%–85% of these patients may develop chronic infection in 6 months which may also be asymptomatic. Therefore, the diagnosis is typically established through routine screening or diagnostic workup of an incidental abnormal finding on routine laboratory tests. Chronic infection is associated with increased risk for hepatic fibrosis, and 10%–20% of patients with HCV infection eventually develop cirrhosis. Currently, effective treatment of hepatitis C is available.

The hepatitis D virus (also called the delta-virus) is a defective pathogen that requires the presence of the hepatitis B virus for infection. Hepatitis D can elicit a specific immune response in the infected host, consisting of antibodies of the IgM and IgG class (antihepatitis D antibodies). Hepatitis E virus is also an enterically transmitted virus. Hepatitis E can also be transmitted by blood transfusion, particularly in endemic areas. Chronic hepatitis does not develop after acute hepatitis E infection, except in the transplant setting and possibly in other settings of immunosuppression. Fulminant hepatitis can occur resulting in an overall case fatality rate of 0.5%–3%. For reasons as yet unclear, the mortality rate in pregnant women can be as high as 15%–25%, especially in the third trimester. The diagnosis of hepatitis E is based upon the detection of hepatitis E virus in serum or stool by polymerase chain reaction (PCR) or by the detection of IgM antibodies against hepatitis E. Antibody tests against hepatitis E alone are less than ideal since they have been associated with frequent false-positive and negative results. In summary, various tests for the laboratory diagnosis of hepatitis include:

- Liver function tests including bilirubin, aspartate aminotransferase (AST), alanine aminotransferase (ALT), alkaline phosphatase (ALP), bilirubin, albumin, and prothrombin time
- Blood test to identify IgM antibody against hepatitis A
- For hepatitis B, testing of hepatitis B surface antigen (HBsAg), hepatitis B core antigen (HBcAg), as well as respective antibodies and viral load
- For hepatitis C, testing of antibodies and hepatitis C viral load
- For hepatitis D, IgG and IgM antibody against hepatitis D

Characteristics of various hepatitis viruses are summarized in Table 4. Because hepatitis B and C infection are the most important components of hepatitis testing, in the following section, laboratory tests for hepatitis B and C are discussed in detail.

**Table 4** Characteristics of various hepatitis viruses.

| Hepatitis virus | Type | Transmission | Comment |
|---|---|---|---|
| Hepatitis A | RNA virus | Fecal-oral | Low mortality, does not cause chronic liver disease |
| Hepatitis B | DNA virus | Parenteral, vertical, sexual transmission | 10% fail to clear virus and may have chronic liver disease/hepatocellular carcinoma |
| Hepatitis C | RNA virus | Parenteral, vertical, sexual transmission | 50%–70% fail to clear virus and may have chronic liver disease/hepatocellular carcinoma |
| Hepatitis D | RNA virus | Parenteral, vertical sexual transmission | Can only infect patients with hepatitis B infection |
| Hepatitis E | RNA virus | Fecal-oral | Low mortality (1%) except pregnant women (10%–20%), does not cause chronic liver disease |

## Testing for hepatitis B

An estimated 350 million people are chronically infected with hepatitis B in the world resulting in an estimated 600,000 deaths per year from cirrhosis, liver failure, and liver carcinoma. In the United States, an estimated 0.8–1.4 million persons are suffering from chronic hepatitis B infection, but with the introduction of hepatitis B vaccine in 1991, now the new infection of hepatitis B in the United States is only 1.6 cases out of 1000,000 people. Hepatitis B is a small-diameter (42 nm) incompletely double-stranded DNA virus with eight distinguishable genotypes (A through H) [12].

Serologic markers available for hepatitis B infection are HBsAg (hepatitis B, surface antigen), HBeAg (hepatitis B, e antigen), anti-HBc (antibody against hepatitis B core antigen; both IgG and IgM), anti-HBs (antibody against hepatitis B surface antigen), anti-HBe (antibody against hepatitis B, e antigen) as well as testing for viral DNA (Table 5).

HBsAg is the first marker to be positive after exposure to hepatitis B virus. It can be detected even before the onset of symptoms. Most patients may clear the virus and HBsAg typically becomes undetectable within 4–6 months. Persistence of HBsAg for more than 6 months implies chronic infection. The disappearance of HBsAg is followed by the presence of anti-HBs. During the window period (after the disappearance of HBsAg and before the appearance of anti-HBs), evidence of infection is documented by the presence of anti-HBc (IgM). Coexistence of HBsAg and anti-HBs has been documented in approximately 24% of HBsAg-positive individuals. It is thought that the antibodies fail to neutralize the virus particles. These individuals should be considered as carriers of hepatitis B virus. There exists a subset of patients who have undetectable HBsAg but are positive for hepatitis B viral DNA. Most of these patients have

| Test | Comments |
|------|----------|
| HBsAg | First detectable agent in acute infection |
| HBcAg | Not tested as it is not detectable in blood |
| HBeAg | Indicates virus is replicating and patient is highly infectious |
| Anti-HBc | First antibody to appear in blood and this test is positive when other tests for hepatitis B are negative during the window period (HBsAg is negative and anti-HBs not yet detectable) |
| Anti-HBe | Indicates virus is not replicating |
| Anti-HBs | Patient has achieved immunity |

**Table 5** Hepatitis B testing.

very low viral load with undetectable levels of HBsAg. Uncommon situations are infection with hepatitis B variants that decrease HBsAg production or mutant strains that have altered epitopes normally used for detection of HBsAg. Time frame of release of various virologic and serological markers in acute hepatitis B infection with recovery is presented in Fig. 1.

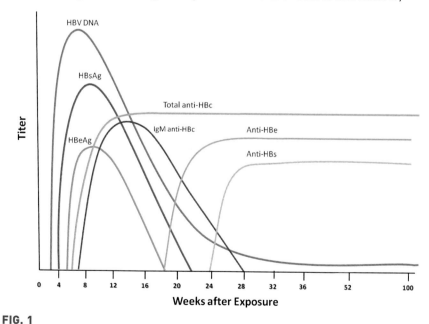

**FIG. 1**

Virological and serological response to acute hepatitis B infection with recovery. *Figure courtesy of Andres Quesda, M.D.*

Individuals with recent infection will develop anti-HBc, IgM antibodies. Individuals with chronic infection and individuals who have recovered from an infective episode will develop anti-HBc, IgG antibodies. However, anti-HBc, IgM antibodies may remain positive for up to 2 years after an acute infection. Levels may also increase and be detected during exacerbations of chronic hepatitis B. This may lead to diagnostic confusion.

In certain individuals, isolated positive anti-HBc antibodies may be detected. The clinical significance of this finding is unclear. Some of these individuals have been found to have hepatitis B DNA by PCR. Transmission of hepatitis B has been reported from blood and organ donors who have isolated positive anti-HBc antibody. On the other hand, there are reports that a certain percentage of these individuals who have anti-HBc are false-positive. Presence of HBeAg usually indicates that the hepatitis B is replicating and the patient is infectious. Seroconversion to anti-HBe typically indicates that the virus is no longer replicating. This is associated with a decrease in serum hepatitis B DNA and clinical remission. In some patients, seroconversion is still associated with active liver disease. This may be due to low levels of wild-type hepatitis B or its variants that prevent or decrease the production of HBeAg.

Testing for HBsAg is most common to investigate a suspected hepatitis B infection but this surface antigen is present for a short time after hepatitis B vaccination because hepatitis B vaccines contain the surface antigen. However, such positive test is unlikely approximately 14 days after vaccination but a weakly positive test result may persist in few individuals beyond 2 weeks [13]. Hepatitis B screening is also important in blood donors and serological tests for HBsAg and anti-HBc are standard screening techniques. However, more recently, nucleic acid-based testing for hepatitis viral DNA has been implemented to avoid transmission-related infection to identify HBsAg-negative window period in a blood donor who may have early acute infection [14]. All major diagnostic companies market immunoassays for testing of serological markers of hepatitis B using various automated analyzers marketed by these companies. Interpretation of hepatitis B serology is summarized in Table 6.

Four types of molecular assays are available for the diagnosis and management of hepatitis B infection: quantitative viral load tests, genotyping assays, drug resistance mutation tests, and core promoter/precore mutation assays. Viral load tests that quantify hepatitis B viral DNA in peripheral blood (serum or plasma fractions) are currently the most useful and most widely used. International practice guidelines also recommend using sensitive nucleic acid amplification techniques for detecting and quantifying hepatitis B viral DNA in clinical practice to determine appropriate therapy. Several commercial nucleic acid amplification tests are currently available for quantifying hepatitis B viral

Table 6 Interpretation of hepatitis B serology.

| Test | Result | Interpretation |
|------|--------|----------------|
| HBsAg | Negative | Susceptible/No infection |
| Anti-HBc | Negative | |
| Anti-HBs | Negative | |
| HBsAg | Negative | Immunity after natural infection |
| Anti-HBc | Positive | |
| Anti-HBs | Positive | |
| HBsAg | Negative | Immunity due to hepatitis B vaccine |
| Anti-HBc | Negative | |
| Anti-HBs | Positive | |
| HbsAg | Positive | Acute infection |
| Anti-HBc | Positive | |
| IgM anti-HBc | Positive | |
| Anti-HBs | Negative | |
| HbsAg | Positive | Chronically infected |
| Anti-HBc | Positive | |
| IgM anti-HBc | Negative | |
| Anti-HBs | Negative | |
| HBsAg | Negative | Unclear interpretation, most likely resolved infection. Other possibilities include false positive anti-HBc, low Level chronic infection or resolving acute infection |
| Anti-HBc | Positive | |
| Anti-HBs | Negative | |

Source: Center for Disease Control and Prevention: division on viral hepatitis.

DNA in serum or plasma. Most hepatitis B viral DNA assays use real-time PCR assays either fully or partly automated. The COBAS AmpliPrep/COBAS TaqMan HBV, version 2.0 and Abbott RealTime HBV assay are widely used. Hologic Aptima HBV Quant Assay is a relatively new test for the determination of hepatitis B viral DNA in serum and plasma samples. It differs from other commercially available test by using target enrichment for nucleic acid purification and using real-time transcription-mediated amplification instead of PCR for nucleic acid amplification and quantification. Furthermore, it incorporates dual targets to improve homogenous quantification of diverse clinical HBV strains [15].

## Testing for hepatitis C

Hepatitis C (HCV) is a single-stranded RNA virus of the genus *Hepacivirus*, and the family Flaviviridae. The central 9.6 kb genome encodes a large polyprotein precursor subject to proteolytic cleavage by viral and host proteases to generate both structural (core, E1 and E2) and nonstructural proteins. It has been estimated that 130–170 million people worldwide are infected with HCV. Most patients with hepatitis C infection (60%–70%) may be asymptomatic but HCV infection is one of the leading causes of chronic liver disease (including liver cirrhosis), the main cause of hepatocellular carcinoma as well as the major indicator for liver transplantation in Western countries. HCV is categorized into seven genetically distinct genotypes; GT1, GT2, GT3, GT4, GT5, GT6, and GT7, as well as 67 confirmed subtypes. In general, GT1 is the most prevalent genotype globally at 49.1%, followed by GT 3 at 17.9%, GT4 at 16.8%, GT2 at 11.0%, and GT5 and 6 at <5% between them [16]. Persons with hepatitis C infection who drink alcohol on a regular basis have higher risk of liver diseases including fibrosis and cirrhosis [17].

Unlike hepatitis B, no vaccine is available for hepatitis C (HCV), but an effective treatment is available. The diagnostics tests available for hepatitis C can be classified under four broad categories:

- Immunoassays for detecting anti-HCV IgG antibodies.
- Recombinant immunoblot assays to detect anti-HCV IgG antibody; these test are more specific than immunoassays for detecting antibodies against HCV.
- Rapid assays for detecting anti-HCV IgG antibodies but these assays may be less specific than laboratory-based immunoassays.
- Detection of viral RNA by polymerase chain or related technique is the gold standard for HCV testing. These tests are in general called nucleic acid testing (NAT).

Commonly used screening assay to detect anti-HCV antibody are immunoassays. The latest third-generation assays have better sensitivity and specificity compared to first- and second-generation assays. In addition, the mean time to detect seroconversion is shortened by 2–3 weeks and now with third-generation assays are capable of detecting HCV infection as early as 10 weeks after exposure and these assays have over 99% specificity. Nevertheless, even third-generation immunoassay can produce false-negative result in patients undergoing hemodialysis or in immunocompromised patients. False-positive results may also occur. In the case of false-positive tests, recombinant immunoblot assays and strip immunoblot assay can be used to detect antibody against HCV. These tests are more specific than immunoassays. For establishing the diagnosis of past infection, recombinant immunoblot assay (RIBA) may also be undertaken. If the anti-HCV test is a false-positive test then recombinant immunoblot assay should be negative. If it is a case of past infection, recombinant immunoblot assay should be positive. There are some individuals who have HCV RNA but anti-HCV antibody testing is negative. This can be seen in the immunocompromised individuals or in the setting of early acute infection. HCV RNA tests are positive earlier than anti-HCV antibody tests. Rapid tests for the detection of anti-HCV also exist but these tests are more expensive than automated immunoassays.

More recently, FDA approved the OraQuick rapid HCV test for use with finger-stick of venous blood specimen. If an individual is positive for anti-HCV then the logical next step is to test for viral RNA. Viral RNA is detectable in serum or plasma as early as 1 week after exposure to HCV. Nucleic acid testing (NAT)-based assays for detecting HCV RNA includes polymerase chain reaction-based assays, branched DNA signal amplification and transcription-mediated amplification. Although both qualitative and quantitative methods are available, quantitative methods are gaining more acceptances [18]. If there is true infection, both anti-HCV antibody and tests for HCV RNA should be positive. If the HCV RNA is negative, then the possibilities include:

- A false-positive anti-HCV antibody test.
- If the individual is a newborn, the anti-HCV may be that of the mother with transfer of antibodies across the placenta.
- Intermittent viremia.
- Past infection.

Interpretation of HCV test results are summarized in Table 7. Blood products are routinely screened for hepatitis C using both immunoassays and NAT method to identify a donor with early infection stage without seroconversion. False-positive anti-HCV test result may also occur with immunoassays and in such cases, viral RNA must be tested. False-positive hepatitis C antibody may be observed in a patient with autoimmune hepatitis but after responding to the therapy, antibody against hepatitis C should be negative.

Autoimmune hepatitis is a periportal hepatitis where human immune system attacks liver cells mistaking them as foreign body and causes increased

**Table 7** Interpretation of HCV test results.

| Immunoassay (HCV antibody) | Recombinant immunoblot (HCV antibody) | HCV RNA (nucleic acid testing) | Interpretation |
| --- | --- | --- | --- |
| Negative | Negative | Negative | Mostly likely no infection |
| Negative | Negative | Positive | Very recent infection (window period of seroconversion) |
| Positive | Positive | Positive | Current infection |
| Positive | Positive | Negative | Past infection or in a newborn antibody from mother |
| Positive | Negative | Negative | False-positive screen |

immunoglobulins and autoantibodies. This disease affects more women than men (70% females) especially young girls and may respond to immunosuppression therapy. It is postulated that a drug, virus, or an environmental agent can trigger a T-cell-mediated cascade directed against liver antigens in genetically susceptible individuals who are prone to autoimmune disease. Autoimmune hepatitis is treated with high doses of prednisolone, which are tapered gradually as azathioprine is introduced. Recent guidelines have described newer treatment using mycophenolate mofetil, calcineurin inhibitors, mTOR inhibitors, and biological agents are potential salvage therapies, but should be reserved for selected nonresponsive patients and administered only in experienced centers. Liver transplantation is a life-saving option for those patients who progress to end-stage liver disease [19].

Antiviral agents are used in treating hepatitis C. Originally hepatitis C was treated with ribavirin and interferon but since 2011, direct acting antivirals (Sofosbuvir, Simeprevir, Daclatasvir, and ledipasvir) that attack hepatitis C virus have been introduced. Moreover, combination drugs such as Sofosbuvir + ledipasvir (Harvoni) and Ombitasvir-Paritaprevir/Ritonavir and dasabuvir (Viekirax) are also approved by the FDA for treating hepatitis C infection. All these drugs can be given orally [20].

## Immunization and false-positive HIV and hepatitis testing

There are reports of false-positive HIV and hepatitis C serological test results following influenza vaccination in some subjects. Although for HIV serology, this phenomenon is transient, for hepatitis C serology, this phenomenon may persist for a longer time. However, Erickson et al. observed false-positive HIV antibody test result in an individual using an enzyme immunoassay 11 days after receiving influenza vaccine but his antibody against hepatitis C was negative. However, Western blot test was indeterminate and no viral load was detected in his blood. The authors considered that his HIV serology test was false-positive and when he was tested after 1 month, his viral load was again undetectable but his Western blot reverted to nonreactive [21]. Rubella vaccination and transitory false-positive results for human immunodeficiency virus Type I in blood donors has also been reported [22].

## Testing for Epstein-Bar virus

Epstein-Barr virus (EBV) targets B lymphocytes and the infected B lymphocytes disseminate the virus throughout the reticuloendothelial system. The T cells mount an immune response against the infected B cells and these T cells which appear as morphologically reactive cells, respond to EBV infection by producing two types/classes of antibodies:

- Antibodies specific to EBV
- Polyclonal antibodies such as heterophil antibodies, cold agglutinins, rheumatoid factors, and antinuclear antibodies (ANA)

EBV-specific antibodies are antibodies to viral capsid antigen (VCA), antibody to early antigen (EA) and antibody to Epstein-Barr nuclear antigen (EBNA). Antibody to VCA can be IgG or IgM. In acute infection, patients are positive for VCA-IgM. Antibody to EA may be positive or negative and antibody to EBNA is usually negative. With past infection, VCA-IgM is negative but VCA IgG is positive. In addition, antibody to EA is negative but antibody to EBNA is positive. Heterophilic antibodies are cross-reacting antibodies to antigens occurring in several species that are not phylogenetically related. Heterophilic antibodies usually include two different types:

- Capable of agglutinating horse red blood cell (this is the basis of Monospot test)
- Do not react with any EBV antigens

Serological markers for diagnosis of EPV infection are summarized in Table 8.

## Summary/key points

- Human immunodeficiency virus (HIV) is a slowly replicating retrovirus responsible for acquired immunodeficiency syndrome (AIDS). HIV infection is sexually transmitted but is also transmitted due to sharing infected needle, blood transfer, and from mother to her newborn. HIV infects vital cells involved in human immune function such as helper T cells especially CD4 + T cells, macrophages, and dendritic cells. Two major types of HIV virus have been characterized: HIV-1 and HIV-2. HIV-1 was initially discovered and is the cause of majority of HIV infection worldwide. HIV-2 has lower infectivity than HIV-1 and is largely confined

Table 8 Serological profile of Epstein bar virus.

| Test | Acute infection | Past infection |
| --- | --- | --- |
| IgM antibody against viral capsid antigen (VCA) | Positive | Negative |
| IgG antibody against viral capsid antigen (VCA) | Positive | Positive |
| Antibody to early antigen (EA) | Positive/negative | Negative |
| Antibody to Epstein-Barr nuclear antigen (EBNA) | Negative | Positive |
| Heterophilic antibody | Positive | Negative |

in West Africa but is spreading slowly in North America. Several groups are classified within HIV-1. HIV-1 M group is the main type of HIV virus seen in clinical practice and M type can be divided into several subtypes.

- After HIV infection, viral RNA can be measured within 10–12 days and viral p24 antigen can be measured a little bit later. In general, IgM is the first antibody against HIV followed by IgG antibodies which appear approximately 3–4 weeks after infection. Moreover, within 1–2 months after infection, HIV antibodies are usually present in almost all of the infected individuals although for few individuals, it may take up to 6 months for antibodies to appear in circulation.

- Testing for HIV can be broadly divided into screening tests and confirmatory tests. Screening tests include standard testing, rapid HIV testing, and combination HIV antibody and antigen testing. Confirmatory testing is performed by Western blot and recombinant immunoblot, or line immunoassay (LIA).

- Standard screen is performed by various immunoassays. The test is based on the detection of IgG antibody against HIV-1 antigens in the serum. HIV antigens include p24, gp 120, and gp 41 and antibodies to gp 41 and p24 are the first detectable serologic markers following HIV infection. IgG antibodies appear as early as 3 weeks but most likely within 12 weeks in the majority of patients and generally persist for life.

- Causes of false-negative HIV screening tests include testing during the window period, fulminant HIV infection, immunosuppression, or immune dysfunction and delay in seroconversion due to early initiation of antiretroviral therapy.

- Since introduction in 1985, ELISA method has been evolved and now third-generation ELISA assays (sandwich format) can detect IgG and IgM antibodies and antibodies to all of the M subtypes as well as N and O group. Antibody testing method capable of detecting O group is important for screening of blood products. More recently, fourth-generation assays have been introduced which simultaneously detect HIV antibodies and p24 antigen reducing the window period to only 13–15 days.

- Rapid HIV antibody tests can provide test result in less than 30 min and can be adopted in point-of-care settings. However, both false-positive and false-negative results may occur with rapid HIV tests and it is important to confirm initial findings with a laboratory-based HIV assay.

- HIV viral load test detects the number of copies of RNA of HIV virus present in per milliliter of serum or plasma and is expressed as copies/mL or in log scale. HIV viral load test indicates viral replication and is often conducted to monitor the progress of antiretroviral therapy. In addition, CD4 lymphocyte counts are also measured to evaluate immune system.

- The nucleic acid tests (NAT) are commercially available tests that can identify HIV nucleic acid (either RNA or proviral DNA) and these testes are based on the principle of polymerase chain reaction (PCR), real-time PCR, nucleic acid sequence-based amplification or ligase chain reaction. NAT assays are useful in special population such as in the window period of infection when antibody against HIV is absent in serum, and newborns of HIV-infected mother where maternal antibody against HIV is present in the serum of the newborn.
- Hepatitis A testing is straightforward where IgM antihepatitis A antibody denotes recent infection and IgG antihepatitis A antibody appears in the convalescent phase of acute hepatitis.
- Hepatitis E virus is also an enterically transmitted virus. Hepatitis E can also be transmitted by blood transfusion, particularly in endemic areas. Chronic hepatitis does not develop after acute hepatitis E infection, except in the transplant setting and possibly in other settings of immunosuppression. The diagnosis of hepatitis E is based upon the detection of hepatitis E virus in serum or stool by polymerase chain reaction (PCR) or by the detection of IgM antibodies against hepatitis.
- The hepatitis D virus (also called the delta-virus) is a defective pathogen that requires the presence of the hepatitis B virus for infection. Hepatitis D can elicit a specific immune response in the infected host, consisting of antibodies of the IgM and IgG class (antihepatitis D antibodies).
- Serologic markers available for hepatitis B infection are HBsAg (hepatitis B surface antigen), HBeAg (hepatitis B, e antigen), anti-HBc (antibody against hepatitis B core antigen; both IgG and IgM), anti-HBs (antibody against hepatitis B surface antigen), anti-HBe (antibody against hepatitis B e antigen) and testing for viral DNA. HBsAg is the first marker to be positive after exposure to hepatitis B virus. It can be detected even before the onset of symptoms. Most patients may clear the virus, and HBsAg typically becomes undetectable within 4–6 months. Persistence of HBsAg for more than 6 months implies chronic infection. The disappearance of HBsAg is followed by the presence of anti-HBs. During the window period (after the disappearance of HBsAg and before the appearance of anti-HBs), evidence of infection is documented by the presence of anti-HBc (IgM). Coexistence of HBsAg and anti-HBs has been documented in approximately 24% of HBsAg-positive individuals. It is thought that the antibodies fail to neutralize the virus particles. These individuals should be considered as carriers of hepatitis B virus. There exists a subset of patients who have undetectable HBsAg but are positive for hepatitis B viral DNA. Most of these patients have very low viral load with undetectable levels of HBsAg. Uncommon situations are infection with hepatitis B variants that decrease HBsAg production or mutant strains that have altered epitopes normally used for detection of HBsAg.

- Individuals with recent infection will develop anti-HBc, IgM antibodies. Individuals with chronic infection and individuals who have recovered from an infective episode will develop anti-HBc, IgG antibodies. However, anti-HBc, IgM antibodies may remain positive for up to 2 years after an acute infection. Levels may also increase and be detected during exacerbations of chronic hepatitis B. This may lead to diagnostic confusion.

- The diagnostics tests available for HCV can be classified under four broad categories: immunoassays for detecting anti-HCV IgG antibodies, recombinant immunoblot assays to detect anti-HCV IgG antibody (these test are more specific than immunoassays for detecting antibodies against HCV), rapid assays for detecting anti-HCV IgG antibodies (these assays may be less specific than laboratory-based immunoassays), and detection of viral RNA by polymerase chain or related technique (the gold standard for HCV testing, these tests in general are called nucleic acid testing: NAT).

- Epstein-Barr virus targets B lymphocytes and the infected B lymphocytes disseminate the virus throughout the reticuloendothelial system. The T cells mount an immune response against the infected B cells and these T cells which appear as morphologically reactive cells, respond to EBV infection by producing two types/classes of antibodies including antibodies specific to EBV and polyclonal antibodies such as heterophilic antibodies, cold agglutinins, rheumatoid factors, and antinuclear antibodies (ANA).

- EBV-specific antibodies are antibodies to viral capsid antigen (VCA), antibody to early antigen (EA), and antibody to Epstein-Barr nuclear antigen (EBNA). Antibody to VCA can be IgG or IgM. In acute infection, patients are positive for VCA-IgM. Antibody to EA may be positive or negative and antibody to EBNA is usually negative. With past infection, VCA-IgM is negative but VCA-IgG is positive. In addition, antibody to EA is negative but antibody to EBNA is positive.

## References

[1] Visseaux B, Le Hingrat Q, Damond F, Charpentier C, Descamps D. Physiopathology of HIV-2 infection. Virologie (Montrouge) 2019;23:277–91.

[2] Luft L, Gill MJ, Church DL. HIV-1 viral diversity and its implications for viral load testing: review of current platforms. Int J Infect Dis 2011;15:e661–70.

[3] Gallant JE. HIV counseling, testing and referral. Am Fam Physician 2004;70:295–302.

[4] Spivak A, Sydnor E, Blankson JN, Gallant JE. Seronegative HIV-1 infection: a review of the literature. AIDS 2010;24:1407–14.

[5] Ling AE, Robbins KE, Brown TM, Dunmire V, et al. Failure of routine HIV-1 tests in a case involving transmission with preseroconversion blood components during the infectious window period. JAMA 2000;284:210–4.

[6] Vardinon N, Yust I, Katz O, Laina A, et al. Anti-HIV indeterminate Western blot in dialysis patients: a long term follow up. Am J Kidney Dis 1999;34:146–9.

[7] Delaney KP, Branson BM, Uniyal A, Phillips S, et al. Evaluation of the performance characteristics of 6 rapid HIV antibody tests. Clin Infect Dis 2011;52:257–63.

[8] Facente SN, Dowling T, Vittinghoff E, Sykes DL, et al. False positive rate of rapid oral fluid HIV test increases as kits near expiration date. PLoS One 2009;4, e8217.

[9] Garcia-Prats AJ, Draper HR, Sanders JE, Agarwal AK, et al. False negative post 18 month confirmatory DNA PCR-positive children: a retrospective analysis. AIDS 2012;26:1927–34.

[10] Drain PK, Dorward J, Bender A, Lillis L, et al. Point-of-care HIV viral load testing: an essential tool for a sustainable global HIV/AIDS response. Clin Microbiol Rev 2019;32(3). e00097-18.

[11] Thuener J. Hepatitis A and B infections. Prim Care 2017;44:621–9.

[12] Wilkins T, Zimmerman D, Schade RR. Hepatitis B diagnosis and treatment. Am Fam Physician 2010;81:965–72.

[13] Rysgaard C, Morris CS, Dress D, Bebber T, et al. Positive hepatitis B surface antigen test due to recent vaccination: a persistent problem. BMC Clin Pathol 2012;12:15.

[14] Kuhns MC, Busch MP. New strategies for blood donor screening for hepatitis B virus: nucleic acid testing versus immunoassay methods. Mol Diagn Ther 2006;10:877–91.

[15] Schønning K, Johansen K, Nielsen LG, Weis N, Westh H. Analytical performance of the Hologic Aptima HBV Quant Assay and the COBAS Ampliprep/COBAS TaqMan HBV test v2.0 for the quantification of HBV DNA in plasma samples. J Clin Virol 2018;104:83–8.

[16] Rabaan AA, Al-Ahmed SH, Bazzi AM, Alfouzan WA, et al. Overview of hepatitis C infection, molecular biology, and new treatment. J Infect Public Health 2020;13:773–83.

[17] Wilkins T, Malcolm J, Raina D, Sachade R, Hepatitis C. Diagnosis and treatment. Am Fam Physician 2010;81:1351–7.

[18] Kamili S, Drobeniuc J, Araujo A, Hayden TM. Laboratory diagnostics for hepatitis C virus. Clin Infect Dis 2012;55(Suppl 1):S43–8.

[19] Liberal R, Krawitt EL, Vierling JH, Manns MP, et al. Cutting edge issues in autoimmune hepatitis. J Autoimmun 2016;75:6–19.

[20] González-Grande R, Jiménez-Pérez M, González Arjona C, Mostazo Torres J. New approaches in the treatment of hepatitis C. World J Gastroenterol 2016;22:1421–32.

[21] Erickson CP, McNiff T, Klausner JD. Influenza vaccination and false positive HIV result. N Engl J Med 2006;354:1422–3.

[22] Araujo PB, Albertoni G, Arnoni C, Ribeiro J, et al. Rubella vaccination and transitory false positive results for human immunodeficiency virus Type I in blood donors. Transfusion 2009;49:2516–7.

# CHAPTER 26

# Autoimmunity, complement, and immunodeficiency

## Introduction to immune system and complement

The immune system is a complex biological structure in an organism capable of protecting the organism from disease by neutralizing the invading organism. Human immune system can detect a wide variety of pathogens including viruses, bacteria, parasitic worms, and even certain cancer cells within human body and is capable of distinguishing such objects from organ's healthy tissue. In autoimmune disease, this distinction is lost, and body's own immune system can target certain cells in an organ or a variety of organs causing disease. Disorder of immune system can also cause inflammatory diseases and cancer. Immunity can be broadly divided into innate (natural) and adaptive (specific) immunity. Innate immunity does not provide defense against any particular pathogen but can be considered as all-purpose immunity and can attack a number of pathogens in a short period of time when challenged. Therefore, innate immunity is the first line of defense. A characteristic feature of innate immunity is that no prior memory of infection by an agent needed for its activation. The components of innate immunity include certain cells and molecules. These cells include granulocytes (neutrophils, eosinophils, basophils, and mast cells), monocytes, dendritic cells, and natural killer lymphocytes. Molecules involved in innate immunity include:

- Complement
- Enzymes (e.g., lysozyme)
- Collectins
- Pentraxin (e.g., C-reactive protein)

The largest group of cells involved in innate immunity is granulocytes. Neutrophils phagocytose microorganisms and possess enzymes such as myeloperoxidase, cathepsins, proteinase 3, elastase, and defensins for killing. Eosinophils play a major role against multicellular parasites (e.g., worms). Eosinophils release toxic proteins against parasites including major basic protein, eosinophilic cationic protein and eosinophil neurotoxin. Basophils and mast cells

**535**

Clinical Chemistry, Immunology and Laboratory Quality Control. https://doi.org/10.1016/B978-0-12-815960-6.00023-6

release chemicals such as histamine which is an inflammatory mediator. Mast cells also release TNF-alpha (tumor narcosis factor-alpha) which recruits and activates neutrophils. Monocytes and macrophages have phagocytic and micro-bicidal properties and also release cytokines as well as clear cellular debris. Monocytes and dendritic cells present antigens to T lymphocytes which can trigger adaptive immunity. Therefore, major steps in achieving adaptive immunity are by phagocytosis, enzyme activation, liberation of cytokines, and activation of complement. Cells use receptors to recognize microbes. Most important of these receptors are known as Toll-like receptors. Complement system consists of plasma proteins which have an important role in immunity and inflammation. The complement proteins are present in plasma in inactive states. These proteins are numbered C1–C9. Many complement proteins once activated in turn can activate other complement proteins.

Specific or adaptive immunity is characterized by greater specificity than innate immunity, but the response is slower than innate immunity. Lymphocytes play important role in adaptive immunity. Three types of lymphocytes, T-helper cells, T-cytotoxic cells, and B-cells mediate specific immunity.

## T lymphocytes and cell-mediated immunity

T cells constitute about two-third of lymphocytes in the peripheral blood. In general T cells are not activated by free or circulating antigens, but such cells can recognize antigens bound to major histocompatibility complex (MHC) molecules. T-cell receptor complex consists of heterodimers of alpha and beta chains, CD3 proteins (gamma, delta, and epsilon proteins) as well as two zeta chains. The antigen associated with the MHC molecule of the cell binds with the alpha and beta chains. The other proteins interact with constant portion of the alpha and beta chains of the T-cell receptors to generate intracellular signals after the alpha and beta chains have recognized the antigen. If the T cell is a CD4 + cell, it will recognize antigens presented in association with Class II MHC molecules. If the T cell is a CD8 + cell, it will recognize antigens presented in association with Class I MHC molecule. There is also a minority of T cells which have gamma and delta chains as the receptors instead of alpha and beta chains. These gamma delta cells do not express either as CD4 or CD8. Typically, more than one signal is required for activation of these T cells. CD8 + lymphocytes differentiate into cytotoxic T lymphocytes that kill cells harboring microbes in their cytoplasm. CD4 + T cells once activated also secrete IL-2which in turn causes proliferation of T lymphocytes. Interferon gamma (IFN-γ) activates macrophages and also stimulates B cells to produce antibodies. Interleukin-4 (IL-4) stimulates B cells to differentiate into IgE producing plasma cells. Some T cells also differentiate into long-living memory cells.

# B cells and humoral immunity

B cells can be directly stimulated by antigens. The antigens are recognized by immunoglobulin molecules on the surface of B cells. These immunoglobulin molecules are either IgM or IgD. A second signal also comes from activated CD4 + T helper cells. Once activated, the B cells differentiate into plasma cells. The first exposure to antigen results in production of IgM class of (primary response) but with subsequent exposure to the same antigen IgG antibodies are produced (secondary response). This ability of changing the antibody class is called "class switching." Immunoglobulins contain both heavy chain and light chain and names of various antibodies are derived from names of the heavy chain, for example, IgG contains gamma heavy chain, hence the name. Various features of immunoglobulins are summarized in Table 1. Characteristics of various immunoglobulins include:

- IgG is the dominant class of immunoglobulins which are produced as a secondary response of adaptive immunity.
- IgM is the primary response of adaptive immunity.
- IgA is found in mucus membrane secretions.
- IgE is required for defense against parasites especially worms.
- IgD is present as B-cell surface receptor.

Many factors can affect immune response. Interestingly acute stress can be associated with adaptive upregulation of some parameters of natural immunity and downregulation of some functions of specific immunity while chronic stress can cause suppression of both cellular and humoral response [1].

# Pathways of complement activation

Complement system was discovered more than 100 years ago by Jules Bordet, and since then the importance of complement system in protecting humans and animals against infections has been well recognized. The name complement

**Table 1** Various classes of immunoglobulins.

| Specific feature | IgG | IgM | IgA | IgE | IgD |
|---|---|---|---|---|---|
| Heavy chain | Gamma (γ) | Mu (μ) | Alpha (α) | Epsilon (ε) | Delta (δ) |
| Molecular type | Monomer | Pentamer | Dimer | Monomer | Monomer |
| Subclass | IgG1–4 | None | IgA1–2 | None | None |
| Interact with | Yes | Yes | Yes | No | No |
| Complements | (Classical) | (Classical) | Alternate | No | No |
| Bind to mast cells | No | No | No | No | No |
| Crosses placenta | Yes | Yes | Yes | No | No |

system is derived from the ability of this system to complement the action of antibodies and phagocytic cells in destroying pathogens from an organism. At present more than 30 complement proteins have been discovered [2]. Complement proteins can interact with each other and are also capable of interacting with cell surface proteins. A key step in the activation of complements is the activation of C3 which can be accomplished through three different pathways:

- Classical pathway: initiated by antigen-antibody complexes fixating with C1.
- Alternate pathway: spontaneous but bacterial polysaccharides and proteins such as properdin, factor B and D are also involved in activating this pathway.
- Lectin pathway: plasma lectin binds to mannose on microbes and activates the pathway.

The classical pathway is triggered by activation of C1 complement. There are three components of C1, which are C1q, C1r, and C1s. IgG or IgM antibodies already bound with antigen can also bind to C1q thus activating complements through the classical pathway. One IgM molecule is capable of activating this pathway, but multiple IgG molecules are needed for activation. Subsequently C1r and then C1s are activated which in turn cleaves C4 and C2 to yield C4a, C4b, C2a, and C2b. Then C4b and C2a form C4b2A (C3 convertase of the classical pathway) that splits C3 into C3a and C3b. At that point C3b combines with C4b2a to form C5 convertase that cleaves C5 into C5a and C5b which eventually leads to the formation of C5b, C6, C7, C8 and C9 membrane attack complex (MAC). C1 is inhibited by C1 inhibitor. C3 convertase is inhibited by decay accelerating factor, a protein that is absent in paroxysmal nocturnal hemoglobinuria (PNH).

The alternative pathway is a low-level activation pathway which is spontaneous but also can be activated by insoluble bacterial polysaccharides, yeast cell walls, etc. in the absence of antibody. This is due to spontaneous hydrolysis of C3 to C3a and C3b which eventually in the presence of factor B and D, forms C3bBb complex which acts as C3 convertase. Properdin stabilizes the C3 convertase by binding to the complex which in turn cleaves other C3 molecules, and the cascade is continued. Factors H and I can inhibit alternative pathway.

In lectin pathway, plasma lectins bind to mannose on microbes, and activation of C4 and C2 takes place as in the classical pathway. Regardless of pathway, the end result of activation of C3 leads to the formation of C3 convertase that breaks down C3 to yield C3a and C3b. C3b can act as an opsonin or bind to C3 convertase to form C5 convertase which breaks down C5 to yield C5a and C5b and subsequently, activation of C6, C7, C8, and C9 are achieved. Thus, complement once activated promotes inflammation, recruits cells, and kills targeted cells. Various activities of complement include:

- Opsonins: C3b and C4b can promote phagocytosis by phagocytic cells.
- Anaphylatoxins: C3a, C4a, and C5a.
- Leukocyte activation and chemotaxis: C5a.
- Cell lysis: Membrane attack complex (C5 to C9, activated).
- Remove circulating antigen-antibody complexes.

Complement deficiency may result in impaired innate immunity (a well-known example is increased susceptibility to *N. meningitidis* infection) and immune complex-mediated inflammation (such as glomerulonephritis, vasculitis, and systemic lupus erythematosus). Deficiency of C1 inhibitor may be inherited, known as hereditary angioedema.

## Immunodeficiency

Immunodeficiency may be primary or secondary. Primary immunodeficiency includes a broad category of diseases including B-cell defect, T-cell defect, and combination of both B- and T-cell defects, complement deficiency, and defective phagocytosis.

### B-cell defect

Patients with B-cell defects are typically susceptible to recurrent bacterial infections, especially respiratory tract infections involving influenzae virus, *Strep pneumonia* and *Staph aureus*. Diarrhea may also be present due to infection due to enterovirus and/or *Giardia lamblia*. However, most viral, fungal, and protozoal infections are cleared due to intact cell-mediated immunity.

X-linked agammaglobulinemia (XLA; also known as Bruton's disease) is a primary humoral immunodeficiency characterized by severe hypogammaglobulinemia and increased risk of infection. The genetic condition results from a mutation in the Bruton tyrosine kinase (BTK) gene located on the X chromosome. This mutation results in arrest in B-cell maturation, from pre-B cells to B cells leading to a near absence of B cells. Patients affected by XLA are most commonly predisposed to frequent and severe bacterial infections. Clinical features are seen soon after birth once maternal immunoglobulins (that have crossed the placenta) start to decline. B cells, plasma cells, and immunoglobulin levels are all decreased. Lymphoid tissues lack germinal centers. Subsequently, there is an increased incidence of leukemia, lymphoma, and autoimmune diseases. This is a rare disease affecting an average of one in 379,000 total births per year or one in 190,000 male births per year in the Unites States. Patients suffering from XLA can be diagnosed in early age and treated with regular intravenous immunoglobulin therapy before the sequelae of recurrent infections. More recently, such patients can also be treated with subcutaneous immunoglobulin [3].

Common variable immunodeficiency (CVID) is a primary B-cell immunodeficiency disorder, characterized by very significant hypogammaglobulinemia. The disease can develop at any age without gender predominance. The underlying causes of CVID remain largely unknown; primary B-cell dysfunctions, defects in T cells, and antigen-presenting cells are involved. Although some monogenetic defects have been identified in some CVID patients, most likely CVID is a polygenic disorder. Patients with CVID suffer from recurrent and chronic infections (e.g., bacterial infections of the respiratory or gastrointestinal tract), autoimmune diseases, lymphoproliferation, malignancies, and granulomatous lesions. The disease can affect multiple organs, including the liver. Replacement therapy with immunoglobulin and antiinfection therapy are the primary treatment regimen for CVID patients [4].

Selective immunoglobulin A deficiency is the most common primary antibody deficiency. Although many patients are asymptomatic, selected patients suffer from different clinical complications such as pulmonary infections, allergies, autoimmune diseases, gastrointestinal disorders, and malignancy. Pathogenesis of this disease is still unknown; however, a defective terminal differentiation of B cells and defect in switching to IgA-producing plasma cells are presumed to be responsible. Some individuals with IgA deficiency are susceptible to develop anaphylactic reactions with blood products. There is no specific treatment for patients with symptomatic IgA deficiency [5].

## T-cell defects

Individuals with T-cell defects have recurrent and persistent viral, fungal, and protozoal infections. Individuals are also at risk for transfusion-associated graft-vs-host disease (GVHD). DiGeorge syndrome is characterized by failure of development of thymus and parathyroid (due to failure of development of the third and fourth pharyngeal pouches). Features of T-cell immunodeficiency include hypoparathyroidism, dysmorphic facies and cardiac defects.

## Both B- and T-cell defects

In severe combined immunodeficiency (SCID), underlying genetic defect may be diverse. About 50% of cases are transmitted as X-linked disorder. These are due to mutation in the gene coding for the gamma chain for receptors for various interleukins (IL) including IL-2, IL-4, IL-7, IL-9, and IL-15. In addition, 40%–50% of cases of SCID are transmitted as autosomal recessive and most common example of this type of SCID is due to mutation in the gene encoding for adenosine deaminase (ADA) enzyme. ADA deficiency causes accumulation of adenosine and deoxyadenosine triphosphate metabolites which are lymphotoxins. In SCID patients, the thymus is hypoplastic, lymph nodes and lymphoid tissues lack germinal centers as well as paracortical T cells. Lack of help from T cells does not allow B cells to be functional and clinical features seen in both B- and T-cell defects are also present in patients with combined B- and T-cell defects.

As mentioned earlier in this chapter, IgM is produced first as a primary immune response and subsequently other classes of immunoglobulins are produced. This is called class switching. One factor involved in class switching is interaction of CD40 molecules on B cells and CD40 ligand (CD40L or CD154) on T helper cells. The most common cause of hyper IgM syndrome is mutation in the gene encoding for CD40L, which is located in the X chromosome. The interaction between CD40 and CD40L is also required for T helper-mediated activation of macrophages. Thus, both humoral immunity and cell-mediated immunity are affected in hyper IgM syndrome, and IgM levels are normal or high with low levels of IgG, IgA, and IgE. Wiskott Aldrich syndrome (WAS) is an X-linked disease characterized by immunodeficiency, eczema, and thrombocytopenia. In this syndrome, the platelets are small in size. The WAS gene codes for the Wiskott Aldrich syndrome protein (WASP). This syndrome is complicated by lymphoreticular malignancies and autoimmune diseases.

Ataxia telangiectasia is an autosomal recessive condition due to mutation of the ataxia telangiectasia mutated gene (ATM gene) which encodes for the ATM protein kinase involved in DNA repair. Clinical features include cerebellar ataxia, oculocutaneous telangiectasia, and lymphoreticular malignancy. EBV (Epstein-Barr virus) associated immunodeficiency (Duncan syndrome or X-linked lymphoproliferative disease) is a disease where individuals develop overwhelming EBV infections, immunodeficiency, aplastic anemia, and lymphomas. Complement deficiency can be associated with deficiency of C3 that leads to increased infections by pyogenic organisms or deficiency of C5–C9, which leads to increased infections by Neisseria (both gonococcus and meningococcus).

Defective phagocytosis can be classified under three broad categories:

- Chronic granulomatous disease (CGD): This disease is characterized by a deficiency of nicotinamide adenine dinucleotide phosphate (NADPH) oxidase resulting in lack of oxidative burst and defective killing of bacteria and fungus that are catalase positive (e.g., staphylococcus and aspergillus). This disorder may be inherited as X-linked recessive and also in autosomal recessive fashion. The gene encoding for the Kx antigen of the Kell blood group system is very close to the gene encoding for NADPH oxidase and if both are affected, Kx antigen may also be lacking. This is the called McLeod phenotype, which is associated with presence of acanthocytes. A screening test used for CGD is the nitro blue tetrazolium test (NBT) where a yellow dye is converted into a blue dye if NADPH oxidase function is intact.

- Chediak Higashi syndrome is an autosomal recessive condition where defective trafficking of intracellular organelles lead to defective fusion of lysosomes with phagosomes. This syndrome is related to mutation in the lysosomal trafficking regulator gene. Granulocytes, lymphocytes, and monocytes exhibit giant lysosomes. Neutropenia, thrombocytopenia, and oculocutaneous albinism are seen in this syndrome along with immunodeficiency.
- Leukocyte adhesion deficiency (LAD): LAD type 1 (LAD-1) is characterized by defective synthesis of LFA-1 and Mac-1, which are integrins. This results in defective leukocyte adhesion to endothelium, impaired leukocyte migration and defective leukocyte phagocytosis. LAD type 2 (LAD-2) is due to absence of sialyl-Lewis X in leukocytes that binds to selectin on endothelium.

## Major histocompatibility complex

Histocompatibility molecules are important for immune response, but these molecules are also responsible for evoking transplant rejections. Histocompatibility molecules bind to peptide fragments of foreign proteins and render them susceptible to attack by specific T cells. The genes encoding for the histocompatibility molecules are clustered on a small segment (small arm) of chromosome 6. The cluster of genes is known as major histocompatibility complex (MHC) or human leukocyte antigen (HLA) complex. The HLA system is highly polymorphic. The proteins encoded by certain HLA genes are also called antigens which are essential elements of immune function and play major roles in histocompatibility during an organ transplant.

Class I MHC molecules are present on all nucleated cells and platelets. There are three different Class I MHC molecules: A, B, and C. Class I molecules are heterodimers of an alpha (or heavy) chain and a smaller beta two microglobulin. The beta two microglobulin molecule is extracellular and the alpha chain has extracellular components as well as parts which traverse the cell membrane and into the cell. The extracellular part of the alpha chain has three domains, alpha1, alpha2, and alpha3. Peptides are able to bind within a groove formed by the alpha1 and alpha2 domains.

Class II MHC molecules are present on B lymphocytes and monocytes. There are three different Class II MHC molecules (DP, DQ, and DR). Class II molecules are also heterodimers of one alpha chain and one beta chain. Both chains have extracellular components with parts that traverse the cell membrane and into the cell. The extracellular portions of both chains have domains alpha1, alpha2 and beta1 and beta 2. The peptide or antigen binding site is formed between the alpha1 and beta 1 domains. Class III MHC molecules are components of the complement system.

Antigens within a cell may bind with Class I MHC molecule which is produced within the cell. This binding takes place in the endoplasmic reticulum. The complex of Class I molecule and the antigen is transported to the cell surface for presentation to CD8+ cytotoxic T lymphocytes. The T-cell receptor (TCR) recognizes and binds with the MHC-peptide complex (MHC molecule-antigen complex). The CD8 molecule also binds with the alpha3 domain of the Class I MHC molecule. The T cell is thus activated. CD8+ T cells are Class I MHC restricted because they can only be activated with antigen which are bound to MHC Class I molecules. Similarly, CD4+ T cells are Class II restricted. A variety of diseases are associated with certain HLA alleles (Table 2). Link between HLA-B27 and ankylosing spondylitis has been well established. Li et al. reported association between HLA-46 with Graves' disease [6].

## HLA testing

HLA test, also known as HLA-typing or tissue typing is used to identify antigens on blood cells that determine compatibility of a recipient with a donor organ. If HLA antigens of the recipient match well with a donor organ, possibility of organ rejection is minimized. However, HLA matching is more complex than blood group matching because there are six loci on chromosome 6 where the genes that code HLA antigens are inherited which include HLA-A, HLA-B, HLA-C, HLA-DR, HLA-DQ, and HLA-DP. In classical serological HLA testing, antibodies are used to distinguish between various variants of HLA antigens. Each antibody is specific for a particular antigen and by using different antibodies HLA serotyping is performed to determine if donor serotype is a good match

Table 2 Association between various HLA alleles and diseases.

| HLA allele | Increased risk of disease |
| --- | --- |
| HLA-B27 | Ankylosing spondylitis<br>Reactive arthritis |
| HLA-B46 | Graves' disease (Asian population)[a] |
| HLA-B47 | 21-Hydroxylase deficiency |
| HLA-DR2 | Systemic lupus erythematosus |
| HLA-DR3 | Type 1 diabetes mellitus<br>Systemic lupus erythematosus<br>Myasthenia gravis |
| HLA-DR4 | Rheumatoid arthritis |

[a]Other HLA types such as HLA-B8 may also be associated with Graves' disease.

for the recipient. However, serological testing is limited by number of antibodies available against specific HLA antigens and more recently, molecular techniques have been used for HLA DNA typing, which is superior to classical serological tests. The HLA Class I genes are by far the most polymorphic genes in the human genome. Current molecular techniques for HLA DNA typing include recombinant DNA technology, chain-termination Sanger sequence, and polymerase chain reaction (PCR)-based amplification. These molecular tests can recognize more alleles than traditional serological testing [7].

HLA typing along with ABO (blood type) grouping is used to evaluate tissue compatibility between a donor and a potential transplant recipient. HLA typing is performed before various transplant types including kidney, liver, heart, pancreas, and bone marrow transplant. The success of a transplant increases with number of identical HLA antigens between a recipient and a potential donor. Major types of HLA tests include:

- HLA antigen typing between donor and recipient: Classically this type of testing is done using serological markers but more recently molecular (DNA) typing, which provides more information, is replacing classical serological testing.
- HLA antibody screening: HLA antibody screening is performed on the recipient in order to determine whether any antibody is present in the recipient that may target against donor organ triggering organ rejection. HLA antibody is not always present in an individual unless the person has received blood transfusion or a woman after pregnancy.
- Lymphocyte crossmatching: This step takes place when a donor is identified, and the objective is to identify whether any antibody is present in the recipient that may be directed against antigen present on the donor's lymphocyte. In this test, serum from the intended recipient is mixed with T and B lymphocytes (white blood cells) of the donor to investigate any potential reaction (positive test result) that will destroy white blood cells of a recipient.

## Transplant rejection

Graft rejection is due to recognition by the host that the graft is a foreign entity. The antigens responsible for graft rejection belong to the HLA system. Rejection involves both the cell-mediated immunity and humoral immunity. In T cell-mediated rejection, individual T cells recognize a single peptide antigen in the graft by two distinct pathways, direct and indirect pathways. In the direct pathway host T cells encounter donor MHC molecules through interstitial dendritic cells that are present in the donor organ and function as antigen presenting cells (APC). These dendritic cells have a high density of MHC molecules capable of directly stimulating the host T cells. The encounter of the host T cells and donor dendritic cells may take place in the graft or when the dendritic cells move out of the graft and migrate to regional lymph nodes. Both CD4+ and CD8+ cells are activated. CD8+ cells are responsible for cell-mediated cytotoxicity. CD4+ cells secrete cytokines which result in accumulation of lymphocytes and macrophages. In the indirect pathway T cells recognize peptides (as antigens) presented by host APCs not donor APCs. The peptides (antigens) are however, derived from the graft tissue. The direct pathway is the major pathway in acute cellular rejection and the indirect pathway is thought to be responsible for chronic rejection.

Antibody mediated rejection is also an important aspect of graft rejection where a host may have developed preformed antibodies to donor antigens even before the transplant. Prior blood transfusion may lead to the development of anti-HLA antibodies because of platelets and white blood cells which are rich in HLA antigens. Multiparous women may also develop anti-HLA antibodies. These antibodies are directed against paternal antigens which are shed from the fetus. Presence of these preformed antibodies can be detected and are referred to as panel reactive antibodies (PRA). High titers of such antibodies are likely going to cause development of hyperacute rejection. In those individuals who are not presensitized, exposure to donor antigens may result in formation of antibodies. These antibodies may cause graft damage by antibody-dependent cell-mediated cytotoxicity, complement-mediated cytotoxicity and inflammation. The primary target of such antibodies is the vessels of the graft. After transplant a patient can be assessed for donor-specific antibodies (DSA). During antibody-mediated rejection, complement is activated and C4 is converted to C4a and C4b. C4b is converted to C4d which can bind to the endothelial and collagen basement membranes. C4d can be detected by monoclonal antibodies on graft biopsies, thus establishing the process of antibody-mediated rejection. Patterns of rejection could be hyperacute, acute (acute cellular rejection, acute humoral rejection), or chronic (GVHD).

## Graft-versus-host disease

GVHD is a major cause of transplant-related morbidity and mortality following allogeneic hematopoietic cell transplantation. HLA matching in bone marrow transplantation reduces chances of GVHD, but subtle difference may be enough to trigger such response. GVHD is clinically described in two forms: acute and chronic. The acute GVHD is primarily induced by T cells (donor T cells recognize the host HLA antigens as foreign entity and become activated) whereas chronic GVHD is induced by both T and B cells, similar in nature to that of autoimmune disorders. Additionally, late acute GVHD, defined as occurring beyond 3 months posttransplant, is associated with high lethality. In acute

GVHD which occurs within days to weeks after transplant is usually accompanied by skin rash, liver dysfunction, and diarrhea. Features of chronic GVHD include dermal fibrosis, cholestatic jaundice, and immunodeficiency. Unfortunately, effective treatment options are very limited beyond steroids. Currently, the only FDA-approved agents for steroid-refractory acute and chronic GVHD are ruxolitinib and ibrutinib, respectively. Even with these new agents, GVHD mortality and its impact on quality of life remains a major clinical challenge [8].

Although GVHD is typically observed in bone marrow transplant recipients where the host is severely immunocompromised (due to underlying disease or due to drugs orrradiation), GVHD is also observed in other solid organ transplants but rarely. The incidence of acute GVHD following orthotopic liver transplantation is 0.1%–1% compared to an incidence of 40%–60% in hematopoietic stem cell transplant recipients [9]. Guo et al. commented that GVHD also occurs after other transplantation such as lung, intestine, and pancreas-kidney. The occurrence of GVHD after kidney transplantation is extremely rare with only five reported cases [10].

## Autoimmune serology

Antinuclear antibodies (ANA) are antibodies directed against various components of the nucleus and ANA test is ordered in patients suspected of autoimmune disease but most commonly ordered in a patient suspected of suffering from systemic lupus erythematosus (SLE). ANA test can be performed by indirect immunofluorescence (IIF) assay on Hep-2 cells or using enzyme-linked immunosorbent assay (ELISA) and result is reported as a titer. In IIF assay patient's serum is incubated with Hep-2 cells (a line of human epithelia cells), followed by adding fluorescein-labeled antihuman globulin (AHG). The serum is serially diluted until the test becomes negative, which provides estimation regarding strength of positivity. Low titers (1:40 to 1:160) are observed in population but titers higher than 1:160 are likely to be significant but titer greater than 1:320 are more likely indicative of true positive results.

ANA shows up on indirect immunofluorescence assay as fluorescent pattern in cells that are fixed to a slide. Therefore, pattern can be further investigated under a microscope. Although there are some overlaps, different patterns can be associated with certain autoimmune diseases. These various patterns include speckled, homogenous, anticentromeric, and peripheral (Table 3).

For the diagnosis of SLE, once ANA is positive further testing for antibodies must be considered. In general ANA testing using Hep-2 cells is very effective in identifying patients with SLE because almost all patients show ANA positivity. In addition, ANA test is also positive in other diseases [11]. Unfortunately, false positive ANA test is common in elderly, and many other conditions

**Table 3** Various patterns of ANA in indirect immunofluorescence assay.

| Pattern | Disease | Further testing/autoantibody |
|---|---|---|
| (Peripheral) | Systemic lupus erythematosus (SLE) | Anti-dsDNA |
| Speckled | SLE, scleroderma, Sjögren's syndrome | Smith antibody |
| | Mixed connective tissue disease | Anti-SAA (Anti-Ro) |
| | | Anti-SSB (Anti-La) |
| | | Anti-topoisomerase I |
| | | (Scl-70) |
| | | U1-RNP antibody |
| | | PCNA antibody |
| Homogenous | SLE, drug-induced SLE | Anti-dsDNA |
| | | Antihistone |
| Nucleolar | Scleroderma, CREST syndrome | RNA-polymerase I |
| | | U3-RNP antibody, PM-Scl antibody |
| Centromere | CREST syndrome, Raynaud's | Anticentromere |
| | | Antibody |
| Diffuse | Nonspecific for any disease | |

U1-RNP antibody, *U1-ribonuclear protein antibody*; PCNA antibody, *proliferating cell nuclear antigen antibody*; PM-Scl antibody, *polymyositis-associated antibody*; CREST syndrome, *limited cutaneous form of systemic scleroderma is often referred as CREST syndrome (acronym of calcinosis, Raynaud's syndrome, esophageal dysmotility, sclerodactyly, and telangiectasia).*

(Table 4). If ANA test is positive, anti-dsDNA (antibody against double stranded DNA) and anti-Smith antibody testing may be undertaken. Tests for anti-dsDNA may be done using the Farr assay or an IIF using Crithidia luciliae. The Farr assay is used to quantify the amount of anti-dsDNA antibodies in serum. Ammonium sulfate is used to precipitate an antigen-antibody complex which is formed if the serum contains antibodies to dsDNA. The quantity of these antibodies is determined by using radioactively labeled dsDNA. Crithidia luciliae is a protozoon containing a kinetoplast which is a mitochondrion rich in dsDNA. Patients serum reacts (if positive) with the kinetoplast and binding is identified with fluorescent antibody.

If ANA test is positive, testing for various other antibodies may be undertaken because in addition to SLE, ANA test is also positive in other disorders. Anti-Smith antibodies were first discovered in the 1960s, when a patient named

**Table 4** Positive ANA tests in disease and false positive ANA test.

| | |
|---|---|
| Positive ANA test | Systemic lupus erythematosus, drug induced lupus, scleroderma, Sjögren syndrome, rheumatoid arthritis, mixed connective tissue disease, polymyositis, dermatomyositis, systemic vasculitis |
| False positive ANA | Elderly, liver disease, HIV infection, multiple sclerosis, diabetes, pulmonary fibrosis, pregnancy and patients with silicone implant |

Stephanie Smith was treated for SLE and a unique set of antibodies against nuclear protein was detected in her blood which were termed as anti-Smith antibodies or in abbreviated form "anti-Sm" antibodies. These antibodies are specific in patients with SPE. Antihistone antibody testing is useful for patients with a positive ANA test and history of exposure to medications associated with drug induced lupus such as procainamide and isoniazid. ELISA assays are available for detecting antihistone antibody or subfractions (H1, H2a, H2b, H3, and H4). However, such antibodies may also be detected in patients with rheumatoid arthritis, localized scleroderma and other diseases characterized by the presence of autoantibodies [12]. Ant-Ro (anti-SSA: anti-Sjögren syndrome A) and anti-La (anti-SSB: anti-Sjögren syndrome B) autoantibodies are usually associated with Sjögren syndrome. Ant-RNP antibody (antiribonuclear protein antibody) can be observed in mixed connective tissue disease. Anti-RNP (ribonuclear protein can be observed in mixed connective tissue disorder. Antitopoisomerase I is also called anti-scl-70. This auto antibody is present in patients with scleroderma. Anticentromere antibody is also found in patients with CREST syndrome and scleroderma. Associations of various antibodies with in ANA testing with diseases are listed in Table 5. Various cytoplasmic antibodies are also associated with different autoimmune diseases. These are summarized in Table 6.

SLE is a multisystem disorder, which can affect the gastrointestinal system in 50% of patients with SLE, these have barely been reviewed due to difficulty in identifying different causes. There are four major SLE-related gastrointestinal system complications: protein-losing enteropathy, intestinal pseudo-obstruction,

**Table 5** Association between various autoantibodies in ANA testing with diseases.

| Antibody | Association | Antigen | Appearance on IFF using Hep-2 cells |
|---|---|---|---|
| Anti-dsDNA | SLE (specific) | DNA backbone | Homogeneous |
| Anti-Smith | SLE (highly specific) | Nonhistone nuclear protein complexed with U1-RNP, involved in mRNA splicing | Speckled |
| Anti-histone | Drug induced SLE, SLE | Histone H1, H2A, H2B, H3, or H4 | Homogeneous |
| Anti-SSA (Ro) | Sjögren (70%), SLE (30%) | Small ribonuclear protein | Speckled |
| Anti-SSB (La) | Sjögren (50%), SLE (15%) | Small ribonuclear protein and/without RNA polymerase III | Speckled |
| Anti-RNP | Mixed connective tissue disease | U1-RNP-associated protein | Speckled |
| Anti-Scl-70 | Scleroderma | DNA topoisomerase I | Finely speckled |
| Anti-centromere | CREST Scleroderma | CENP B | Anti-centromere |

IIF, *indirect immunofluorescence assay.*

**Table 6** Association of cytoplasmic antibodies with various autoimmune diseases.

| Antibody | Disease |
| --- | --- |
| Antismooth muscle (SMA) against actin | Autoimmune hepatitis |
| Antimitochondrial against various mitochondrial antigens (M2 most specific) | Primary biliary cirrhosis |
| Anti-Jo-1 against histidyl tRNA synthase | Polymyositis, dermatomyositis (interstitial lung disease) |
| Antiparietal cell | Pernicious anemia |
| Antiendomysial | Celiac sprue, dermatitis herpetiformis (specific) |
| Antimicrosomal | Hashimoto disease |
| Antithyroglobulin | Hashimoto disease |

hepatic involvement and pancreatitis. SLE-related protein-losing enteropathy is characterized by edema and hypoalbuminemia. The most common site of protein leakage is the small intestine and the least common site is the stomach. More than half of SLE-related intestinal pseudo-obstruction patients had ureterohydronephrosis, and sometimes they manifested as interstitial cystitis and hepatobiliary dilatation. Lupus hepatitis and SLE accompanied by autoimmune hepatitis share similar clinical manifestations but may have different autoantibodies and histopathological features. Moreover, positive antiribosome P antibody is a good indication of diagnosis of lupus hepatitis. Lupus pancreatitis is usually accompanied by high SLE activity with a relatively high mortality rate. Early diagnosis and timely interventions with administration of corticosteroids and immunosuppressants are effective for most of the patients [13].

## Antineutrophil cytoplasmic antibodies

Antineutrophil cytoplasmic antibodies (ANCAs) are autoantibodies mainly of IgG type directed against antigens present in cytoplasmic granules of neutrophils and monocytes. ANCA may recognize multiple antigens, but antibodies against only two antigens (proteinase 3 and myeloperoxidase) have clinical significance. There are two main types of ANCA: c-ANCA (cytoplasmic-ANCA) and p-ANCA (perinuclear-ANCA). Immunofluorescence on ethanol fixed neutrophils is used for detection of ANCA. When serum is incubated with alcohol fixed neutrophils, two different types of reactivity may be observed in individuals with ANCA. If c-ANCA is present, cytoplasmic granular immunofluorescence activity is observed where c-ANCA has specificity against proteinase-3 and is seen in Wegener granulomatosis. The other type of reactivity is where perinuclear immunofluorescence pattern is observed if p-ANCA is present, which has specificity against myeloperoxidase (MPO). This pattern is observed in patients with microscopic polyarteritis nodosa, polyarteritis nodosa, and Churg-Strauss syndrome.

## Hypersensitivity reactions and hypersensitivity-mediated disorders

Hypersensitivity reactions are generally of four types including immediate or type I, antibody-mediated or type II, immune complex-mediated or type III and T cell-mediated or type IV.

In immediate or type I hypersensitivity reaction, when a host is exposed to an antigen, IgE antibodies are produced that are bound to the surface of mast cells, which triggers mast cell degranulation. Mast cell products are responsible for the subsequent clinical manifestations such as allergic rhinitis, bronchial asthma, and even anaphylactic reactions. Vaccination may also cause type 1 hypersensitivity reaction, which may occur immediately after vaccination to up to 4 h but delayed reactions may also occur. Serious hypersensitivity reactions after influenza vaccines especially in people with severe egg allergy are particularly important because of the large number of persons vaccinated annually. The most common symptoms of acute-onset IgE-mediated hypersensitivity range from urticaria to angioedema to anaphylaxis. Serious anaphylactic (1.3 cases per million vaccine doses) or cutaneous adverse reactions do occur but are extremely rare [14].

The antibody-mediated or type II hypersensitivity mediated disorder is due to antibodies directed against antigens, which are components of cells. Sometimes the antigen is exogenous in nature, being adsorbed onto cell surfaces (e.g., drug or their metabolites). Examples of this type of hypersensitivity disorder include Myasthenia Gravis, Goodpasture syndrome, autoimmune hemolytic anemia, and autoimmune thrombocytopenia. In Graves' disease the antibody binds to the thyroid-stimulating hormone (TSH) receptor and stimulates it resulting in hyperthyroidism. Therefore, Graves' disease is an example of type II hypersensitivity reaction; however, some authors prefer to put this disease in a different category, type V hypersensitivity reaction.

In immune complex-mediated or type III hypersensitivity mediated disorder, large amounts of antigen-antibody complexes are formed which especially if persist in the circulation may deposit in various tissues causing inflammatory response. Common sites of immune complex deposition are kidneys, joints, and skin. Examples of disease states due to this mechanism are SLE, polyarteritis nodosa, post streptococcal glomerulonephritis, and serum sickness.

T cell mediated or type IV reactions can be subclassified into delayed type hypersensitivity (DTH) and T cell-mediated cytotoxicity. The classical example of delayed type hypersensitivity is the tuberculin reaction. Upon first exposure to tubercle bacilli, macrophages take up bacteria, process their antigen and present them on their surface. This antigen in association with Class II MHC molecules are recognized by CD4+ T lymphocytes which can remain as memory CD4+ lymphocytes. However, during subsequent exposure (tuberculin testing) these CD4+ memory T cells gather at the site of inoculation. Gamma interferon is secreted which recruits macrophages that are the major mediator of DTH. Prolonged DTH reactions yield a granulomatous inflammation causing accumulation of macrophages. Some of these macrophages may be converted to epithelioid cells and some into giant cells. The macrophages maybe surrounded by lymphocytes and even a rim of fibrous tissue. In T cell-mediated cytotoxicity CD8+ T cells are responsible for killing of antigen bearing target cells. This type of cytotoxicity is important against viral infections and tumor cells.

## Summary/key points

- Immunity can be broadly divided into innate (natural) and adaptive (specific) immunity. Innate immunity is nonspecific all-purpose immunity, but adaptive immunity is characterized by greater specificity than innate immunity but response slower than innate immunity.
- T cells constitute about two-third of lymphocytes in the peripheral blood. T-cell receptor complex consists of heterodimers of alpha and beta chains, CD3 proteins (gamma, delta, and epsilon proteins) as well as two zeta chains. The antigen associated with the MHC (major histocompatibility complex) molecule of the cell binds with the alpha and beta chains. If the T cell is a CD4+ cell, it will recognize antigens presented in association with Class II MHC molecules. If the T cell is a CD8+ cell, it will recognize antigens presented in association with Class I MHC molecule. There is also minority of T cells which instead of alpha and beta chains possess gamma and delta chains as the receptors. These gamma delta cells do not express either as CD4 or CD8. Typically, more than one signal is required for activation of these T cells.
- B cells can be directly stimulated by antigens which are recognized by immunoglobulin molecules on the surface of B cells. These immunoglobulin molecules are either IgM or IgD. A second signal also comes from activated CD4+ T helper cells. Once activated, B cells differentiate into plasma cells. The first exposure to antigen results in production of IgM class of (primary response) but subsequently IgG antibodies are produced (secondary response). This ability of changing the antibody class is called class switching.

- Activation of complements involves activation of C3 which can be accomplished through various pathways including classical pathway (initiated by antigen-antibody complexes fixating with C1), alternate pathway (spontaneous but bacterial polysaccharides and proteins such as properdin, factor B and D are also involved in activating this pathway) and lectin pathway (plasma lectin binds to mannose on microbes and activates the pathway).

- Various activities of complement include opsonins (C3b and C4b can promote phagocytosis by phagocytic cells), anaphylatoxins (C3a, C4a, and C5a), leukocyte activation and chemotaxis (C5a), cell lysis (membrane attack complex C5–C9, activated), and removing circulating antigen-antibody complexes.

- Complement deficiency may result in impaired innate immunity (a well-known example is increased susceptibility to *N. meningitidis* infection) and immune complex mediated inflammation: glomerulonephritis, vasculitis, and systemic lupus erythematosus. In addition, deficiency of C1 inhibitor may be inherited, known as hereditary angioedema.

- Patients with B-cell defects are typically susceptible to recurrent bacterial infections, especially respiratory tract infections involving influenzae virus, *Strep pneumonia* and *Staph aureus*. Diarrhea may also be present due to infection caused by enterovirus and/or *Giardia lamblia*. However, most viral, fungal, and protozoal infections are cleared due to intact cell mediated immunity.

- Bruton's disease (X-linked agammaglobulinemia) is due to a mutation on chromosome Xq22 affecting the gene for a tyrosine kinase known as Bruton tyrosine kinase or B-cell tyrosine kinase. This mutation results in arrest in B-cell maturation, from pre-B cells to B cells.

- Individuals with T-cell defects have recurrent and persistent viral, fungal, and protozoal infections. Individuals are also at risk for transfusion associated graft vs host disease.

- DiGeorge syndrome is characterized by failure of development of thymus and parathyroid glands (due to failure of development of the third and fourth pharyngeal pouches). Features of T-cell immunodeficiency include hypoparathyroidism, dysmorphic facies, and cardiac defects.

- In severe combined immunodeficiency (SCID), underlying genetic defect may be diverse. About 50% of cases are transmitted as X-linked disorder. These are due to mutation in the gene coding for the gamma chain for receptors for various interleukins (IL) including IL-2, IL-4, IL-7, IL-9, and IL-15. In addition, 40%–50% of cases of SCID are transmitted as autosomal recessive and most common example of this type of SCID is due to mutation in the gene encoding for adenosine deaminase (ADA) enzyme. ADA deficiency causes accumulation of adenosine and deoxyadenosine triphosphate metabolites which are lymphotoxins. In SCID patients the thymus is hypoplastic, lymph nodes and lymphoid tissue lack germinal centers as well as paracortical T cells. Lack of help from T cells does not allow B cells to be functional and clinical features related to both B- and T-cell defects are present in the patient.

- One factor involved in class switching is interaction of CD40 molecules on B cells and CD40 ligand (CD40L or CD154) on T helper cells. The most common cause of hyper IgM syndrome is mutation in the gene encoding for CD40L, which is located in the X chromosome. The interaction between CD40 and CD40L is also required for T helper-mediated activation of macrophages. Thus, both humoral immunity and cell mediated immunity is affected in hyper IgM syndrome and IgM levels are normal or high with low levels of IgG, IgA, and IgE.
- Wiskott Aldrich syndrome (WAS) is an X-linked disease characterized by immunodeficiency, eczema, and thrombocytopenia. In this syndrome the platelets are small in size. The WAS gene codes for the Wiskott Aldrich syndrome protein (WASP). This syndrome is complicated by lymphoreticular malignancies and autoimmune diseases.
- Ataxia telangiectasia is an autosomal recessive condition due to mutation of ataxia telangiectasia mutated gene (ATM gene) which encodes for the ATM protein kinase involved in DNA repair.
- Defective phagocytosis can be classified under three broad categories:
- Chronic granulomatous disease (CGD): This disease is caused by a deficiency of NADPH oxidase resulting in lack of oxidative burst and defective killing of bacteria and fungus that are catalase positive (e.g., staphylococcus and aspergillus). This disorder may be inherited as X-linked recessive and also in autosomal recessive fashion. The gene encoding for the Kx antigen of the Kell blood group system is very close to the gene encoding for NADPH oxidase and if both are affected, Kx antigen may also be lacking, it is called the McLeod phenotype, which is associated with presence of acanthocytes.
- Chediak Higashi syndrome is an autosomal recessive condition where defective trafficking of intracellular organelles lead to defective fusion of lysosomes with phagosomes. This syndrome is related to mutation in the lysosomal trafficking regulator gene. Granulocytes, lymphocytes, and monocytes exhibit giant lysosomes. Neutropenia, thrombocytopenia, and oculocutaneous albinism are seen in this syndrome along with immunodeficiency.
- Leukocyte adhesion deficiency (LAD): LAD type 1 (LAD-1) is due to defective synthesis of LFA-1 and Mac-1, which are integrins. This results in defective leukocyte adhesion to endothelium, impaired leukocyte migration and defective leukocyte phagocytosis. LAD type 2 (LAD-2) is

due to absence of sialyl-Lewis X in leukocytes which binds to selectin on endothelium.

- The genes encoding for the histocompatibility molecules are clustered on a small segment (small arm) of chromosome 6. The cluster of genes is known as major histocompatibility complex (MHC) or human leukocyte antigen (HLA) complex. The HLA system is highly polymorphic. Class I MHC molecules are present on all nucleated cells and platelets. There are three different Class I MHC molecules: A, B, and C. Class I molecules are heterodimers of an alpha (or heavy) chain and a smaller beta two microglobulin. The beta two microglobulin molecule is extracellular and the alpha chain has extracellular components as well as parts which traverse the cell membrane into the cell. The extracellular part of the alpha chain has three domains, alpha1, alpha2, and alpha3. Class II MHC molecules are present on B lymphocytes and monocytes. There are three different Class II MHC molecules (DP, DQ, and DR). Class II molecules are also heterodimers of one alpha chain and one beta chain. Both chains have extracellular components with parts that traverse the cell membrane and into the cell. The extracellular portions of both chains have domains alpha1, alpha2 and beta1 and beta 2.

- Antigens within a cell may bind with Class I MHC molecule which is produced within the cell. This binding takes place in the endoplasmic reticulum. The complex of Class I molecule and the antigen is transported to the cell surface for presentation to CD8 + cytotoxic T lymphocytes. The T-cell receptor (TCR) recognizes and binds with the MHC-peptide complex (MHC molecule-antigen complex). The CD8 molecule also binds with the alpha3 domain of the Class I MHC molecule. The T cell is thus activated. CD8 + T cells are Class I MHC restricted because they can only be activated with antigen which are bound to MHC Class I molecules. Similarly, CD4 + T cells are Class II restricted.

- Patterns of transplant rejection could be hyperacute, acute (acute cellular rejection, acute humoral rejection), or chronic (graft vs host disease).

- Graft vs host disease is typically observed in bone marrow transplant recipients where the host is severely immunocompromised (due to underlying disease or due to drugs or irradiation) and the donor tissue has fully immunocompetent cells. The donor T cells recognize the host HLA antigens as foreign entity and become activated.

- ANA test can be performed by indirect immunofluorescence (IIF) assay on Hep-2 cells or using enzyme-linked immunosorbent assay (ELISA) and result is reported as a titer. In IIF assay patient's serum is incubated with Hep-2 cells (a line of human epithelia cells), followed by adding fluorescein-labeled antihuman globulin (AHG). The serum is serially diluted until the test becomes negative which provides estimation regarding strength of positivity. Low titers (1:40 to 1:160) are observed in population but titers higher than 1:160 are likely to be significant but titers greater than 1:320 are more likely indicative of true positive results.

- ANA shows up on indirect immunofluorescence assay as fluorescent pattern in cells that which are fixed to a slide. Therefore, pattern can be further investigated under a microscope. Although there are some overlaps, different pattern can be associated with certain autoimmune disease. These various patterns include speckled, homogenous, anticentromeric and peripheral. If ANA test is positive, anti-dsDNA (antibody against double stranded DNA) and anti-Smith antibody testing may be undertaken. Tests for anti-dsDNA may be done using the Farr assay or an IIF using Crithidia luciliae.

- Antineutrophil cytoplasmic antibodies (ANCA) are autoantibodies mainly of IgG type directed against antigens present in cytoplasmic granules of neutrophils and monocytes. ANCA may recognize multiple antigens, but antibodies against only two antigens (proteinase 3 and myeloperoxidase) have clinical significance. There are two main types of ANCA, c-ANCA (cytoplasmic-ANCA) and p-ANCA (perinuclear-ANCA). Immunofluorescence on ethanol fixed neutrophils is used for detection of ANCA.

- If c-ANCA is present, cytoplasmic granular immunofluorescence activity is observed where c-ANCA has specificity against proteinase-3 and is seen in Wegener granulomatosis. The other type of reactivity is where perinuclear immunofluorescence pattern is observed if p-ANCA is present which has specificity against MPO. This pattern is observed patients with in microscopic polyarteritis nodosa, polyarteritis nodosa and Churg-Strauss syndrome.

- Hypersensitivity reactions are generally of four types including immediate or type I, antibody mediated or type II, immune complex-mediated or type III, and T cell-mediated or type IV.

- In immediate or type I hypersensitivity reaction when a host is exposed to an antigen, IgE antibodies are produced which are bound to the surface of mast cells that triggers mast cell degranulation.

- The antibody-mediated or type II hypersensitivity mediated disorder is due to antibodies directed against antigens which are components of cells.

- In immune complex mediated or type III hypersensitivity mediated disorder, large amounts of antigen-antibody complexes are formed which especially if persist in the circulation may deposit in various tissues causing inflammatory response. Common sites of immune complex deposition are kidneys, joints, and skin.

- T cell-mediated or type IV reactions can be subclassified under two categories: delayed type hypersensitivity (DTH) and T cell-mediated cytotoxicity. The classical example of delayed type hypersensitivity is the tuberculin reaction. In T cell-mediated cytotoxicity CD8+ T cells are responsible for killing of antigen bearing target cells. This type of cytotoxicity is important against viral infections and tumor cells.

## References

[1] Segerstrom SC, Miller GE. Psychological stress and the human immune system: a meta-analytical study of 30 years of inquiry. Psychol Bull 2004;130:601–30.

[2] Glovsky MM, Ward PA, Johnson KJ. Complement determinations in human disease. Ann Allergy Asthma Immunol 2004;93:513–23.

[3] Arroyo-Martinez YM, Saindon M, Raina JS. X-linked agammaglobulinemia presenting with multiviral pneumonia. Cureus 2020;12, e7884.

[4] Song J, Lleo A, Yang GX, Zhang W, et al. Common variable immunodeficiency and liver involvement. Clin Rev Allergy Immunol 2018;55:340–51.

[5] Yazdani R, Azizi G, Abolhassani H, Aghamohammadi A. Selective IgA deficiency: epidemiology, pathogenesis, clinical phenotype, diagnosis, prognosis and management. Scand J Immunol 2017;85:3–12.

[6] Li Y, Yao Y, Yang M, Shi L, et al. Association between HLB-B*46 allele and Graves' disease in Asian population: a meta-analysis. Int J Med Sci 2013;10:164–70.

[7] Erlich H. HLA DNA typing: past, present and future. Tissue Antigens 2012;80:1–11.

[8] Betts BC, Xue-Zhong Y. Editorial: Pathogenesis and therapy of graft-versus-host disease. Front Immunol 2019;10:1797.

[9] Bitar C, Olivier K, Lee C, Vincent B, Martin J. Acute graft-vs-host disease following liver transplantation. Cutis 2019;103(6):E8–E11.

[10] Guo Y, Ding S, Guo H, Li S, et al. Graft-versus-host-disease after kidney transplantation: a case report and literature review. Medicine (Baltimore) 2017;96(26):e7333.

[11] Fritzler MJ. Choosing wisely: review and commentary on anti-nuclear antibody (ANA) testing. Autoimmun Rev 2016;15:272–80.

[12] Hasegawa M, Sato S, Kikuchi K, Takehara K. Antigen specificity of antihistone antibodies in systematic sclerosis. Ann Rheum Dis 1998;57:470–5.

[13] Li Z, Xu D, Wang Z, Wang Y, et al. Gastrointestinal system involvement in systemic lupus erythematosus. Lupus 2017;26:1127–38.

[14] McNeil MM, DeStefano F. Vaccine-associated hypersensitivity. J Allergy Clin Immunol 2018;141:463–72.

# CHAPTER 27

# Effect of herbal supplements on clinical laboratory test results

## Use of herbal remedies in the United States

Throughout the history of mankind, herbal remedies were the only medicines available. However, when a pharmaceutical is prepared from a plant source, the active ingredient is sold in the pure form following extensive steps of extraction, purification, and standardization. In contrast, herbal remedies are crude extract of plant products and may contain active ingredients along with other active components which may cause toxicity. According to WHO, approximately 80% of world population relies on herbal medicines. In addition, many patients take herbal medicines concurrently with conventional drugs. For example, in the United States, the concurrent use of herbals and conventional drugs occurs in 20%–30% patients [1]. As a result, a relatively safe herbal product such as St. John's wort may cause treatment failure due to drug-herb interactions [2].

In the United States, herbal remedies are sold according to 1994 Dietary Supplement Health and Education Act where herbals are classified as food supplements. In addition to herbal remedies, food supplements, such as vitamins, minerals, amino acids, extracts, metabolites, etc. are also sold under this law. Although manufacturers of herbal remedies are not allowed by law to claim any medical benefits from using such products, but at the same time, they are not under surveillance by the Food and Drug Administration (FDA). Therefore, dietary supplements can be manufactured and sold without demonstrating safety and efficacy to FDA as required for drugs. Unlike drugs which are highly regulated by the FDA, for herbal supplements, FDA cannot take steps against a toxic supplement unless patients have already been harmed.

The popularity of using herbal supplements is steadily increasing among the general population in the United States where it has been estimated that roughly 20,000 herbal products are available. The 10 most commonly used herbal supplements are echinacea, ginseng, ginkgo biloba, garlic, St. John's wort, peppermint, ginger, soy, chamomile, and kava [3]. Population survey

**557**

also indicates that one-third to one-half of Americans take dietary supplements. Users are more likely to be women, non-Hispanic whites, and more financially secure than nonusers [4].

## How herbal remedies affect clinical laboratory test results

An herbal remedy may affect clinical laboratory test results by one of the following mechanisms:

- Herbal remedy may produce an unexpected test result by direct physiological effect of the herbal remedy on human body. For example, use of a hepatotoxic herb such as kava can cause elevated liver function tests due to hepatotoxicity.
- Herbal supplement may interact with a therapeutic drug causing clinically significant drug-herb interactions. For example, St. John's wort induces liver enzymes that metabolize cyclosporine and tacrolimus thus reducing its blood level. Reduced blood cyclosporine or tacrolimus level may cause treatment failure or even possibility of organ rejection.
- An ingredient of herbal supplement can cross-react with assays antibody causing interference. This has been reported only in therapeutic drug monitoring of digoxin using immunoassays where Chinese medicine such as Chan Su may cause falsely elevated digoxin concentration (see Chapter 15).
- Herbal product may contain undisclosed drugs as adulterants and an unexpected drug level (such as phenytoin in a patient who never took phenytoin but took a Chinese herb) may confuse the laboratory staff and the clinician.
- Herbal supplement may be contaminated with a heavy metal or a heavy metal may be an active ingredient of the herbal supplement, such as in Indian Ayurvedic medicine. Heavy metal toxicity may occur after use of such herbal supplements.

## Liver damage as reflected by abnormal liver function test after using certain herbals

The best documented organ toxicity due to use of certain herbal supplements is liver toxicity, and abnormal liver function tests are the first indications of such toxicity. Measurements of the serum or plasma activities of the enzymes aspartate aminotransferase (AST), alanine aminotransferase (ALT), γ-glutamyl transferase (GGT), and alkaline phosphatase (ALP) are routinely performed to assess liver function. In general, abnormal liver function tests in the absence of any hepatitis infection is a strong indication of liver damage due to use of an herbal supplement. Important points regarding herbal-induced liver injury include:

- Abnormal liver function tests such as elevated liver enzymes and possibly bilirubin with negative serological tests for hepatitis or related viral infection is a strong indication of liver toxicity due to use of an herbal supplement.
- Most common herbal supplement associated with liver damage is kava, an herbal sedative and anxiolytic agent.
- Other hepatotoxic herbals are chaparral, comfrey, germander, and pennyroyal oil. Prolonged use of a hepatotoxic herb (3 months or more of continuous use) may cause irreversible liver damage and even death. Various hepatotoxic herbs are summarized in Table 1.

There is a case of a 42-year-old healthy white male who showed highly elevated liver enzymes 3 weeks after taking a trip to Samoan Island for 20 days and repeatedly participated in kava ceremonies consuming a total of 2–3 L of traditional kava preparation. He was admitted to the hospital for suspected liver injury and was discharged after 19 days. It was determined that his liver toxicity was due to ingestion of kava drinks [5]. Becker et al. reported a case of a patient who, after using kava for 52 days, developed acute liver failure requiring liver transplantation [6]. There are more than 100 cases of hepatotoxicity due to prolonged exposure to kava. Coingestion of alcohol may potentiate the hepatotoxicity. In one case that resulted in death, the individual was reported to have consumed a standardized extract containing 30%–70% kava lactones [7]. In addition to kava, use of chaparral, comfrey, and germander may also cause severe hepatotoxicity, and even death, but these herbs are encountered less frequently than kava in clinical practice. Key points regarding hepatotoxicity of comfrey include

**Table 1** Common herbal supplements that may cause liver damage.

| Herbal supplements | Indication for use | Death associated with use |
|---|---|---|
| Kava | Herbal sedative/anxiolytic agent | Yes |
| Chaparral | Antioxidant, anticancer, anti-HIV | No |
| Comfrey | Repairing broken bone, gout, arthritis | Yes |
| Germander | Herbal weight loss | No |
| Mistletoe | digestive aid, heart tonic | No |
| Lipokinetix | Herbal weight loss product | No |
| Pennyroyal | Aromatherapy, inducing abortion | Yes |
| Noni juice | Stimulating immune system | No |

- Pyrrolizidine alkaloids found in comfrey are responsible for liver damage.
- Russian comfrey is more toxic than European or Asian comfrey.

LipoKinetix has been promoted as a weight loss aid and an alternative to exercise that increases metabolism. This product contains phenylpropanolamine, caffeine, yohimbine, diiodothyronine, and sodium usniate. Both phenylpropanolamine (a banned drug) and sodium usniate may be responsible for liver damage due to use of LipoKinetics. Sodium usniate is derived from usnic acid which is also present in Kombucha Tea (also known as Manchurian Mushroom or Manchurian Fungus tea), prepared by brewing Kombucha mushroom in sweet black tea. Acute liver damage due to drinking of Kombucha tea has been reported. Herbalene, promoted for weight reduction, may also cause liver injury. Fatal hepatitis in a 68-year-old woman was related to use of the herbal weight loss product Tealine which contained hepatotoxic germander extract [8].

Pennyroyal (*Mentha pulegium*) is a plant in the mint genus whose leaves release a spearmint-like fragrance when crushed. Portions of the plants as well as the essential oil are used for a variety of purposes including as an additive to bath products and in aromatherapy. Traditionally, pennyroyal has been brewed as a tea to be ingested in small amounts as an abortifacient and emmenagogue. Ingestion of as little as 10 mL of pennyroyal oil may cause severe toxicity. Death has been reported from ingestion of pennyroyal oil. Interestingly, the antidote used in acetaminophen overdose, *N*-acetylcysteine, has been used successfully in treating pennyroyal toxicity. Noni juice which is prepared from noni fruits that grow in Tahiti is indicated for stimulating heart and also used as a digestive aid. There are case reports that noni juice may cause hepatotoxicity but such effects are usually reversed after discontinuation of noni juice.

Epigallocatechin 3-gallate (EGCG), the major catechin present in green tea, is an effective antioxidant. Consumption of green tea is not associated with liver damage in humans, and green tea infusion and green tea-based beverages are considered safe. The tolerable upper intake level of 300 mg EGCG/person/day is proposed for food supplements. However, taking green tea extract where EGCG intake level is much higher than 300 mg/day (especially exceeding 600 mg/day), liver toxicity may be encountered [9]. Since 2006, there have been more than 50 reports in the medical literature of clinically apparent acute liver injury with jaundice attributed to green tea extracts [10].

## Kidney damage and herbal supplements

In 1993, rapidly progressing kidney damage was reported in a group of young women who were taking pills containing Chinese herbs while attending a

| Table 2 Common herbs associated with kidney damage. |
| --- |
| Aristolochic acid containing Chinese herbs |
| Wormwood plant |
| Sassafras |
| Horse chestnut |
| Kava |
| Calamus |
| Chaparral |
| Wormwood oil |
| White sandalwood oil |

weight loss clinic in Belgium. It was discovered that one prescription Chinese herb has been replaced by another Chinese herb containing aristolochic acid, a known toxin to kidney [11]. Later there were many reports of kidney damage due to use of herbal supplements contaminated with aristolochic acid in the medical literature. There are several herbal supplements which are known to cause hematuria and proteinuria. Examples of these herbs are kava, calamus, chaparral, horse chestnut seed, and wormwood oil. Common herbs associated with kidney damage are listed in Table 2.

## Kelp and thyroid function

Kelp (seaweed) is a part of natural diet in many Asian countries. The popular Japanese food sushi is wrapped with seaweed. In addition, kelp extracts are also available in the form of tablets in health food stores and are used as a thyroid tonic, antiinflammatory, and metabolic tonic and also as dietary supplement. Kelp tablets are rich in vitamins and minerals but also contain substantial amounts of iodine. Usually, eating sushi or Japanese food should not cause any problem with thyroid although some Asian seaweed dishes may exceed tolerable upper iodine intake of $1100\,\mu g$/day [12]. However, taking kelp supplements on a regular basis for a prolonged time may cause thyroid dysfunction, especially hyperthyroidism, due to high iodine content of kelp supplements. Some kelp preparation may also contain arsenic. There is a report of a 39-year-old woman who showed laboratory test results indicative of hyperthyroidism with elevated free T3 and free T4 but suppressed level of TSH ($<0.01\,$mU/L). The patient admitted taking a Chinese herbal tea prescribed by a Chinese herbal specialist. The preparation of tea revealed large amount of kelp. The patient was advised to discontinue the tea and was treated with antithyroid drug (40 mg thiamazole) and 40 mg propranolol daily. After 7 months her free T4 and T3 returned to normal value but her TSH was slightly decreased (0.14 mU/L). Because her hyperthyroidism was resolved clinically, her thiamazole dosage was reduced to 20 mg/day. The iodine-induced thyrotoxicosis in this patient was due to ingestion of kelp-containing herbal tea [13].

## Miscellaneous abnormal test results due to use of certain herbals

Various abnormal test results may also be encountered due to use of certain herbal supplements. Although measuring hypertension is not a clinical laboratory test, blood pressure is one of the first few parameters measured when a person is presented to a clinic or emergency department. Although use of ephedra in weight loss products is banned in the United States, infrequently ephedra is encountered in weight loss products imported to the United States from various Asian countries. A popular example is Ma-huang. Hypertension is common after the use of ephedra-containing products.

Another relatively safe herbal product licorice, which is also used in candies as a flavoring agent, may further increase blood pressure in a person suffering from hypertension. In addition, these patients are also vulnerable of developing hypokalemia and possibly pseudohyperaldosteronism from regular use of licorice. Glycyrrhizic acid found in licorice is possibly responsible for increasing blood pressure after licorice use.

Many herbal supplements such as ginseng, fenugreek seed, garlic, bitter melon, bilberry, dandelion, burdock, and prickly pear cactus are indicated for lowering blood glucose. In addition, dietary supplement chromium is also capable of lowering serum glucose level. Patients suffering from diabetes mellitus and taking oral hypoglycemic agent should not use any such herbals without the approval of their physicians because severe hypoglycemia may occur due to interaction of these herbals with oral hypoglycemic agents. Patients suffering from insulin-dependent diabetes should also refrain from using such herbal supplements.

## Drug-herb interactions

Although many drug-herb interactions have been reported in the literature, clinically significant drug-herb interactions are more commonly encountered in clinical situation involves St. John's wort, an herbal antidepressant and Western drug. In addition, warfarin also interacts with many herbal supplements. In general, it has been recommended that the following groups of patients should not take any herbal supplement because they are very susceptible to drug-herb interactions:

- Organ transplant recipients must not take any herbal supplements because immunosuppressants, especially cyclosporine and tacrolimus, are susceptible to various interactions with herbal supplements. Clinically significant interaction between St. John's wort and

cyclosporine or tacrolimus may cause potential rejection of transplanted organ due to increased clearance of both drugs as a result of pharmacokinetic interaction with St. John's wort.
- Patients taking warfarin should avoid herbal supplements because many clinically significant interactions have been reported between warfarin and various herbal supplements.
- Patients suffering from HIV infection and being treated with HAART (highly active antiretroviral therapy) should avoid all herbal supplements due to potential treatment failure as a result of interaction of antiretroviral agents with certain herbs.

St. John's wort is a popular herbal antidepressant which is composed of dried alcoholic extract or alcohol/water extract of hypericum, a perennial aromatic shrub with bright yellow flowers that bloom from June to September. The flowers are believed to be most abundant and brightest around June 24, the day traditionally believed to be the birthday of John the Baptist. Therefore, the name St John's wort became popular for this herbal product. Active components of St. John's wort, hypericin and hyperforin, are responsible for pharmacokinetic interactions between many Western medications and St. John's wort. Although St. John's wort interacts with most drugs pharmacokinetically, pharmacodynamic interaction of St. John's wort with several drugs has also been reported. Key points involving interaction of St. John's wort with various drugs are as follows:

- Hyperforin, an active component of St. John's wort, induces cytochrome P 450 mixed function oxidase, the major liver enzymes responsible for the metabolism of many drugs and thus increases clearance of many drugs which may result in treatment failure.
- Hypericin, another active component of St. John's wort modulates P-glycoprotein pathway thus affecting clearance of drugs that are not metabolized by liver enzymes such as digoxin.
- Pharmacodynamic interaction of St. John's wort with various SSRI such as paroxetine, sertraline, or venlafaxine may produce life-threatening serotonin syndrome.

The most important pharmacokinetic interaction of St. John's wort with various drugs includes its interaction with immunosuppressants (reduced efficacy of cyclosporine and tacrolimus but no interaction with mycophenolic acid), warfarin (reduced efficacy) and various antiretroviral agents (reduced efficacy). Hyperforin concentration in St. John's wort is a variable factor for drug-herb interactions involving St. John's wort. Herbal preparations with low hyperforin content may not show significant drug-herb interactions [14]. Pharmacokinetically important drug interactions with St. John's wort are summarized in Table 3.

**Table 3** Important pharmacokinetic interactions between various drugs and St. John's wort that may cause treatment failure.

| Drug class | Comments |
| --- | --- |
| Immunosuppressant agents | Reduced levels of cyclosporine and tacrolimus<br>No interaction with mycophenolic acid |
| Antiretroviral agent | Reduced levels of indinavir, saquinavir, atazanavir, lopinavir, lamivudine, and nevirapine |
| Anticoagulant | Reduced level of warfarin |
| Anticancer agents | Reduced levels of imatinib and irinotecan |
| Cardiovascular drugs | Reduced levels of digoxin, verapamil, and nifedipine |
| Benzodiazepines | Reduced levels of alprazolam, midazolam, and quazepam |
| Hypoglycemic agents | Reduce level of gliclazide |
| Antiasthmatic agent | Reduced levels of theophylline |
| Statins | Reduced efficacy of simvastatin and atorvastatin |
| Oral contraceptives | Failure of contraception by ethinyl-estradiol and related compounds |
| Antidepressant | Reduced level of amitriptyline |
| Synthetic opioid | Reduced levels of methadone and oxycodone |

Therapeutic drug monitoring is very useful in determining certain drug-herb interactions including drug-herb interactions involving St. John's wort. A 65-year-old patient who received a renal transplant in November 1998 had a trough whole blood level tacrolimus concentration between 6 and 10 ng/mL which was within the therapeutic range. The patient experienced depression in July 2000 and started self-medication with St. John's wort (600 mg/day). In August 2000, the patient showed an unexpected low tacrolimus concentration of 1.6 ng/mL. When the patient stopped taking St. John's wort following medical advice, the tacrolimus level returned to the previous range without any dosage increase [15].

Many herbal supplements are known to potentiate effect of warfarin and may produce excessive anticoagulation causing bleeding problems. In such cases, increased INR with no change of dosage may be an early indication of such warfarin-herb interactions. In general, it is assumed that angelic root, anise, bogbean, borage seed oil, bromelain, capsicum, chamomile, clove, fenugreek, feverfew, garlic, ginger, ginkgo biloba, horse chestnut, licorice root, meadowsweet, passionflower herb, red clover, turmeric extract, and willow bark potentially increase the effectiveness of warfarin, thus increasing the risk of bleeding in a patient taking warfarin and one of these supplements. In contrast, green tea extract and St. John's wort reduces the efficacy of warfarin.

## Herbs adulterated with western drugs and herbs contaminated with heavy metals

Sometimes herbal medicines manufactured in various Asian countries are contaminated with Western drugs but the product labels do not mention the presence of such drugs. Of 2069 samples of traditional Chinese medicines collected from eight hospitals in Taiwan, 23.7% contained pharmaceuticals most commonly caffeine, acetaminophen, indomethacin, hydrochlorothiazide, and prednisolone [16]. Lau et al. reported a case of phenytoin poisoning in a patient after using Chinese medicines. This patient was treated with valproic acid, carbamazepine, and phenobarbital for epilepsy but was never prescribed phenytoin [17]. Heavy metal contamination is another major problem with Asian medicines. Ko reported that 24 of 254 Asian patent medicines collected from herbal stores in California contained lead, 36 products contained arsenic and 35 products contained mercury [18]. Lead and other heavy metal contaminations (cadmium and mercury) are common in Indian Ayurvedic medicines. Unfortunately, some Ayurvedic medicines contain heavy metals as a part of the active ingredient.

## Grapefruit juice-drug interaction

The first grapefruit juice-drug interaction was reported in 1989 where significantly increased felodipine bioavailability was observed after drinking grapefruit juice. Pharmacokinetic studies have demonstrated that intake of grapefruit juice can increase the bioavailability of many drugs classes including some calcium channel blockers, benzodiazepines, and statins. The mechanism is irreversible inhibition of cytochrome P450 (CYP) 3A by furanocoumarins present in the juice. Unlike other known CYP3A inhibitors, normal consumption of grapefruit juice only inhibits CYP3A in the enterocyte cells lining the small intestine but hepatic CYP3A activity remains unaffected, except with unrealistically large ingestion of grapefruit juice. Grapefruit juice can also decrease the bioavailability of some drugs (e.g., fexofenadine) and such interactions are due to reduction in drug uptake transport via inhibition of organic anion transporting polypeptides (OATPs) by grapefruit juice flavonoids. In addition, grapefruit juice may be capable of inhibiting P-glycoprotein (P-gp), esterase, and sulfotransferases but effects are usually modest. Currently, the interaction of 85 drugs and grapefruit juice has been reported and such effects can last up to 12 h after drinking grapefruit juice [19].

Lilja based on a study of 10 healthy volunteers observed that when simvastatin was taken with grapefruit juice, the mean peak serum concentration and the mean area under the serum concentration-time curve of simvastatin were increased 12.0-fold and 13.5-fold, respectively, compared with control. This is equivalent to take 12 tablets at one time. When simvastatin was administered 24 h after ingestion of the last dose of grapefruit juice, the Cmax and AUC (0-infinity) were increased 2.4-fold ($P < 0.01$) and 2.1-fold ($P < 0.001$), respectively, compared with control. Therefore, taking simvastatin and grapefruit juice together may cause severe toxicity such as liver damage and muscle damage due to toxic simvastatin level [20].

The FDA has required that some drugs taken by mouth should include warnings against drinking grapefruit juice or eating grapefruit while taking the drug. These drugs include [20]:

- Statins such as simvastatin and atorvastatin where blood concentrations are increased causing drug toxicity such as muscle pain and potential liver damage.
- Certain blood pressure medication, such as nifedipine, where blood pressure may decrease due to increased drug level in blood.
- Immunosuppressants, such as cyclosporine, where toxicity may be increased due to increased blood levels.
- Cardioactive drug amiodarone where blood level is increased if taken after drinking grapefruit juice.
- Some antianxiety drugs, such as buspirone, where blood level is increased.
- Corticosteroid, such as budesonide, where blood level is increased.
- Some antihistamines, such as Allegra (fexofenadine), where blood level is decreased and may cause therapeutic failure.

Important drug interactions with grapefruit juice are summarized in Table 4.

## Summary/key points

- Abnormal liver function tests such as elevated liver enzymes and possibly bilirubin with negative serological tests for hepatitis or related viral infection is an indication of liver damage due to use of herbal remedies most commonly kava. Other hepatotoxic herbals are chaparral, comfrey, germander, and pennyroyal oil.
- Chinese herbs used for weight loss may contain aristolochic acid, a known toxin to kidney causing nephrotoxicity.
- Kelp (seaweed) is rich in iodine and taking kelp supplement on regular basis may cause thyroid dysfunction.
- Weight loss product such as Ma-huang may contain ephedra which may cause hypertension and even damage to the heart.
- St. John's wort, an herbal antidepressant, interacts with many drugs causing treatment failure due to reduced concentration of a particular

**Table 4** Important interaction of grapefruit juice with drugs.

| Drug class | Specific drug | Comments |
|---|---|---|
| Calcium channel blockers | Felodipine, manidipine, nicardipine, nifedipine, nimodipine, nitrendipine, pranidipine, verapamil | Increased bioavailability |
| Statins | Atorvastatin, simvastatin, lovastatin | Increased bioavailability |
| Benzodiazepines | Alprazolam, diazepam, midazolam, triazolam, quazepam | Increased bioavailability |
| Immunosuppressants | Cyclosporine, tacrolimus | Increased bioavailability |
| Cardioactive | Amiodarone | Increased bioavailability |
| Anticonvulsants | Carbamazepine | Increased bioavailability |
| Antibiotics | Erythromycin, clarithromycin | Increased bioavailability |
| Proton pump inhibitors | Lansoprazole, omeprazole | Increased bioavailability |
| Narcotic analgesic | Methadone, oxycodone | Increased bioavailability |
| Steroids | Methylprednisolone, prednisolone | Increased bioavailability |
| Protease inhibitors | Amprenavir, saquinavir | Increased bioavailability |
| Antianxiety | Buspirone | Increased bioavailability |
| Antihistamine | Fexofenadine | *Significantly reduced bioavailability and AUC* |

drug in blood. Hyperforin, an active component of St. John's wort induces cytochrome P 450 mixed function oxidase, causing increased clearance of many drugs. Hypericin, another active component of St. John's wort, modulates P-glycoprotein pathway thus affecting clearance of drugs that are not metabolized by liver enzymes.

- Clinically significant interaction between St. John's wort and cyclosporine or tacrolimus may cause potential rejection of transplanted organ due to increased clearance of both drugs as a result of pharmacokinetic interaction with St. John's wort. Patients taking warfarin should avoid St. John's wort because St. John's wort significantly reduces the efficacy of warfarin by increasing its clearance. Patients suffering from HIV infection and being treated with HAART (highly active

antiretroviral therapy) should avoid St. John's wort because St. John's wort reduces the efficacy of many protease inhibitors.
- Pharmacodynamic interaction of St. John's wort with various SSRI such as paroxetine, sertraline, or venlafaxine may produce life-threatening serotonin syndrome.
- Indian Ayurvedic medicines and herbal supplements manufactured in Asian may be contaminated with heavy metals most commonly lead, mercury, and arsenic. In addition, certain herbal supplements manufactured in Asian countries may be contaminated with Western drugs.
- Pharmacokinetic studies have demonstrated that intake of grapefruit juice can increase the bioavailability of many drugs classes including some calcium channel blockers, benzodiazepines, and statins. Furanocoumarins present in the juice only inhibits CYP3A in the enterocyte cells lining the small intestine, but hepatic CYP3A activity remains unaffected. Therefore, grapefruit juice-drug interaction is only observed if a medication is taken orally after drinking grapefruit juice but effect may last 12 h. Therefore, drinking grapefruit juice should be avoided by people taking certain medications.

## References

[1] Choi JG, Eom SM, Kim J, Kim SH, Huh E, Kim H, et al. A comprehensive review of recent studies on herb-drug interaction: a focus on pharmacodynamic interaction. J Altern Complement Med 2016;22:262–79.

[2] Madabushi R, Frank B, Drewelow B, Derendorf H, Butterweck V. Hyperforin in St. John's wort drug interactions. Eur J Clin Pharmacol 2006;62:225–33.

[3] Bent S. Herbal medicine in the United States: review of efficacy, safety and regulation. J Gen Intern Med 2008;23:854–9.

[4] Navarro VJ, Khan I, Bjornsson E, Seeff LB, Serrano J, Hoofnagle JH. Liver injury from herbal and dietary supplements. Hepatology 2017;65:363–73.

[5] Christl SU, Seifert A, Seeler D. Toxic hepatitis after consumption of traditional kava preparation. J Travel Med 2009;16:55–6.

[6] Becker MW, Lourençone EMS, De Mello AF, Branco A, et al. Liver transplantation and the use of KAVA: case report. Phytomedicine 2019;56:21–6.

[7] Denham A, McIntyre MA, Whitehouse J. Kava-the unfolding story: report on a work-in-progress. J Altern Complement Med 2002;8:237–63.

[8] Mostefa-Kara N, Pauwels A, Pines E, Biour M, Levy VJ. Fatal hepatitis after herbal tea. Lancet 1992;340:674.

[9] Dekant W, Fujii K, Shibata E, Morita O, Shimotoyodome A. Safety assessment of green tea based beverages and dried green tea extracts as nutritional supplements. Toxicol Lett 2017;277:104–8.

[10] Mazzanti G, Di Sotto A, Vitalone A. Hepatotoxicity of green tea: an update. Arch Toxicol 2015;89:1175–91.

[11] Vanhaelen M, Vanhaelen-Fastre R, Nut P, et al. Rapidly progressive interstitial renal fibrosis in young women: association with slimming regimen including Chinese herb. Lancet 1993;341:387–91.

[12] Teas J, Pino S, Critchley A, Braverman LE. Variability of iodine content in common commercially available edible seaweeds. Thyroid 2004;14:836–41.

[13] Mussig K, Thamer C, Bares R, Lipp HP, et al. Iodine induced thyrotoxicosis after ingestion of kelp containing tea. J Gen Intern Med 2006;21:C11–4.

[14] Chrubasik-Hausmann S, Vlachojannis J, McLachlan AJ. Understanding drug interactions with St John's wort (*Hypericum perforatum* L.): impact of hyperforin content. J Pharm Pharmacol 2019;71:129–38.

[15] Bolley R, Zulke C, Kammerl M, Fischereder M, Kramer BK. Tacrolimus induced nephrotoxicity unmasked by induction of CYP3A4 system with St. John's wort. [Letter]. Transplantation 2002;73:1009.

[16] Huang WF, Wen KC, Hsiao ML. Adulteration by synthetic therapeutic substances of traditional Chinese medicine in Taiwan. J Clin Pharmacol 1997;37:344–50.

[17] Lau KK, Lai CK, Chan AYW. Phenytoin poisoning after using Chinese proprietary medicines. Hum Exp Toxicol 2000;19:385–6.

[18] Ko RJ. Adulterants in Asian patent medicines. N Engl J Med 1998;339:847.

[19] Hanley MJ, Cancalon P, Widmer WW, Greenblatt DJ. The effect of grapefruit juice on drug disposition. Expert Opin Drug Metab Toxicol 2011;7:267–86.

[20] Lilja JJ, Kivistö KT, Neuvonen PJ. Duration of effect of grapefruit juice on the pharmacokinetics of the CYP3A4 substrate simvastatin. Clin Pharmacol Ther 2000;68:384–90.

# Index

Note: Page numbers followed by *f* indicate figures, and *t* indicate tables.

Printed in the United States
by Baker & Taylor Publisher Services